Functional and Dysfunctional Sexual Behavior

Functional and Dysfunctional Sexual Behavior

A Synthesis of Neuroscience and Comparative Psychology

Anders Ågmo

Department of Psychology
University of Tromsø
Norway

ELSEVIER

AMSTERDAM • BOSTON • HEIDELBERG • LONDON • NEW YORK • OXFORD
PARIS • SAN DIEGO • SAN FRANCISCO • SINGAPORE • SYDNEY • TOKYO
Academic Press is an imprint of Elsevier

Academic Press is an imprint of Elsevier
84 Theobald's Road, London WC1X 8RR, UK
30 Corporate Drive, Suite 400, Burlington, MA 01803, USA
525 B Street, Suite 1900, San Diego, CA 92101-4495, USA

First edition 2007

British Library Cataloguing in Publication Data
A catalogue record for this book is available from the British Library

Library of Congress Cataloging in Publication Data
A catalog record for this book is available from the Library of Congress

ISBN: 978-0-12-370590-7

For information on all Academic Press publications
visit our web site at http://books.elsevier.com

Printed and bound in Great Britain

07 08 09 10 11 10 9 8 7 6 5 4 3 2 1

Contents

Preface

For a long time I was fascinated by trying to understand the behavioral, endocrine and neural mechanisms controlling rat sexual behavior, independently of whether such an understanding was of any use or not. My favorite afternoon distraction was to walk over to the lab and see how the experiments were going, and if possible I made some observations of a couple of copulating rats. Although I have had this habit for many years, I have never got bored of seeing the same motor patterns repeated over and over again. There is always something new to discover. There may be some unexpected effect of an experimental treatment, a rat that behaves in an unusual way, or just a funny conversation with a student observer, who I had judged as very boring until then. However, with the passage of time and the accumulation of knowledge and experience, it was impossible to avoid getting some thoughts concerning the meaning and even the real interest of experimental studies of rodent sexual behaviors. These thoughts partly originated in discussions with people outside the field of rat sex research, who candidly asked if the rat data had any relevance for humans. My standard answer was that I didn't care, since my interest was focused on scientific rather than utilitarian problems. If that answer were satisfactory or not was of no concern to me. At that time another issue appeared more disturbing than questions regarding applicability of rat research to humans. The discovery that once glorious notions slowly were forgotten and replaced by other ones, and that dominant research themes and strategies demised and were replaced by other ones troubled me a lot. Still worse, the replacements were not necessarily new, they were often only different. Like we can observe at the seasonal shows at Paris' haute couture houses, fashions come and go, and some come back again in an eternal vicious circle. The ephemerality of many hypotheses and research questions as well as the reappearance of once rejected notions seemed to me far more appalling than potential applicability of the findings.

Despite the transitory nature of many hypotheses, some notions survive all challenges and remain unchanged. They are what I like to call elements of solid knowledge, entirely different from the momentarily fashionable concepts appearing and disappearing in unending succession. The meaningful and really worthwhile in the research effort must undoubtedly be these elements of solid knowledge. I am convinced that knowledge concerning cause–effect relationships normally belongs to the category of solid, while all other knowledge usually belongs to the category of ephemeral and sometimes to the field of metaphysics. However, much research pretending to address cause–effect relationship is based on assumptions going far beyond available data, and belongs in reality to metaphysics. Tragically enough, this kind of research is almost always presented as science and not as speculation as it should be.

It is not easy and perhaps not always possible to distinguish solid from ephemeral or metaphysical knowledge. Philosophers concerned with theory of knowledge have made many suggestions, but none of them is easily employed in scientific practice. In my ignorance I decided to believe that findings based on sound experimental procedures and that are replicated in many laboratories over many years are solid and true. Having solved, at least in my imagination, the problem of stability and truthfulness of knowledge I started to consider that of the usefulness and the applicability of the rodent data to human behavior. This is a question that has been discussed by many of my colleagues during many years, but I could not find any answer that was really convincing. The simple fact that the discussion continues is by itself an excellent sign of the absence of a universally acceptable and compelling answer.

This book is an attempt to provide a picture of mammalian sexual behaviors based on solid rather than on ephemeral or metaphysical knowledge. It so happened that when I started to expand my interest from rodents to other mammals, including the human, I found that some behavioral, neural, and endocrine mechanisms of sexuality not only had been solidly established in several species but also that they were astoundingly similar among species. At the same time, it became clear that the human is not just another mammal. In the same way as human language has no correspondence in other animals, some human sexual behaviors are quite unique. This uniqueness appeared for some time as most irritating, but it turns out that it can easily be explained within a theoretical framework that also explains non-human sexual behaviors. Based on this theoretical framework, a comprehensive analysis of mammalian sexual behaviors accommodating the vast majority of known facts came almost by itself. It appears on the following pages.

Writing a book produces an enormous intellectual satisfaction and a substantial amount of fatigue. For example, I have described a large number of experimental and some non-experimental studies. In order to provide an accurate description, I was forced to carefully read all the original reports. That was a main source of fatigue. Some very complicated studies required an immense effort just to find out what had been done, and with whom. In the end, though, the satisfaction associated with writing largely outweighs the unpleasantness of fatigue. I do hope that the reader will share the experience of enormous satisfaction when having finished the book.

I decided not to include any acknowledgements in this book. Not because I consider that there is nothing to acknowledge, but because I fear that it is far too much. The list would never end, but despite that a lot of people who have contributed to my points of view in important ways would certainly have been forgotten. I will make one exception, though. It would be very unfair not to mention the sufferings endured by my wife, Claudia, during many months. She had most of our social life and summer vacations ruined by a constantly unavailable husband, and when he was available he insisted on talking about themes of little interest, such as the lives and thinking of Saint Augustine or Origen, or about men's behavior when seeking sexual contacts in a public convenience. I cannot refrain from expressing my deepest and most humble gratitude to her for not throwing me out.

Introduction

The first experimental study of sexual behavior that I have been able to localize was published around 1780. The Italian priest and naturalist, Lazzaro Spallanzani, had made a series of experiments with the aim of elucidating the stimulus control of sexual behavior in frogs. The original report was published in an inaccessible source. I have been able to find a translation of this report into French, printed in Geneva in 1784 (Spallanzani, 1784). It seems likely that the original was published somewhat before that. Therefore, it is reasonable to date the initiation of *experimental* studies of sexual behavior to the late 18th century. This does not mean, in any way, that sexuality had not been the subject of inquiry long before. Historians could certainly present evidence for accounts of sexuality dating from the remote past. The novelty in Spallanzani's contribution was the employment of experimental techniques and the choice of a non-human organism to experiment on. In fairness, it must be pointed out that Spallanzani was not at all interested in human sexuality and he did not for a moment make generalizations from his frogs to any other animal. He became interested in finding out the stimulus control of the motor patterns characteristic of copulatory behavior in frogs as an add-on to his work on how and when fertilization takes place in this species. Linné had exposed a hypothesis on that subject and Spallanzani showed that Linné was wrong. A sad thing to say for a Swede.

It appears that Spallanzani's pioneering analysis of copulatory behavior in male frogs never turned into a scientific success story. Indeed, the scientific enthusiasm for the study of sexual behavior was not awesome for the hundred years following the publication of his experimental data. It was not until the late 19th century that the German endocrinologist Steinach continued Spallanzani's work on the male frog's copulatory behavior and showed that castration eliminated it. In the same paper, Steinach introduced the rat as subject in experimental studies of sexual behavior and showed that castration strongly reduced that behavior (Steinach, 1894). During the first half of the 20th century, there was a surge in interest in sexual behavior, but it was not until the 1940s that research really took off. While the early work was done mainly in Germany and France, American scientific giants like Frank Beach and Alfred Kinsey spurred interest on the west side of the Atlantic Ocean. During the decades after 1950, there was a rapid expansion in sex research

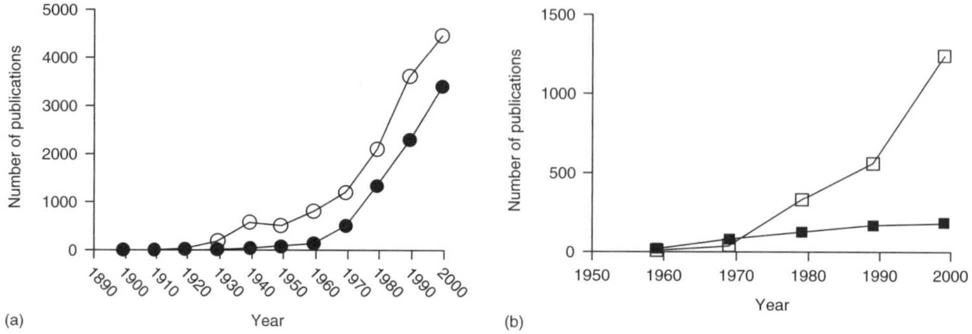

FIGURE I.1 (a) Number of publications per 10-year period (1890–1899, etc.) obtained from the PsycLIT database. ○ , total number of registered publications from the period. Note that the numbers given on the ordinate should be multiplied by 100. ● , number of publications when using the search term 'sexual behavior'. (b) Number of publications per 10-year period (1950–1959, etc.) when crossing the search terms 'sexual behavior' and 'human', □ . The same number when crossing 'sexual behavior' and 'rat', ■ . The data points are located at the last year of a decade, i.e. 1899, 1969, etc. (Reprinted from Ågmo and Pfaff, 1999. Copyright 1999, with permission from Elsevier.)

in America. Non-human mammals were the favorite subjects and, despite Kinsey's groundbreaking studies of human sexuality, it was not until the 1960s that human copulatory behavior was analyzed in the laboratory. From about 1970, scientific research on human sexual behavior increased rapidly and that increase was accentuated when AIDS started to attract attention. An illustration of the number of publications on sexual behavior during the period from 1890 to 1999 is found in Figure I.1. It can easily be seen that the expansion in the number of research publications on sexuality closely follows the expansion of the total scientific output, suggesting that sexuality did not increase its share of attention within the scientific community. The enhanced scientific effort in the field was simply a result of the increasing resources allocated to science in general during the postwar period. What is noteworthy, though, is that research on non-human sexual behaviors has stagnated. The increased number of publications during the last decade of the last century is entirely accounted for by the increase in research on the human. Most unfortunately, the bulk of that research is focused on the behavior patterns relevant to the spread of sexually transmitted diseases or unwanted pregnancies. One favorite theme has been the circumstances determining the use of condoms, for example. Although this information might have some marginal relevance for public health, it has none at all for our understanding of the basic mechanisms guiding human sexual behaviors.

Despite its shortcomings, the considerable research effort has produced much important knowledge. Since the research on non-human animals mostly employs experimental procedures and is performed under strictly controlled laboratory conditions, progress in the non-human areas of sexual behavior research has been much faster and much more solid than it has been in the field of human sexual behaviors. Basic research in non-human animals has elucidated the role of gonadal hormones and their actions within the brain are beginning to be understood. At the same time, reproductive physiologists have learned a lot about the actions of these hormones in peripheral tissues. Some fundamental behavioral principles have been unequivocally established in non-human mammals and we now begin to have some ideas regarding how external stimuli affect the nervous system when determining the likelihood of activation of sexual behaviors. When we turn to human sexual behavior, the employment of experimental techniques is far more restricted for obvious reasons. Many manipulations that are routinely performed on non-human animals are unacceptable for human use. One consequence of this is that most of the human data stem from non-experimental sources. In many studies of human sexual behavior information is obtained by asking people to produce self-reports, in the form of diaries or answers to questionnaires. We know that introspection is a most uncertain procedure, because there is no way to determine the truthfulness of the data obtained through it. As a consequence, knowledge about human sexual behavior is frequently contradictory and almost always fragmentary. In fact, a substantial portion of that knowledge is nothing more than informed speculation.

Besides the limited availability of experimental data on human sexual behaviors, there are other circumstances that have hampered progress in our understanding of it. Human sexuality is embedded in a large amount of social and religious rules and it is surrounded by and associated with many myths. Scientists are not insensitive to dominating social prejudice and much of the research on human sexuality has been heavily influenced by different kinds of bias originating in preconceived notions of its inner nature. Likewise, some myths have been and are so pervasive that they occasionally are confounded with established facts. One consequence of this is that some of the myths rarely, if ever, have been confronted with reality. Another aspect of the social context is that some societies do not consider research on human sexuality useful or even desirable, thereby making the obtention of research funds unnecessarily challenging. In sum, the social context in which research is performed poses many difficulties for the scientist interested in human sexuality and this has certainly contributed to slow down the advance of our understanding of it.

A comparison between the amount of presently existing solid knowledge regarding non-human and human sexuality would reveal that we know far more about the former. In this situation, a reasonable question is whether some of the knowledge obtained in non-human animals could inform us about fundamental aspects of human sexuality. My personal conviction is that it could. I would even dare to propose that it could inform us a lot. That conviction was, in fact, the main reason for taking on the considerable effort to write this book.

Scientists have a tendency to specialize. There is no reason to criticize this necessity, but with regard to sexual behavior it has had the consequence that those working with non-human sex behavior are only remotely interested in and informed about human sex, and vice versa. Furthermore, there is a generally held belief that the mechanisms controlling non-human sexual behavior are essentially different from those controlling human sexual behavior. Due to that belief, non-human data are usually ignored by those interested in human sexuality and human data are usually ignored by those interested in non-human sexuality. There are many reasons for the notion that sex is entirely different in mice and men. Some of these reasons will be the subject of detailed analyses in this book. However, I will not discuss them here. Instead, I will focus on the similarities between human and non-human sexual behaviors for a moment. We now know that sexual behavior in all mammals, including the human, is dependent on gonadal hormones. The mechanisms of hormone actions are, as far as can be judged from available data, similar if not identical in all mammals. The modulatory actions of some neurotransmitters are, as far as can be judged from available data, similar if not identical in all mammals. Still more important is the fact that the behavioral principles, and the theoretical models putting these principles into a system, can be applied to non-human as well as to human sexuality without the slightest modification. While the expression of sexual behavior and the specific motor patterns involved are different in different species, the basic functioning of the behavioral systems controlling these varying expressions is always the same.

This book will try to demonstrate the usefulness of a unitary approach to the study and analysis of sexual behavior in all mammals. Unitary approaches to behavioral mechanisms have demonstrated their utility in some fields of behavioral science for a long time. A good example could be Skinnerian learning principles. Skinner's schedules of reinforcement work in exactly the same way in all mammalian species and subtleties like the partial reinforcement effect have been found in every mammalian species studied. The application of Skinner's theoretical framework to human and non-human behavior alike has proved to be very successful and allows for an understanding of behavioral phenomena that otherwise would have been unexplainable. Likewise, Pavlov's classical conditioning has been found to follow the same principles in all mammals and even the effects of minor variations in interval between the appearance of conditioned and unconditioned stimulus affect learning equally in the human and rat, just to take one example. Although sexual behavior might seem to be more complex than the learning mechanisms I mentioned, the basic principles guiding it are not necessarily so complex. There is no reason to doubt their applicability to all mammalian species.

In the case of sexual behavior, there is an additional and important advantage with a unitary approach. Almost every human has some direct or indirect knowledge about sexuality, based on personal experience, social and moral conventions and age-old mythology. Much of that knowledge is probably false. Not long ago, we all knew that masturbation produced dramatic effects, like emptying the skull of its substance, or producing general weakness, or outright insanity. Likewise, not long ago, we knew that premarital sex results in later sexual perversion, or that homosexuals transform heterosexual youths by simply looking at them in public toilets. When we turn to non-human animals, the plethora of myths is not as abundant. There are still many misconceptions diffused by Discovery Channel or National Geographic, but their impact is not as devastating as the myths about human sexuality are. Furthermore, the many moral principles perverting views on human sexuality do not apply to non-human mammals. The reduced number of popular misconceptions regarding non-human sexual behavior and the absence of moral rules make it possible to approach sexual behavior in other mammals without too many preconceptions. A unitary approach to human and non-human sexual behavior could strongly benefit from this.

I have systematically written about non-human *mammals* rather than non-human *animals*. This is not an accident. The ambition of proposing a unitary approach to the analysis and understanding of sexual behaviors would probably be unrealistic if all animals were to be included. First of all, we do not know enough about the sexual behaviors of invertebrates and some vertebrates. In the absence of a reasonable amount of behavioral knowledge, it is completely impossible to judge whether the proposed model also can account for the behavior of these animals. Furthermore, the behavioral model I will propose can be anchored to reality through reference to specific sensory, central nervous and endocrine processes. All available evidence suggests that the endocrine mechanisms controlling sexual behaviors are quite similar in most mammals, while this not necessarily is the case with regard to reptiles or insects. Most of the sensory organs are also similar in mammals, but they are quite different in fishes and arachnoids. The brain structures involved in sexual behaviors appear to be equivalent in all mammals, but they are by necessity different in molluscs and plathyhelminths. The genitals have the same basic structure in most mammals, while they are of a different kind in birds and arthropods. All these arguments made it look completely inconvenient to extend the model to include non-mammalian sexual behavior.

In the first chapter, I will present the philosophical bases that have guided the writing of this book. Essentially, I insist on the paramount importance of providing scientific explanations in terms of cause–effect and never in terms of imposed purpose. This has some importance for the content we give to the term sexual behavior and for how we can proceed for understanding it. I also examine the origin of popular (and scientific) notions about sexuality. As we will see, these arbitrary notions have penetrated Western thought to such a degree that most of us consider them as established truths. Knowing from where these notions come makes it easier to get rid of them. This chapter is quite fundamental for an adequate appreciation of the rest of this book and I recommend even the most hurried reader to invest a couple of minutes in reading it.

The second chapter begins with a presentation of the theoretical model providing the basis for the organization of current knowledge. Existing scientific information about sexuality is so vast that it is impossible to grasp its essence without having a theoretical structure that can impose order upon it. I will outline a model that describes how internal states and external stimuli interact in the control of sexual behaviors. Then I will proceed with a description of these behaviors in humans and in non-human mammals. The theoretical model is indispensable reading, while the descriptions of behavior can be considered a luxury. Nevertheless, there are some comments on the differences between the copulatory motor patterns in humans and non-humans that are quite important for an appropriate understanding of many arguments appearing throughout the book.

In Chapter 3, I make a painstaking analysis of how external stimuli control sexual behaviors in non-human mammals. One purpose with this is to make evident that much of our knowledge of this basic aspect of sexuality remains on a level of informed speculation. Another is to illustrate how researchers in non-human sexual behavior proceed when determining whether stimuli within a particular sensory modality influence sexual behaviors. This is a long procedure and all the intricacies involved are not always apparent. However, I consider it of utmost importance to show that, despite a large amount of research, there are still many questions waiting for an answer. Chapter 3 also illustrates that work with non-human animals is not always and not necessarily associated with greater clarity of results than human studies can offer and that most results in fact are open to different interpretations. In this chapter, as well as in the rest of the book, I describe many experiments in some detail, including intricacies of the methods employed. I do that because there is an obligatory and fundamental relationship between the methods employed and the results obtained. If methods are inadequate, then the results are ambiguous or, in the worst of cases, completely meaningless. The careful reading of the methods section of a research report is essential for our possibility to judge its significance. I realize that the sometimes long-winded discussions of methods may appear fastidious, but science is a fastidious enterprise. There is no way around that fact.

In Chapter 4, we turn to the stimulus control of human sexual behaviors. This is a fascinating field of inquiry and we will find out that much knowledge is lacking. We will also find out that we actually know a lot about certain elements. This latter is a reassuring discovery.

In Chapter 5, we turn to the control of sexual behavior by events inside the organism. The role of gonadal hormones is the subject of this chapter. In the following chapter, the brain structures involved are examined. This is to give some background for many themes that will appear in later chapters. We then turn from the fundamental, general aspects of mechanisms controlling sexual behaviors to the operation of these mechanisms. In Chapter 7, I will describe how sexual behaviors can be modified by experience. Above all, we will see that the execution of sexual responses is rewarding and reinforcing in the same way as food or drink are. Both human and non-human data will be covered. At the end of the chapter, we will be convinced that any stimulus may acquire sexual significance through classical conditioning and that operant and social learning are fundamental factors in the shaping of human sexual behavior. Some crucial differences between humans and other mammals will be reiterated here. We will then leave themes concerning sexual behavior in general and focus on a few issues that I consider of particular relevance and well suited for an analysis in terms of the theoretical model proposed in Chapter 2.

In Chapter 8, we will take a look at same-sex behaviors in non-human animals. We will see that copulatory motor patterns and sexual preferences must be clearly differentiated. We will also see that copulatory motor patterns rarely are sexually dimorphic and that all mammals, including the human, are basically bisexual as proposed by Freud some time ago. In Chapter 9, we turn our attention to homosexuality in the human. The main thrust of that chapter is to convince us that the notion of homosexuality is a quite meaningless social construction. Indeed, from a scientific point of view, the sex of the partner a human chooses for displaying sexual responses is quite uninteresting. I make some comments on why some homosexuals favor a biological cause for a social construct and then proceed to criticize the data presented in support of that hypothesis.

A direct continuation of the conceptual discussion in Chapter 9 follows in Chapter 10. There I try to come to grips with the notions of functional and normal. It is concluded that both are arbitrary concepts, but that it is difficult to avoid them. This, however, is not of much concern as long as we realize that they are arbitrary, that they can be changed at any moment and that normal in one place does not necessarily mean the same in another place. In Chapter 11, we turn to a sexual dysfunction, hypoactive sexual desire disorder. A rather exhaustive analysis of diagnosis, prevalence, etiology and potential treatments is provided. I insinuate that the theoretical framework employed throughout the book can be applied to that disorder and that it can provide valuable suggestions for treatment, in addition to being able to account for the origin of the disorder. In support for that proposal I describe the only psychotherapeutic treatment with some empirical foundation in certain detail.

In Chapter 12, the opposite condition, hyperactive sexual desire, is discussed. I show that the notion in itself is arbitrary and perhaps meaningless but, nevertheless, I propose diagnostic criteria and suggestions for treatment. After that section, there is a rather extensive discussion of some of the paraphilias.

Arriving to the end of the book, we will find that much material has been covered. Many research reports and many reviews have been presented, discussed and summarized. Many experiments have been described in some detail. Unfortunately, I had to conclude that some studies employed sloppy methods. Other studies were based on very ingenious ideas and this was also acknowledged. We have also seen examples of the most extravagant transformations between the results section of a paper and the ensuing discussion of these results. There were occasions where scientists interpreted their results in a way diametrically opposed to what they actually found. Conclusions going far beyond the data were not unusual either and far-fetched speculations were ingredients of many research reports. I have tried to point out all the elements that can make a study's conclusions doubtful whenever necessary. It must absolutely be added that, in spite of the occasional deficiencies in the research, the overwhelming majority of the research reports contribute to an enhancement of our understanding of sexuality.

I have tried to present evidence both in favor and against most of the hypotheses we have encountered in the fairest possible way. It has not been exceptional to find conflicting results and, in these cases, I have arrived either at the conclusion that we do not know or at a cautious suggestion or prediction with regard to what future research may reveal. Some of these predictions may turn out to be false, but only the future can tell us. Nevertheless, the account of sexual behaviors given in this book, and the theoretical approach employed, offers a new and integrated framework for the comprehension and analysis of human sexual behaviors and those of other mammals devoid of much of the moral and social and scientific prejudices that have infected so many earlier analyses of these behaviors.

In mentioning an experiment or another kind of research report, I now and then allow myself to manifest my appreciation of it. Some people maintain that value statements are unacceptable in serious writings. I disagree with this and without any trace of shame I can write that a study was elegant or sloppy, whichever applies according to my judgment. In the same spirit of rebelliousness, I treat myself to a candid and discreet expression of humor here and there. There is no need for a serious text to be consistently boring. In fact, I have the impression that charlatans always write in a very boring way so that no one will take the trouble to read all their nonsense and discover its emptiness. Finally, as a modest European intellectual, I do not prohibit myself from using words and expressions that the hypocrites would consider inconvenient. Freedom of speech is, after all, a fundamental human right.

References

Ågmo, A. and Pfaff, D.W. (1999). Research on the neurobiology of sexual behavior at the turn of the millennium. *Behav Brain Res* **105**, 1–4

Spallanzani, L. (1784). *Expériences pour servir à l'histoire de la génération*. Barthelemi Chirol: Geneva

Steinach, E. (1894). Untersuchungen zur vergleichenden Physiologie der männlichen Geschlechtsorgane insbesondere der accessorischen Geschlechtsdrüsen. *Pflügers Arch ges Physiol* **56**, 304–338

CHAPTER

1

On the purpose of sex and some notes on scientific explanations

1 On the definition of sexual behavior

Any discussion of a scientific subject can benefit from clear definitions of the basic terms employed. Many misunderstandings, conceptual confusions and sterile arguments may thereby be avoided. Since this is a book on sexual behavior, it would, therefore, be convenient to start out with a definition of that term. Some people use the expression 'reproductive behavior' instead of 'sexual behavior', either because they want to avoid the vulgarity or stigma sometimes associated with the word sex, or because they want to emphasize that the behavior has something to do with perpetuation of the individual and with the species. The expression 'mating behavior' is frequently employed, particularly when referring to non-human animals, but I have been unable to find any difference between this term and the term 'reproductive behavior'. It will consequently be ignored.

Having eliminated the expression 'mating behavior', we can now start to look at some definitions of 'sexual behavior' and 'reproductive behavior' appearing in non-specialized sources. The online version of the *Encyclopedia Britannica* defines *sexual* behavior as: 'any activity – solitary, between two persons, or in a group – that induces sexual arousal' (retrieved September 20, 2005, from *Encyclopedia Britannica Online* http://search.eb.com/eb/article-9110426). On the other hand, *reproductive* behavior is defined as: 'any activity directed toward perpetuation of a species' (retrieved September 20, 2005, from *Encyclopedia Britannica Online* http://search.eb.com/eb/article-9110426). *Wikipedia* defines sexual behavior (human) thus: 'Sexual behavior is a form of physical intimacy that may be directed to reproduction (one possible goal of sexual intercourse), spiritual transcendence, and/or to the enjoyment of any activity involving sexual gratification' (retrieved September 20, 2005, from *Wikipedia*, the free encyclopedia http://en.wikipedia.org/wiki/Human_sexual_behavior). This encyclopedia does not mention reproductive behavior.

These definitions are probably useless in a scientific context. For example, the notion that sexual behavior is *any activity that induces sexual arousal* has several interesting implications, some of which are counterintuitive. Imagine a woman, comfortably seated on her favorite couch reading an old novel by David Herbert Lawrence. It so happens that she gets sexually aroused by a passage in *The Plumed Serpent*. Would that mean that reading is sexual behavior? I could easily invent dozens of examples equally absurd, but this one should be sufficient to illustrate the point. Likewise, reproductive behavior is said to be *any activity directed toward perpetuation of a species*. Does this mean that the veterinarian inseminating a cow is displaying reproductive behavior? Or the technician who diluted the sperm directly after collection from the bull, did she show reproductive behavior while doing so? Yet both the veterinarian and the technician performed activities directed toward perpetuation of a species. It appears that the definitions provided by at least some electronic resources do not really clarify the issue. Perhaps they even facilitate some misunderstandings.

Turning to more specialized publications, like the The *Blackwell Encyclopedia of Social Psychology* (Manstead and Hewstone, 1995), we continue to find the same lack of clarity as in the non-scientific sources. Here, the heading sexual behavior is followed by this statement:

> The field encompasses biological and behavioral substrates of genitally oriented arousal and reproduction, which among humans includes the study of gender, close relationships, sexual orientation, deviance, antisocial behavior, and dysfunction' (p. 524).

Not only does the preceding sentence fail to provide anything resembling a definition, but it also expands the term in many surprising directions. It would perhaps be unfair to suggest that social psychology is a field lacking in rigor, but it is embarrassingly evident that at least some of its 'definitions' are close to meaningless. Obviously, the *Blackwell Encyclopedia of Social Psychology* does not bother to mention 'reproductive behavior'. Other publications, like the multivolume work *The Corsini Encyclopedia of Psychology and Behavioral Science* (Craighead and Nemerofff, 2001) or its abbreviated counterpart (Corsini and Auerbach, 1996), does not mention sexual behavior, although obscure subjects like sexual abstinence and sex therapies are worthy of inclusion, in addition to the more traditional themes of sex differences, sex roles and sexual deviation. The rather limited concern with definitions of sexual or reproductive behaviors in psychology publications is repeated in biology. The *Encyclopedia of Biology* (Rittner and McCabe, 2004) mentions neither sexual nor reproductive behavior. As could be expected in a biology text, though, reproduction cannot be entirely ignored and it is even somehow associated with sex. We find the heading *sexual reproduction* followed by the definition: 'the creation of a new individual following the union of two gametes . . . ' (p. 306). It is difficult to object to this definition, but it has no relevance for those concerned with behavior.

Simple examples like those just mentioned illustrate at least two things. First, it is not easy to define sexual or reproductive behavior. It seems to be so difficult that most texts do not even try to do so. Alternatively, the meaning of sexual and/or reproductive behavior is considered so evident that no definition is needed. Another thing that can be learned from the few available definitions is that there might be some conceptual difference between reproductive behavior on one side and sexual behavior on the other. This is nicely illustrated in *The Encyclopædia Britannica*. There, reproductive behavior is associated with perpetuation of the species, while sexual behavior is not limited to that. Remember that any activity inducing sexual arousal is included under the term sexual behavior. No reference to perpetuation of the species is made. *Wikipedia* makes implicitly the same distinction, when it is mentioned that sexual behavior can either be directed to reproduction (one possible goal of sexual intercourse) or to the enjoyment of any activity involving sexual gratification. The latter seems to be independent of reproduction. The possible difference between sexual and reproductive behavior will be discussed shortly. I will eventually come to the conclusion that sexual behavior needs to be conceived as a behavior entirely unrelated to reproduction if we ever want to be in the position of understanding human sexual behavior and probably also that of other animals, at least mammals. Before being able to do so I need, however, to describe how reproductive behavior frequently is looked upon among biologists. Within the biological tradition, the notion of reproductive behaviors is usually strictly limited to behaviors leading to the successful union of gametes, sperm and ovum, although paternal behaviors sometimes are included among the reproductive behaviors. It is evident, though, that the word reproductive always refers to the function or end result or purpose of the behavior, namely reproduction. This, in turn, means that the term is based on a teleological approach to behavior. The implications and consequences of that will be analyzed in the following section.

Before ending this first section of the first chapter and in agreement with the idea that basic definitions should come at the outset, I will immediately venture into an attempt to define sexual behavior: *Sexual behavior is any action leading to sexual reward. Sexual reward is a state of positive affect activated by physical stimulation of the genitalia or mental representations of such stimulation*. This definition may seem to be lacking in many respects but, during the next few chapters, I will present a substantial amount of data in support of it. Long before the end of this book, I hope to have been so convincing that it will appear to be the best definition available.

2 The relationship between biological function and the causation of behavior

2.1 Teleology

As mentioned a few lines above, sexual/reproductive behavior is sometimes analyzed in terms of its biological purpose, the union of gametes and ensuing procreation. The sterility of such an approach can be made evident through a short discussion of the history and philosophical underpinnings of what are normally called teleological explanations. Then, I will briefly contrast that kind of futile explanation with the fruitfulness of a causal approach.

Teleology (from Greek telos, *end*, and logos, *reason*) is normally understood as a doctrine that maintains that phenomena can be explained by reference to their purpose or end. The purpose or end is described as *final cause* by Aristotle. For example, if we use Aristotelian thinking to explain why an egg turns into a chicken, the answer would be that this is what the egg is for. Within the teleological tradition, this explanation would be satisfactory. Consideration of the processes in the egg that lead to the formation of a chicken would be unnecessary. The fundamental role of final causes in Aristotelian teleology makes it enter into what we call ontology or metaphysics. Many explanations in contemporary biology are essentially teleological in the way that physiological processes are understood in terms of what they do in and for the organism. The resort to teleological thinking has become still more pronounced when biologists, and many psychologists too, try to understand behavior. As an illustration to this, we can simply look at teleological explanations of why a rat engages in sexual activities or why it eats. A teleologist would say that the rat copulates because it needs to reproduce and it eats because it needs food for survival. This is certainly true, but at the same time it does not explain anything at all about the circumstances under which sexual behavior or eating occurs. Likewise, we get no idea as to the mechanisms determining the intensity of sexual behavior or the amount eaten. Another teleologist, under the cover name of a sociobiologist, may say that men like to have sex with women with a slim waist because such women are more fertile than those with a wide waist. Here, we get a step farther in sophistication, because it is not only assumed that reproduction is the cause of sexual activity, but also that we want to maximize the efficiency of the behavior in terms of its biological function. Again, this does not say anything at all about the mechanisms that make some men like a slim waist.

Even some biologists have expressed dismay over the frequent use of teleological explanations with regard to phenomena of great relevance to biology. The following quote is a splendid example of this:

> Biology is sick. Fundamentally unscientific modes of thought are increasingly accepted, and dominate the way the subject is explained to the next generation. The heart of the problem is that we persist in making (literally) sense of a world that we now know to be senseless by attributing subjective values to the objects in it, values that have no basis in reality (Hanke, 2004, p. 143).

Function or purpose are, according to the line of reasoning presented in the preceding paragraph, not entities that exist by themselves but impositions of the human mind; human inventions to make a complex world comprehensible. The cost associated with this is that a real understanding of the world is replaced by a fictitious understanding, based on our prejudices or on subjective values.

As pointed out many years ago (Monod, 1970), the basic process of biology, evolution, is driven by chance events, namely, random recombination of genes in organisms with sexual reproduction and fortuitous mutations in such and other organisms. Chance events can have no purpose or function. The accidental modifications in genes are brought into some order by natural selection, but that process can have no more purpose or goal than mutations have. Selective pressures are spatially and temporally unstable. At one time and at one place a certain characteristic may be favorable, but at the same time and at another place this same characteristic may be unfavorable. At the same place, but at different times, a constant characteristic can be either favorable or unfavorable. Selective pressures may favor variety and change, but they certainly do not impose purpose on evolution. If the basic mechanisms underlying evolutionary processes do not need any purpose in order to function quite well, then behavior does not need any purpose in order to function well, or in order to be adaptive, to put it in words dear to biologists. However, the habit of talking about behavior as purposeful is so firmly rooted that we need to pay some attention to it. It could even be useful to subject descriptions and explanations of behavior in terms of purpose to a rather detailed analysis. To that end, or better, with this purpose in mind, we need some help from philosophers of science.

A splendid analysis of the philosophical bases for teleological explanations and the costly consequences of them was presented in a little book a few years ago (McLaughlin, 2001). Of particular interest in the present context is, obviously, the discussion of function or purpose or goals as causes for, among other things, behavior. Concerning human behavior, it is not unreasonable to maintain that it sometimes is goal directed or purposeful. For example, I might subject myself to a long walk to the city center with the purpose of having a beer with some friends. In this case, my behavior was activated and directed by a purpose, it had a goal and it had a function (allow me to have some beers with friends). The purpose, goal and function can each legitimately be maintained to be the cause of my long walk. The legitimacy of this teleological explanation of behavior rests on the assumption that I was able to have a mental representation of drinking beer with friends. It is the *representation* of the goal or purpose, someplace in my brain, that activated and directed my behavior, not the goal or purpose itself. Thus, in any organism able to have a representation of the goal or the purpose of its own behavior, that representation can be considered a cause of the behavior. A consequence of this reasoning is that teleological explanations are legitimate for organisms in which such representations exists, and only for such organisms. This excellent analysis solves probably many problems regarding the explanation of human behavior. There are not many reasons to doubt that humans are able to have representations of the goals of their behavior. However, the problem with teleological explanations remains for all species where we do not know whether individuals really have representations of their goals or not. It would be entirely impossible, basically because of my ignorance, to enter the discussion on the possible existence of goal representation in non-human animals, and it is perhaps not even necessary to do so with regard to sexual behavior. As it turns out, the most likely representation of the goal of sex behavior in the human, which can function as an efficient cause for activating sex-related behaviors, operates in exactly the same way as certain external stimuli in other animals. More on this will follow in Chapter 2. For the moment, it is sufficient to say that we do not need any notion of a representation of a goal for explaining non-human sexual behavior. Thereby, there is no need to speculate about possible mental representations in non-human animals, hence the issue is of no relevance for us. In order to avoid misunderstandings, though, I should point out that the existence of mental representations of many kinds, even among invertebrates, is beyond doubt (see e.g. Fyhn *et al.*, 2004, for a hypothesis concerning the representation of space in rats, or Lloyd, 1987, for an analysis of representation in *Aplysia*). However, the existence of representations of some physical aspects or relationships in the external world does not suggest representations of goal or purpose. It might be adventurous, but I would like to propose that goal representations become possible only through the human language.

2.2 Cause–effect relationships

The preceding discussion should have made clear that the notion of final cause or purpose as an explanatory element is lacking in many respects. It does not offer any convincing and scientifically useful explanation for most phenomena of nature. One exception, perhaps the only one, is some aspects of human behavior as mentioned above. Despite the logical deficiencies inherent in teleological explanations, they persist in many areas of biology and in some areas of psychology. On the contrary, in physics and chemistry, explanations in terms of purpose or final cause were already beginning to be abandoned in classical Greece. Even the champion of teleological thought, Aristotle, realized that many phenomena in physics required other kinds of explanations. By the 17th century, teleology had essentially disappeared from physics. In Newton's mechanics, there is no teleology left. For example, the movement of bodies can be explained in terms of cause instead of in terms of purpose. The impact of a moving body A on another moving body B can be entirely explained considering their masses, speeds, directions of movement, elasticities and so on. The importance of each of these factors can be precisely expressed in an equation and accurate predictions can be made. Instead of explanations in terms of function or purpose, Newtonian mechanics offers explanations in terms of causes and ensuing effects. Modern science is essentially based on this kind of explanation and it is quite likely that the change from an emphasis on function to an emphasis on causes has been a major factor in scientific progress. Questions like 'why?' and 'what for?' have been replaced by questions of 'how?'. The 'how?' question can be answered by causal explanations, but it can also be taken a step further in the way that we can explain 'how' without considering a cause. As an example, we could use Einstein's law of gravitation. It simply states that the path and speed of a planet are so adjusted that the total separation, measured over any section of its course, shall be a maximum. This law explains *how* a planet moves, but it does not say a word of *why* the planet follows a course of maximum separation or the cause of its movement (Hull, 1959). It offers a pure description of a phenomenon, but that description is enough to allow us to predict the planet's path, even in a distant future. The accuracy of the predictions can in no way be improved by references to metaphysical questions of 'why?' and 'what for'?

The spectacular progress that followed the disappearance of teleology from physics and chemistry is now repeated in molecular biology, a branch of biology where teleology has become efficiently eradicated. This is also the case in the neurosciences. Most of contemporary neuroscience offers strictly mechanical, causal explanations of the mysteries of nerve cell function (Revonsuo, 2001). One immediate consequence is that nervous functions have become less mystical than ever. The intimacies of the sodium channel or of transmitter release can be entirely explained in causal, mechanistic terms without any reference to purpose. According to Revonsuo (2001), within the neurosciences there is only the subfield of cognitive neuroscience that still persists in the employment of non-mechanistic, non-causal explanations. The consequence is that no single cognitive component has been discovered for which there is general agreement among investigators (Van Orden and Paap, 1997). It is my firm belief that behavioral sciences could also benefit from abandoning teleological explanations and concentrate on causes rather than on purposes. The potential fruitfulness of such an approach will, hopefully, become apparent during the reading of the remaining parts of this chapter.

Before returning to the subject matter of this book, it can be useful to make a short comment on what is meant by explanation in terms of cause–effect. It appears that one of the first comprehensive analyses of this was presented by David Hume in his *An Enquiry Concerning Human Understanding* (Hume, 1758). Briefly, there are three components of all causal relations: contiguity of time and place, temporal priority of the cause, and constant conjunction. If we imagine that C causes E, then C and E must occur at the same place and about the same time (contiguity of time and place), but C must always precede E (temporal priority). Furthermore, if C is present, then E must invariably occur. There is nothing inherent in C that determines E, so a relation cause–effect can only be discovered by experience. Hume himself mentions that there is no logical reason for a moving billiard ball to make a second ball start moving upon contact with the first. However, since this always occurs we can conclude that the first ball *causes* the second to move. An interesting consequence of this argument is that reason is not the basis for discovering any cause–effect relationship. It is only experience that can allow us to discover such relations. If we want to apply this to our search for understanding sexual behavior, it must be concluded that we should not think of the behavior but observe it. Experiment, not reason, will illuminate us.

There is one additional issue that needs to be addressed before we can leave the discussion of cause. In teleology, as well as in biology, it has been customary to divide causes into two kinds, efficient cause versus final cause in classical teleology, and ultimate versus proximate cause in biology. Let us start with teleology. The efficient cause is the immediate cause of an event while the final cause is the purpose. If eggs are being fried, the final cause may be my breakfast, but the efficient cause is the heat applied to the frying pan. In this example, the coagulation of the proteins in my egg is caused by heat and nothing else. The conformational changes in the proteins underlying coagulation can be entirely attributed to increasing temperature. The efficient cause explains completely the phenomenon of an egg being fried, that is the 'how?' question. The answer to the 'why?' question, I will have fried eggs for breakfast, does not contribute in the slightest to an understanding of what is happening in the eggs. Furthermore, I may apply heat to eggs for many reasons in addition to having them for breakfast, meaning that the final cause of the event 'eggs being fried' can be variable and it cannot be inferred from the event itself. The ambiguity or even arbitrariness inherent in the process of assigning a final cause to an event becomes particularly evident in the behavioral sciences, where it normally is an observer and not the behaving individual who will have opinions about the purpose of the observed behavior. Let us for a moment return to my walk to the city center. I know that the purpose, or final cause, is to have beer with friends. However, another friend who saw me walking in the direction of the city center may well imagine that I did that because I was on my way to a movie theater. She came to that conclusion because she had heard that an excellent movie was playing that week. A colleague who also saw me might have concluded that I was going to the long distance bus terminal to pick up a visitor. He had done that himself a couple of days ago. As soon as an observer imputes a purpose to another individual's behavior, he is likely to do that according to his personal inclination. This is far more evident when we try to give a purpose to the behavior of individuals belonging to a species different from our own. It is quite sufficient to hear someone describing the extremely purposeful behavior of his dog to discover the absurdity of the reasoning behind. It seems, then, that explanations of behavior in terms of purpose, i.e. teleological explanations, are extremely risky, perhaps with the exception of accounts of our own behavior. Personally, though, I find that my own behavior normally defies any rational explanation.

In biology, the term 'efficient cause' has been replaced by 'proximate cause', and 'final cause' with 'ultimate cause'. The adjective final comes from the Latin word *finalis*, meaning *of or pertaining to an end*, while the adjective ultimate is derived from the Latin word *ultimus*, meaning *last, final*. It seems that final and ultimate causes are not only conceptually but also semantically equivalent. The reason why biologists have replaced final with ultimate is perhaps to make the expression's teleological foundations less evident and the association with metaphysics less salient. Having said this, it should also be emphasized that many biologists are quite comfortable with teleological explanations. It is particularly, or perhaps even only, in evolutionary biology that these kinds of explanations are legion. The prime example, however, comes from the popular quasiscientific discipline of evolutionary psychology, where teleology is not only accepted but also venerated (see e.g. Wakefield, 2005). Some of the excesses of evolutionary psychology will be illustrated in a few pages. The more sensate biologists will, nevertheless, maintain that the complete explanation of a phenomenon includes an answer to the question 'why?'. This 'why?' does not necessarily mean 'what for?', a reference to the purpose. It can also mean 'how come?', which is a reference to the evolutionary processes that underlie the phenomenon (Mayr, 1961).

In this latter question, there is no reference to purpose and an answer to it would not belong to teleology. Although the distinction between the questions 'what for?' and 'how come?' could easily be maintained, many biologists fail to do that and explanations in evolutionary biology are frequently offered in teleological terms (see Mayr, 1997). When it comes to behavior, it can be maintained that the evolutionary processes responsible for it, if any, are of no immediate interest. If we understand how the behavior is controlled by internal and external causes we have achieved a complete explanation. The evolutionary history of these causes is entirely irrelevant. It is likely that this principle could apply to most organismic phenomena, in addition to behavior.

Let us take a simple example. For the sake of clarity, I use the behavior of an inanimate entity. My car is equipped with ABS brakes. I happen to know the history and purpose of this kind of brake. One morning I discovered a serious malfunction in them. So serious that I almost ended in the ditch. That event made it evident that they needed repair. Here, my knowledge of the history of the ABS brakes was of no more use than my knowledge of their purpose. The mechanic who fixed them, a young woman, had never heard of their history and she had very vague ideas about their purpose. However, she knew how they worked. Such a simple example should be enough to illustrate that knowledge of evolutionary history or purpose is of little practical value. It may satisfy or thrill our intellect and constitutes, perhaps, a healthy entertainment. Nothing more.

To summarize, it can be maintained that metaphysical explanations, including teleological, have no place in science. Consequently, explaining behavior in terms of its function should be abandoned. On the contrary, any meaningful analysis of behavior requires the establishment of cause–effect relationships. The only way to do so is by careful observation and, whenever possible, with the aid of rigorously designed experiments.

Within the behavioral sciences, the principles outlined in the preceding paragraphs have been dominant in the behaviorist, neobehaviorist and radical behaviorist traditions. The splendid learning theories of Hull (1943) and Spence (1956) are good examples of how behavior has been carefully analyzed in terms of cause–effect. Skinner has taken that tradition a step further in his insistence of act – consequence (cause–effect) as the basic principle determining most of human and non-human behavior. I maintain that sexual behavior in mice and men also can be analyzed and understood in terms of cause–effect relationship. I also insist on Hume's criteria of contiguity and conjunction. Since reproduction (birth of the young) is non-contiguous with sex behavior, it can have no causal relationship to it. This argument lacks force, though, if with reproduction we understand fertilization. Although not simultaneous, fertilization is always temporally associated with sex behavior. What is far more important is that there is no constant conjunction insofar as sex is not always associated with reproduction. In the human, the vast majority of sexual activity has no reproductive consequences and in other animals the association is uncertain. Data from an old Swedish survey of sex behaviors show that there are about 1100 copulations per birth, making an association between sexual intercourse and reproduction something quite exceptional. Likewise, female mice, rats and rabbits copulate with a large number of males during the period of estrus, yet only one of the males will sire offspring. The display of copulatory behaviors by all the other males had no reproductive consequence.

The preceding examples illustrate that there is no constant conjunction between sexual behavior and reproduction. Thus, according to the criteria of causality established by one of the most respected philosophers of all times, David Hume, sexual activity is not causally related to reproduction. I hope that the claim I presented several pages ago, that sex behavior is independent of reproduction, thereby have been substantiated. A caveat needs to be presented immediately in order to avoid exaggeration. Reproduction is not possible without sex behavior and, whenever reproduction occurs in animals with sexual reproduction, sex behavior has been executed at one moment or another. This means that the relation sex behavior – reproduction is asymmetrical. It is very well possible to have sex without reproduction, but reproduction is not possible without preceding sex behavior. This asymmetry is unproblematic and the easiest way to handle it is by considering reproduction as an occasional, physiological side effect of sex behavior, without any importance for the mechanisms determining that behavior.

The asymmetric relation between sex behavior and reproduction can probably explain why some biologists have insisted on a necessary association between the two. This can easily be illustrated by an example where reproduction (or better fertilization) is the event R and sex behavior is the event S. We can formulate the law that if R, then it was preceded by S. The principle of constant conjunction is therewith satisfied. The principles of temporal priority and contiguity are also satisfied, since sex behavior preceded reproduction (fertilization) and reproduction (fertilization) was contiguous with sex behavior. Thus, S is the cause of R, in other words sex behavior causes fertilization. According to a biological point of view, then, it is entirely legitimate to associate sex with reproduction. On the contrary, if the starting point is S, then we can formulate a law saying that if S, it is not always followed by R. From the behavioral point of view, as mentioned, there is no causal relationship between S and R. It appears, then, that if our focus of attention is on mechanisms of reproduction, then it is reasonable to maintain an obligatory relationship between sex behavior and reproduction, while if our focus is on behavior, then it is reasonable to reject such a relationship. Since the subject of the present discussion is behavior, I consider sex behavior as a non-reproductive behavior.

3 The coincidence between Darwinian and Judæo-Christian views on human sexual behavior

3.1 The biological analysis of sexual behavior in the context of evolution and natural selection

As mentioned in the preceding section, there is a long tradition for teleological explanations in parts of biology and biological accounts of sexual behavior are no exception. The reproductive consequences of sexual acts are thus paramount in any analysis of sexual behavior as soon as a biologist is involved. It is tacitly or explicitly assumed that evolution has assured that sexual behavior is as efficient as possible for fertilization to occur.

An analysis of sex behavior in terms of biological function and evolution rests on some or all of the following assumptions:

1 The purpose of sexual behavior is fertilization. The purpose is also the cause.
2 Evolution has maximized the ability of sexual behavior to achieve fertilization.
3 If two or more different behavior patterns existed, then evolution has assured to eliminate the less efficient ones.

This means that all kinds of sexual behavior patterns not leading to fertilization should have been eliminated through the course of evolution. Furthermore, no one would engage in sexual activity if fertilization were not at least potentially possible. Non-fertile sex carries a cost to the individual but no benefit and any tendency to engage in such behavior should have been disadvantageous and eliminated through the process of natural selection. Finally, the behavior patterns expressed should be efficient for assuring the union of gametes with the least cost possible. That probably implies that copulation should be as short as feasible and the least physically exerting as feasible.

Let us briefly examine each of these suppositions. We have already discussed the problems with teleology quite extensively, so I will not do it again. The second and third assumptions are worthy of some analysis, though. If evolution had maximized the ability of sexual behavior to achieve fertilization, it should have assured that sex behavior only occurs when the individual is fertile. Any individual showing behavior outside of fertile periods should have been forced to carry extra costs and should have been extinguished. If we accept assumption 2, we must also accept that women represent one of evolution's failures. Females of some primate species in addition to *Homo sapiens* have also the privilege to share this position. Men are not better, in fact. They accept copulation with unfertile women, some even with insistence and pleasure. Males of many other species have the same bad habit of trying to copulate with unfertile females. A sexually experienced male rat may be just as insistent and persistent as any man when trying to copulate with an unwilling female. A female rat may be unwilling because she is prepubertal or because she is not in estrus. In both cases she is unfertile. I will not multiply the examples, but it is quite evident that evolution has failed to assure that sex behavior maximizes the possibility of fertilization. Having sex when not fertile is incompatible with the basic criteria of natural selection and sex with an unfertile individual is equally futile in an evolutionary perspective. We have at least three alternative solutions to get away with these imperfections. The first is to accept that natural selection is far from perfect and has allowed many inadaptive behaviors to persist. This, in turn, means that it is risky to assume that behavior is always adaptive. The second is an extension of the first, in the way that it is maintained that evolution has not shaped the intricacies of sexual behavior because natural selection has failed to exert systematic pressures on specific behavior patterns. The third is to invent elaborate and often quite esoteric arguments as to the adaptive advantages of showing inadaptive behavior. Personally, I favor the second alternative.

Considering the third assumption, the evidence in favor is mixed. Sometimes evolution seems to have been quite efficient. A rabbit, for example, shows a very efficient sexual behavior – one single intromission, always associated with ejaculation. It is all a question of about one second. Very efficient in time and in effort. Moreover, the doe is a reflex ovulator. She displays lordosis, making vaginal penetration possible, only when in estrus and when in estrus she is always fertile. Shortly after copulation, the female will ovulate and fertilization is almost certain since the male deposited semen a few hours earlier. After ovulation, the propensity to show lordosis will soon be reduced and finally absent. It would seem that evolution has done a good job, assuring that sex behavior in rabbits is fast, almost effortless, and frequently fertile. It is not so easy, though. A male rabbit may spend hours mounting an unfertile female if given the opportunity to do so. Although his sex behavior seems very efficient, it is not certain that it really is. It is not known how much time a wild male rabbit engages in mounting unfertile females, but it is not unlikely that it is far superior to the time he spends copulating with fertile females. If this indeed were the case, then his behavior would be quite inadaptive.

A male rat needs to make a large number of vaginal penetrations, intromissions, before ejaculating. The number varies often between 6 and 12, and the time investment can be from under a minute to up to half an hour or more. Since the female has the habit of running back and forth, and indeed *has* to run back and forth in order to incite the male to mount and intromit, it becomes a rather exhausting exercise. A poor rat needs to expend a lot of time and energy in order to deposit sperm. Extremely inefficient compared to a rabbit. Sexually experienced male rats also spend some time mounting infertile females, when given the opportunity, making their sex behavior still more inefficient. The inefficiency of the male rat's sex behavior is often explained as a consequence of the fact that the female rat requires several intromissions for successful implantation to occur (Adler, 1969) and that his behavior in fact is very adaptive. This is pure nonsense, for there is no *a priori* reason for the female rat to require many intromissions for subsequent successful implantation of the fertilized ova, thereby being far less efficient than a female rabbit.

There are innumerable examples of inefficient sexual behaviors among mammals and many of very efficient behaviors. The question is why the inefficient behaviors have not been eliminated through natural selection. They seem rather to have been rewarded. The rat is a real cosmopolitan. In fact, just as cosmopolitan as the human. Rabbits much less so. Many explanations have been advanced for the existence of dramatically inefficient sexual behaviors, all having in common that they try to invent adaptive advantages of clearly inadaptive behavior through long-winded and often surprisingly creative teleological arguments. Another explanation for the persistence of inefficient sex behavior is that there are so many characteristics and combinations of characteristics that determine the overall fitness of an individual that details of sexual behavior have a most limited impact. Natural selection operates, probably in a most efficient way, on all characteristics having significant influences on fitness, but far less so or not at all on characteristics having only marginal impact. This reasoning could explain the great variety of sexual behaviors that can be found in mammals, not to mention in other vertebrates. This variability stands in sharp contrast to the limited variability in other systems. The cardiovascular system, for example, is essentially identical in all mammals, both with regard to anatomy and physiology. The reason is probably that cardiovascular functioning is essential for survival and even small variations in the efficiency of this system are associated with substantial differences in fitness.

The preceding paragraphs should have illustrated that none of the basic assumptions in the traditional biological analysis of 'reproductive behavior' is supported by fact. They can be saved, though, by making still more assumptions about adaptiveness, by giving a purpose to inadaptive behaviors. When these additional assumptions are challenged, then additional assumptions can be made in order to save the previous ones and so on in eternity. The most reasonable conclusion is that it is unknown how much selective pressure has been put on sexual behaviors. The margin for random variation is, consequently, unknown. My guess is that evolution, through natural selection, has been quite generous in allowing sexual behavior to be quite inefficient and highly variable between one species and another. A wide margin of random variation, unavailable for natural selection, implies that speculations about evolutionary advantages are basically meaningless. A way to make sense of a senseless (random) world.

I have many times mentioned that biology likes to offer teleological explanations of sex behavior and, consequently, that the purpose and cause must be reproduction. I have also analyzed how a Darwinian analysis of sex behavior suggests that the only adaptive sex is that associated with reproduction. Teleology and evolutionary theory have merged in the conclusion that sex must be associated with reproduction and, if it is not, it is inadaptive. The next step is to say that inadaptive sex is anti-natural. This kind of reasoning has, in some societies and at some points in time, led to the tragic notion that humans displaying non-reproductive sexual behaviors are inferior to those that display sexual behaviors that potentially can be reproductive. Among the many behaviors considered anti-natural or inadaptive are sexual activities with individuals of the same sex, cunnilingus, fellatio or anal sex. All these activities were and still are, at some places, not only regarded as inferior but also severely punished either by social opprobrium, penal action or both. Biologists have provided a scientific rationale for such actions through their use of evolutionary theory.

3.2 Saint Augustine and sex: the *poena reciproca* and the role of love

As soon as we turn from non-human to human sexual behavior, an enormously difficult complication presents itself. Unlike sexual behaviors in non-human animals, human sexual behaviors are, to a substantial degree, socially constructed. For an excellent analysis of the social construction of sexuality, the reader is referred to the brilliant classic by Gagnon and Simon (2002). Here, I will only briefly introduce the notion of human sexuality as a social rather than biological entity.

Many human behavior patterns, particularly those pertaining to the category of precopulatory behaviors, have not by themselves any specific meaning. They acquire a sexual connotation through social learning and they may be highly variable between one culture and another and they can also change their meaning within a culture over time. One pre-copulatory behavior common in Europe and some other places, mutual tactile stimulation of the rostral end of the digestive canal (an activity vulgarly known as kissing), has no sexual meaning at all in other cultures. Its capacity to enhance sexual arousal efficiently in most Europeans is a consequence of social learning. Likewise, where, with whom and when to execute copulatory behaviors are determined by social learning in the human. I will not pursue this discussion here, but we must have clear right from the outset that human sexual behavior is determined by social conventions far more than by biological factors. Any discussion of human sexual behaviors ignoring the role of social learning is doomed to failure. An immediate consequence of this statement is that we must add a social perspective to sexuality whenever discussing human sexual behaviors.

Biologists have probably had a rather limited influence on the way human societies regard sexual activities. Biology is rather young as a science and has only had a couple of generations to influence human thought. There are other, much more powerful influences that have determined most of our views on sexuality. In Europe and the Americas, Christianity has undoubtedly exerted an enormous influence. In other parts of the world and in pre-Columbian America, other religions have been and are tremendously influential. Despite the importance and the large number of people adhering to these other religions, they have had a most limited influence on the scientific study of sexuality. For a long time, science as we know it was mainly a European undertaking. More recently, some former European colonies have acquired scientific predominance. This does not change much, since they are extensions of European culture. However, as an act of courtesy it has become habitual to talk of Western rather than European culture. Independently of the name we put on it, most of the scientific study of sexuality has been performed in this Western culture.

Scientists are part of the societies in which they live. They share their societies' ideals, beliefs, prejudices and values. When entering the laboratory, they carry with them all these ideals, beliefs, prejudices and values. When they decide which problems to study and, particularly when interpreting their data, all these ideals, beliefs, prejudices and values will be present, exerting implicit or explicit, conscious or unconscious influences. This is probably of little concern if the subject is quantum mechanics or non-Euclidean geometry, because Western society does not attach too many ideals, beliefs, prejudices and values to quantum mechanics or non-Euclidian geometry. It is a completely different thing with regard to sexuality.

Western society has very clear opinions on that matter and some of these opinions are profoundly rooted in a very long history. Most of them are probably so firmly rooted in history that they are no longer questioned. They have become what we may call facts of nature. These ideals, beliefs, prejudices and values have had an enormous impact on research pertaining to the domain of sexuality. In order to be able to estimate the magnitude of that impact it could be helpful to try to make them explicit. A short incursion into history might be necessary for this.

Sexuality has traditionally been a subject for all kinds of moral analyses. In Christianity, there seems even to have existed, and continues to exist, something close to an obsession with the morals of sex, perhaps also with sex itself. There is a good reason for this.

One of the church fathers, Saint Augustine (Augustinus Aurelius, bishop of Hippo, doctor of the Roman Catholic Church, AD 354–430) is regarded by many as the most influential of Christian thinkers. He wrote quite extensively on sexuality and he wrote quite explicitly. His luminous analyses have become the basis for the views on sexuality still ferociously defended by the Holy Roman Catholic and Apostolic Church as well as by the majority of Lutheran churches. The following summary of Augustine's views on sexuality is essentially based upon three excellent works: Brown (1983, 1988) and Schmitt (1983). To the specialist, it will certainly appear superficial and uninformed, but it should be acceptable for the purpose of showing how his analyses of human sexuality still are the bases for most contemporary opinions, both among men of faith and among scientists. Those especially interested in the subject may want to venture into the original writings, particularly volume 14 of the City of God, and Sermon 51, a Christmas sermon dedicated to the issue of sex.

Augustine knew that sex is a prerequisite for reproduction. He also knew that reproduction is necessary because the human is mortal. And he is mortal because Adam and Eve lost their eternal life and were thrown out of paradise because they had insurrected against God's will. This means that the necessity of the sexual act for the persistence of the human species is a direct consequence of the original sin, of a hiatus between the will of man and the will of God. As soon as Adam and Eve had made their own wills independent of the will of God, parts of them became resistant to their own conscious will. Their bodies were touched with a disturbing new sense of the alien, in the form of sexual sensations that escaped their control. Here we have the essence of the problems with sexuality according to the saint: it is outside the control of the will. For example, Augustine observed that orgasm was a surge of sensation over which the mind had no final control, either to summon it, or to cancel its quality once experienced. In the same way, erection may occur whether we want it or not and it may also refuse to occur when we want it. The uncontrollable physical responses to desire became the appropriate punishment for Adam's sin, a *poena reciproca* according to Augustine and many others.

The uncontrollable elements in sexual desire reveal the working in the human person of a *concupiscentia carnis*, of a permanent flaw in the soul that tilted it irrecoverably toward the flesh and away from God's will. Each time sexual desire (a manifestation of *concupiscentia carnis*) is activated, so is the *poena reciproca*. Every manifestation of sexual desire is a reminder of the original sin and, as such, a divine punishment for the most fateful flaw of human nature. As a result of these considerations, sexuality would never be trivial in the West.

There are two conditions that must be satisfied before a human being can engage in sexual activity without offending God: that the aim is procreation and that the participants are united in love between themselves and in love of God. Only through a limitless love of the next (the partner) and of God could the *concupiscentia carnis* and the associated *poena reciproca* become compatible with the Christian faith and any hope of a post-mortem eternal life. Needless to say, once a small number of children had been begotten, any decent human should be continent. In fact, Augustine considered the perfect Christian marriage a marriage of continence, for a simple reason. The justification for procreation was to increase the number of inhabitants in the city of God, but that number was sufficient already at the time of Augustine, so he preferred to insist on moderation of procreation under all circumstances and on continence as the ideal. It may be interesting to note that Augustine did not have the slightest interest in the preservation of the species. To him, procreation had a religious, not biological, aim. Nevertheless, if, by some default of character, sexual activity had to be displayed, it should be with procreation in mind. If not, it was a sin. Augustine, knowing the force of the flesh and of lust, the famous *concupiscentia carnis*, realized that people sometimes would succumb to their sexual urges and have sex for pleasure. If it occurred in a loving couple, he was generous enough to consider it as a venial sin.

What we can observe here is that the sexual act should be performed with the purpose of reproduction if it should be performed at all. Although the arguments employed by Augustine are principles of faith rather than of biology, the conclusion is the same as the one reached in Darwinian biological thinking. The question that immediately surfaces is whether the Darwinian notion of sex for, and only for, reproduction is a necessary corollary of evolutionary theory or whether it is a consequence of the fact that most biologists have grown up with the religious idea that sex is for reproduction. Most of those biologists were probably not aware of the fact that Saint Augustine's reasoning provides complete justification for that opinion. Perhaps they felt that evolutionary theory offered a justification for something they had been taught but which they did not understand why they had been taught. There is probably not any final answer to this question, but the possibility that evolutionary theory was and is used to justify an otherwise unfounded, social and/or religious convention, should not be overlooked. What is clear, though, is that there is no necessity to impose an association between sex and reproduction in order to explain how evolution has succeeded in assuring that most individuals will engage in sexual activities. This issue will be discussed at length in Chapter 2.

Another important point within Augustine's reasoning that persists in contemporary society is that sexual activity is only acceptable when the participants are in love with each other. This religious ideal seems to be firmly anchored in Western societies, but few of those confessing their adherence to it are aware of its origin and the reasons that led Saint Augustine to propose it. Although numerous survey studies reveal that many Europeans and Americans attach a great importance to being in love with the sexual partner (see e.g. Aron and Aron, 1991; Hendrick and Hendrick, 2002), very few, if any, of these surveys have ever asked their subjects *why* it is important or *how* it has become important. Again, we have all been taught that it is, but only some rare inquisitorial minds take the trouble to ask why. When they do, explanations like natural, the way it must be and the like appear. They are probably as well founded as the notion that sex is for reproduction, but none of them appears particularly compelling. In this context it might be illuminating to observe that the importance of love as justification for sexual activities is much more evident in Western cultures than in, for example, India or Japan (Jankowiak and Fischer, 1992; Levine *et al.*, 1995; Medora *et al.*, 2002). It is not impossible that the Augustinian notions on sexuality have been more successful in penetrating the minds of Europeans than those of Indians or Japanese.

Love will not be a major theme in this book. It will, in fact, not be a theme at all. The concept of love suits extremely well in literature and in the other fine arts, and some not as fine like the movie industry, but it is uncertain whether love has any place as a scientific entity or object of study. I would support the notion presented in an extraordinary little paper, almost never cited, by Beall and Sternberg (1995). They consider love to be a social construction and the definition as well as the emotional experience associated with it are, consequently, contextually determined. They change between cultures and between times within a culture. Thus, the idea that love is something intrinsic to human nature is entirely false. Notwithstanding, there are some scientists who have succeeded in localizing love to the medial insula, the anterior cingulate cortex, the putamen and the caudate nucleus with the aid of functional magnetic resonance imaging (fMRI) procedures (Bartels and Zeki, 2000). Reassuring to learn that love has such a wide distribution. Another fMRI study of this phenomenon of such a paramount importance for human health found that love is localized to the right ventral tegmental area and the caudate nucleus (Aron *et al.*, 2005). This is a far more limited distribution than the first study mentioned here, but at least one structure coincides between studies, the caudate nucleus. The obvious conclusion, which the authors had overlooked, must be that love is an extrapyramidal event. It would perhaps be too optimistic to expect some of these scientists to solve another fundamental problem in neuroscience: the localization of the superego, the entity determining the permissible expressions of love. Only time will show if they will be successful. A reasonable guess would be that the pineal turns out to be the site. It is, after all, the spot where body and soul interact according to Descartes. Independently of the results of that future study, the notion of love does not seem particularly useful in a scientific analysis of sexuality and will therefore be respectfully ignored.

After the preceding parenthesis we can now return to the real subject of this section. I proceed immediately to the conclusion, which is quite simple. Some of the Christian views on sexuality coincide with, or are the origin of, the biological notion of sex as a behavior in the service of reproduction. Some others of the

Christian analyses of sexuality have not had any immediate impact on scientific hypotheses concerning sex, but they have determined or influenced society's norms concerning sexuality.

4 If sex is not for reproduction, what is it for?

In the preceding sections, I invested many words in an effort to convince the reader that questions like 'why?' and 'what for?' have no place in science. This truth has not changed, so the question posed in the title must be reformulated. One possible and entirely acceptable reformulation is: 'under which circumstances is sexual behavior activated and what are the consequences of executing the behavior?' We can immediately see that this question does not refer to function but to cause and consequence. Nevertheless, and as mentioned many times, it is undoubtedly so that fertilization is preceded by sexual behavior. No biologist, and I assume also no psychologist, would like to dispute that. However, even if this is perfectly true, it does not explain, in any way, the circumstances under which sexual behavior occurs, with whom it occurs, how intense the behavior is, or how frequently it is executed. Furthermore, and which has also been mentioned many times, there is no cause–effect relationship between sexual behavior and reproduction. If we are interested to know the mechanisms, behavioral and neural, that control this behavior, we must forget its biological function and start to look for events that activate the behavior. This assertion is not new or unusual in any way. Thirty years ago, the founder of behavioral neuroendocrinology and a giant in the study of sexual behavior, Frank Beach, wrote the following brilliant passage:

> Since no animal mates in order to reproduce, but animals must mate in the service of species survival, we are faced with the problem of identifying the source of reward or positive reinforcement which impels individuals to copulate. The problem has scarcely been recognized as far as sexual activity is concerned (Beach, 1976, p. 471).

Frank Beach clearly took the position that sexual behavior cannot be understood through an analysis of its biological function. In other words, he rejected teleology. Moreover, he realized that sexual behavior is rewarding and reinforcing. This insight provides the basic clue for any serious effort to study sexual behavior. As we will see in Chapter 2, sex in itself is indeed a rewarding activity, just like eating or drinking. A determination of the circumstances, internal and external, under which the behavior occurs and reward is obtained is all we need in order for a complete understanding. In that endeavor, there is not the slightest need for any mention of function.

If individuals engage in sexual behavior not because they feel an urge to reproduce or to obey the religious principle of 'multiply' but because they want to experience the positive affect inherent in sexual activities, then sex ceases to be a reproductive activity and turns into a recreative activity like going to the theater, riding a roller coaster or playing bridge. This notion could have tremendous implications for society's views on sexuality but none for sexual behavior. In fact, humans and other animals have always had sex for the fun of it. Undoubtedly, many humans have made and make efforts to obey the many rules established by society, biology, and doctrines of faith, but they seem to be in a minority and their success is uncertain. Although the Holy Roman Catholic and Apostolic church does not accept contraception, a vast majority of Catholics use unauthorized methods to dissociate sex from reproduction. Although many Lutheran groups reject homosexuality, there is no indication that homosexuality is less frequent in countries where such groups are influential. While individual sexual behavior always seems to have been directed towards the obtention of positive affect, the norms supposedly controlling that behavior have established otherwise. One of the reasons for the excitement produced by the Kinsey reports (Kinsey *et al.*, 1948, 1953) was, in fact, that they made evident the huge discrepancy between official norms and actual behavior.

The recognition of the fact that individuals engage in sexual activities for recreational purposes facilitates our understanding of sexual behavior in humans and other animals, but it has no necessary implications for society's views on sexuality. Most social rules have their origin in religious notions and such notions are not appropriate subjects for scientific analysis.

5 The high cost of the imposed association between sexual behavior and reproduction

Most of those interested in human sexual behavior realize that biological mechanisms are important. At the same time, it is accepted that the expression of sexual activities is socially rather than biologically determined. This applies to the choice of partner, to the place and time chosen for the performance of sexual behaviors, as well as for the specific behaviors displayed. Considering that procreation for a long time has been regarded as the only legitimate reason for engaging in sexual behaviors, it becomes immediately evident that the only socially acceptable sexual activity is penile–vaginal intercourse. Any other activity leading to genitally induced positive affect is sinful, anti-natural, asocial and possibly also criminal. Such activities include masturbation (solitary or mutual), fellatio, cunnilingus, anal sex, and whatever other way the human mind can figure out for obtaining sexual reward. Sex between individuals of the same sex can never lead to procreation and is, therefore, by definition sinful, anti-natural, asocial and frequently also criminal.

If the notion that the only legitimate reason for engaging in sex is procreation had not been dominating Western societies for centuries, a tremendous amount of human suffering would have been avoided. The most obvious is certainly that gays and lesbians would have been able to lead a decent and peaceful life. Instead of that, they have been and are frequently still victims of persecution on the part of the forces of repression, discrimination or rejection by religious communities and sometimes by conservative civil societies and even by family members.

Homosexuals are not allowed to formalize their relationships in the same way as heterosexuals are with the exception of a handful of advanced countries accepting homosexual marriage. All this inhumane treatment of fellow humans is justified by the notion of sexual activity for and only for procreation.

The emphasis on penile–vaginal procreative sex as the only acceptable sexual activity has led to a reduction in the quality of life for numerous groups besides lesbians and gays. One of the largest is menopausal women. In menopause, reproduction is not possible and sexual activity cannot be legitimate. Thus, those menopausal women taking either religion or biology seriously felt obliged to abandon sex and move out of the conjugal bedroom. Their poor husband was also forced to abandon sex, since it would be sinful to have sex with a non-fertile woman. He could at least find a solution to this dilemma by taking a younger woman as mistress, but the solution could only be partial since he then committed the mortal sin of adultery. In fact, having sex with his wife would have had a lesser cost, since Saint Augustine determined that sexual activity without the goal of reproduction is only a venial sin. The husband's predicament is probably that his wife refuses to commit any sin at all, forcing him to choose the mistress solution. This choice is not unproblematic in terms of collateral damage, to use a term popular within the American military establishment. The husband will probably suffer feelings of guilt, the mistress will face similar feelings and, in addition, both of them will experience the many inconveniences of secrecy and the accompanying fear of discovery. In the case of secrecy failure, the wife and other close relationships may experience distress. This simple example of one possible consequence of the belief that sexual activity needs to be directed towards reproduction should be enough to illustrate that this idea may be responsible for significant reductions in the quality of life of many individuals.

Sexual activity among lesbians and gays as well as that of menopausal women becomes illegitimate as a result of the obligatory association between sex and reproduction. A corollary of this principle is that only penile–vaginal penetration is a morally and biologically acceptable sex behavior. Indeed, social learning has convinced most humans that this sexual activity not only is morally superior to other sexual activities, but also that the satisfaction it produces is superior to that offered by any other sexual activity. Both these notions contribute to deprive several human groups besides gays, lesbians and menopausal women of sexuality. One such group is men suffering from impotence. They could easily satisfy their female partners with masturbation or cunnilingus and their male partners with masturbation and fellatio. They could receive sexual satisfaction by being masturbated or fellated. Both activities can easily induce ejaculation/orgasm in the absence of erection. In fact, impotence would probably be a rather non-significant problem if penile–vaginal intercourse were not the only acceptable sexual activity and, in popular as well as in expert belief, the most satisfying. Another group deprived of sexual activity consists of victims of physical handicaps making the exercise of penile–vaginal intercourse exceedingly difficult or outright impossible.

Another important sequel of insisting on the association between sex and reproduction concerns the attitude towards the use of contraceptives. It is well known that the Holy Roman Apostolic and Catholic Church has rejected the employment of contraceptives for more than 1500 years, following the principles of Saint Augustine. Many Lutheran churches and the American federal government presently do the same. The amount of human suffering caused by these attitudes is difficult to estimate. What is clear, though, is that they have brought no benefit to mankind. It is not only religious organizations and fundamentalist governments that reject the use of contraceptives because of doctrines of faith. Many individuals are also influenced by these attitudes and refrain from the use of contraceptives from fear of appearing immoral. It has been argued that many unwanted pregnancies are a result of the strange idea that sex without contraception is less immoral than sex with contraception (see Gudorf, 1994, for an excellent discussion of this and other consequences of what she calls 'procreationism'). Tragedy, both for mother, father and child, is a not uncommon result of these pregnancies. Again, the notion that sex is for reproduction takes a high toll on human well-being.

Before leaving this section, there are three things that must be made clear. One is that religious beliefs cannot be criticized on scientific grounds. If some religious group wants to maintain that copulation without the explicit purpose of reproduction is a sin, it is certainly entitled to do so and its members have certainly every right to respect that principle if they would like to. What becomes problematic is the habit of some religious groups imposing their beliefs on others. The extreme expression of this is the incorporation of religious principles into common law. Even though this is a scientific text on sexual behavior, I feel morally obliged to make explicit my absolute rejection of that habit.

Another thing that must be made clear is that biologists' affirmation that the purpose of sex is reproduction is a metaphysical notion and can therefore be criticized as any other philosophical principle. My own opinion is that metaphysics has no place in science and it is consequently not even necessary to criticize metaphysical ideas on philosophical grounds in order to dismiss them.

Lastly, it must be mentioned that the widely held supposition that penile–vaginal intercourse is the most satisfying sexual activity has no scientific support. There are no objective measurements of the intensity of positive affect produced by different kinds of genital stimulation, but there are many studies reporting introspective data. These data are almost always in the form of responses to questionnaires or rating scales of different kinds. It appears that women find cunnilingus offering a higher degree of satisfaction than penile–vaginal intercourse (Hurlbert *et al.*, 1993). Orgasm is also more frequent during cunnilingus than during any other kind of genital stimulation. Furthermore, women find the orgasm that is readily associated with masturbation as more intense than the rather infrequent orgasm associated with traditional intercourse. Men find often fellatio more satisfying than penile–vaginal intercourse and fellatio is the most requested service at massage parlors (Bryant and Palmer, 1975; Armstrong, 1978). Although the reliability of these data are difficult to determine and although I have a profound skepticism with regard to data from questionnaires, I feel tempted to accept that penile–vaginal intercourse is not necessarily perceived as superior to other sexual activities, contrary to what good morals would suggest. The fact that questionnaire data suggest that non-vaginal sex may be more satisfying than vaginal sex is remarkable for two reasons. First, strongly dominant religious norms stipulate vaginal sex and condemn everything else. Thus, social desirability, which, as we will see in later chapters, is an important determinant of questionnaire responses, would favor penile–vaginal intercourse. Second, to the general public, the Freudian notion that vaginal orgasm is deep and mature compared to the immature and superficial clitoral orgasm is widely known and believed in. Here, the prejudice created by this notion should likewise have influenced responding in the way of favoring penile–vaginal intercourse. With these arguments in mind, it is indeed surprising that cunnilingus, an immoral sexual act for the religious and an act producing clitoral orgasm for the Freudians, turned out to be favored by women. With regard to men, their favored activities are also classified as immoral by the religious. Freud was not as fascinated by male orgasm as he was by the female and did not, to my knowledge, explicitly compare different means of obtaining male orgasm, perhaps with the exception of masturbation. His influence on popular beliefs should be far less evident for men than for women, leaving us mainly with religious notions favoring penile–vaginal intercourse. It is, consequently, less surprising that penile–vaginal intercourse neither is the favorite for men.

From a comparative perspective we must immediately accept that penile–vaginal copulation is the dominant and usually the only behavior displayed by non-human animals. Rats are sometimes found to lick their penis or their vaginal opening, but whether this response is similar or at all related to masturbation is uncertain. Females may also drag their hindquarters, including the vaginal opening, against the floor, but there is no evidence showing that this is a sexual response related to self-stimulation or masturbation. Male primates are known to masturbate the erect penis in response to the sight or odor of a female and this can certainly be regarded as a sexual response directly equivalent to masturbation in the human. Some female primates stimulate their clitoris with the tail, again a response most likely equivalent to human masturbation. Despite these few examples of non-vaginal sexual activities in primates, it is undoubtedly so that they are very rare in non-human mammals. I will not here speculate about the reasons for this. Such speculations will be presented in a more appropriate context (Chapter 2). Suffice to repeat the extremely trivial statement that there are a huge number of behavioral differences between humans and other mammals and the fact that a human behavior is not found elsewhere in the animal kingdom does not make it anti-natural or immoral. Writing or reading a book, for example, is neither of the two.

6 Differences and similarities between sex and other primary needs like food and drink

Sex is normally included on the list of primary needs or drives. Likewise, sex is frequently considered a primary reward and reinforcer (the difference between reward and reinforcer will become obvious within a couple of chapters). However, it is not evident that sex should be lumped together with food, drink, need for oxygen, adequate temperature, defecation, micturition and avoidance of pain. The lack of food or drink for a prolonged period of time eventually becomes incompatible with the survival of the organism. An organismic need is being built up and the individual will feel clear manifestations of the need in the form of a sensation of hunger or thirst. This sensation of need is called the drive stimulus in Hullian theory (Hull, 1943) and simply motivation in many other contexts. A fundamental feature of this motivation is that it increases with increasing deprivation. It is generally so that the amount eaten or the latency to start eating are directly and inversely, respectively, related to the length of deprivation and hence to the intensity of motivation. After a very long deprivation, the individual is weakened because of the organismic need and the motivation may fail to manifest itself in behavior. After a still longer deprivation the individual is transformed into a corpse.

Events like these never occur with sex. There is no need building up, no sensation of a need and no death waiting at the end of deprivation. In fact, it has never been observed that abstinence from sexual activity has any deleterious consequences for the organism. Many humans abstain voluntarily from sexual activity without suffering any deterioration of health. Sex is a luxury, not a necessity. Thus, sex is very different from food or drink.

It was, again, Frank Beach who first realized that sexual motivation is not activated by an internal process but by an external stimulus, a potential mate (Beach, 1956). In most adult individuals, the sight, smell or sound of a potential mate activates sexual motivation and the motivation then activates approach behaviors and, if the mate is in an appropriate state, copulatory behavior. In the case of a male rat, the sight and smell of a female might activate sexual motivation and approach behavior and, if the female happens to be in estrus, she will be attracted to the male and invite him to initiate copulatory behavior. A potential mate functions as a positive incentive and, since the incentive is necessary for the activation of motivation, the kind of motivation activated by it is called incentive motivation. The important point here is that motivation is dependent on an external stimulus.

Although food and drink generate internal stimuli that activate motivation independently of any external event, there are also alimentary stimuli having incentive properties, that is they can activate approach behavior and food consumption in the absence of an internal drive stimulus. I might have had an abundant dinner before going to a late reception, and not have the slightest intention to eat anything during that reception. However, discovering that they serve canapés topped with an abundant amount of the most exquisite Iranian Beluga caviar, I might feel an intense hunger and eat as many canapés as my level of shame allows me. If nothing else, this example shows that the motivation to eat indeed can be activated by external stimuli, incentives, even in the absence of an internal drive stimulus. This in turn means that food and sex both obey incentive motivational mechanisms while only food is sensitive to deprivation and the ensuing internal drive stimulus.

A few words on an important difference between humans and other animals must be added. Any human can have a mental representation of a sexual incentive at any time. She/he may even have a representation of being engaged in sexual activity. We have already learned that such representations can function as causes of human behavior. This means that a human, at any time, can activate sexual motivation through a representation of a sexual incentive. Moreover, it is known that at least some women can have mental representations of sexual activity so potent that they reach orgasm (Sizemore, 1976; Loewenstein, 1978; Whipple *et al.*, 1992). This fact very nicely illustrates the force of mental representations as causes of behavior and even of physiological reactions. Women obtaining orgasm by mental representation of sexual activity show the same hormonal and circulatory changes as women obtaining orgasm as a result of physical stimulation of the genitalia, either by coitus or by masturbation (Whipple *et al.*, 1992; Exton *et al.*, 1999, 2001). Furthermore, the likelihood of obtaining orgasm by masturbation is related to the appearance of representations of intercourse (Lentz and Zeiss, 1983). Orgasm through representations is also rewarding (induces a state of positive affect), further illustrating the power of mental representations. For some reason, those working in the field of human sexual behavior do not like the word representation and replace it with the unscientific expressions of 'fantasy' or 'imagery'. This latitude in the use of language may confound some readers. Nevertheless, we have here eloquent examples of how representations can function as causes of human behavior. Whether representations of sexual incentives and/or of sexual activities exist in non-human animals is not known. The most parsimonious approach to this issue is imagining that they do not exist and conclude that non-human animals are not sexually motivated in the absence of a sexual incentive perceived by the senses.

Another short comment is needed before ending this section and it has to do with what is frequently called popular science or cultural prejudice. Some people are completely convinced that sexual motivation builds up over time. We have all been exposed to stories of the fantastic exploits of sailors at the whore-houses in distant harbors, or of soldiers showing far more heroism at the brothels than in the trenches. I know that these kinds of stories are frequent at men-only gatherings and, although I never have been able to witness it in person, I would bet that they are equally frequent at women-only gatherings. There is no reason to believe that this anecdotical evidence has any correspondence in reality. No one has ever tried to gather reliable human data on this subject and the evidence from animal studies is clear-cut. If absence of sexual activity for a long time has any effect on indices of sexual motivation and performance, it is in the opposite direction. We have data from rats that were tested for sex behavior three times a week for several weeks and then forgotten for more than two months. After rediscovery, they were again tested for sex behavior. It was far inferior to the behavior displayed at the last test of the previous series of tests, despite the fact that they had copulated quite frequently for some time at that earlier point. Their level of 'deprivation' was very low then, but it should have been quite high after the two months of inactivity. Instead, the intromission and ejaculation latencies were prolonged and the intromission ratio was somewhat reduced. All this was indicative of reduced sexual performance, exactly the opposite to what any notion of deprivation would predict. It is my impression that most of my colleagues working with sex behavior have made similar observations, that is that animals regularly engaged in sexual activity show more intense behavior than animals having suffered long periods of abstinence.

I do not know the origin of the myth that abstinence enhances sexual motivation, but it is entirely unfounded. If not, the pope and other clerics would be real sex bombs.

Another popular and occasionally not so popular myth is that good health requires a certain amount of sexual activity. Abstinence or an unsuitably low frequency of sexual intercourse is deleterious to health, according to this myth. Some support can be mobilized from an extremely vicious reading of the clinical literature. One of the supporters of the notion of a minimum healthy frequency of sexual intercourse (Brody, 1997), obviously understood as penile–vaginal intercourse, cites some data purportedly showing that a too low sexual activity is unhealthy. Among the evidence cited is a rather exotic proposal regarding the factors causing Down's syndrome (Martin-DeLeon *et al.*, 1987). Since old sperm show a higher incidence of trisomies than young sperm, it is suggested, it would be convenient to have intercourse very frequently in order to assure that the spermatozoon fertilizing the ovum is very young. A problem with this idea is that the frequency of intercourse has only a minor or no influence on the speed of passage of sperm through the epididymis. Furthermore, Down's syndrome is not overwhelmingly common. Additional, equally convincing, support comes from a report showing that sexually active menopausal women had less vaginal atrophy than inactive women (Leiblum *et al.*, 1983). This is certainly an important aspect of health, but it is not evident that a high frequency of penile–vaginal intercourse impeded atrophy. Perhaps women suffering from vaginal atrophy do not like to have sex as frequently as those without that problem. More support for a healthy minimum level of sexual activity was mustered from a study of men with and without prostatic cancer in the Minneapolis-St Paul area (Mandel and Schuman, 1987). Men with cancer reported a lifetime frequency of intercourse slightly lower than men without cancer. This was taken as indicating that sex is good for health. However, the cancer patients had more experiences with prostitutes, had a larger number of homosexual partners and they had suffered more venereal disease than controls. Moreover, some of these other differences were more related to the incidence of prostate cancer than the frequency of intercourse. I would, actually, conclude that these data show that abstinence from sexual activity is far healthier than sexual indulgence. In fact, any argument as to beneficial effects of sexual activity needs to show that such effects are more important than the costs associated with it. One of the major costs is, without doubt, infection with sexually transmitted diseases. There is no need to explain that such infections are extremely rare in the abstinent and there is no need to explain that the probability of infection is directly related to the intensity of sexual behaviors. Other well-known costs of sexual activity are an increased incidence of cancer of the cervix. The earlier sexual activity is initiated, the more partners a woman has, the more venereal diseases she suffers, the more likely is cancer of the cervix. Even the side effect of sexual activity, pregnancy, is a potential contributing factor to the likelihood of cervix cancer (Kjær *et al.*, 1992; International Collaboration of Epidemiological Studies of Cervical Cancer, 2006). I would assume that the costs of sexual activity largely outweigh its rather uncertain benefits.

The notion of a healthy level of sexual activity is scientifically unfounded. Instead it seems to be determined by social conventions. Brody (1997) suggests that one penile–vaginal intercourse per week should be the lower limit. In the 19th century, experts considered that one intercourse per week should be the upper limit. More activity than so was deleterious to health (reviewed in Fellman and Fellman, 1981).

Summary

1 It is difficult to define sexual behavior. The most adequate definition seems to be one referring to sexual reward more than to particular motor patterns or, still worse, specific goals or functions.
2 Explanations of behavior in terms of its purported function or final cause belong to teleology and are entirely unacceptable with one exception: humans may have a mental representation of the goal or purpose of their behavior and this representation may act as a cause. In other animals, we do not know if representation of goals exists. The consequence is that teleological explanations can only be applied to human behavior and only when the causative representation can be unequivocally known.
3 Explanations of behavior should always be given in terms of cause–effect. There is no exception to this. Other explanations are quasiscientific at the best and meaningless in most cases.
4 Biology has a long tradition of explaining phenomena teleologically. So with sexual behavior. It is performed for the sake of reproduction. Thereby, the alleged purpose is confounded with the cause, as always in teleology. Any sexual activity not leading to fertilization is inadaptive from an evolutionary point of view and should have been eliminated through natural selection.
5 Within the Christian tradition, sexual behavior must be performed with the explicit goal of reproduction, otherwise it is a venial sin. Saint Augustine wrote brilliant analyses of the inevitable association between the original sin and the sexual act. He also established the connection between sex and love and ended up maintaining that sexual acts should not only have the purpose of reproduction but also be executed by loving partners. These notions have still profound influences in contemporary Western society.
6 Sexual behavior is intrinsically rewarding or 'pleasurable'. It is the representation of sex-induced pleasure that moves humans to search for a sexual partner and eventually engage in copulatory behavior. In other animals, some stimuli are unconditioned sexual incentives, activating approach and copulatory reflexes.

7 Sexual motivation is not built up as a result of deprivation. It is activated by a sexual incentive or the representation of a sexual incentive or of sexual acts. Abstinence from sexual activity has no known deleterious consequences for health and does not threaten the individual's survival. Sex is a luxury. This statement does not, in any way, imply that human sexual activity is without importance for feelings of well-being and satisfaction with life. Likewise, appropriate sexual activity may be essential for the maintenance of certain close interpersonal relationships.

8 The motivation to eat and drink has an endogenous component, caused by deprivation. Absence of food or drink for a long time threatens the individual's survival. Food and drink are necessities.

9 The human is probably unique among animals in her capacity to evoke mental representations of sexual incentives or sexual acts in the absence of external, sexually relevant stimuli.

10 The human, at difference to other mammals, may obtain sexually induced positive affect through a variety of behavior patterns in addition to penile–vaginal intercourse.

References

Adler, N.T. (1969). Effects of the male's copulatory behavior on successful pregnancy in the female rat. *J Comp Physiol Psychol* **69**, 613–622

Armstrong, E.G. (1978). Massage parlors and their customers. *Arch Sex Behav* **7**, 117–125

Aron, A. and Aron, E.N. (1991). Love and sexuality. In *Sexuality in close relationships* (K. McKinney and S. Sprecher, eds), pp. 25–48. Lawrence Erlbaum: Hillsdale

Aron, A., Fisher, H., Mashek, D.J., Strong, G., Li, H. and Brown, L.L. (2005). Reward, motivation, and emotion systems associated with early-stage intense romantic love. *J Neurophysiol* **94**, 327–337

Bartels, A. and Zeki, S. (2000). The neural basis of romantic love. *Neuroreport* **11**, 3829–3834

Beach, F.A. (1956). Characteristics of masculine 'sex drive'. In *Nebraska symposium on motivation* (M.R. Jones, ed.), pp. 1–32. University of Nebraska Press: Lincoln

Beach, F.A. (1976). Cross-species comparisons and the human heritage. *Arch Sex Behav* **5**, 469–485

Beall, A.E. and Sternberg, R.J. (1995). The social construction of love. *J Soc Pers Relatsh* **12**, 417–438

Brody, S. (1997). *Sex at risk. Lifetime number of partners, frequency of intercourse, and the low AIDS risk of vaginal intercourse.* Transaction Publishers: New Brunswick

Brown, P. (1983). *Augustine and sexuality.* The Center for Hermeneutical Studies in Hellenistic and Modern Culture: Berkeley

Brown, P. (1988). *The body and society. Men, women and sexual renunciation in early christianity.* Columbia University Press: New York

Bryant, C.D. and Palmer, C.E. (1975). Massage parlors and hand whores – some sociological observations. *J Sex Res* **11**, 227–241

Corsini, R.J. and Auerbach, A.J. (1996). *Concise encyclopedia of psychology*, 2nd edn. John Wiley: New York

Craighead, W.E. and Nemerofff, C.B. (2001). *The Corsini Encyclopedia of Psychology and Behavioral Science*, 3rd edn. John Wiley: New York

Exton, M.S., Bindert, A., Krüger, T., Scheller, F., Hartmann, U. and Schedlowski, M. (1999). Cardiovascular and endocrine alterations after masturbation-induced orgasm in women. *Psychosom Med* **61**, 280–289

Exton, M.S., Krüger, T.H. C., Koch, M. *et al.* (2001). Coitus-induced orgasm stimulates prolactin secretion in healthy subjects. *Psychoneuroendocrinology* **26**, 287–294

Fellman, A.C. and Fellman, M. (1981). The rule of moderation in late nineteenth century American sexual ideology. *J Sex Res* **17**, 238–255

Fyhn, M., Molden, S., Witter, M.P., Moser, E.I. and Moser, M.B. (2004). Spatial representation in the entorhinal cortex. *Science* **305**, 1258–1264

Gagnon, J.H. and Simon, W. (2002). *Sexual conduct: the social sources of human sexuality*, 2nd edn. AldineTransaction: New Brunswick

Gudorf, C.E. (1994). *Body, sex, and pleasure. Reconstructing Christian sexual ethics*. The Pilgrim Press: Cleveland

Hanke, D. (2004). Teleology: the explanation that bedevils biology. In *Explanations: Styles of explanation in science* (J. Cornwell, ed.), pp. 143–155. Oxford University Press: Oxford

Hendrick, S.S. and Hendrick, C. (2002). Linking romantic love with sex: development of the perceptions of love and sex scale. *J Soc Pers Relatsh* **19**, 361–378

Hull, C.L. (1943). *Principles of behavior: an introduction to behavior theory*. Appleton-Century-Crofts: New York

Hull, L.W.H. (1959). *History and philosophy of science*. Longmans, Green and Co.: London

Hume, D. (1758). *An enquiry concerning human understanding*. Reprinted 1999. Oxford University Press: Oxford

Hurlbert, D.F., Apt, C. and Rabehl, S.M. (1993). Key variables to understanding female sexual satisfaction – an examination of women in nondistressed marriages. *J Sex Marital Ther* **19**, 154–165

International Collaboration of Epidemiological Studies of Cervical Cancer (2006). Cervical carcinoma and reproductive factors: collaborative reanalysis of individual data on 16 563 women with cervical carcinoma and 33 542 women without cervical carcinoma from 25 epidemiological studies. *Int J Cancer* **119**, 1108–1124

Jankowiak, W.R. and Fischer, E.F. (1992). A cross-cultural perspective on romantic love. *Ethnology* **31**, 149–155

Kinsey, A.C., Pomeroy, W.B. and Martin, C.E. (1948). *Sexual behavior in the human male*. Saunders: Philadelphia

Kinsey, A.C., Pomeroy, W.B., Martin, C.E. and Gebhard, P.H. (1953). *Sexual behavior in the human female*. Saunders: Philadelphia

Kjær, S.K., Dahl, C., Engholm, G., Bock, J.E., Lynge, E. and Jensen, O.M. (1992). Case-control study of risk factors for cervical neoplasia in Denmark. II. Role of sexual activity, reproductive factors, and venereal infections. *Cancer Causes Control* **3**, 339–348

Leiblum, S., Bachmann, E., Kemmann, E., Colburn, D. and Swartzman, L. (1983). Vaginal atrophy in the postmenopausal woman. The importance of sexual activity and hormones. *J Am Med Assoc* **249**, 2195–2198

Lentz, S.L. and Zeiss, A.M. (1983). Fantasy and sexual arousal in college women: an empirical investigation. *Imagination Cognition Personal* **3**, 185–202

Levine, R., Sato, S., Hashimoto, T. and Verma, J. (1995). Love and marriage in 11 cultures. *J Cross-Cult Psychol* **26**, 554–571

Lloyd, D. (1987). Mental representation from the bottom up. *Synthese* **70**, 23–78

Loewenstein, S. (1978). An overview of some aspects of female sexuality. *Social Casework* **59**, 106–115

Mandel, J.S. and Schuman, L.M. (1987). Sexual factors and prostatic cancer: results from a case-control study. *J Gerontol* **42**, 259–264

Manstead, A.S.R. and Hewstone, M. (1995). *The Blackwell encyclopedia of social psychology*. Blackwell: Oxford

Martin-DeLeon, P.A., Williams, M.B. and Reynolds, J.F. (1987). Sexual behavior and Down syndrome: the biological mechanism. *Am J Med Genet* **27**, 693–700

Mayr, E. (1961). Cause and effect in biology. *Science* **134**, 1501–1506

Mayr, E. (1997). *This is biology. The science of the living world*. Harvard University Press: Cambridge

McLaughlin, P. (2001). *What function explains: functional explanations and self-reproducing systems*. Cambridge University Press: Cambridge

Medora, N.P., Larson, J.H., Hortacsu, N. and Dave, P. (2002). Perceived attitudes towards romanticism: a cross-cultural study of American, Asian-Indian, and Turkish young adults. *J Comp Fam Stud* **33**, 155–178

Monod, J. (1970). *Le hasard et la nécessité: essai sur la philosophie naturelle de la biologie*. Editions du Seuil: Paris

Revonsuo, A. (2001). On the nature of explanation in the neurosciences. In *Theory and method in the neurosciences* (P.K. Machamer, R. Grush and P. McLaughlin, eds), pp. 45–69. Pittsburgh University Press: Pittsburgh

Rittner, T. and McCabe, T.L. (2004). *Encyclopedia of biology*. Facts on File: New York

Schmitt, E. (1983). *Le mariage chrétien dans l'oeuvre de saint Augustin : une théologie baptismale de la vie conjugale*. Études Augustiniennes: Paris

Sizemore, L.F. (1976). *Female orgasm and erotic fantasy*. PhD, Northwestern University

Spence, K.W. (1956). *Behavior theory and conditioning*. Yale University Press: New Haven

Van Orden, G.C. and Paap, K.R. (1997). Functional neuroimages fail to discover pieces of mind in the parts of the brain. *Philos Sci* **64**, S85–S94

Wakefield, J.C. (2005). Biological function and dysfunction. In *The handbook of evolutionary psychology* (D.M. Buss, ed.), pp. 878–902. John Wiley: Hoboken

Whipple, B., Ogden, G. and Komisaruk, B.R. (1992). Physiological correlates of imagery-induced orgasm in women. *Arch Sex Behav* **21**, 121–133

CHAPTER

2

An incentive motivational framework and the description of sexual behaviors

1 Sexual motivation: theoretical framework

As was pointed out in Chapter 1, sexual motivation is not activated by endogenous events but by an external stimulus. This is not the whole story, though. The external stimulus must initiate some processes within the central nervous system and these internal processes must sooner or later translate themselves into observable behavior, which means activity in skeletal muscles. This overt activity may or may not be associated with responses in the autonomous nervous system. In order to understand the mechanisms of motivation, it is essential to have some basic knowledge of the interactions between certain kinds of stimuli, central nervous activities and behavior. To that end, the incentive motivation theory elaborated by Dalbir Bindra (Bindra, 1969, 1974, 1976, 1978) is most useful. I will start this chapter with an introduction to that theory.

Environmental stimuli can be divided into three classes. Those that have no effect on an organism's behavior are neutral. The vast majority of stimuli belong to this class. Some stimuli may activate approach behaviors and once approached they may be consumed or interacted with in a variety of forms. An example is food, which is not only approached but also eaten if the organism is in an appropriate state. Another example could be a conspecific of the opposite sex. If the organism is in an appropriate state, it approaches a conspecific and, if that conspecific also is in an appropriate state, sexual interactions may be initiated. If not, social instead of sexual interactions may occur. Any stimulus inducing approach behavior in the organism is a positive incentive. Finally, some stimuli produce withdrawal. Examples are excessive heat, like that coming from a forest fire, or an intense smell of putrefaction, like that coming from a forgotten garbage bag, or some painful stimulus like electric shock. Stimuli producing withdrawal are negative incentives.

In addition to producing the behavioral responses of approach and withdrawal, incentive stimuli may activate visceral responses mediated by the autonomous nervous system or by the endocrine system. Examples of such actions of incentive stimuli may be the autonomous response of erection produced by an appropriate sexual incentive. Men and male primates and even male rats frequently respond with erection to stimuli emitted by a sexually receptive female. Similarly, negative incentives like an aggressive intruder may produce an endocrine response such as release of corticosteroids from the adrenal gland cortex as well as sympathetic responses such as release of catecholamines from the adrenal medulla and increases in heart rate and blood pressure. Thus, besides producing organized activity in skeletal muscles incentives may, or may not, produce visceral responses of many kinds.

One very important characteristic of incentives, positive or negative, is that the approach or avoidance behaviors they activate are arbitrary in the sense that there is no obligatory relationship between one incentive stimulus and a motor pattern. A steak may activate intense approach behaviors in a dog, but the specific motor patterns executed depend on whether the steak is made available in the dog's food bowl or on the dish of an honored guest. In the first case, the dog would run to the food bowl and ingest the steak with little decorum, while in the second case she would politely sit on the floor beside the guest, staring at him with begging eyes as he put a piece of steak to his mouth. If that failed to produce any result, the dog may start making discreet sounds or lightly scratch the guest's leg, or whatever the dog's experience tells her is the most efficient way to get hold of the alimentary incentive. In the same way, a male rat may press the lever in a Skinner box in order to reach a receptive female, or run over an electrified grid, or overcome whatever obstacle an experimenter may invent in order to humiliate his rodent subject. The point should be clear: there is no obligatory association between a particular incentive (like food or a female) and a particular motor pattern for approaching the incentive. The subject chooses the motor pattern most likely to lead to success according to the specific situation and his previous experience with this and similar situations.

A positive incentive will not always produce approach and a negative incentive will not always produce withdrawal. While an adult male rat will approach a sexually receptive female and eventually initiate sexual interaction with her, a prepubertal male will show much weaker approach behaviors and very little sexual interaction will occur. Likewise, a rat that has been deprived of food for 24 hours will rapidly approach and start to eat its normal food pellets when given the opportunity to do so, while a rat that just has finished a large meal of delicious chocolate pellets will pay no attention to her normal pellets. These petty examples show that a constant stimulus (a sexually receptive female or standard food pellets) may or may not have effects on the organism. Obviously, there is no necessary dichotomy effect – no effect. In most cases there are simply variations in the magnitude or intensity of response activated by the incentive stimulus. Nevertheless, it is clear that some internal mechanism that can account for the varying impact of constant incentives needs to be included in any theoretical model of motivational mechanisms. In incentive motivation theory this internal mechanism is called the *central motive state*.

The expression *central motive state* was probably introduced by Morgan in his famous 1942 textbook (Morgan, 1942). He ascribes four characteristics to it:

1 It is, at least partly, self-perpetuating. This merely means that it is not spontaneously fluctuating from one moment to another but shows a continuous and maintained functioning until some event reduces or terminates it. A food-deprived rat running in a maze for food reward at the end does not run in fits. It runs quite smoothly at each trial, but the intensity of approach behavior as well as of food consumption will be reduced when satiety is approached.
2 General activity increases with increasing intensity of the central motive state.
3 The central motive state activates specific behavior patterns. Supposedly, these behavior patterns should be related to the origin of the central motive state. If, again, food deprivation is the cause for the activity of the central motive state, behavior patterns likely to be activated are those related to search for food and eating.
4 The central motive state induces 'a set or potentiality for presenting various patterns of behavior when the appropriate stimulus conditions in the external environment are available. This is the priming property of the c.m.s.' (Morgan, 1942, p. 461). In other words, the central motive state primes the individual for reacting to certain stimuli, again supposedly, stimuli that have some relationship to the origin of the central motive state.

Points 1 and 2 were certainly most important in the kind of motivational discussions that were common in the early 1940s, but seem to be of slight interest in a contemporary analysis of motivation. Points 3 and 4, though, are rather central to modern incentive motivation theory. They imply, among other things, that there are many different central motive states, several of which may be active at the same time. The one that will translate into overt behavior is the one corresponding to the stimuli present in the environment. Any reader interested in motivation may find it intellectually rewarding to compare these statements with the views on response hierarchies, drive stimuli, external stimuli and incentive value presented in the writings of Clark Hull (Hull, 1943, 1951, 1952).

Bindra systematically avoids speculations concerning the intimate nature of the central motive state. This is probably a wise attitude, since it should be regarded and treated as an abstract concept without any necessary correspondence in reality. It is, nevertheless, a useful concept that allows us to explain exactly what it is intended to explain, namely the variations in reaction to a constant stimulus between different situations. Having said this, it may be appropriate to point out that we *can* make some speculations about the nervous bases of the central motive state associated with sexual behaviors. We even have data allowing us to go from speculations to specific hypotheses and occasionally from hypothesis to established fact. More about that will come further down and in other chapters.

Figure 2.1 illustrates what an incentive motivational model of sexual behaviors could look like.

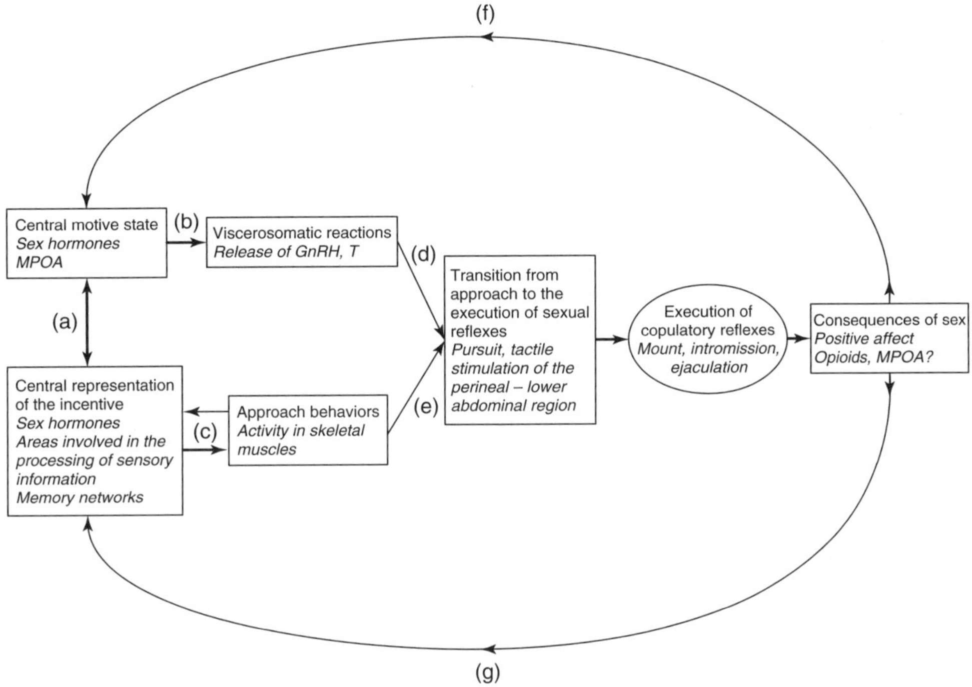

FIGURE 2.1 A model for sexual incentive motivation. The text in *italics* represents the example of the male rat. (a) A reciprocal excitatory relationship functioning in such a way that the central motive state enhances the sensory system's sensibility to stimuli with sexual significance. When such stimuli are perceived, the sensory system excites the central motive state which, in turn, further sensibilizes the sensory system, i.e. the relationship is one of reciprocal positive feedback. (b) At a certain threshold level of activity, the central motive state engages a series of viscerosomatic activities preparing the subject for sexual interaction. (c) The appropriate environmental stimuli activate motor patterns that bring the subject in contact with the source of stimulation. During approach, additional incentive stimuli may be encountered. These will be centrally represented and enhance the central motive state through (a). (d) and (e) Provided that approach behaviors have been successful and that appropriate viscerosomatic reactions are being accomplished, the subject's behavior *may* change from unconditioned or conditioned instrumental responses to the execution of sexual reflexes. These are activated by tactile stimulation of the perineal or lower abdominal region. If the subject is sexually inexperienced, such stimulation is obtained accidentally. If the subject already has acquired sexual experience, then conditioned instrumental responses may facilitate the attainment of tactile stimulation necessary for activation of sexual reflexes. At the point of transition from approach to execution of copulatory reflexes, the behavioral sequence is aborted in the absence of tactile stimulation. In the case when sexual reflexes indeed are activated, sex behavior will normally continue until ejaculation. (f) The positive affect induced by ejaculation will feed back to the central motive state where a short-lasting inhibitory system is activated. (g) At the same time, the positive affect and associated processes of reinforcement will strengthen learning of associations between itself and environmental cues. These cues will acquire incentive properties in relation to the intensity of the positive affect that is experienced. MPOA, medial preoptic area; GnRH, gonadotropin-releasing hormone; T, testosterone. (Reprinted from Ågmo, 1999. Copyright 1999, with permission from Elsevier.)

When there is an active central motive state and an appropriate stimulus, behavior is usually activated. As always, behavior is activity in skeletal muscles. This means that the incentive stimulus impinging on the active central motive state in the end must activate motor neurons. The process could be conceived as a kind of automatism where efferent stimulation excites the central motive state which, in turn, excites motor neurons, in the same way as a sensory neuron excites a motor neuron in a reflex arch. In motivated behavior, large amounts of sensory neurons, perhaps from several modalities, vastly distributed in the brain, may connect to large numbers of motor neurons, vastly distributed in the brain. These connections might involve several short interneurons and perhaps also longer projection neurons. Both kinds of neurons may be modulated by the central motive state, making the connection between sensory input and organized motor output either impossible or possible with varying probability, in the same way as other polysynaptic reflexes are modulated by interneurons of several kinds. The point here is that motivated behavior, like sex, can be regarded as a sequence of automatic reactions to environmental stimuli. This allows us to understand and explain the behavior without any reference to goal or purpose. An automatic response occurs as a consequence of a specific pattern of stimulation, whether we want it or not, and without any purpose. An erection in response to a beautiful student during a lecture has certainly no purpose, except for reminding us of the *poena reciproca*, the humiliation of not having control over our organism with our will, as so eloquently exposed by Saint Augustine.

The preceding discussion is applicable to both human and non-human sexual incentive motivation. However, as repeatedly pointed out in Chapter 1, the human may replace an external incentive stimulus with the mental representation of such a stimulus. The requirement that an incentive be present in the environment does not, therefore, apply to the human. As we will see, this difference has some important consequences. For example, while male primates almost never masturbate in the absence of sexually relevant external stimuli, men frequently do so. They may intentionally produce external sexual incentive stimuli, by opening a pornographic magazine or turning on one of the many TV channels offering pornographic movies, but masturbation may also be activated solely by mental representations of sexual incentives. Female primates rarely masturbate and they never do so in the absence of adequate sexual incentive stimuli. Most women masturbate, and they usually do it in the absence of external sexual incentives. Like men, they may produce such incentives on purpose, but again pure mental representation may be enough for activating the behavior. Thus, while non-human sexual behaviors are never or almost never performed in the absence of sexual incentives, humans may engage in such behaviors at any moment.

If we conceive the scheme incentive stimulus (real or in the form of a mental representation) – central motive state – motor output as an automatism, we gain a lot in conceptual power and scientific rigor. Not only do we get rid of purpose, hence teleology, but we can also propose mechanistic cause–effect explanations. This can be done at the purely behavioral level. If we maintain the central motive state constant, we can determine the relationships between incentive stimuli and ensuing responses in the way behaviorists do, without any consideration of processes within the nervous system. If, on the contrary, we maintain the incentive stimuli constant, we can manipulate the nervous system and determine changes in motor output. This would eventually allow us to understand the cellular processes underlying the abstract concept of central motive state. Both approaches are currently used, and have been used for a long time, but they have not been integrated into a coherent theoretical framework. The next few chapters should illustrate the fruitfulness of the conceptual framework offered by incentive motivational theory in the form it has been outlined here.

Once approach behaviors have been successful, copulation may be initiated. In contrast to the arbitrary approach behaviors, copulation in non-human mammals consists of a series of stereotyped motor patterns. These motor patterns are essentially somatic (thrusting, contraction of penile muscles, ejaculation in the male or lordosis in the female of many species) or autonomous (erection, seminal emission) reflexes or fixed action patterns, to employ a term popular in classical ethology. Some scientists are of the opinion that the motivational mechanisms determining the likelihood of appearance of the sexual reflexes or fixed action patterns are best described in terms others than those employed in incentive motivational theory. The motivational model proposed by Konrad Lorenz many years ago (Lorenz, 1950) is a viable, or perhaps even excellent, option. Likewise, the motivational framework in Clark Hull's learning theory is perfectly adequate. In fact, Hull used some of Frank Beach's work on female rat lordosis as basis for his theoretical analysis (Hull, 1943). Both the Lorenz and Hull models posit that there is an inverse relationship between the intensity of the stimulus required to activate a behavior pattern and the intensity of motivation to display the pattern. Thus, a weak stimulus will activate behavior only if motivation is strong, whereas a weakly motivated animal will respond only to a powerful stimulus. Interestingly enough, both models concentrate on the probability of response activation and not on the intensity of response. The reason is certainly that the intensity of a reflex response varies little, while the probability of occurrence may vary from 0 to 1. I have earlier suggested that the incentive motivational framework is essential for an understanding of the mechanisms controlling the arbitrary approach behaviors, while the Hullian or Lorenzian frameworks may be better for understanding the occurrence or non-occurrence of copulatory reflexes (Pfaff and Ågmo, 2002). My opinion has not changed.

2 Copulatory behaviors

I will present the fundamental elements of sexual behavior in reverse order. First we will examine copulatory behavior, the series of reflexes or fixed action patterns that usually are activated or executed in association with sperm deposit in the female's vagina or uterus. Please note that this statement refers to copulatory behaviors in non-human mammals. We have already learned that human sexual interactions do not necessarily end in intravaginal sperm deposit. We have also, at least implicitly, noted that human sexual behaviors are far less stereotyped than those found in other mammals. The discussion beginning here and continuing over the next several pages refers, consequently, mainly to non-human copulatory behavior. In fact, due to the rather important difference between human and non-human copulatory activities with regard to variability and flexibility of behavior patterns, I have found it convenient to analyze human behavior in a separate section.

In the male mammal, the reflexes or fixed action patterns executed during copulation ultimately lead to vaginal penetration and eventually ejaculation. In the female mammal, the reflexes or fixed action patterns essentially allow the male to achieve vaginal penetration. In addition to these fundamental aspects of copulatory behavior, males and females of many species display a varying number of other behaviors during the course of sexual interaction. Most of these other behaviors have unknown or little known consequences for the end point of sperm deposit. Furthermore, there are many instances in which elements of copulatory behavior are displayed in contexts where no sperm deposit is possible. Usually, these behavioral elements are identical to some behaviors normally displayed during the sequence of events leading to sperm deposit. Frequently, these behaviors are heterotypical, i.e. a female displaying elements of typically male copulatory behavior or vice versa. Many examples of this will appear later on. I will include all reflexes or fixed actions patterns of this kind under the label of copulatory behavior.

2.1 On the importance of the distinction between approach behaviors and the execution of copulatory reflexes

Sexual interaction, potentially leading to sperm deposit, is impossible at a distance. Two or more individuals need to be in close proximity before sperm transfer can occur. The establishment of the required proximity must inevitably precede the activation of copulatory behavior. The ways in which humans and other animals succeed in establishing this proximity is extremely variable. In fact, it is safe to propose that proximity is established by a series of entirely arbitrary behaviors as already pointed out in the previous section. Whereas copulatory behavior is very stereotyped in all mammals, with the exception of the human, behaviors leading to proximity are extremely varied. They are determined by an interplay between the specific context and individual experience. The importance of proximity as a requisite for copulation cannot be overestimated. For many animals, and perhaps particularly for the human, copulation is a far lesser problem than finding with whom to copulate. Despite this rather obvious fact, the stereotyped copulatory behavior has attracted scientists' attention far more than the intriguing and fundamental problems of the arbitrary behaviors employed for the establishment of proximity. One consequence is that we know far more about the regulation of copulatory reflexes than we know about the mechanisms determining behaviors leading to proximity. This state of affairs makes it possible to give a quite clear picture of copulatory behavior, whereas the present picture of behaviors associated with the establishment of proximity is confusing. That is one reason for beginning with an analysis of copulatory behavior. An old didactic principle is to begin with the simple and proceed with the complex. Another reason for beginning with copulatory behavior is that our discussion of proximity establishment frequently will refer to some or another aspect of copulatory behavior. Since most scientists have ignored the proximity establishment, we often need to make inferences based on observations of copulatory behavior and, to make these inferences meaningful, we need to have some basic knowledge of that behavior. However, it is also very important to keep in mind that there are fundamental differences between approaching and establishing contact with a potential mate and the ensuing (if all goes well) execution of copulatory reflexes.

One powerful reason for maintaining that sexual approach behaviors need to be classified and analyzed in a category different from the category of copulatory behaviors is the obvious and repeatedly mentioned distinction between arbitrary and stereotyped. This distinction has many consequences. One is that the arbitrary approach behaviors may be fundamentally molded by learning, basically operant conditioning. The stereotyped copulatory behaviors are far less moldable by experience and, when learning occasionally affects them, it appears to be because of classical rather than operant conditioning. A more extensive discussion of the role of learning is found in Chapter 7, but I find it necessary to anticipate this subject here because of its importance for keeping approach behaviors separate from copulation. Besides the differential role of learning, there are many other good reasons for the distinction between copulation and approach. These will appear in the following paragraphs.

Others may argue that the distinction between the establishment of proximity and actual copulatory behavior is artificial and unfounded in reality. It could be maintained, for example, that both aspects are necessarily associated with reproduction and represent, therefore, inseparable parts of a behavioral sequence. This argument loses its force when we consider the dichotomy stereotyped versus arbitrary behaviors and the consequences of this outlined in the preceding paragraph. Furthermore, it is likely that the nervous processes involved in approach are different from those involved in the execution of copulatory reflexes. Illustrations of this difference are the fact that some individuals will try to establish proximity while never initiating copulatory behavior (Stone *et al.*, 1935) and that experimental treatments may affect proximity establishment but not copulatory behavior and vice versa. Examples of this will be found throughout this chapter.

Another argument against the distinction between proximity establishment and copulatory behavior could be that the establishment of proximity with a potential mate is not a sexual behavior at all. Animals may approach conspecifics in many non-sexual situations and the behavioral and neural mechanisms determining the approach behavior leading to the establishment of proximity may be common for non-sexual and sexual contacts. In keeping with this reasoning, there would be no need for including the proximity establishment in the category of sexual behaviors.

In fact, mechanisms responsible for this kind of behavior could be better understood if they were analyzed independently from sexual behaviors, according to this point of view. Since the actions or motor patterns of which approach behavior consists are extremely variable, there is no way to distinguish patterns leading to sexual proximity from patterns leading to other kinds of proximity. As examples of situations where proximity of different kinds are established and maintained for some time through the action of forward locomotion, I can mention social and sexual approach between rats. Laboratory rats are gregarious and they like to be close to other rats when given the opportunity (Latané and Glass, 1968; Latané, 1969; Latané *et al.*, 1973; Ågmo *et al.*, 2004) even though no sexual interaction will take place. This kind of proximity may conveniently be called social. In this context, one rat typically reduces the distance to another rat by forward locomotion and, since the other rat may be moving around, this forward locomotion may continue for a rather long period of time. Likewise, forward locomotion is involved in sexual approach and it is also used for maintaining contact during actual copulation. The muscular activity, i.e. the motor patterns, involved in both social and sexual approaches may be identical. Considering the impossibility of distinguishing behaviorally one kind of approach from another, it is impossible to consider the establishment of sexual proximity as anything different from the establishment of other kinds of proximity. To make the point in slightly different language: animals, including humans, may search to establish proximity with conspecifics without any ensuing copulatory behavior employing exactly the same motor patterns as those used for the establishment of sexual proximity, making it ludicrous to propose a particular behavioral category of 'the establishment of sexual contact'. Moreover, social approach behaviors may be modified by operant learning in exactly the same way as sexual approach behaviors are.

The reasoning exposed above could easily be rejected by stating that the finality or purpose or function of the behaviors is different when sexual and social proximity are sought. According to my own opinion, expressed quite clearly before, this argument has no sense. Fortunately, there are scientifically sound arguments for maintaining that the establishment of sexual contact is determined by factors different from those involved in the establishment of social contact. Instead of advancing a teleological argument, I advance empirical observations concerning the causation of these approaches. They go like this:

1 If sexual approach were *dependent* on gonadal hormones in the same way as copulatory behaviors are, and if social approach were *independent* of such hormones, then it could be concluded that sexual and social approach are at least partially controlled by different central nervous mechanisms. In fact, castrated males approach a sexually receptive female less than intact males and ovariectomized females approach an intact male less than sexually receptive females. At the same time, social approach is not modified by gonadal hormones. Intact and castrated males approach an ovariectomized female (a social incentive) with the same intensity and sexually receptive and non-receptive females approach

a castrated male (another social incentive) with equal intensity (Ågmo, 2003; Ågmo *et al.*, 2004). If the establishment of proximity to a potential mate can be experimentally separated from other kinds of proximity establishment, then it seems most useful to consider it a unique behavioral category. Furthermore, since it is dependent on gonadal hormones, it seems reasonable to consider it as part of sexual behavior.

2 We have shown that continuous sexual interaction with a sexually receptive female for 4 hours significantly reduces the intensity of the males' approach towards a sexual incentive while leaving that towards a social incentive unaffected. Likewise, a female rat shows reduced approach towards an intact male immediately after having received three ejaculations while approach towards a castrated male did not differ from that shown when the females were unmated (Ågmo *et al.*, 2004; Spiteri and Ågmo, 2006). Similar data have been obtained in an entirely different procedure (Sloan and Latané, 1974). All these observations convincingly show that the mechanisms determining the intensity of sexual approach are different from those determining social approach.

I would like to propose that the central motive state determining the behavioral impact of sexual incentives (conspecifics with which sexual interactions may occur) is different from that determining the impact of social incentives (conspecifics with which social but not sexual interaction may occur). We have already seen that some behavioral data support this proposal. Concerning the neural control of social versus sexual approach, it has been found that preoptic lesions eliminating copulatory behaviors and reducing approach to a sexually receptive female fail to modify social interactions (e.g. Paredes *et al.*, 1993a). Finally, some drugs modify a sexual incentive approach while not affecting, or having opposite effects on, a social incentive approach (Viitamaa *et al.*, 2006). There is, then, ample evidence for a behavioral, anatomical and neurochemical separation of mechanisms determining sexual approach and mechanisms determining social approach.

The preceding arguments are entirely based on data from rats. However, as we will see in the appropriate section, sexual proximity establishment has been experimentally studied in a few other species. Results from these studies support the notions outlined above. For the moment we will abandon the question of how to understand this part of sexual behavior and proceed with an introduction to the scientific study of copulatory behavior. This may prove to be more exciting than would be expected.

2.2 Male copulatory behavior

Here I will present a description of the behavior. The neural and endocrine mechanisms involved are the subjects of a later chapter. Male copulatory behavior is a sequence of stereotyped motor patterns ending in ejaculation, the expulsion of seminal fluid. Some species may make several ejaculations in a row. During the time when I worked with rabbits, our champion once made 17 ejaculations during a 30-minute test. He had received no treatment whatsoever. It was his spontaneous level of sexual activity he had shown for us. Rats are less active and normally become unresponsive to a female after 5–10 ejaculations. They normally require several hours to achieve that. The human is still less active. In fact, most men stop copulatory activity after the first ejaculation. Occasionally, a second or even a third ejaculation may be achieved before abandoning copulatory activity, but that seems to be quite unusual. Although there may be a substantial variability between species with regard to the number of ejaculations that are normally achieved during a sexual encounter, the basic behavior pattern is rather constant. We will, therefore, concentrate on copulatory behaviors leading up to the first ejaculation. As always, the rat will be used as our model animal. This is not only because I have some kind of emotional attachment to rats, but also because the rat is by far the best known animal. The principles that have been derived from studies of male rats are applicable to most other species, so there is no waste of time and effort to focus on rats.

The male rat's copulatory behavior starts with a mount. A sexually experienced male almost always mounts the female from behind. He places his forequarters on the female's back and presses his abdominal/preputial region against her rump. This physical contact usually activates a rhythmic anterior-posterior movement of the male's pelvic region, copulatory thrusting. The motor pattern of pelvic thrusting has been carefully described. The frequency of the anterior-posterior movement is around 18 Hz and the duration of thrusting is about 400 ms (Beyer *et al.*, 1982; Moralí and Beyer, 1992). It appears that the generation of this highly stereotyped movement is independent of testicular hormones. Male rats will continue to display exactly the same thrusting pattern as long after castration as they will continue to mount (Beyer *et al.*, 1981). The stimulus triggering pelvic thrusting is unknown, but there are some data suggesting that thrusting is a tactile reflex activated by mechanical stimulation of the preputial region (Contreras and Ågmo, 1993). This is a region surrounding the base of the penis. Anesthesia of the preputial region makes it difficult for the male to initiate pelvic thrusting. Tactile stimulation of other parts of the perineal and inguinal regions may also activate thrusting (Maillard and Edwards, 1990), but this is less clear. Nevertheless, it is quite remarkable that we still ignore the stimulus control of pelvic thrusting, the basic element in male rat copulatory behavior.

Mounting is associated with enhanced intracavernous pressure. Telemetric recordings have shown an increase of around 35 mmHg during mounting (Giuliano *et al.*, 1994). This rather modest increase clearly suggests that penile erection is far from complete, but it should be sufficient for making the penis protrude from the preputial sheath. The duration of the pressure increase was found to be around 0.5 seconds, which means that erection lasts longer than the mount itself. It is, in fact, relatively common to see a moderately erect penis during the male's dismount. A more recent study offered quantitatively different results (Bernabé *et al.*, 1999). In this study, peak intracavernous pressure during mount was reported to be almost 300 mmHg and the duration of enhanced pressure was over 3 seconds.

Some mounts will end in penile insertion. If mounting is associated with erection and appropriate activity in the penile striated muscles, then it is possible for the penis to penetrate into the vagina. Intracavernous pressure is always larger during vaginal penetration, called an intromission among the rat professionals, than during mount. It reaches a level of about 60 mmHg according to old data (Giuliano *et al.*, 1994), suggesting that erection is more complete. However, this is not enough for a successful intromission. The male also needs to contract the ischiocavernosus muscles in order to achieve the necessary forward orientation and elevation of the penis. Extirpation of these muscles virtually eliminates a male's capacity for intromission (Sachs, 1982) and electromyographic recordings *in copula* have established that enhanced activity in the ischiocavernosus precedes vaginal penetration (Holmes *et al.*, 1991). A more recent study of intracavernous pressure (Bernabé *et al.*, 1999) reported that peak pressure during intromission was above 700 mmHg and the duration of the pressure increase was, just as in the case of a mount, around 3 seconds. The mean systemic blood pressure in a rat is around 130 mmHg. An intracavernous pressure above that cannot be caused by vascular mechanisms alone. This means that the pressures recorded during both mount and intromission in the Bernabé *et al.* (1999) study must be caused by some additional mechanism. It was proposed that contractions of the ischiocavernosus muscles contribute to the high intracavernous pressure. This probably means that these muscles are important not only for the orientation of the penis, but also that they enhance penile rigidity. Therefore, they contribute to successful intromission in two ways.

Behaviorally, a vaginal penetration starts with the male in the same position he assumes in any mount and pelvic thrusting is initiated. Suddenly the male makes a strong forward thrust, which immediately is followed by rapid backwards withdrawal. Pelvic thrusting has a shorter duration when a mount ends in vaginal penetration. This is due to a rapid cessation of thrusting immediately upon penile insertion. Intravaginal thrusting never occurs in rats during intromission and the time the penis remains in the vagina is very short, a couple of milliseconds. The stimulus control of vaginal penetration is not better known than that of pelvic thrusting, although it is generally believed that tactile stimulation from the glans penis is essential. The forward thrust causing vaginal penetration is supposed to be activated when the penis makes contact with the vaginal orifice. This hypothesis is strengthened by the fact that experimental manipulations reducing sensory feedback from the penis, for example section of the dorsal penile nerve or the application of local anesthetics to the penile epithelium, abolish intromission without reducing mounting or affecting the thrusting pattern (Larsson and Södersten, 1973; Contreras and Ågmo, 1993). One question that has not been solved is how the male can distinguish penile contact with the vaginal orifice from contacts with the surrounding fur. Temperature or texture may be critical. Although not of direct relevance for the rat issue discussed here, I cannot refrain from mentioning that observations from rabbits would suggest that temperature rather than texture is critical for the forward thrust.

Reproductive physiologists now and then want to collect semen and when they want to collect semen from rabbits they employ a hand-held artificial vagina. This object is nothing exotic. It is simply a glass tube about 15 cm long, with open ends. Inside the tube there is a condom with the top fixed to a rubber stopper at the bottom of the tube. The condom's opening is turned inside out around the tube's opening and fixed with a rubber band. Before use, the rubber stopper at the bottom is removed and the tube is filled with water. Then, the semen collector puts a rabbit fur over her right arm, takes the artificial vagina in the right hand and introduces the ensemble in the rabbit's cage. He will immediately mount and the collector can feel the rabbit penis stroking her hand and the artificial vagina with the typical thrusting frequency of rabbits, about 12 Hz (reviewed in Moralí and Beyer, 1992). The water temperature is the critical thing in all this process. If it is too cold, the rabbit will never insert his penis into the artificial vagina. If the temperature is around 39°C, he will do it immediately when his penis touches it. During semen collection the texture of the vaginal opening is always kept constant but temperature can accidentally vary. Thus, texture cannot be the factor determining if there will be intromission or not. These extremely informal rabbit data may perhaps not be generalized to rats, but if I were forced to make a guess as to the critical stimulus provoking the forward trust leading to vaginal penetration, I would place my bet on temperature. The glabrous skin around the vaginal opening is certainly warmer than the surrounding fur.

After several intromissions, the rat will ejaculate. The number of intromissions required for activating ejaculation is fairly variable. Rats ejaculating after less than four intromissions are rare and those requiring more than 20 are equally rare. The usual number is between 6 and 10. Ejaculation starts exactly like a mount. Then it is transformed into an intromission, which does not seem to end. In fact, the vaginal penetration coupled with ejaculation may last a second or more. During ejaculation the male appears to embrace the female's anterior back with his forelegs. Rhythmic abdominal contractions are also observed. In contrast to intromission, ejaculation is associated with intravaginal thrusting. The frequency of the intravaginal thrusting is much lower than that of the thrusting preceding penetration (Beyer *et al.*, 1982). Finally, the male opens his arms and slowly rises on his hind legs. At that moment, the female runs away. Following the ejaculation the male will rest for several minutes.

The intracavernous pressure during ejaculation is above that observed during intromission, around 100 mmHg according to old data and almost 800 mmHg according to more recent data. This very high pressure is probably caused by contraction of another of the striated penile muscles, the bulbospongiosus. In rats, contraction of this muscle is important for the proper placement of semen. Extirpation of the bulbospongiosus muscles leads to reduced fertility (Sachs, 1982).

Before leaving ejaculation, I will just point out that in rats, as in other mammals, this is a double process, one controlled by the autonomous nervous system and the other by the somatic nervous system. The first process consists of the transport of semen from the epididymis through the vas deferens to the urethra and contraction of the accessory sexual glands, notably the seminal vesicles and the prostate. Physiologists consider this as a process under the control of the sympathetic nervous system and call it emission. Then the seminal liquid must be expelled from the urethra, preferably to the outside. This is achieved by contraction of skeletal muscles, evidently under the control of the somatic nervous system, in a process called ejaculation. Because emission is an unobservable process both for rats and humans while ejaculation is quite notorious, we have a tendency to forget about the preceding emission. This tendency is so strong that we, in vulgar language, call the entire process ejaculation. The somatic contractions enhance intraurethral pressure, causing flow of liquid in the direction of least resistance. Since the vasa deferentia void close to the bladder end of the urethra, the principle of least resistance would direct the seminal liquid to the bladder. Therefore, a sphincter muscle, with the rather poetic name of *musculus sphincter vesicae*, around the bladder opening has to be contracted at the appropriate moment. Failure to contract the sphincter produces a retrograde ejaculation, which sometimes is a problem in men. This, however, is of no concern for us.

The stimulus triggering ejaculation (here used in the vulgar sense) is certainly mechanical stimulation of penile mechanoreceptors. The amount of such stimulation needed for activation of the autonomous reflex of emission and the subsequent somatic reflex of ejaculation is apparently variable between individuals and within an individual from one day to another. The causes of this variability are unknown, as are the exact mechanisms of the two reflexes. A discussion of these issues does not seem to be warranted here.

The basic unit of male rat copulatory behavior is the mount, with or without intromission/ejaculation. If a mount is not leading to an intromission, most males will shortly perform another mount, and perhaps a third and a fourth. Between these mounts the male will do nothing but groom his genitals or pursue the female. Such a group of mounts has been defined as a mount bout, 'a sequence of mounts (one or more) with or without intromission, uninterrupted by any behavior (other than genital autogrooming) that is not oriented toward the female' (Sachs and Barfield, 1970, p. 360). If a mount leads to intromission, it is frequently but not always the last in a bout. Between mount bouts there is a far longer interval than between each mount within a mount bout. The intermount bout interval is independent of whether the last mount in the preceding bout was a mount with or without intromission. This was elegantly shown in a brilliant paper published many years ago (Sachs and Barfield, 1970). That paper is undoubtedly one of the most important papers that ever has been published in the field of sexual behavior, yet it has been cited only 85 times, at least according to the electronic version of the Science Citation Index at the beginning of 2006. What Ben Sachs and Ron Barfield did was surprisingly simple. Rats copulated either with normal females, with whom they could display intromissions as often as they could or would, or with females with a closed vagina. In this latter case, their copulatory behavior was forcibly limited to mounting. Then they compared the intermount bout intervals between these two conditions. There was no difference. What this simple study shows is that the mount is the basic unit of male rat sexual behavior. Intromission only has a consequence as the interruption of an ongoing mount bout. The implication of this is that there is, somewhere in the brain, a kind of 'mount generator' with an oscillating excitability.

The execution of a mount produces a temporary inhibition of the mount generator and, when this inhibition is large enough, mounting stops. One, two, or three mounts in a row might be necessary for arriving at the level of inhibition that stops mounting. An intromission produces a larger inhibition than a mount, normally making one intromission sufficient to stop mounting. However, it occurs now and then that a rat makes an intromission that is rapidly followed by another mount or intromission. As soon as mounting has stopped, the inhibition starts to disappear and within some time (habitually 30–120 seconds) mounting is reinitiated. Ejaculation causes a more profound inhibition of the mount generator, which requires several minutes for recovery.

The concept of a mount generator may seem hopelessly old fashioned, and perhaps it is. Nevertheless, the basic idea, that male rat copulatory behavior is an orderly series of excitation–inhibition processes, coincides with everything we know about the temporal pattern of the behavior. An important contribution to our understanding of copulatory behavior would be to characterize the neural mechanisms responsible for the oscillating excitability. Regrettably, I do not think that it will ever be done. Any practical application of such knowledge seems remote, to say the least, making the subject uninteresting to funding agencies.

Before leaving male sexual behavior, I will briefly mention the procedure normally used for the quantitative description of it. A male and a female are put together in a rather small observation arena and a standard set of behavioral measures is recorded in one way or another. These behavioral measures will be mentioned frequently in this book and it is almost a necessity to know them. They are also part of the basic knowledge you can expect from a reasonably well-cultivated human. First we record the time the male needs to perform the first mount, that is the time that passes between the moment the couple gets together until the first mount. This is the mount latency. Please do not forget that when I talk of a mount it always requires the presence of pelvic thrusting. Then we record the time to the first mount with vaginal penetration, the intromission latency. The time between the first intromission and ejaculation is the ejaculation latency. The duration of the period of inactivity following ejaculation is the post-ejaculatory interval. Normally, the sexual inactivity is considered ended when the male makes his first post-ejaculatory mount with vaginal penetration. Occasionally, the criterion for the end of inactivity is the first post-ejaculatory mount independently of whether it is associated with intromission or not. In addition, the number of mounts without intromission as well as the number with it are counted. It is also common to calculate the proportion of mounts ending in intromission (the number of mounts with intromission divided by the number of mounts with intromission + the number of mounts without intromission). The result is frequently called the intromission ratio, or more vulgarly the hit rate. A more complete description of the male's behavior can be given by calculating the interval between intromissions, that is the ejaculation latency divided by the number of intromissions. Likewise, the copulatory rate may be determined by adding the number of mounts with and without intromission and dividing it by the time from the first mount until ejaculation. Other measures of copulatory behavior have now and then been used. In fact, it is only the scientist's creativity that limits the number of possible descriptors of male copulatory behavior.

How all these measures should be interpreted is a matter of eternal discussion. Some like to imagine that the mount latency somehow represents the intensity of the male's sexual motivation. Others consider that the post-ejaculatory interval is a better indicator of motivation. The number of pre-ejaculatory intromissions may depend on the amount of sensory stimulation required for activation of the reflexes of emission and ejaculation, or on some other factor. The intromission ratio is often interpreted as an indicator of erectile function. If erection is deficient, then few mounts will end in intromission. This is a perfectly logical reasoning, but it is complicated by the fact that a reduced intromission ratio can be a consequence of deficient contraction of the ischiocavernosus muscles. The dramatic consequences of extirpation of these muscles were mentioned above, but I will further reinforce the argument by an additional example. Certain drugs facilitating γ-aminobutyric acid- (GABA) ergic neurotransmission have as a major effect on copulatory behavior a dramatic reduction of the intromission ratio. This reduction is unrelated to deleterious effects on pelvic thrusting. Drug treated animals show a thrusting pattern identical to that of controls (Ågmo and Contreras, 1990). On the contrary, electromyographic recordings of activity in the ischiocavernosus muscles showed that they contracted much less in drug treated rats than in controls (Paredes *et al.*, 1993b). At the time, we concluded that the reduced intromission ratio was an immediate result of the deficient ischiocavernosus contraction leading to insufficient elevation of the penis. However, we now know that ischiocavernosus contraction contributes to intracavernous pressure in a significant way, most likely enhancing penile rigidity. Thus, the deficient contraction of the muscles not only altered the elevation of the penis, a prerequisite for vaginal penetration, but it also affected erection in a deleterious way. The real cause of a diminished intromission ratio could, then, be either deficient elevation of the penis (a direct consequence of drug effects on ischiocavernosus contraction) or a deficient erection (an indirect effect of weakened ischiocavernosus contraction) or a combination of both. It is not easy to figure out how an experiment that offered unequivocal data as to which alternative is the correct one should be designed. The conclusion we can draw from this example is that it is safest to limit our interpretation of the behavioral descriptors to very simple statements of what actually was observed.

I will not contribute to the never-ending discussion of the meaning of the many descriptors of male copulatory behavior. A minimalist point of view is more in agreement with my personal taste. Thus, I prefer to maintain that they are descriptors with an unknown meaning. The effects of experimental manipulations, either behavioral, physiological, pharmacological or genic can be carefully described and these descriptions can provide us with an understanding of the basic mechanisms involved in copulatory behavior without any need for far-fetched interpretations. The point can be illustrated by a straightforward example. An experimental treatment, for example a drug, leads to reduced mount latency. The minimalist interpretation would be that the drug made the animals initiate mounting faster, which is an interpretation strictly tied to the observational data. A more fanciful conclusion would be that the drug enhanced sexual motivation, and that in turn made the males mount faster. This is certainly a possible explanation, but there are many alternatives that could equally well account for the behavioral effect. The rats may mount faster because their general activity was increased, or their perceptual system identified the partner as a receptive female faster after drug treatment, or the sensitivity of the male's abdominal skin was enhanced making tactile stimulation from the female's back more efficient, and so on. None of these explanations refer to any change in motivation, but to changes in general arousal, or in perceptual and sensory systems. The fact that the mount latency was reduced does not tell us anything about the cause of that reduction and we should not make false acclaims of knowing something that we cannot know. However, we could obviously proceed with all kinds of control experiments ruling out all alternative explanations and, in the end, we *could* be entitled to state that the reduced mount latency was caused by enhanced sexual motivation. In scientific practice, all these control experiments are almost never performed before launching the conclusion they could justify. This practice, as widespread as unfortunate, has led to innumerable myths concerning the effects of many experimental manipulations on male copulatory behavior and associated phenomena, like sexual motivation. The interested reader can find an extensive analysis of one of these myths in a review published some time ago (Paredes and Ågmo, 2004).

With this outcry for modesty in our interpretations we will almost leave male copulatory behavior. Before doing so, however, I must add some comments concerning mammalian species other than rats. Otherwise it may be thought that the account of copulatory behavior given in the preceding paragraphs is of limited interest. This is not the case. One or another form of mounting is extremely common among mammalian males and penile insertion associated with mounting is equally common. Although the exact motor patterns coupled to mounting are very stereotyped within a species, they may differ considerably between species, but there is no reason for believing that the neural and behavioral mechanisms are equally diverse. As will be seen in a later chapter, the neural control of copulatory behavior seems to have been highly conserved. The penile structure may vary and the mechanisms of erection and insertion are not necessarily identical among species. Humans and rodents and many other mammals have penile corpora cavernosa, hence a vascular erection, while still others do not have corpora cavernosa in the penis and, consequently, employ other mechanisms for erection. In rats, the ischiocavernosus muscles are important for penile insertion, while these muscles, although present also in the human, do not contribute, in the slightest, to this act. Instead, the partner's hand may be employed, something never happening among rodents. Some species, like mice and men, normally show intravaginal thrusting, while rats do so only at ejaculation. Hamsters display intravaginal thrusting at some intromissions but not at others. Rats and hamsters need several intromissions before ejaculating, while rabbits, mice and men only need one. The rabbit ejaculates within milliseconds of penile insertion, without any trace of intravaginal thrusting, while mice make a few thrusts and ejaculate within a couple of seconds. Men are usually somewhat slower than mice and perform more extensive intravaginal thrusting. Despite all these apparent differences, the neural and behavioral mechanisms associated with intromission and ejaculation seem to be remarkably similar in different species. I insisted on rats not only because I like them but also because we know much more about rat copulatory behavior than we know about any other species, as mentioned before. Moreover, a large number of experiments employing rats will be presented in the next few chapters. Most of them would not be adequately understood without some basic knowledge of rat copulatory behavior. I guarantee that we will discover how useful these rat data indeed are for the understanding of human sexual behavior during the last few chapters of this book. Nevertheless, if someone would like to insist on an analysis of copulatory behavior in a wider selection of species I can suggest some first class reading. There are two brilliant reviews that come to mind, one dealing with most male mammals from which there are usable data (Dewsbury, 1972) and one concentrating on primates (Dewsbury and Pierce, 1989). Both reviews are rather old, but any better analysis of male copulatory behavior from the viewpoint of comparative psychology has not been published since.

Now I will turn to females. Not surprisingly, the discussion will have a slight positive bias toward rats.

2.3 Female copulatory behavior

A sexually receptive female rat will quietly approach a male and within a rather short distance of him (about one body length) she will turn away and make a short, fast run and then suddenly stop with her hind quarters raised. This behavior pattern is called solicitation (McClintock *et al.*, 1982; McClintock and Anisko, 1982). It is a very important part of the female's copulatory behavior in seminatural conditions, like those used by Martha McClintock in her very elegant studies. A most interesting observation is that more than 90% of sexual interactions are initiated by a female solicitation (McClintock and Adler, 1978). The male initiated only 3% of the interactions. In the remaining 7% it was unclear who the initiator was. This is true not only for the initial mounting episode, but also for the entire sequence of mounts with and without intromissions leading to ejaculation. It appears, then, that the female is in control of the pace of copulatory interactions from the first mount until the end of copulation, in addition to being in control of the initiation of copulatory behavior. I occasionally give talks on sex behavior to lay audiences and in these talks I find it amusing to emphasize this fact. It is not unusual that a male listener spontaneously and loudly remarks that it is not only among rats that the female is in command.

In the standard mating test arena there is not room enough for the female to display solicitation behavior and she loses most of her control of the copulatory interaction. This is very important to remember, because most studies of female rat sexual behavior have been performed in small arenas. In small arenas, where there is no way for the female to run away from the male, it appears that the male controls the interaction. Under these circumstances, the female's behavior is limited to responding to the male's mounts with lordosis, a concave dorsiflexion associated with raised hindquarters. This is achieved by extension of the hind legs. Simultaneously, the tail is turned sideward. The resulting body posture exposes the vaginal opening and makes it possible for the male to achieve intromission. In addition to lordosis, the female may respond to the male's approaches with short runs followed by a sudden stop, or she may even make short hops. These behaviors are identical to parts of the solicitation behavior described above. What is usually missing in small spaces is the initial approach to the male and the turning away from him at some distance. Nevertheless, the short runs and hops, called darting or hop-darting in the scientific literature, are frequently confounded with and used as synonyms for the entire sequence of solicitation. Another frequent behavior displayed by the female is ear wiggling. She is, in reality, not wiggling the ears but shaking and turning the head at a rather high frequency (Pfaff *et al.*, 1973). To the human observer it looks like she wiggles her ears, therefore the name given to the behavior. Ear wiggling is often associated with hop-darting, but may also occur independently of this behavior. All three behaviors were termed proceptive in a classic 1976 paper by Frank Beach (Beach, 1976) and that name has been universally used ever since.

A female displaying lordosis in response to a male's mount or an experimenter's fingers is said to be sexually receptive. A huge amount of data has established that the display of lordosis is dependent on ovarian hormones. It would not be unreasonable to pose that, for all practical and scientific purposes, it is absolutely hormone-dependent. Furthermore, in 1958, it was shown that the ovarian hormones act in the brain when predisposing the female for the behavior. George Harris and Richard Michael implanted minute amounts of estrogen into the hypothalamus of ovariectomized cats and found that the implanted females responded with lordosis to a male's mounts while control females did not (Harris *et al.*, 1958). They also presented data showing that the hormone had not leaked out from the site of implantation to the circulation (Harris and Michael, 1964). The unequivocal conclusion was that ovarian hormones act within the brain to predispose the female for displaying lordosis in response to mount. This was the first clear-cut demonstration of a central nervous action of hormones in the control of a complex behavior. Data from rabbits (Davidson and Sawyer, 1961) and ewes (Signoret, 1970) soon confirmed that lordosis and sexual receptivity are under the control of estrogens acting in the hypothalamus. The fact that lordosis is a behavior controlled by hormones acting in the brain has made it attractive for those interested in central nervous hormone actions. In the early 1970s, Don Pfaff at the Rockefeller University in New York set out on a series of brilliant studies aiming at a careful description of the lordosis in female rats. He analyzed the stimulus control and found that tactile stimulation of the flanks and perineal region triggered lordosis. He employed electrophysiological techniques in order to determine the sensory pathways involved in the transmission of the tactile stimulation, determined the exact brain sites for estrogen actions and described the pathways for the output from these sites to the muscles in the back underlying the lordosis posture. Summaries, some pretty extensive, of this extraordinary work have been published at irregular intervals and I strongly encourage readers to treat themselves to the intellectual delight provoked by reading these (Pfaff *et al.*, 1973; Pfaff, 1980, 1999). That the processes underlying lordosis in rats are not unique for this species has been demonstrated in a series of studies in cats (van der Horst and Holstege, 1998). Here, I will not give any detailed picture of the intricacies of lordosis. It is sufficient to mention that one or another kind of lordosis posture as a fundamental element of copulatory behaviors is widespread among female mammals. A comparative analysis of lordosis among rodents was published many years ago by Don Dewsbury (Diakow and Dewsbury, 1978) and it is still unsurpassed in its clarity of description, although some teleology has infiltrated the conclusions.

Lordosis is not a major component of copulatory behavior in women. In fact, it is not a component at all. Likewise, some ungulates like the ewe does not display lordosis in response to male mounts. A sexually receptive ewe displays an extremely quiet copulatory behavior. She just stands still when the male mounts. When non-receptive, ewes always run away from the male and he never gets an opportunity to perform a mount. The point I would like to make here is that, although the stimulus control as well as the endocrine mechanisms and the brain sites crucial for lordosis are well known, this knowledge may not be of vast usefulness when we want to talk about copulatory behavior in women and perhaps not even when our interest is focused on some species where lordosis is not part of female copulatory behavior.

Lordosis is not the only behavior characteristic of a sexually receptive female rat. I already mentioned that the female controls the pace of sexual interaction when given the opportunity to do so. This opportunity is, as I also mentioned, lost in small spaces like those habitually employed in the laboratory. Observations of sexual interaction among wild rats in natural environments, for example garbage deposits, have shown that a receptive female is pursued by a band of male rats. One of the pursuing males may suddenly be the object of a female solicitation, which normally is followed by a mount with or without intromission/ejaculation. After such an interaction the female frequently makes herself unavailable for the pursuing males in some way or another, for example by disappearing in a burrow (Steiniger, 1950; Calhoun, 1962; Robitaille and Bouvet, 1976). The males will wait outside until the female comes out again. Another copulatory interaction will take place and the female will again disappear for a while. This appears to go on until the female's sexual receptivity vanishes.

A situation similar to the wild can easily be created in the laboratory. All that is needed is an observation arena arranged so that there is a section to which only the female has access. One solution is to put an elevated platform somewhere in the arena to which the female has free access while the male is prevented from acceding to it. Another is to tether the male in one corner while the female is left free to move around. A third is to introduce a physical barrier dividing the arena. If the barrier has an opening and if the female is smaller than the male, then the opening can be adjusted so that only the female can pass. The bigger male gets stuck if he tries to penetrate the opening. In all these arrangements, the female is guaranteed to be able to escape from the male whenever she likes. At the same time, the male is always accessible in case the female would like to return to him. In situations of this kind, it is typical to see the female approach the male, who rapidly mounts with or without intromission. Now and then the female disappears to her private area. She soon returns, there is another mount with or without intromission, the female escapes again, she returns after a while, there is another mount with or without intromission, and so on. This come-and-go behavior on the female's part is called pacing. Paced mating tests have become rather popular and quite a lot is known about the factors controlling the female's behavior. Paramount among these factors is the amount of sensory sexual stimulation received by the female. After a mount without intromission, it is rather unusual for the female to escape. In fact, the probability of escape is no larger than it would have been if no sexual interaction had occurred (Ellingsen and Ågmo, 2004). After a mount with intromission it is more likely that the female escapes and after an ejaculation she almost always escapes. If sensory feedback from the vagina is eliminated by section of the pudendal and pelvic nerves, then there is no difference between a mount with and without intromission and there is no difference between an intromission and an ejaculation (Erskine, 1992).

Besides determining the probability of female escape, the amount of sensory stimulation also determines the duration of the female's escapes. In the case of escape after a mount, she will return very fast. If she escaped after an intromission, she will take some more time before returning. The longest absence follows an ejaculation. The proportion of mounts with and without intromission/ejaculation followed by escape as well as the duration of escapes have been attributed many meanings. Some would say that the smaller the proportion of escapes and the faster the returns are, the more sexually motivated is the female. Both effects may be interpreted as indicating that the female wants to be close to the male in case he would like to mount. This, in turn, must be indicative of high sexual motivation. On the contrary, if the proportion of escapes is high and the return is slow, then we may say that the female does not want to be close to the male, hence she is not sexually motivated. All this sounds very nice, but before making claims of this kind it would be convenient to try to figure out the mechanisms determining the female's escape.

We already know that the amount of sensory stimulation is the critical factor. What we do not know is which kind of sensory stimulation the female responds to. It is most likely that penile insertion causes distension of the vagina for the duration of the insertion. It is also likely that the ejaculatory plug causes vaginal distension that outlasts the penile insertion that was associated with ejaculation. The rat's semen coagulates within seconds of ejaculation and forms a solid plug that is somehow attached to the cervix. This plug is big enough to distend at least the deeper parts of the vagina. One possible cause for the females' habit of escaping from the male after intromissions and ejaculation may be this vaginal distension. The more intense distension caused by ejaculation (penis + ejaculatory plug enhance the *degree* of distension and the continued presence of the ejaculatory plug enhances the *duration* of distension) can easily explain the increased likelihood of escape after ejaculation compared to intromission, and also the longer duration of absence from the male after ejaculation. If this reasoning were correct, than stretch receptors in the vaginal wall would be critical. There is some evidence for the existence of such receptors. Inflation of an intravaginal balloon activates sensory neurons in the pudendal nerve in cats (Cueva-Rolón *et al.*, 1994). The firing of these neurons declines very slowly during sustained vaginal distension, suggesting that the sensory receptors do not adapt to this kind of stimulation. On the contrary, receptors responding to a moving probe adapted rather fast. Similar results have been reported from rats (Katter *et al.*, 1996). Ejaculation should cause sustained vaginal distension while the intromitting penis should be similar to a moving probe. Thus, ejaculation should produce long-lasting activity in sensory receptors while the activity should be short lasting after intromission. The duration of sensory stimulation thereby seems to be associated with the duration of the female's escape from the male. The electrophysiological data nicely substantiate the hypothesis that vaginal distension can be the stimulus causing female withdrawal from the male.

Another hypothesis is that penile insertion is painful to female rats. The argument for this is that vaginocervical stimulation produces an analgesic response (Komisaruk and Wallman, 1977; Komisaruk and Whipple, 1995) and analgesic responses are typically associated with painful events. According to this point of view, the female escapes from the male because of vaginal pain. This is not an unlikely hypothesis. Most of those who have spent some time observing copulating rats have noticed that the female often vocalizes when being penetrated and she may also turn against an insistent male and reject his approaches with a certain lack of politeness or even with outright violence if the male should dare to try to mount shortly after having performed an intromission. In addition to this kind of anecdotal observation, there are experimental data suggesting that vaginal distension is aversive to rats. In a series of quite ingenious studies, rats were first subjected to tail pinch. They could terminate this noxious stimulation by introducing the head into a small tube located in front of them. All the rats learned to do so after a few trials. The interpretation of this was that the rats had learned to stop an aversive stimulus by the operant response of introducing the head into the tube. In the next phase, tail pinch was replaced with uterine or vaginal distension produced by a gradual increase of pressure in an intrauterine or intravaginal balloon. Exactly as in the first phase of the experiment, the stimulus could be stopped by introducing the head into the tube. All the rats responded to even moderate increases in pressure in the intravaginal balloon while the intrauterine balloon produced a much weaker response (Berkley *et al.*, 1995). The conclusion drawn, that vaginal distension is aversive, does not seem unwarranted. In a subsequent study, it was reported that the response to intravaginal pressure increases depended on the estrous cycle. It was at a minimum around estrus/proestrus, exactly the period in which a cycling female is sexually receptive (Bradshaw *et al.*, 1999). The conclusion was that vaginal distension is least aversive at the moment in which it is most likely to occur, that is during copulation, a behavior strictly limited to the period of proestrus/estrus.

Whether we want to interpret the female's response to the balloon as a pain response or not is probably a matter of taste. Personally, I prefer to call it a response to an aversive stimulus. According to basic concepts of learning theory, the head-introducing response was acquired because it stopped an aversive stimulus, tail pinching or distension of internal genitals. If these stimuli had not been aversive, a response stopping them would not have been learned. The fact that the females learned constitutes the proof that these stimuli are aversive. I will not enter into an analysis of the apparent circular nature of this reasoning. Learning theorists have solved the problem long ago and any good textbook in the field of learning will provide convincing arguments for the validity of the relationship between the acquisition of responses ending a stimulus and an aversive internal state produced by such a stimulus. The intimate nature of the emotional experience associated with the presence of the aversive stimulus is impossible to determine, but since the stimuli employed appear painful to a human observer, we imagine that they are also painful for rats.

A completely different explanation for the copulatory behavior of the female rat has been proposed in an extraordinary well-written chapter by Komisaruk and Whipple (2000). They suggest that a sexual interaction produces a momentary reduction of sexual motivation. The degree of reduction is determined by the intensity of vaginocervical stimulation, being least after a mount and largest after an ejaculation. If sexual motivation is momentarily reduced, then the female has no need to remain in proximity of the male and could as well escape. When sexual motivation spontaneously recovers after a few moments, she will return to the male for another sexual interaction. This intriguing proposal accounts for all known experimental facts in a very elegant way. Furthermore, it is no way incompatible with the vaginal distension hypothesis presented above. Indeed, it can be maintained that the origin of the reduction of sexual motivation is the sensory stimulation received from vaginal stretch receptors. For the sake of fairness, it must be mentioned that Komisaruk and Whipple (2000), at difference to myself, regard the aversive stimulation/pain hypothesis as a more viable alternative.

Other behaviors, like ear wiggling, displayed by female rats during copulation are poorly understood. For example, the potential effects of ear wiggling on the male have never been studied. One imaginative hypothesis saying that the vestibular stimulation produced by the head shaking and turning underlying ear wiggling facilitates lordosis has been put forth (Pfaff, 1999). Some convoluted electrophysiological evidence was presented in favor of that hypothesis, but it is far from compelling. Personally, I find it safest to propose that ear wiggling is an entirely meaningless behavior, affecting copulatory behaviors neither of the female displaying it nor of the male watching it. The proposal of a meaningless, reflex-like behavior is certainly offensive to those believing that nature is perfect. However, nature may be perfect only if so designed by a perfect creator. Evolution, through natural selection, may favor or disfavor alternative solutions, but there is no reason to believe that evolution ever has arrived at a perfect solution. Only a creationist conviction could justify the idea that behavior cannot be meaningless. Those of us believing in evolutionary theory have no problems with imperfections.

I have given a rather extensive account of the female rat's sexual behavior. As was the case with the male, the main reason is that the female rat is the only animal for which we have a considerable amount of excellent experimental data. The essential characteristic of female rat sexual behavior is the sequence of excitation–inhibition, manifested as approach–withdrawal behaviors. Incidentally, this aspect of female sexual behavior had been ignored for many years. It was not so long ago that it was discovered that the female likes to escape from the male after some sexual interactions when given the opportunity to do so (Peirce and Nuttall, 1961). By a simple change of observation procedure, the before then ignored aspects of female copulatory behavior were suddenly seen. Here we have an eloquent example of how the choice of behavioral procedure can determine what we will observe. It is also an excellent example of how decisive it may be to evaluate ecological validity of our procedures. With that I mean that the possibile behavioral repertoire expressed in the laboratory should be as similar as possible to that expressed in the wild. Please note that this reasoning applies to the study of spontaneous behaviors like sex, but it is perhaps entirely inadequate for other kinds of behaviors. The importance of strictly controlled environments for learning experiments has been eloquently expressed by one of the giants in that field (Spence, 1956). After this little digression we should return to our subject matter.

The sequence excitation–inhibition, approach–avoidance seen in female rats is identical to what we see in the male. After a mount, there is a short period of inactivity, after an intromission it is slightly longer and after an ejaculation it is longer still. The similarity in the temporal organization of sexual behavior in male and female rats is striking. There are, in fact, only two consistent, but most insignificant, differences: the period of inactivity in the male is shorter than in the female following an intromission and it is longer in the male than in the female following an ejaculation. We will later see that the similarities between males and females extend from behavior to neural control and endocrine regulation of sex behavior.

Female copulatory behavior has been described in many species other than rats, but in none of those does the precision and depth of analysis approach, in the slightest, those found in the rat. There is no need to make a detailed comparative analysis. Nevertheless, I would like to point out that copulatory behavior in the human female is fundamentally different from that of the female rat. This also applies to men and rats, so I will talk about both sexes at the same time. Copulatory behavior in women and men is not characterized by constant interchanges excitation–inhibition, approach–avoidance. In the human, copulatory behavior is a continuous process, as so elegantly described by Masters and Johnson in their now classical work (Masters and Johnson, 1966). This, in fact, applies to all primates and many other species. It would appear that rats and some other species have a rather peculiar behavior. Despite the indisputable difference between rat and human copulatory behavior, we will discover that many experimental manipulations have similar effects in rats and humans. There are, for example, some drugs that reliably enhance the ejaculation latency in rats. They do exactly the same in men. More about the many similarities between rats and humans will appear in future chapters.

A crucial thing to observe here is that, when writing these accounts of copulatory behavior, I have made no mention of function. There is no teleology in this chapter.

The attentive reader should have noted that I have simply described behavior patterns and tried to show how these behavior patterns are under the control of specific stimuli. When trying to determine the stimulus control of a particular behavior we determine its external causation. As shown in the section on incentive motivational theory, the external stimulus must act on something in the central nervous system when producing behavior. It can, therefore, easily be argued that a complete understanding of a behavior is impossible until we have elucidated these central nervous mechanisms. However, there is no way to determine how incentive stimuli act in the central nervous system when controlling a particular behavior without knowing which these stimuli are. Moreover, we cannot unravel how an incentive stimulus acts without knowing the exact kind of behavioral responses it controls. Thus, without a careful description of behavior and the stimulus control of its occurrence and patterning, progress in neuroscience will be slow and precarious. These comments are obviously not limited to the neuroscience of sexual behavior but apply equally well to all areas of behavioral neuroscience.

On the preceding pages, there is a simple outline of copulatory behavior. We know a lot more than I have cared to describe, but I see no point in making everyone an expert in that behavior. Not even scientifically speaking. Likewise, I see no point in making everyone an expert in comparative psychology. I have tried to concentrate on some basic aspects, all of which will be important to master in order to appreciate many of the arguments that will appear throughout this book. We will now turn to copulatory behaviors in the human.

2.4 Copulatory behavior in men and women

Throughout history there have been many descriptions of human copulatory behavior. Most of them belong to the fine arts, some to pornography and a few to the medical sciences. Descriptions satisfying some minimum requirements of objectivity and accuracy are more rare. Without discrediting the early pioneers in sex research, I will start my analysis of copulatory behavior with the extraordinary data collected by Alfred Kinsey and colleagues and published in two landmark books (Kinsey *et al.*, 1948, 1953). Those interested in research on human sexuality prior to the Kinsey studies can find an excellent summary in a charming book (Bullough, 1994). However, there is one extremely interesting text preceding the Kinsey era that I cannot avoid mentioning. In 1926, the Dutch gynecologist, Theodoor Henrik van de Velde, published an account of human sexual behavior (van de Velde, 1965; I have had access only to a reprint of the 1926 edition. All following citations refer to the 1965 reprint), including plenty of advice as to how to increase the pleasure derived from sexual activities. Human sexual interactions were presented as a sequence of stages, beginning with prelude, continuing with love-play, followed by coitus and finally the epilogue or postlude. The progression between stages and the associated level of sexual excitation in a few cases of sexual intercourse is illustrated in Figure 2.2.

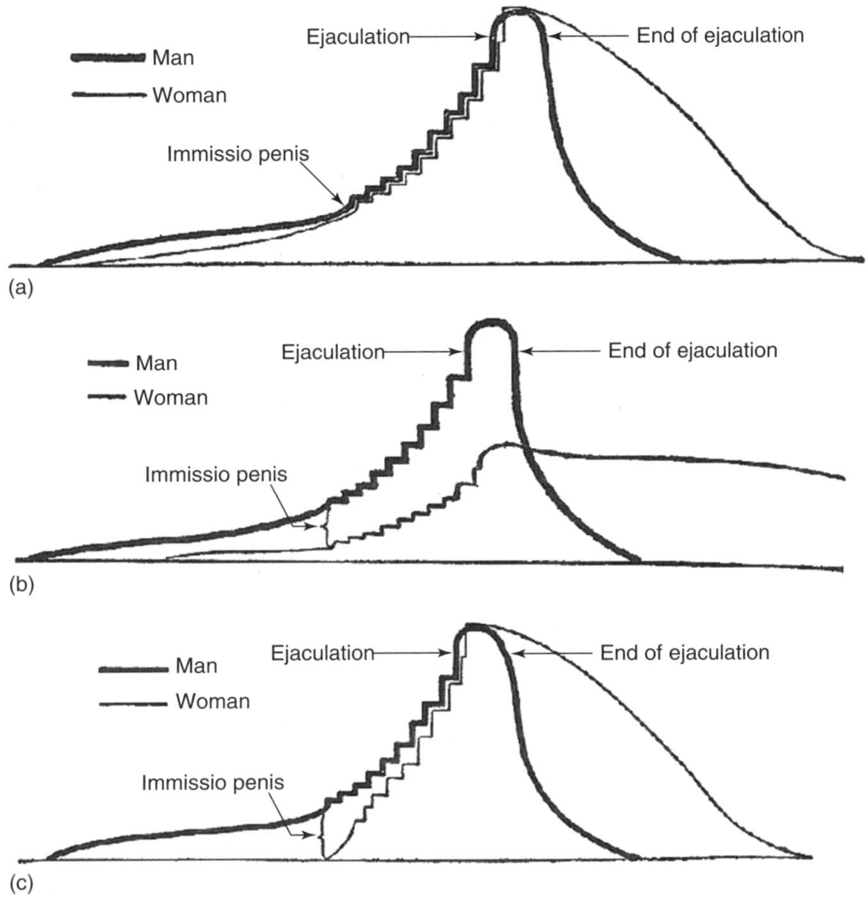

FIGURE 2.2 van de Velde's illustration of human coital interactions. (a) The changes in sexual excitation during an ideal copulatory encounter. With ideal is understood that the man and the woman reach orgasm at the same time. Excitation is defined as the summation of sexual desire and pleasure, bodily and psychic. (b) Similar to panel (a), but here the sexual interaction occurs with an inexperienced woman without adequate coital stimulation. (c) Similar to panel (a), but now the woman is sexually experienced. The prelude was omitted, but the woman's low initial excitation was compensated for by her experience. (From van de Velde, 1965. Copyright 1926. Reprinted with permission.)

The prelude is described in this way: 'As soon as the first stirrings of the impulse of approach are perceptible, the prelude to sexual union begins' (van de Velde, 1965, p. 102). Further: 'In the prelude the impulse of approach works through the three senses of sight, hearing and smell. Taste and touch come into play at closer range, and it may be said that touch, once in action, soon becomes paramount' (van de Velde, 1965, p. 104). Although this was written long before incentive motivational theory existed, van de Velde's description of the initial stage of sexual interaction is stated in exactly the same way as incentive motivation theory would have it, that is the senses register certain incentive stimuli which activate sexual approach behaviors.

The love play arises from the impulse for sexual gratification and it almost always begins, in Western cultures, with the erotic kiss, according to van de Velde. It proceeds with tactile stimulation of various parts of the body, eventually concentrating on the genitals. During the love play, sexual arousal (erection in the man and vaginal lubrication in the woman) needs to be produced. If the woman shows signs of insufficient lubrication, the man is recommended to ameliorate the situation through repeated mechanical stimulation of the vulva. 'And this may best, most appropriately, and most expeditiously be done without the intermediary offices of the fingers, but through what I prefer to term the kiss of genital stimulation or genital kiss: by gentle and soothing caresses with lips and tongue' (van de Velde, 1965, p. 118). Likewise, if the man's erection is insufficient, the woman may provide oral stimulation to the penis. These suggestions must have been far ahead of their time. In 1926, the practice of cunnilingus or fellatio, even as preparations for penile–vaginal intercourse, was not socially acceptable and even constituted a felony in many places.

When erection and vaginal lubrication have been obtained, the third stage of sexual interaction, coitus, may begin. This stage starts with the insertion of the penis into the vagina, reaches culmination in the ejaculation of semen and in the orgasm in both partners, and finally ends with penile withdrawal. The 'sexual tension' increases stepwise from the introduction of the penis until orgasm in both the man and the woman. van de Velde illustrates the degree of 'sexual tension' in a series of figures, which 40 years later were employed by Masters and Johnson (1966) in their important book on human sexuality.

The postlude or epilogue is not a stage that attracts much of van de Velde's attention. He briefly mentions that it may be entirely absent in couples who do not feel love for each other. In the ideal case, 'it should echo, and vibrate, and reawaken in the preliminaries of a new act of sexual communion' (van de Velde, 1965, p. 102).

An interesting aspect of van de Velde's book is that he does not mention the factual basis for his analysis of human sexual behavior. I assume that his clinical experience forms the basis for all his assertions, but the more detailed nature of this experience is never mentioned. This is a serious limitation, but despite this limitation we need to accept that his discussion of human sexual behavior has had considerable impact. His book is said to have sold more than 800 000 copies and, as we will see in a few pages, the notion of human sexual behavior as characterized by stages still survives. What also survives is the insistence on penile–vaginal intercourse. No other form of sexual behavior is considered by van de Velde. He makes it explicitly clear that he refers exclusively to normal intercourse between opposite sexes and this has continued to be the case in almost all subsequent studies of human sexual behavior. One noteworthy exception is found in the work of Masters and Johnson (1966), which also includes data obtained during solitary masturbation.

van de Velde described and gave advice concerning the perfect penile–vaginal intercourse. Although extremely interesting and influential, his book did not contain any data as to what humans actually do when they have sex, with whom they have sex, at what age they start and stop to have sex, and so on. All these and many other questions remained unanswered until the Kinsey group published their reports on human sexual behavior. It must immediately be mentioned, though, that they did not report data on human copulatory behavior based on direct observation. Almost all their data stem from interview responses. In these interviews, people were asked to recall and tell their sexual history. Questions like the age at the first masturbatory experience, at the first intercourse, kinds of copulatory behavior displayed and with whom, number of partners, homosexual activities and so on were asked. The veracity of the answers was checked in several ways and data from interviewees showing signs of giving false answers were discarded. However, it is well known that the tendency to provide false information is largest when sensitive questions are asked (Coxon, 1999; Jagannathan, 2001). Most people seem to consider that questions about their copulatory behavior are sensitive and the likelihood of obtaining answers based on imagined social desirability rather than on facts is not negligible. Most unfortunately, available data do not allow for any direct evaluation of the coincidence between self-reported sexual behaviors, be it in the form of written questionnaire answers, face-to-face or telephone interviews, or computer assisted procedures, and the actual behaviors displayed. In fact, some authors suggest that self-report data are essentially useless (Brody, 1995; Lewontin, 1995). I tend to agree with this proposal, but at the same time I have to recognize that extensive reviews of the vast literature suggest that some useful information *might* be obtained through self-reports (Catania *et al.*, 1995; Weinhardt *et al.*, 1998). Nonetheless, Kinsey intuitively figured out that people are prone to lie about their sexual activities when answering a questionnaire. Lying to an interviewer staring into your eyes is much more difficult, thought Kinsey, and abandoned written questionnaires very early in his research. Although interviews performed by extraordinary interviewers like Kinsey and associates may be superior to other forms of data collection, they are not perfect. The fact that people have to rely on their memory when answering most, if not all, of the questions inevitably leads to some uncertainty as to the correspondence between what actually occurred and what the person remembers occurred. Nevertheless, despite all their defects, the Kinsey data are probably the most reliable we have concerning human copulatory behavior. Later studies, based on written questionnaires or telephone interviews, provide data that probably are far less reliable than the Kinsey data. I will, in fact, avoid such data whenever possible. They are, anyway, well known from front-page headlines in newspapers and long articles in men's magazines.

The main findings reported in both Kinsey books were that human copulatory behavior is more variable than expected and less dependent on social rules than hoped. Many behavior patterns, like cunnilingus (stimulation of the vulva with the mouth and tongue), fellatio (stimulating the glans penis or the penile body with the mouth and tongue), anal intercourse or the display of copulatory behavior with individuals of one's own sex, were all far more common than ever imagined. These facts were still more remarkable considering that these behaviors were illegal in most states of the USA and severely punished if prosecuted. In the late 1940s, the penal codes of many states established maximal prison terms of around 20 years for sodomy or fellatio, for example. Some other American and many European countries had similar laws. The discovery that many otherwise exemplary citizens engaged in sexual activities 'contrary to nature' as they were regarded at the time, produced a profound social shock and a violent outcry from the guardians of morals, like Anglican bishops, cardinals of the Holy Catholic and Apostolic Church of Rome, members of the United States Senate's Government Operations Committee, Lutheran pastors and many others. They tried to combine their moral complaints with scientific criticisms, with the obvious aim of reducing the credibility of the Kinsey data. The rather unorthodox sampling procedures employed were one of the main targets of attack. It was argued that the estimates of the frequency of some copulatory behaviors were much exaggerated because of biased samples and that, in reality, the copulatory behavior of Americans was much closer to the social norms than the Kinsey data suggested. This point may have been reassuring for many conservatives, but the data showing that copulatory behavior was far more diverse than previously believed do not appear to be solely a consequence of biased sampling. In fact, a group of respected statisticians, including John Tukey and William Cochran, were asked to examine the procedures behind the data presented in the book on male copulatory behavior and they concluded that the sampling procedures employed by Kinsey probably were the best available. The statisticians found some minor errors in the data presentation, but basically the results were accepted as scientifically sound (Cochran *et al.*, 1954).

An important innovation by the Kinsey group was to quantify the intensity of sexual activity by estimating the number of orgasms reached per unit time. This was called the total sexual outlet and usually expressed as the number of orgasms achieved per week. Earlier research had preferred to express the intensity of sexual activity as the number of coitions per unit time. The underlying supposition was that the only sexual activity in which people engage and which, therefore, is the only activity worth studying, is penile–vaginal intercourse. The Kinsey data show that this is not the case. Penile–vaginal intercourse is only one of many ways in which women and men may obtain orgasm, and this important fact was first acknowledged by Kinsey. Among the many consequences of this notion is that sexual activities must be regarded as a means to obtain orgasm and the specific sexual activity that leads to that end is uninteresting. Such a conception of sexuality is totally different from the point of view of Saint Augustine and of most other Christians, and it is also contrary to the opinion of biologists. To these, the end point of sexual activity is procreation as mentioned several times already. Here we have another source of criticism of the Kinsey approach to sexuality: its lack of moral considerations and the absence of condemnations of non-procreative sex. Those reading Swedish would likely enjoy a funny little book illustrating the fury provoked by the scientific rather than moral analysis of human sexuality characterizing Kinsey and his collaborators (Sundström, 1949). Similar books were certainly published in many other languages.

The Kinsey studies of copulatory behavior in the USA had clearly shown that human sexual behavior is most variable within a country. Others had taken an anthropological approach and studied copulatory behavior in different cultures

(Ford and Beach, 1951). Again, diversity seems to be the best descriptor. However, anthropological data may not always be entirely reliable, since most descriptions of sexual acts are based on hearsay and very few on actual observations. Nevertheless, it does not appear to be too risky to suggest that human copulatory behavior is far more variable than in any other primate. I even dare to suggest that no other mammal shows the amount of intraspecies variation as the human.

The Kinsey studies as well as Ford and Beach's (1951) account are basically descriptions of general motor patterns of copulatory behavior. The information contained in terms like intercourse with rear entry, anal intercourse etc. is not sufficient if we are looking for quantitative information on aspects of copulatory behavior. Questions like the latency from penile insertion to ejaculation, or from the initiation of masturbation until orgasm, or the mean duration of cunnilingus or fellatio are not easy to answer from the kind of descriptive data offered. Furthermore, on the few occasions when quantitative data on some of these aspects are presented, they are based on verbal reports of memories of copulatory activities. The degree of coincidence of these memories with reality is difficult to evaluate. Laboratory observations of these and many other copulatory behaviors could provide the required information, but such studies were unconceivable for a long time. It appears that Kinsey had started to make direct observations of copulatory behavior (Jones, 1997), but at his death in 1956 nothing had been published.

It was not until the early 1950s that scientists began to observe human copulatory behavior in a laboratory setting. The groundbreaking work performed at Washington University by William Masters and Virginia Johnson from 1954 until the early 1960s is still unsurpassed in quantity and probably also in quality. In their classical description of human copulatory behavior, Masters and Johnson (1966) distinguished four arbitrary stages: arousal, plateau, orgasm and resolution. The notion of stages or phases was explicitly taken from van de Velde, although the phases were slightly modified and renamed. The world famous figures used by Masters and Johnson (1966) to illustrate 'the sexual response cycle' were also adjustments of van de Velde's (1965) figures. Figure 2.3 illustrates the Masters and Johnson (1966) curve. Its similarity to the curves shown in Figure 2.2 is probably not a pure consequence of coincidence. The most noteworthy modifications were that van de Velde's first phase, the prelude, was omitted and that orgasm was added as a phase of its own. van de Velde regarded orgasm as the end of the coition phase. It is of some importance to keep in mind that the four stages, perhaps with the exception of orgasm, do not correspond to qualitatively different events occurring during human sexual intercourse. Rather they reflect the chronological progress of human sexual activity, from erection and vaginal lubrication (arousal) through penile insertion and intravaginal thrusting in men and the consequent mechanical stimulation of the vulva, clitoris and vagina in women (plateau) to ejaculation and the experience of orgasm in men and rhythmic contractions of the perineal musculature and the experience of orgasm in women. The resolution phase corresponds to detumescence in men and to the cessation of vaginal exudation in women. While ejaculation (orgasm) is almost always followed by the resolution phase and frequently the end

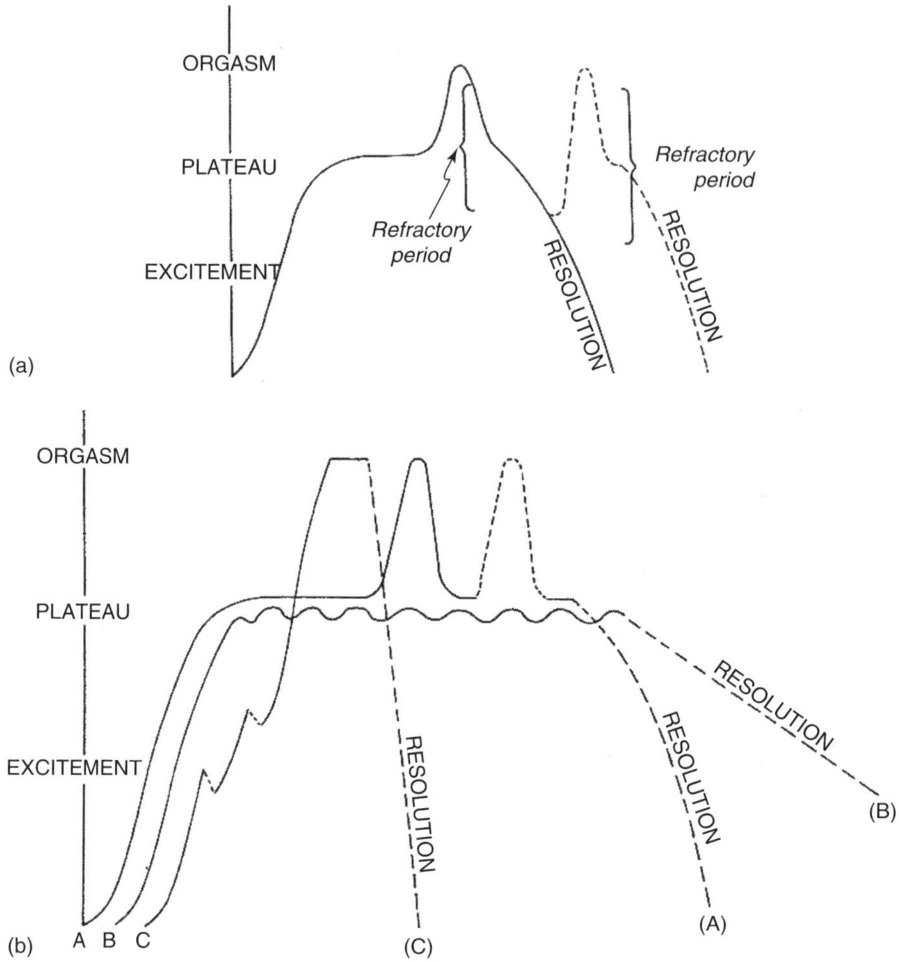

FIGURE 2.3 The Masters and Johnson curve illustrating the phases of human coital interaction. (a) The male sexual response cycle; (b) the female sexual response cycle. (From Masters and Johnson, 1966. Reprinted with permission from Lippincott Williams & Wilkins.)

of copulatory activity in men, women may experience two or more orgasms in a row before entering the resolution phase. It is possible to interrupt this latter phase and initiate a new response cycle with arousal, plateau and so on, but this is rather uncommon. Most humans, at least in Western cultures, do not reinitiate copulatory activity until after a rather long and very variable interval, ranging from hours to weeks.

Logically, we begin our passage through the phases of copulatory activity with the arousal stage. According to the Master and Johnson model, the arousal or excitement phase is characterized by erection in men and vaginal lubrication in women. Clitoral tumescence is also common. All these responses are consequences of enhanced genital blood flow. There is no mention whatsoever of any motivational causes or consequences of arousal in the Masters and Johnson description. This is obviously a serious limitation. The intensity of the visceral response to an incentive stimulus must be determined by the intensity of motivation in exactly the same way as intensity of responses in skeletal muscles are determined by motivational factors. This fact was neglected by Masters and Johnson. The likely reason is that they were not particularly interested in the psychological processes behind the physiological responses that were their object of study. This is by no means a criticism of the very important contribution they made. It is not possible to cover all aspects of such a complex phenomenon as human copulatory behavior at the same time. Nevertheless, it was soon recognized that the arousal phase ignored a critically important portion of sexual interaction. Enhanced genital blood flow does only occur if the individual feels a desire to experience sexual stimulation, provided either by another individual or by himself or herself. In other words, the arousal phase follows an initial phase of sexual desire (Kaplan, 1979). Interestingly enough, the desire phase had been presented in van de Velde's (1965) model of human sexual behavior under the name of prelude, but it was suppressed by Masters and Johnson (1966). In another chapter (Chapter 4) we will analyze the sexual incentives that may give rise to this desire or initiate the prelude. The addition of a desire phase to the Masters and Johnson model has been almost universally accepted. Furthermore, the Masters and Johnson plateau phase has been collapsed with their arousal phase into something now known as the excitation phase (Robinson, 1976). Some scientists have further sophisticated the analysis of the sexual response cycle and divided the desire phase in two components, one endogenous, and one activated by sexual stimulation (Levin, 2000). The endogenous phase is supposed to be created spontaneously, meaning that there is no physical stimulation of the genitalia or other evident physical events that can explain the activation of desire. The second phase is associated with the presence of exogenous sexual stimuli such as mechanical stimulation of the genitals or other parts of the body. This second phase of desire is simultaneous with the excitation phase.

An interested reader may find it rewarding to compare the notion of endogenous desire phase with Freud's *Ichlibido*, and the second, exogenous desire phase, with Freud's *Objektlibido* (Freud, 1905). Some striking similarities are evident. It appears, then, that Helen Kaplan's (1979) addition of a desire phase to the Masters and Johnson (1966) model brought us some 40 years back to the van de Velde model, originally published in 1926. Levin, who published his division of the desire phase into endogenous and exogenous in 2000, brought us back to a conceptual framework from 1905. This nicely illustrates that scientific progress sometimes consists of the fruitful union of almost forgotten concepts with newly acquired data.

If the copulatory behavior consists of penile–vaginal intercourse, then the excitation phase sooner or later will be manifest as vaginal penetration and thrusting. The moment of the excitation phase at which penetration occurs is highly variable and probably determined by own experience or social learning or a combination of both. Manuals of copulatory behavior, in the Western hemisphere, hypocritically called marriage manuals, give usually extensive advice as to when penetration is convenient, desirable or permissible. Common sense suggests that penetration is problematic before the penis has attained an adequate erection and before vaginal lubrication is sufficient to make penetration possible. It might even be regarded as desirable to assure that vaginal lubrication has reached a level where penetration can be made without causing pain to the woman. Most other suggestions as to the ideal instant for penetration are most likely based on uninformed speculation. The moment of penetration does not seem to be of any particular interest and I will refrain from any further discussion of it and, above all, I will give no advice. Moreover, I will not discuss the many bodily orientations that may allow for vaginal penetration. Only the creativity of the human mind and the acrobatic capacities of the partners pose limits to these orientations. Again, manuals of copulatory behavior as well as women's and men's magazines offer abundant advice. Rather than dwelling on this vulgar subject I want to insist on the fact that penile–vaginal intercourse is only one of many possible copulatory behaviors in the human. Other bodily openings of sufficient size can be employed for penetrative intercourse, and these other openings have the advantage of being available in both sexes. Thus, men fond of penetrative copulatory behavior can have the choice between heterosexual and homosexual copulatory behavior as soon as they free themselves from the inevitable limitations inherent in the insistence on penile–vaginal intercourse. Behaviors executed during the excitation phase may also be non-penetrative. Mutual masturbation is one example, cunnilingus another. Just as is the case with non-vaginal penetration, non-penetrative copulatory behaviors are equally possible as homosexual or heterosexual activities. All copulatory behaviors other than vaginal intercourse and masturbation were ignored by Masters and Johnson (1966) as well as by most others who recently have studied human copulatory behavior. Consequently, we do not know much about the physiological reactions occurring during these alternative behaviors. It may be reasonable to assume that they are not fundamentally different from those occurring during penile–vaginal intercourse.

The excitation phase may or may not lead to the point of orgasm. In men, the sensation of orgasm is normally associated with seminal emission and ejaculation. The stimulus triggering ejaculation is normally mechanical stimulation of the glans penis. Sensory receptors in the glans penis were described in the mid-1800s and they have been termed genital end bulbs. Unlike glabrous skin, such as the finger tips, or the lips and palate, there are no Merkel endings in the glans and few Pacinian and Ruffino corpuscles. On the contrary, there is an abundance of free nerve endings. A peculiar type of corpuscular receptor has been described (Halata and Munger, 1986). It seems to consist of axon terminals resembling a tangled skein of free nerve endings. Somatosensory information is relayed from the glans penis to the spinal cord by the dorsal penile nerve (Yang and Bradley, 1999). Blockade of nerve impulse conduction in the dorsal penile nerve eliminates ejaculation (Wieder *et al.*, 2000), making evident the crucial importance of sensory stimulation for ejaculation associated with copulatory behaviors. Parenthetically, it must be mentioned that ejaculation can occur in the absence of or with minimal sensory stimulation, as happens in 'wet dreams' in young men and at 'spontaneous ejaculation' in rats. Nevertheless, ejaculation in conjunction with copulatory behavior is always a consequence of mechanical stimulation. Returning to the dorsal penile nerve, it must be mentioned that this nerve innervates the penile shaft in addition to the glans (Yang and Bradley, 1998). This opens up the interesting possibility that mechanical stimulation of the shaft may trigger ejaculation. I know of no experimental study evaluating the possibility of triggering ejaculation through stimulation of the penile shaft without concomitant stimulation of the glans penis, but this possibility cannot be excluded. Anyway, this question can have no practical meaning since it is difficult, if not impossible, to imagine any copulatory behavior involving the penile body while leaving the glans penis completely without stimulation.

The male orgasm/ejaculation is associated with bodily reactions and psychic experiences. The bodily reactions were described by Masters and Johnson (1966) and I see no reason to repeat their description here. The psychic activities, frequently described in terms of emotion, have been the subject of myriads of comments from all kinds of experts, true and false, including cardinals, sexologists, psychoanalysts, charlatans of all sorts and innumerable scientists. There is no need to comment further on this issue. I refer the curious reader to the safe hands of any newsstand attendant.

Orgasm in women is sometimes considered to be less known than ejaculation/orgasm in men. There may be many obscure reasons for this and at least one reason of some conviction: female orgasm is not as apparent as the male orgasm. This, together with the fact that female orgasm is not a requisite for fertilization, made it a rather uninteresting topic for centuries. We shall not forget, though, that in the times of Saint Augustine it was believed that the experience of orgasm was necessary for both the copulating man and woman for the act to be fertile. Since non-fertile copulation was a sin, we may conclude that sex without orgasm was sinful. This notion of Saint Augustine was lost sometime during the Dark Ages and for centuries European men did not pay much attention to orgasm in women. Whether women did or did not was of little or no concern. Nevertheless, female orgasm has been the subject of a respectable amount of research during most of the 20th century. In addition, and as is the case with orgasm in men, there is an enormous amount of popular descriptions of the orgasmic experience. I should not, in principle, waste words on commenting on this subject, but there are so many misconceptions about the female orgasm circulating in popular and some not so popular writings that I cannot resist the temptation to present my own parsimonious view as a counterpoise. At difference to most of the popular opinions, I try to base my point of view on empirical data. As with so many other aspects of human copulatory behavior, Masters and Johnson (1966) were the first to provide a rather detailed description of the physiological responses accompanying orgasm in women. These responses were quite similar to those found in men, i.e. a series of rhythmic genitopelvic and anal contractions. Other responses typically seen at orgasm in both men and women are body rigidity, muscle spasms, hyperventilation, sweating and vocalizations. Master and Johnson (1966) considered the orgasm experience to be more variable in women than in men, although its basic structure was the same in both sexes.

The similarity of the experience of orgasm in men and women was elegantly shown in a study where medical students, obstetricians-gynecologists and psychologists were asked to determine the sex of the author of 48 anonymous, written descriptions of orgasm. Both groups of professionals as well as the medical students failed on the task (Vance and Wagner, 1976). It appears, then, that women and men produce indistinguishable accounts of their experience of orgasm. A more complex study asked almost a thousand students to portray their orgasm experience by rating how well 60 adjectives described it. A Likert scale ranging from 'does not describe it at all' to 'describes it perfectly' was used for the rating of adjectives like incredible, euphoric, shuddering, relaxing, spurting or throbbing, just to give a few examples. Two orgasmic contexts were rated: solitary masturbation and sex with a partner. The partner could be of the same or of the opposite sex and orgasm could be obtained through coital (supposedly penetrative) or non-coital activities. Results were evaluated in the complicated ways that psychologists like, employing procedures such as principal component analysis, multivariate analysis of covariance, confirmatory factor analysis, and so on. The conclusion was, nevertheless, that there was no qualitative difference between sexes with regard to the description of orgasm while there were some marginal differences in intensity ratings of the adjectives. For example, men considered 'shooting sensations' more intense than women did (Mah and Binik, 2002). The authors point out that the minor sex differences perhaps can be explained by sex stereotypes of emotional responsiveness or simply by differential social desirability responding. Women have a tendency to report more intense emotions than men and they also show more intense physiological reactions to emotional stimuli than men (Grossman and Wood, 1993). This does not reflect any inherent sex difference. The larger emotional responses displayed by women are usually interpreted as socially constructed rather than biologically determined. Likewise, women have higher scores on scales evaluating social desirability and they are less comfortable than men answering questions about their sexuality (Plaud *et al.*, 1998). These socially determined sex differences can and probably do account for the small discrepancies observed between women's and men's descriptions of the orgasm experience. It seems rather safe to posit that the experience of orgasm is similar, if not outright identical, in both sexes. The importance of this assertion will become evident in future chapters.

We saw a couple of paragraphs ago that ejaculation/orgasm in men is triggered by tactile stimulation of the penis. The stimulus triggering orgasm in women seems also to be tactile stimulation. The question of some importance here is tactile stimulation of what. Everyone familiar with classical psychoanalysis knows that Freud considered clitoral stimulation to be of importance for orgasm but, albeit important, clitoral orgasm represents a kind of primitive or immature sexual experience when compared to vaginal orgasm. This latter was supposed to be triggered by tactile stimulation of the vagina. Empirical studies, again performed by Masters and Johnson (1966), have failed to detect any difference between orgasm induced by the exclusive stimulation of the clitoris and orgasm induced by stimulation of the vagina. However, it is most likely that vaginal stimulation is associated with some degree of clitoral stimulation because of the morphology of the female genitalia. Vaginal stimulation caused by the thrusting penis during penetrative intercourse is always associated with indirect stimulation of the clitoris because of the back and forth movement of the labia minora. As expressed by Masters and Johnson (1966, p. 59):

> A mechanical traction develops on both sides of the clitoral hood of the minor labia subsequent to penile distension of the vaginal outlet. With active penile thrusting, the clitoral body is pulled downward the pudendum by traction exerted on the wings of the clitoral hood. . . . The rhythmic movement of the clitoral body in conjunction with active penile stroking produces significant indirect or secondary clitoral stimulation.

The same series of events will obviously also occur if the penis is replaced by another thrusting object. An inevitable consequence of this is that it is technically difficult to determine the role of exclusive vaginal stimulation. I know of no study of 'vaginal orgasm' where clitoral stimulation has been convincingly excluded. Perhaps a purely vaginal orgasm does not exist. In any case, Masters and Johnson (1966) concluded that there is no difference between clitoral and vaginal orgasm. Moreover, the bulk of available data suggest that clitoral stimulation is crucial for orgasm in most women (see, for example Mah and Binik, 2001 for a discussion). This assertion does not exclude other means for women to obtain orgasm. Masters and Johnson (1966) mention examples of women who can obtain orgasm through the exclusive stimulation of the nipples, and we have already learned that some women may reach orgasm through mental representations alone, in complete absence of tactile stimulation of any kind. Despite these facts, it is no exaggeration to state that orgasm induced by copulatory behaviors normally requires tactile stimulation of the clitoris as a triggering stimulus.

Anatomical studies of the human clitoris have revealed that its general structure is rather similar to that of the penis. It consists of erectile tissue, corpora cavernosa and bulbs, as well as of a non-erectile glans, similar to the glans penis. The bulbs extend laterally around the urethral and vaginal openings. It has for a long time been discussed whether the bulbs are part of the clitoris or not, but recent anatomical studies seem to have settled the issue with the unequivocal conclusion that they are (O'Connell *et al.*, 2005). A very important consequence of this is that the ventral side of the vaginal opening is bordered by clitoral tissue. In fact, there is such an intimate relationship between the ventral vaginal opening, the urethral opening and the clitoris that these three structures can be considered a functional unit. Indeed, branches of the cavernous nerve originate in the ventrolateral wall of the vaginal opening and join the dorsal nerve of the clitoris (Yucel *et al.*, 2004). This nerve is equivalent to the main sensory nerve of the penis, the dorsal nerve. In addition to containing fibers from the vaginal opening, the dorsal nerve of the clitoris contains fibers from the urethral sphincter complex and an abundant supply from the glans clitoris. The glans contains a large number of sensory receptors, just as the glans penis does. The nature of these sensory receptors is not entirely clear. Pacini's corpuscles have been described, but the abundant free nerve endings and the particular receptor-like structure formed by them that has been observed in the glans penis have not been found in the glans clitoris. One reason is that the only detailed study of clitoral sensory receptors was performed more than 50 years ago (Yamada, 1951), with rather primitive histological techniques. In the absence of more data, I simply suggest that the sensory characteristics of the glans clitoris are not much different from those of the glans penis. In the same way as there is a structural and functional basis for the idea that tactile stimulation of the penis is crucial for ejaculation/orgasm in male copulatory behaviors, we have now shown that a similar structural and functional basis exists for the notion that tactile stimulation of the clitoris is essential for female orgasm produced by copulatory behaviors.

Before ending the discussion of female orgasm, I consider it useful to mention two very neat studies giving support to the conclusion presented in the preceding paragraph. In one, the vaginal wall was stimulated by a device delivering electric currents of variable intensity. The device was rather small and it was kept in a fixed position during the periods of stimulation, thereby avoiding most mechanical stimulation of the vaginal opening. Sensitivity to the electrical stimuli was generally low. In fact, it was lower than that of the back of the hand (Weijmar-Schultz *et al.*, 1989). There was only one stimulus location that revealed high sensitivity to stimulation. Curiously enough, this was the ventral vaginal wall close to the introitus. In the other study, women were asked to stimulate the vaginal walls with their fingers and rate the 'erotic sensitivity' (Alzate and Londoño, 1984). Results showed that stimulation of a site described as 'the upper anterior wall' was the most efficient for attaining orgasm. If we assume that the upper anterior wall is equivalent to the ventral wall close to introitus, then we must conclude that both studies showed that the most sensitive part of the vagina is exactly where the clitoral bulbs are found, which is on the ventral part of the vaginal opening. Thus, without knowing it, these studies support the notion that clitoral stimulation is crucial for female orgasm. Just for the fun of it, I might add that none of the scientifically sound studies mentioned here (as well as many others that have not been mentioned) has provided any support for the existence of a site known as the G spot, or the Grafenberg spot, famous from popular magazines and some bestselling books.

After orgasm there are two possible ensuing events: continuation of copulatory activity, or the end of it. As already mentioned, the latter alternative is the most common in men, while many women may experience repeated orgasms before ending copulation. Regardless of the number of orgasms, the cessation of copulatory activity leads to the resolution phase. During that phase, physiological parameters return to baseline. In later chapters, we will analyze some of the behavioral consequences of sexual stimulation, particularly of orgasm, but the resolution phase, as described by Masters and Johnson (1966), will not attract our attention.

This short description of copulatory behavior in men and women should satisfy the minimum requirements for an adequate appreciation of the following chapters. Before closing this section, though, we might venture into a short comparison of copulatory behavior displayed by non-human animals and those displayed by women and men. A salient difference is that copulatory behavior in non-humans is essentially limited to penile–vaginal intercourse, and occasionally masturbation, while it may be extremely variable in the human. The reasons for this dramatic difference are not completely clear. It is easy to suggest that non-human animals are not creative enough, or do not have the motor capacities needed for the enormous variation typical of the human. For example, it may be motorically impossible for a male rat to achieve anal or oral penetration of a female. Likewise, the rats' creativity may not be sufficient for getting the idea that prolonged licking of the genitals may be enough for triggering ejaculation and orgasm. However, the rat genital organs have structural and functional bases for activating orgasm rather similar to what I have described for the human (see Marson, 1995 for an example and some references). Considering that, it might be a mystery that rats do not masturbate. Likewise, a male chimpanzee would certainly be motorically able to stimulate a female's clitoris until orgasm, but for some reason he has never chosen to do so. Perhaps he does not find any pleasure in it and then he would have no reason to do it. More surprisingly is that the female chimp does not stimulate her own clitoris to orgasm. That would probably cause her pleasure, so it is difficult to see why she does not. Of course, one explanation can be that she has never discovered how pleasurable it is, perhaps because female chimps have an anatomy such that the clitoris is not readily stimulated except by a male's vaginal penetration. This hypothesis may be reinforced by the fact that male chimps do masturbate, and it is obvious even for a human observer that the chimp's penis is easily accessible for manual exercises. I will not penetrate the problem of stereotyped versus highly variable copulatory behaviors here, but it is important to keep this difference between men and mice in mind.

At the same time as there are fundamental differences between human and non-human animals with regard to copulatory behaviors, there are also basic similarities. The basic model of copulatory behavior posing the existence of an ordered sequence of desire – excitation – ejaculation/orgasm – return to baseline is applicable to males and females of any mammalian species. The basic behavioral and neural processes involved are probably also quite similar across species. Likewise, the fact that the model applies both to females and males is probably also true in all mammalian species. However, when we turn to a more detailed examination of each of the phases, we will find some differences in addition to similarities. Salient among the similarities is the fact that male ejaculation/orgasm in association with copulation is always triggered by mechanical stimulation of the penis. As far as we know, this is the case for all mammals and probably also for other classes of vertebrates. Most unfortunately, this clear-cut statement cannot be immediately applied to women. The origin of orgasm, and even the existence of it, in non-human females is unclear. We know that copulatory behaviors are associated with positive affect in females of all species where this question has been studied, but we do not know if this positive affect is associated with the physiological reactions typical of orgasm in women, notably the rhythmic contractions of pelvic muscles. Thus, while the mechanisms of ejaculation/orgasm in males seem to have been conserved during evolution, the mechanisms of orgasm in females may have been modified. The modifications may be limited to the physiological manifestations of orgasm but they may also involve the stimuli triggering it. It is only to be hoped that some day we will have enough data to answer this question. Nevertheless, the universal utility of the sequence desire – excitation – ejaculation/orgasm – return to baseline seems to be beyond doubt.

2.5 The relationship between sexual desire (motivation) and excitation (arousal)

Desire is a term popular in the literature on human sexual behavior. Due to some undoubtedly unfounded prejudice against fashionable terms, I strongly dislike it. Its precise meaning is unclear and it is extremely difficult to use it in a comparative perspective. Proposing that a female rat desires to engage in copulatory activity is extremely repulsive. I have elsewhere argued, with convincing arguments, that the word 'desire' needs to be replaced with the word 'motivation' in any scientific text (Ågmo *et al.*, 2004). Motivation is a well established concept in the behavioral sciences and it can easily be integrated into a large body of theoretical constructions, where clear definitions are possible. The replacement of desire by motivation allows me, for example, to apply an incentive motivational analysis to the problems of approach to a mate and to the visceral reactions produced by incentive stimuli. Erection or vaginal lubrication can, for example, be analyzed and understood as visceral responses to sexual incentive stimuli.

Arousal or excitation are also imprecise terms. Some would use either of these terms as equivalent to motivation. The statement 'I am so aroused' could mean that the person making it feels an immediate urge to engage in sexual activity and/or that she perceives abundant vaginal lubrication. The same reasoning can be applied to the term 'excited'. Here I will not suggest a replacement term. Instead, I want to make clear that arousal, in the context of human sexual behavior, refers exclusively to enhanced genital blood flow. This is eloquently expressed in the *Diagnostic and Statistical Manual of Mental Disorders*, 4th edition (DSM IV, American Psychiatric Association, 2000). The definition of female sexual arousal is: 'The arousal response consists of vasocongestion in the pelvis, vaginal lubrication and expansion, and swelling of the external genitalia' (p. 543). There is no explicit definition of arousal in men, but male erectile disorder appears as a subcategory of sexual arousal disorders. In men, erection is the main manifestation of enhanced genital blood flow. This means that the DSM IV equates sexual arousal with genital blood flow also in men. By now it should be clear that arousal, as employed by clinicians, has nothing to do with mental events, such as wanting sex, or anything of the kind. It refers strictly to an autonomous response, a visceral reaction.

The preceding clarification of terms can immediately be applied to a motivational analysis of the first phases of copulatory behavior, desire and excitation. I mentioned above that the presently most popular description of these phases divides the desire phase into endogenous and exogenous. The endogenous phase of desire, characterized by 'specific sensations which move the individual to seek out or become receptive to sexual experiences' (Kaplan, 1979, p. 10), sounds rather similar to what I have called sexual incentive motivation in its first phase, approach to a mate. The desire phase as defined above does not, actually, belong to what we have called copulatory behavior. We will return to that phase and its stimulus control in the chapter on human sexual incentives. For the moment it is sufficient to say that the addition of the desire phase, the phase where sexual motivation is activated, solves a fundamental problem that was not addressed at all in the Masters and Johnson model of human copulatory behavior. I think of the problem of the motivational background for enhanced genital blood flow during the arousal phase, the phase we now call excitation. By adding desire (motivation), activated by an external or internal incentive stimulus, we can explain how the enhancement of genital blood flow comes under the control of motivational mechanisms. Without a motivational mechanism, the origin of vaginal lubrication or erection would be unexplainable. This would also be the case with variations in the *intensity* of lubrication or erection. Having introduced the motivational basis, we can now use *degree* of lubrication or erection as sensitive indicators of sexual motivation. Prior to these responses, however, sexual motivation must have been activated by external or internal (mental representations of sexual activities) incentive stimuli. The recognition that sexual motivation, or in incentive motivational terms, the activation of a central motive state by sexual incentives, needs to precede excitation and, furthermore, has to continue to be active during the entire excitation phase until orgasm allows for an elegant and comprehensible analysis of human sexual interactions.

Having such a fine model of human copulatory behavior, it must be regretted that no one has profited from the rather detailed descriptions of physiological arousal (excitation) to pursue an analysis of its control by motivation. In fact, the only study to have addressed the relationship between sexual motivation and intensity of the arousal response, which I know of, was not performed in humans but in male chimpanzees (Nadler and Bartlett, 1997). As a measure of arousal the authors employed the magnitude of erection during a sexual encounter with a receptive female. Erection was quantified simply by measuring the penis length at various moments. The momentaneous length measurements were then expressed as a proportion of length measured at maximum erection. This ingenious procedure made it possible to estimate the relative magnitude of erection throughout the sexual encounter. The relative magnitude of erection was then correlated with several measures of sexual performance, thought to depend on the degree of motivation. The results suggested that the relative magnitude of erection indeed is determined by motivational mechanisms similar or identical to those determining the intensity of other aspects of copulatory behavior. It is somewhat unfortunate that the extremely interesting conclusion is obscured by the use of a confusing theoretical framework. Nevertheless, this single study is all we have of experimental data concerning the relationship between the intensity of arousal (defined as intensity of genital blood flow) and motivation.

We know a lot more of the stimulus control of the arousal response than we know about its control by motivational mechanisms. There are diverse reasons for this. The most important is that lack of enhanced genital blood flow in some situations is a rather common clinical problem. In men, it leads to erectile dysfunction, a condition called impotence in ordinary language. Impotence has been of much concern to medical practitioners for many years and much basic research concerning the physiological mechanisms of erection has been prompted by this interest. Concepts like motivation do not form part of a physiologist's conceptual framework and is rarely, if ever, mentioned in their analyses of erectile mechanisms. Although motivation is of no concern to the physiologist, she/he needs to have some idea of the stimuli that activate the physiological response under scrutiny. The clinicians, for their part, need some reliable diagnostic procedure for determining the severity and/or the characteristics of the erectile dysfunction in their patients. A standard procedure is to expose the patients to some kind of erotic stimulus, like a pornographic movie, and measure the arousal response to that stimulus. This is easily done by measuring penile circumference with a strain gauge. Clinicians, like physiologists, are not particularly interested in psychological concepts like motivation and they use non-motivational concepts when trying to understand and treat impotence. This is even the case with impotence treatments centered on manipulations of motivational mechanisms. The famous sensate focus training described by Masters and Johnson (1970) is essentially a procedure leading to enhanced sexual motivation by the generous employment of several sexual incentive stimuli, including mechanical stimulation of the genitalia. However, the lack of a coherent theoretical framework adequate for sexual functions impeded the recognition of this fundamental fact and the term motivation is not used at all in the discussion of treatment of impotence. In fact, the only time the word motivation appears in Human Sexual Inadequacy is when describing the patient's reasons for seeking treatment. These reasons were called motives, and the single use of the term motivation was awkward indeed: '. . . intense motivation for reversal of symptomatology . . .' (Masters and Johnson, 1970, p. 357). Perhaps this little example illustrates the fact that clinicians have their worries far away from motivational mechanisms as understood in motivation theory.

Our discussion of the very basic question of the relationship between motivation (desire) and the following phases of human copulatory behavior is far shorter than its theoretical and clinical importance merit. The reason for this brevity is that very little is known. Factual knowledge is frequently supplanted with more or less founded speculations and I will finish this section with some speculative proposals.

First, by applying the incentive motivational framework outlined earlier, we can establish that sexual incentive stimuli have the dual effect of activating motivation, behaviorally manifested as approach to the sexual incentive, and visceral reactions, notably enhanced genital blood flow. The sexual approach is a paramount feature of van de Velde's (1965) first phase of human sexual interactions, the prelude, and it can be assumed to coincide with the desire phase in more recent models of human sexual behavior. An extremely important question here is whether the two aspects of sexual motivation, approach and enhanced genital blood flow, are always activated simultaneously or not. The answer to that question becomes crucial to our understanding of deficient sexual arousal, called arousal disorder in women and erectile dysfunctions in men. We will return to this important problem in the chapters dedicated to these disorders. Before that, we have to acquire some knowledge about sexual incentives, which will be the subject of the two following chapters. Here I only want to make the reader cognizant of the existence of the question.

Second, some sexual incentive stimuli, be they proximal or distant, may not be able to activate sexual motivation in a particular situation and/or at a particular moment. Other stimuli may, however, be effective in that situation. Caressing the face or lower arms or kissing may be inefficient for enhancing sexual motivation in a woman at a particular moment, but mechanical stimulation of the clitoris might be highly effective. The question is whether the woman would allow this stimulation to occur if the previous stimuli had failed to activate sexual motivation. This is another issue that will become of some importance when we arrive at problems with hypoactive sexual desire disorders. Are these disorders caused by insufficient exposure to efficient sexual incentive stimuli, or are they caused by hypoactivity of the central motive state, making the individual unresponsive to such stimuli? For the moment I will not consider this fundamental topic. Nevertheless, a general principle, maintaining that the intensity of stimulation determines the probability of activation of motivation, and consequently also the degree of arousal, can be reiterated. This principle is not new and surprising in any way. As mentioned in the first section of this chapter it was established, in general terms, more than 50 years ago, and it has had a wide application in ethology and related fields (Lorenz, 1950). It also coincides with the characteristics of the concept of drive stimulus as employed in Hullian theory (Hull, 1943).

Third, 'spontaneous' desire, or 'endogenous' desire, meaning the sexual motivation that is activated without any known reason, is certainly not spontaneous. Some external stimuli may activate sexual desire and there is no reason to assume that all such stimuli must be available for the conscience. Any stimulus may have become a conditioned sexual incentive and it is quite possible that the conditioning process never formed part of our conscious experience. When and if stimuli of this kind activate sexual incentive motivation, we would call it a case of spontaneous or endogenous desire activation since we are not conscious of the connection between the stimulus/stimuli and the central motive state.

Summary

1 According to incentive motivation theory, environmental stimuli are divided into three classes. Neutral stimuli do not affect an organism's behavior. A positive incentive activates activity in skeletal muscles leading to approach to the incentive. A negative incentive activates activity in skeletal muscles leading to withdrawal from the incentive.

2 In addition to the effects on motor patterns, positive and negative incentives activate visceral reactions. A sexual incentive may cause release of hormones or enhanced genital blood flow, for example.

3 The impact of an incentive is determined by the central motive state. An incentive will affect behavior only when the central motive state is in an appropriate condition.
4 In non-human mammals, copulatory behaviors are a sequence of somatic reflexes. In males, the autonomic reflex of erection is required for the full expression of copulatory behavior.
5 The mount is the basic unit of male rat copulatory behavior. It appears that the mount is a somatic reflex activated by tactile stimulation of the perineal and ventral skin.
6 In female rodents, lordosis is the motor pattern associated with sexual receptivity. It is a somatic reflex activated by tactile stimulation of the lower back and flanks.
7 Females of some species display not only lordosis during copulatory interaction but also proceptive behaviors, notably hop-darting and ear wiggling.
8 When given opportunity to do so, the female rat controls the pace of sexual interactions.
9 In the human, copulatory behavior is usually divided into four phases. First comes the desire phase, then follows the phase of excitation (arousal) culminating in orgasm and finally appears the resolution phase.
10 Human copulatory behaviors are far less stereotyped than copulatory behaviors in non-human mammals. Penile–vaginal intercourse is only one of many forms a human can employ for the obtention of sexual reward.
11 In the human, the relationship between sexual desire and sexual arousal is somewhat complicated.

References

Ågmo, A. (1999). Sexual motivation. An inquiry into events determining the occurrence of sexual behavior. *Behav Brain Res* **105**, 129–150

Ågmo, A. (2003). Unconditioned sexual incentive motivation in the male Norway rat (*Rattus norvegicus*). *J Comp Psychol* **117**, 3–14

Ågmo, A. and Contreras, J.L. (1990). Copulatory thrusting pattern in the male rat after acute treatment with GABA transaminase inhibitors. *Physiol Behav* **47**, 311–314

Ågmo, A., Turi, A.L., Ellingsen, E. and Kaspersen, H. (2004). Preclinical models of sexual desire: Conceptual and behavioral analyses. *Pharmacol Biochem Behav* **78**, 379–404

Alzate, H. and Londoño, M.L. (1984). Vaginal erotic sensitivity. *J Sex Marital Ther* **10**, 49–56

American Psychiatric Association (2000). *Diagnostic and statistical manual of mental disorders*, 4th edn. American Psychiatric Association: Washington, DC

Beach, F.A. (1976). Sexual attractivity, proceptivity, and receptivity in female mammals. *Horm Behav* **7**, 105–138

Berkley, K.J., Wood, E., Scofield, S.L. and Little, M. (1995). Behavioral responses to uterine or vaginal distention in the rat. *Pain* **61**, 121–131

Bernabé, J., Rampin, O., Sachs, B.D. and Giuliano, F. (1999). Intracavernous pressure during erection in rats: an integrative approach based on telemetric recording. *Am J Physiol Regul Integr Comp Physiol* **45**, R441–R449

Beyer, C., Contreras, J.L., Larsson, K., Olmedo, M. and Moralí, G. (1981). Effects of castration and sex steroid replacement on the motor copulatory pattern of the male rat. *Physiol Behav* **27**, 727–730

Beyer, C., Contreras, J.L., Larsson, K., Olmedo, M. and Moralí, G. (1982). Patterns of motor and seminal vesicle activities during copulation in the male rat. *Physiol Behav* **29**, 495–500

Bindra, D. (1969). The interrelated mechanisms of reinforcement and motivation, and the nature of their influence on response. In *Nebraska symposium on motivation* (W.J. Arnold and D. Levine, eds), pp. 1–33. University of Nebraska Press: Lincoln

Bindra, D. (1974). A motivational view of learning, performance, and behavior modification. *Psychol Rev* **81**, 199–213

Bindra, D. (1976). *A theory of intelligent behavior*. Wiley: New York

Bindra, D. (1978). How adaptive behaviour is produced: a perceptual-motivational alternative to response reinforcement. *Behav Brain Sci* **1**, 41–52

Bradshaw, H.B., Temple, J.L., Wood, E. and Berkley, K.J. (1999). Estrous variations in behavioral responses to vaginal and uterine distention in the rat. *Pain* **82**, 187–197

Brody, S. (1995). Patients misrepresenting their risk factors. *Int J STD AIDS* **6**, 392–398

Bullough, V.L. (1994). *Science in the bedroom. A history of sex research*. Basic Books: New York

Calhoun, J.B. (1962). *The ecology and sociology of the Norway rat*. US Government Printing Office: Washington, DC

Catania, J.A., Binson, D., van der Straten, A. and Stone, V. (1995). Methodological research on sexual behavior in the AIDS era. *Ann Rev Sex Res* **6**, 77–125

Cochran, W.G., Mosteller, F. and Tukey, J.W. (1954). *Statistical problems of the Kinsey report on sexual behavior in the human male. A report of the American Statistical Association committee to advice the National Research Council Committee for Research in Problems of Sex*. American Statistical Association: Washington, DC

Contreras, J.L. and Ågmo, A. (1993). Sensory control of the male rat's copulatory thrusting patterns. *Behav Neural Biol* **60**, 234–240

Coxon, A.P.M. (1999). Parallel accounts? Discrepancies between self-report (diary) and recall (questionnaire) measures of the same sexual behavior. *AIDS Care* **11**, 221–234

Cueva-Rolón, R., Munoz-Martínez, E.J., Delgado-Lezama, R. and Raya, J.G. (1994). The cat pudendal nerve: afferent fibers responding to mechanical stimulation of the perineal skin, the vagina or the uterine cervix. *Brain Res* **655**, 1–6

Davidson, J.M. and Sawyer, C.H. (1961). Effects of localized intracerebral implantation of oestrogen on reproductive function in the female rabbit. *Acta Endocr* **37**, 385–393

Dewsbury, D.A. (1972). Patterns of copulatory behavior in male mammals. *Quart Rev Biol* **47**, 1–33

Dewsbury, D.A. and Pierce, J.D. Jr (1989). Copulatory patterns of primates as viewed in broad mammalian perspective. *Am J Primatol* **17**, 51–72

Diakow, C. and Dewsbury, D.A. (1978). A comparative description of the mating behavior of female rodents. *Anim Behav* **26**, 1091–1097

Ellingsen, E. and Ågmo, A. (2004). Sexual incentive motivation and paced sexual behavior in female rats after treatment with drugs modifying dopaminergic neurotransmission. *Pharmacol Biochem Behav* **77**, 431–445

Erskine, M.S. (1992). Pelvic and pudendal nerves influence the display of paced mating behavior in response to estrogen and progesterone in the female rat. *Behav Neurosci* **106**, 690–697

Ford, C.S. and Beach, F.A. (1951). *Patterns of sexual behavior*. Harper & Row: New York

Freud, S. (1905). *Drei Abhandlungen zur Sexualtheorie*. F. Deuticke: Leipzig

Giuliano, F., Bernabé, J., Rampin, O., Courtois, F., Benoit, G. and Rosseau, J.P. (1994). Telemetric monitoring of intracavernous pressure in freely moving rats during copulation. *J Urol* **152**, 1271–1274

Grossman, M. and Wood, W. (1993). Sex differences in intensity of emotional experience: a social role interpretation. *J Personal Soc Psychol* **65**, 1010–1022

Halata, Z. and Munger, B.L. (1986). The neuroanatomical basis for the protopathic sensibility of the human glans penis. *Brain Res* **371**, 205–230

Harris, G.W. and Michael, R.P. (1964). The activation of sexual behaviour by hypothalamic implants of oestrogen. *J Physiol* **171**, 275–301

Harris, G.W., Michael, R.P. and Scott, P.P. (1958). Neurological site of action of stilboestrol in eliciting sexual behaviour. In *Ciba Foundation Symposium on the Neurological Basis of Behaviour* (G.E.W. Wolstenholme and C.M. O'Connor, eds), pp. 236–251. Churchill: London

Holmes, G.M., Chapple, W.D., Leipheimer, R.E. and Sachs, B.D. (1991). Electromyographic analysis of male rat perineal muscles during copulation and reflexive erections. *Physiol Behav* **49**, 1235–1246

Hull, C.L. (1943). *Principles of behavior: an introduction to behavior theory*. Appleton-Century-Crofts: New York

Hull, C.L. (1951). *Essentials of behavior*. Yale University Press: New Haven

Hull, C.L. (1952). *A behavior system: an introduction to behavior theory concerning the individual organism*. Yale University Press: New Haven

Jagannathan, R. (2001). Relying on surveys to understand abortion behavior: some cautionary evidence. *Am J Public Health* **91**, 1825–1831

Jones, J.H. (1997). *Alfred C. Kinsey: a public/private life*. W.W. Norton: New York

Kaplan, H.S. (1979). *Disorders of sexual desire and other new concepts and techniques in sex therapy*. Simon and Schuster: New York

Katter, J.T., Dado, R.J., Kostarczyk, E. and Giesler, G.J. (1996). Spinothalamic and spinohypothalamic tract neurons in the sacral spinal cord of rats. 2. Responses to cutaneous and visceral stimuli. *J Neurophysiol* **75**, 2606–2628

Kinsey, A.C., Pomeroy, W.B. and Martin, C.E. (1948). *Sexual behavior in the human male*. Saunders: Philadelphia

Kinsey, A.C., Pomeroy, W.B., Martin, C.E. and Gebhard, P.H. (1953). *Sexual behavior in the human female*. Saunders: Philadelphia

Komisaruk, B.R. and Wallman, J. (1977). Antinociceptive effects of vaginal stimulation in rats: neurophysiological and behavioral studies. *Brain Res* **137**, 85–107

Komisaruk, B.R. and Whipple, B. (1995). The suppression of pain by genital stimulation in females. *Ann Rev Sex Res* **6**, 151–186

Komisaruk, B.R. and Whipple, B. (2000). How does vaginal stimulation produce pleasure, pain, and analgesia? In *Sex, gender, and pain* (R.B. Fillingim, ed.), pp. 109–134. IASP Press: Seattle

Larsson, K. and Södersten, P. (1973). Mating in male rats after section of the dorsal penile nerve. *Physiol Behav* **10**, 567–571

Latané, B. (1969). Gregariousness and fear in laboratory rats. *J Exp Soc Psychol* **5**, 61–69

Latané, B., Edwards, J., Steele, C. and Walton, D. (1973). Social attraction among and between albino and hooded rats. *Bull Psychonom Soc* **2**, 20–22

Latané, B. and Glass, D.C. (1968). Social and nonsocial attraction in rats. *J Personal Soc Psychol* **9**, 142–146

Levin, R.J. (2000). Sexual desire and the deconstruction and reconstruction of the human female sexual response model of Masters and Johnson. In *Sexual appetite, desire and motivation: Energetics of the sexual system* (W. Everaerd, E. Laan and S. Both, eds), pp. 63–93. Royal Netherlands Academy of Arts and Sciences: Amsterdam

Lewontin, R.C. (1995). Sex, lies, and social science. *NY Rev Books* **42**, 24–29

Lorenz, K. (1950). The comparative method in studying innate behavior patterns. *Symp Soc Exp Biol* **4**, 221–268

Mah, K. and Binik, Y.M. (2001). The nature of human orgasm: a critical review of major trends. *Clin Psychol Rev* **21**, 823–856

Mah, K. and Binik, Y.M. (2002). Do all orgasms feel alike? Evaluating a two-dimensional model of the orgasm experience across gender and sexual context. *J Sex Res* **39**, 104–113

Maillard, C.A. and Edwards, D.A. (1990). Cutaneous desensitization of the abdomen and inner thighs disrupts copulation in male rats. *Soc Neurosci Abstr* **16**, 50.8

Marson, L. (1995). Central nervous system neurons identified after injection of pseudorabies virus into the rat clitoris. *Neurosci Lett* **190**, 41–44

Masters, W.H. and Johnson, V.E. (1966). *Human sexual response*. Little, Brown & Co.: Boston

Masters, W.H. and Johnson, V.E. (1970). *Human sexual inadequacy*. Little, Brown & Co.: Boston

McClintock, M.K. and Adler, N.T. (1978). The role of the female during copulation in wild and domestic Norway rats *(Rattus norvegicus)*. *Behaviour* **67**, 67–96

McClintock, M.K. and Anisko, J.J. (1982). Group mating among Norway rats. I. Sex differences in the pattern and neuroendocrine consequences of copulation. *Anim Behav* **30**, 398–409

McClintock, M.K., Anisko, J.J. and Adler, N.T. (1982). Group mating among Norway rats. II. The social dynamics of copulation: competition, cooperation, and mate choice. *Anim Behav* **30**, 410–425

Moralí, G. and Beyer, C. (1992). Motor aspects of masculine sexual behavior in rats and rabbits. *Adv Study Behav* **21**, 201–238

Morgan, C.T. (1942). *Physiological psychology*. McGraw-Hill: New York

Nadler, R.D. and Bartlett, E.S. (1997). Penile erection: a reflection of sexual arousal and arousability in male chimpanzees. *Physiol Behav* **61**, 425–432

O'Connell, H.E., Sanjeevan, K.V. and Hutson, J.M. (2005). Anatomy of the clitoris. *J Urol* **174**, 1189–1195

Paredes, R.G. and Ågmo, A. (2004). Has dopamine a physiological role in the control of sexual behavior? A critical review of the evidence. *Progr Neurobiol* **73**, 179–226

Paredes, R.G., Highland, L. and Karam, P. (1993a). Socio-sexual behavior in male rats after lesions of the medial preoptic area: evidence for reduced sexual motivation. *Brain Res* **618**, 271–276

Paredes, R.G., Holmes, G.M., Sachs, B.D. and Ågmo, A. (1993b). Electromyographic activity of rat ischiocavernosus muscles during copulation after treatment with a GABA transaminase inhibitor. *Behav Neural Biol* **60**, 118–122

Peirce, J.T. and Nuttall, R.L. (1961). Self-paced sexual behavior in the female rat. *J Comp Physiol Psychol* **54**, 310–313

Pfaff, D.W. (1980). *Estrogens and brain function: neural analysis of a hormone-controlled mammalian reproductive behavior*. Springer: New York

Pfaff, D.W. (1999). *Drive. Neurobiological and molecular mechanisms of sexual motivation*. MIT Press: Cambridge

Pfaff, D.W. and Ågmo, A. (2002). Reproductive motivation. In *Steven's Handbook of Experimental Psychology. Volume 3, Learning, Motivation, and Emotion*, 3rd edn (H. Pashler and R. Gallistel, eds), pp. 709–736. Wiley: New York

Pfaff, D.W., Lewis, C., Diakow, C. and Keiner, M. (1973). Neurophysiological analysis of mating behavior as hormone-sensitive reflexes. *Progr Physiol Psychol* **5**, 253–297

Plaud, J.J., Gaither, G.A. and Weller, L.A. (1998). Gender differences in the sexual rating of words. *J Sex Marital Ther* **24**, 13–19

Robinson, P. (1976). *The modernization of sex: Havelock Ellis, Alfred Kinsey, William Masters and Virginia Johnson*. Harper & Row: New York

Robitaille, J.A. and Bouvet, J. (1976). Field observations on the social behaviour of the Norway rat, *Rattus norvegicus* (Berkenhout). *Biol Behav* **1**, 289–308

Sachs, B.D. (1982). Role of striatal penile muscles in penile reflexes, copulation, and induction of pregnancy in the rat. *J Reprod Fert* **66**, 433–443

Sachs, B.D. and Barfield, R.J. (1970). Temporal patterning of sexual behavior in the male rat. *J Comp Physiol Psychol* **73**, 359–364

Signoret, J.P. (1970). Action d'implants de benzoate d'oestradiol dans l'hypothalamus sur le comportement d'oestrus de la brebis ovariectomisée. *Annales Biol Anim Biochimie Biophys* **10**, 549–566

Sloan, L. and Latané, B. (1974). Sex and sociability in rats. *J Exp Soc Psychol* **10**, 147–158

Spence, K.W. (1956). *Behavior theory and conditioning*. Yale University Press: New Haven

Spiteri, T. and Ågmo, A. (2006). Modèles precliniques du désir sexuel. *Sexologies*, **15**, 241–249

Steiniger, F. (1950). Beiträge zur Soziologie und sonstigen Biologie der Wanderratte. *Zeitschr Tierpsychol* **7**, 356–379

Stone, C.P., Barker, R.G. and Tomilin, M.I. (1935). Sexual drive in potent and impotent male rats as measured by the Columbia obstruction apparatus. *J Genet Psychol* **47**, 33–48

Sundström, E. (1949). *Uppgörelse med Kinsey-rapporten*. Norlin: Stockholm

van de Velde, T.H. (1965, originally published in 1926). *Ideal marriage. Its physiology and technique*. William Heinemann Medical Books: Westport

van der Horst, V.G.J.M. and Holstege, G. (1998). Sensory and motor components of reproductive behavior: pathways and plasticity. *Behav Brain Res* **92**, 157–167

Vance, E.B. and Wagner, N.N. (1976). Written descriptions of orgasm: a study of sex differences. *Arch Sex Behav* **5**, 87–98

Viitamaa, T., Haapalinna, A. and Ågmo, A. (2006). The adrenergic α_2 receptor and sexual incentive motivation in male rats. *Pharmacol Biochem Behav* **83**, 360–369

Weijmar-Schultz, W.C.M., van de Wiel, H.B.M., Klatter, J.A., Sturm, B.E. and Nauta, J. (1989). Vaginal sensitivity to electrical stimuli: Theoretical and practical implications. *Arch Sex Behav* **18**, 87–95

Weinhardt, L.S., Forsyth, A.D., Carey, M.P., Jaworski, B.C. and Durant, L.E. (1998). Reliability and validity of self-report measures of HIV-related sexual behavior: Progress since 1990 and recommendations for research and practice. *Arch Sex Behav* **27**, 155–180

Wieder, J.A., Brackett, N.L., Lynne, C.M., Green, J.T. and Aballa, T.C. (2000). Anesthetic block of the dorsal penile nerve inhibits vibratory-induced ejaculation in men with spinal cord injuries. *Urology* **55**, 915–917

Yamada, K. (1951). On the sensory nerve terminations in clitoris in human adult. *Tohoku J Exp Med* **54**, 163–174

Yang, C.C. and Bradley, W.E. (1998). Peripheral distribution of the human dorsal nerve of the penis. *J Urol* **159**, 1912–1916

Yang, C.C. and Bradley, W.E. (1999). Innervation of the human glans penis. *J Urol* **161**, 97–102

Yucel, S., de Souza, A. Jr and Baskin, L.S. (2004). Neuroanatomy of the human female lower urogenital tract. *J Urol* **172**, 191–195

CHAPTER

3

The stimulus control of sexual incentive motivation: non-human mammals

1 Introduction

I will now proceed with a discussion of sexual incentives. This very fundamental issue will occupy most of this chapter. The central nervous and endocrine mechanisms associated with the central motive state will be mentioned only marginally at this stage. They will, in fact, be the subjects of later chapters. Moreover, the present discussion will be heavily concentrated on simple incentive stimuli of possible importance for sexual behavior in non-human animals. Over the years, we have learned quite a lot about the kind of stimuli that may have sexual incentive properties in these animals. We have some ideas of how and under which circumstances certain stimuli are produced by an individual, how these stimuli affect sensory mechanisms in other individuals, where and through which pathways the sensory neurons project in the brain and what these projections may do in the projection areas. In short, we have some idea of the anatomy and physiology of the neural circuitry involved in the processing of sexually relevant stimuli in some non-human animals. This knowledge is of utmost importance if we ever are to understand the central nervous control of sexual behavior.

I will start the analysis of the sexual incentive motivational mechanisms with a short introduction to the comparative aspect. Non-human animals sometimes have sensory organs which react to different stimuli from those to which human sense organs react. One consequence is that the representation of the external world in some animals may be completely different from our own and the most surprising stimuli may have incentive properties. After a few comments on this I will abandon the truly comparative perspective and in painstaking detail analyze two groups of stimuli that are supposed to be of substantial importance for rodent sexual behavior, namely high frequency sounds and non-volatile chemicals. These two groups of stimuli have attracted much attention and, fortunately for our purposes, we have enough data to present a fairly complete picture of their actions. I will profit from the context and also try to show how cautious we need to be when ascribing a sexual function to a stimulus and how we need to proceed in order to establish such a function beyond reasonable doubt. This second aim of my analysis is close to my heart. If successful, at the end of this chapter I will have convinced you that we should always be humble when interpreting our data and, above all, avoid going beyond what we have actually observed. Moreover, before insisting on one explanation, we should have ruled out all alternative explanations. Humility and the consideration of alternative explanations, particularly those we dislike for one reason or another, are not overly common in the scientific enterprise.

The human will be almost absent from this chapter. This is perhaps not a consequence of a rational decision, but it seemed convenient to keep the human out of the discussion of sexual incentives at this stage. There are, in fact, many reasons for believing that human sexual incentives are rather different from those acting in other animals. I will not go into that issue here. It will be thoroughly discussed in the next chapter. I am convinced, though, that any discussion of human sexual incentives becomes far more understandable and interesting with the background offered by data from non-human mammals.

2 Detection of a potential mate and the activation of sexual incentive motivation: behavioral considerations

2.1 The incentive stimulus

No stimulus can affect an organism's behavior if it is not captured by its sense organs. In the case of the human, this basic fact does not constitute any immediate problem, because most of us are well aware of our own sensory capabilities. Many other animals, though, have sensory capacities that are quite different from our own. Everyone knows, or believes to know, that dogs or rats can perceive odors with a sensitivity far beyond ours. Here we have an example of different sensory threshold, which is a quantitative difference. There are also qualitative differences in sensory capacity, even among mammals. Some species may have receptors sensitive to light that others do not see, or to sounds that others do not hear or to tastes that others do not feel. What is perhaps still more disturbing is the existence of entirely different sense organs or sensory receptors. The electric organ of some fishes could be an example of this. The nocturnal fish *Gnathonemus petersii* detects objects and recognizes the surrounding world through the use of receptors sensitive to changes in the self-generated, electrical field (von der Emde, 2004). Such an animal lives in a sensory world very different from our own. A consequence of this is that the most unexpected stimulus, which may never be noticed by a human observer, can function as positive or negative incentive. However, since the discussion of sexual incentives in this book will be limited to mammals, we do not need to discuss this issue here. It becomes of great importance, though, as soon as we want to make comparative studies in non-mammalian vertebrates and in invertebrates.

Although there are few qualitative differences in sensory capacities between mammals, we need to take into account some diversity. For example, it is not certain that color vision is universal among mammals (Jacobs, 1993). Even if that would turn out to be the case, the excellent trichromatic color vision of humans and other primates is different from the dichromatic color vision of most other mammals. Furthermore, some rodents, including rats and mice, may have a rather unique dichromatic vision. One photopigment has maximal sensitivity to light with a wavelength around 510 nm (green) and the other has its peak sensitivity around 360 nm (ultraviolet) (Peichl, 2005). Whether the existence of these two pigments is associated with color vision or not is presently unknown. Independently of this, it can be guaranteed that the rodents possessing these two pigments have an image of their surroundings quite different from ours and probably also different from most other mammals. In fact, ultraviolet-sensitive cones have only been identified in marsupials and bats in addition to some rodents. The importance of this is not overwhelming, because color does not seem to be a basic quality of mammalian sexual incentives. Indeed, a mammal's sexual incentive properties are rarely dependent on visual stimuli. The only reasonably well-established exceptions to this assertion are found among the primates, including the human.

In contrast to the modest importance of visual stimuli in non-human mammals, auditory stimuli appear to contribute to the sexual incentive properties of some species. In fact, many rodent species and several species in a couple of other mammalian orders perceive and produce sounds of frequencies above 20 kHz, the upper limit for human hearing. These sounds have been reasonably well studied in mice and rats, even though some aspects are still unclear. I will discuss these high frequency sounds and their possible functions in some detail. One reason for this is that these sounds are good examples of a potentially important incentive stimulus which the human is unable to detect without adequate instruments. There seems to be something inherently attractive or exotic with sensory experiences unavailable to the scientist himself. At least this is the explanation I propose for the rather large amount of research dedicated to high frequency sounds. The substantial amount of data gathered with regard to these sounds is the second reason for choosing them for a closer analysis. There is a sufficiently large number of studies to illustrate how research should proceed when we want to determine whether a stimulus is a sexual incentive or not and, if it is, whether it is unconditioned (functions without learning) or conditioned (previous experience with the stimulus is obligatory). Before starting this discussion we will take a very short look at some quite obscure events between the sense organs and the central motive state. If not done here, it might be forgotten.

The fact that a stimulus can be registered by the sensory system does not by itself imply that it will affect behavior. The stimulus must also be perceived and correctly identified. Perception is a very complex process and I will not venture into a discussion of how physical energy acting on a receptor is transformed into a perceived object and how a perceived object is either given a meaning or ignored. Incentive motivational theory is not a theory of perception. Nevertheless, a prerequisite for a stimulus to have incentive properties is that it has been correctly identified. A food object acquires appropriate incentive properties if and only if it is correctly identified as a food object, and a mate acquires appropriate incentive properties only when identified as a mate. If not, absurd behaviors would occur, like trying to copulate with a food pellet or ingesting the mate rather than copulating with him. In some way or another, the stimulus must be connected with its corresponding central motive state. Otherwise behavior would be completely disorganized, indistinguishable from random activity.

Almost all natural incentives are compound stimuli, in the way that they may activate more than one sensory modality. Furthermore, their distinctiveness may depend on more than one dimension within a modality. These facts make it quite difficult to determine exactly which modality/dimension carries the incentive properties. For example, a female rat in estrus is a powerful sexual incentive for a male rat. This is simple enough. However, if we try to determine exactly which characteristic of the female is necessary for her incentive property we immediately run into huge problems. A female has a particular body shape, perhaps she moves in particular ways, perhaps she produces noises and sounds, certainly she smells and she has a more or less smooth fur and her movement of the ears might be quite sensual. More scientifically: a female emits many different kinds of visual stimuli, many different olfactory stimuli and perhaps also auditory stimuli. We have, then, at least three sensory modalities involved and many different stimuli and configurations of stimuli within each. Which of all this is used by the male when he identifies her as a potential mate? The answer to that question would also tell us which stimuli carry the female's sexual incentive properties. The incentive properties of the female must reside in exactly those stimuli used for her correct identification. A similar comment can be made regarding an alimentary incentive, a food pellet. Is it the color, shape or odor or a combination of these that allows for correct identification and hence carries the incentive properties of the food pellet? I will not try to answer these questions now. I will, however, illustrate the point with an anecdote.

A hungry rat is walking around on the sidewalk of the East 70th street on Manhattan's Upper East Side, just outside the post office between the 2nd and 3rd avenues. Outside that post office, a lady is always depositing grains, normally corn, for the pleasure of the many pigeons squatting in the area. Our rat is also there and, as soon as the lady throws her handful of grains to the ground, it immediately approaches them and starts to eat. The rat has correctly identified the objects as food. My scientific curiosity sometimes made me wait for the grain lady and when she came I walked accidentally by and threw a cigarette butt at the same time and place as she threw the corn. The rat never approached the butt, showing that it had correctly identified it as a non-healthy object or at least as a non-food object. I do not know if it was a difference in size, shape, odor or brightness that allowed the rat to distinguish the butt from the grains, but it did so with efficiency.

The expression 'correct identification' that I have used repeatedly in the preceding paragraphs needs to be clarified. For a human, identification would probably mean that we perceive the object as a member of a class of objects and we give it the label that corresponds to that class. A tennis ball would be perceived as an object pertaining to the same class of objects as footballs, golf balls, ping pong balls and so on, and it would be labeled ball. Correct would mean that the label we choose coincides with what most other humans would choose. In other words, we identify and give meaning to objects through our language. Non-human animals do not use language in that way, probably because they have none. They must identify and give meaning to objects in a different way. One process that could replace the human language is that the central nervous system of non-human animals simply connects specific types of sensory input to a particular central motive state. The sensory input excites the central motive state which, in turn, excites the appropriate motor output. This simple scheme has a substantial advantage: we can entirely dispose of meaning, or say that giving meaning to an object is nothing more than making it excite the appropriate central motive state. Thus, we can explain behavior by exclusively mechanistic cause–effect relationships. I believe that this kind of explanation allows for an understanding of most aspects of sexual behavior in rodents. In the human, however, the possession of language has complicated the meaning of meaning. Although normally the domain of philosophers, I will try to grasp some consequences of human language in later chapters. One consequence was, in fact, already noted in Chapter 1: the frequent use and power of mental representations in the human.

Having disposed of the term 'meaning' in our analysis of incentives in non-human animals, we can now return to the incentive stimuli. The ambiguity, or difficulty in identifying the exact sensory dimension or combination of dimensions of a stimulus that carry the incentive properties is most annoying. I will illustrate this through a discussion of the role of sounds in rodents. Not because rodents are particularly important among mammals, but because we know a lot more about auditory sexual incentive stimuli in some rodent species than in any other mammal. Then I will proceed with an analysis of olfactory stimuli. Concerning odors we have, in addition to a wealth of facts from rodents and some information from other species, data from some non-human primate species. Short mention will also be made of visual stimuli. A few primate studies show that such stimuli may be of some importance in non-human primates. We already know that they are of much significance in the human.

2.2 Auditory sexual incentive stimuli in rats and mice

Ultrasonic vocalizations in mice

High-frequency sounds emitted by rats were first described in 1954 (Anderson, 1954). In that first report, it was suggested that these sounds might serve for communication between individuals and/or for orientation in space comparable to the echolocation used by bats. The echolocation hypothesis was subjected to a few experimental tests (Rosenzweig *et al.*, 1955; Riley and Rosenzweig, 1957; Bell *et al.*, 1971), with negative results and then it was abandoned. However, while high frequency sounds do not seem to be employed by rats for orientation in space, audible sounds may be used (Rossier *et al.*, 2000). Nevertheless, what remains as a viable hypothesis concerning the function of high frequency sounds and what is frequently regarded as a fact more than a hypothesis, is a communicative role of these sounds.

In rats and mice, high frequency sounds have been extensively studied during the 50 years that have passed since their discovery and we now possess a substantial amount of experimental data. Individuals of both species produce sounds with frequencies from 22 kHz up to 70 kHz in addition to audible sounds. Among the many questions scientists have asked are, how are the sounds produced, in which situations are they produced and what effect may they have on potential listeners? We will briefly examine each of these questions.

Forcing air between the vocal cords produces audible vocalizations in both rats, mice and humans. The flow of air makes the cords vibrate and, by adjusting the degree of contraction of the cords, it is possible to modulate the frequency of the vibrations, hence of the tone, produced. However, neither the vocal cords nor the contraction of them is perfect. The consequence is that the tone produced is not pure. It consists rather of a major frequency plus a number of harmonically related frequencies. The high frequency sounds produced by rats and mice are often pure tones without harmonics and discontinuous changes can occur within a single sound pulse. This is not possible to do with vibrating vocal cords. Because of these and other reasons it was concluded that vibrating vocal cords did not produce the high frequency sounds (Roberts, 1975b). The problem was, then, to determine the source of the sound. Mice, rats and men can produce a lot of sounds without the aid of vocal cords, for example in the mouth or in the nose. It was first shown that high frequency sound production is associated with expiration (Roberts, 1972), an observation that was confirmed many years later (Jourdan *et al.*, 1997). The usual pathway of high frequency sound emission is the mouth, but the nose can also be used (Roberts, 1975a). Thus, rats and mice may produce high frequency sounds with the mouth shut, a kind of rodent ventriloquism. It was also found that the source of the sound is the larynx and, finally, it was determined that the vocal cords do not vibrate but function as a whistle when they produce the high-frequency sounds (Sanders *et al.*, 2001). The absence of vibration is thought to be due to a strong constriction of the vocal cords. In view of the fact that forcing air between the vocal cords, albeit strongly contracted, indeed produces the sound in the same way as audible vocalizations are produced, it is reasonable to call the high frequency sounds vocalizations. To distinguish them from audible vocalizations, it has become customary to denominate them ultrasonic vocalizations.

The nucleus ambiguus is a central nervous structure necessary for ultrasonic vocalizations. Lesion of this structure eliminates them and neurons within it fire in tonic bursts immediately before ultrasound emission (Yajima *et al.*, 1982). It may be observed that the nucleus ambiguus is also involved in the production of speech in the human (reviewed in Jürgens, 2002). It receives numerous afferents from the forebrain and is also associated with neurons controlling respiration. This means that the motor neurons in the nucleus ambiguus are controlled by respiratory activity in the way that they are prone to fire only at expiration. Their firing rate may be modulated by activity at more distant brain sites as well.

A fundamental issue is whether the ultrasonic vocalizations communicate something to conspecifics or if they simply are noises without communicative function. The simple fact that they exist is not a sufficient basis for assuring that they are used to convey messages of any kind. Any hypothesis as to a communicative function of ultrasonic vocalizations needs to be substantiated by experimental data. In order to provide those data, it may be useful to start out with a determination of the circumstances under which these sounds are emitted. Once we know where and when rodents produce ultrasonic vocalizations, we could try to find out if and how these vocalizations affect the behavior of other individuals. If there were no effect we would be forced to conclude that they have no communicative function at all. On the other hand, if they altered the behavior of other individuals, we could describe these alterations and based on that description we could try to infer a meaning. This latter process is uncertain and mostly based on intuition, and it is perhaps not necessary at all. A reliable description of the effects of a vocalization on the receiving individual may make an attribution of meaning unnecessary. I will try to show this in a few paragraphs. Before that, though, it is most important to address other issues, those of intentionality and purpose.

In the context of vocalizations, it is easy to succumb to the temptations of anthropomorphism. The human is heavily dependent on sounds, generally in the form of speech, to communicate with other individuals and it is quite natural for us to assume that other species use sounds in the same way as we do. One of the most irresistible temptations seems to be to attribute a purpose to the vocalization in the same way as much of our own speech has a purpose. The purpose we impose on other animals is normally based on our intuitive knowledge of what a human would do in a situation equivalent to the one our experimental animal is subjected to. Eloquent examples of this are found in the abundant literature on vocalizations observed in infant rats when they are separated from their mother. In such a situation, the pups will soon start to vocalize and the dam will react to the sounds by picking the pups up and putting them back into the nest. It is rather common to maintain that the pups, in this situation, vocalize *because* the separation from the mother has induced emotional distress. These 35 kHz vocalizations are, in fact, frequently named distress calls (Winslow and Insel, 1991). The emotion of distress, or separation anxiety as some more extravagant scientists say, is supposed to be the *cause* of the vocalizations. Furthermore, their *purpose* is to let the mother know the tragic emotional state of her offspring and incite her to come to the rescue. According to this reasoning the pups vocalize intentionally and purposefully.

Here we have a beautiful example of teleology combined with anthropomorphism. We have all observed that some babies cry when taken away from their mother and we usually explain the crying as a result of emotional distress. If a rat pup produces sounds in a similar situation, we immediately imagine that the cause must be the same as in the human. Hence we consider the rat vocalizations as distress calls. Furthermore, when the baby cries we think that she expresses a desire for being reunited with the mother, in other words the cry is intentional and has a purpose. Hence we consider the pup's vocalizations as directed by the same intentionality and purposefulness. This reasoning intuitively makes sense. The only problem is that it might not correspond to reality as manifested in experimental facts. Some scientists maintain that the vocalizations emitted after separation from the mother are a by-product of abdominal contraction in response to reduced blood pressure caused by lowered body temperature (Blumberg and Alberts, 1990; Kirky and Blumberg, 1998; Blumberg and Sokoloff, 2001). Increasing expiratory resistance by intense contraction of the vocal cords should, according to this point of view, be a kind of reflex subserving the cardiovascular system and the sounds produced should simply be an acoustic by-product. In that way it is similar to a sneeze in humans, a reflex serving a physiological function and accidentally producing a lot of sound. These are very intriguing proposals and they are supported by a substantial amount of empirical observations. Thus, separation-induced vocalizations can be entirely explained by a purely mechanistic cause, without any mention of intention or purpose. Without any teleology. Such a mechanistic explanation has obviously less poetic or romantic appeal than stories about separation distress or anxiety, which probably is the reason why it has not become popular and is never mentioned in the science programs of the BBC. It is very important to observe that this mechanistic explanation of the vocalizations in no way challenges the data on their effects on the mother's behavior. On the contrary, it nicely illustrates the efficiency of evolutionary mechanisms. Most pups will sooner or later find themselves outside of the nest. If not picked up by the mother, the survival of these pups is uncertain. In consequence, maternal behaviors assuring efficient pick up of pups outside of the nest can be expected to be strongly favored by natural selection. Females responding to the accidental sounds from cooling pups (cooling being the main short-term consequence of being outside of the nest) should be favored over females not responding. We have here a nice example of how evolution *could* have utilized a sonorous by-product, without any function, and transformed it into a positive incentive. This combined with the pups' capacity to activate the fixed action pattern of 'transport back to the nest' would drastically enhance the likelihood of survival. These conjectures have no scientific value whatsoever. I just include them for the pleasure of those insisting on answers to the unscientific question of 'why?'. Moreover, this long paragraph has no factual content of importance for sexual behavior. Nevertheless, the principles stressed here, the uselessness of anthropomorphisms and of explanations in terms of purpose applied to ultrasonic vocalizations are of utmost relevance to our discussion on vocalizations as sexual incentives.

The huge literature on ultrasonic vocalizations induced by several kinds of stressful situations will be ignored. The following account will be strictly limited to ultrasonic vocalizations in the context of sociosexual behavior. First, we need to establish if ultrasonic calls are emitted at all during the execution of such behaviors. If they are, then it could be useful to describe them, at least in terms of frequency and duration. We would also need to try to find out which are the external stimuli that trigger the vocalizations, and under which circumstances these stimuli are effective. Questions like hormone-dependency of the emission of the calls are certainly of interest. Finally, we need to establish whether the calls affect the behavior of the receiving organism and, ideally, we should also determine under which conditions the receiving individual is sensitive to the calls.

Since the now classic papers on ultrasonic vocalizations during sexual interaction in mice (Sales, 1972) and during the post-ejaculatory interval in male rats (Barfield and Geyer, 1972) were published in the early 1970s, much knowledge has accumulated. As we will see, though, not enough for giving a clear picture of the role of these vocalizations in sexual behavior and particularly not enough for determining their incentive motivational importance. It is clear, though, that ultrasounds do not modify non-sexual, social interaction in rats according to a well designed study (Borden *et al.*, 1977). The conclusion drawn by the authors of that study was simply that the function of rats' ultrasonic vocalizations in social interactions may be more expressive than communicative. It would seem, then, that if these vocalizations have any function at all, they must be related to sexual rather than to social interactions.

Rats and mice emit quite a lot of ultrasonic sounds before and during copulation. An early study in mice reported that the majority of 70 kHz vocalizations are emitted by the male during the early part of sexual interaction, for example when sniffing the female anogenital region and when mounting (Sales, 1972). When the female is anesthetized, the amount of calling is unchanged, while there are no calls when the male is anesthetized. The obvious conclusion is that calls are emitted only by the male (Whitney *et al.*, 1973). A later study of the temporal distribution of vocalizations during copulation in mice (White *et al.*, 1998) obtained slightly different results. The calling rate remained almost constant throughout sexual interaction, with some reduction during intromission and ejaculation. The highest call rate was actually found before and during the ejaculatory mount, i.e. before the penile insertion that eventually led to ejaculation. This is certainly not at the beginning of sexual interaction. In a second part of the experiment, either the male or the female was devocalized. It was confirmed that almost all 70 kHz calls were emitted by the male. This was also the case for calls around 40 kHz. Interestingly, the sexual interaction was not modified by devocalizing the males, suggesting that the vocalizations have no role in copulatory behavior. Similar data and a similar conclusion had been published several years before (Nunez *et al.*, 1985). With regard to copulatory behavior in mice they may constitute an epiphenomenon without behavioral function.

There is a large discrepancy between the number of studies that has recorded the amount of ultrasonic vocalizations during copulation in mice and the number of studies that has been undertaken with the purpose of determining whether these vocalizations have any function or not. It may seem curious that so many scientists have been willing to invest time and other resources in the description of a phenomenon with unknown importance. It would seem reasonable to proceed in the opposite direction, that is to begin with the question of whether these calls are worth studying at all rather than first studying them and then ask if it was worth it. As we have seen, the original assumptions concerning a role of these vocalizations in the regulation of copulatory behavior turned out to be false. This means that the propensity of female mice to exhibit lordosis, and perhaps also proceptive behaviors, is not affected by the male's vocalizations. Please remember that only the male vocalizes during copulation. It can easily be argued that the idea that copulatory behavior should be affected by sounds is a strange one. In the first place, it is known that lordosis is a tactile reflex and there is very little or no reason to imagine that tactile reflexes are modulated by sound. Proceptive behaviors are also activated by tactile or olfactory stimulation, so the same argument holds for them. Furthermore, sounds are most efficient for communication over some distance, but during copulation the distance between the participating individuals is inevitably quite short. In conclusion, there has never been much of a rational basis for the assumption that male vocalizations should modify copulatory behavior.

Before sexual interactions can be initiated, it is necessary to be located in the proximity of a partner. An important aspect of any analysis of sexual behavior must be, then, how the necessary proximity to a partner is achieved. As I have already mentioned, a partner must first be detected. In addition, the partner must emit some stimulus or stimuli that may activate approach behaviors in another individual. This other individual may also emit stimuli with incentive properties for the partner. Thus, while being *attracted by* the partner it may also be *attractive for* the partner. In that way we get a kind of mutual approach. One rather obvious possible function of ultrasonic vocalizations is that of being sexual incentives. Sounds are probably more efficient than odors for acting at a distance in the attraction of potential mates. Odors are carried by the wind and only individuals located downwind from the emitter will be exposed to them. Furthermore, it is difficult to localize precisely the source of odorous stimuli. However, sounds are carried by the air in all directions from the emitter, also upwind, and are much easier than odors to locate in space. A simple tropotaxis is sufficient for finding the source of sound while a much more unreliable anemotaxis is required for localizing odors. The advantage of sound over odor holds for any sound, so there is no particular reason for rodents to emit ultrasounds rather than audible sounds. Those wanting to explain everything, that is to find a rational cause for every phenomenon of nature, would argue that an advantage with ultrasounds in relation to audible sounds is that they are rapidly attenuated. It is believed that they do not surpass a distance of about 10 m in the wild (Smith, 1979). This means that the emission of such sounds does not reveal your position to animals at a distance superior to that. Crepuscular or nocturnal animals, like most rodents, are frequently victims of flying predators, but most of the predators fly at an altitude superior to 10 m. Ultrasounds would not reach them. Therefore, many rodents emit ultrasounds rather than audible sounds. These arguments make up a nice and convincing story but, in reality, they carry no conviction. The distance at which sounds are attenuated does not only depend on their frequency but also on their intensity. Thus, rodents could have limited the range of audible sounds by reducing their intensity. Evolution could have made them employ a low voice rather than a high pitch. There is no way to figure out why one option was favored over another. Here we have another example of the futility of the question 'why?'.

Ultrasonic vocalizations may indeed have positive incentive properties in mice. The only study that evaluated this rather obvious prediction found that female mice approach a vocalizing male more than a devocalized male (Pomerantz *et al.*, 1983). Moreover, if intact females were given the opportunity to approach two devocalized males, in a situation where synthetic 70 kHz sounds were presented right behind one of them, the females approached the male combined with the sound more than the male without sound. Similar tests were performed with ovariectomized subjects. After treatment with oil, the females did not distinguish between a vocalizing and a devocalized male or between two devocalized males, one presented together with synthetic vocalizations and the other without. After treatment with estradiol benzoate + progesterone, in doses sufficient to induce full sexual receptivity, the females again spent more time with the vocalizing male. This very elegant experiment clearly shows that ultrasonic vocalizations actually enhance a male's incentive value. It also shows that the impact of the vocalizations is dependent on the female's hormonal condition. Vocalizations are only effective when the female has been exposed to ovarian hormones, that means when she is in a state where sexual interactions with a male are likely to occur. This makes it possible to suggest that the vocalizations have no social incentive value, because if they had, then the females should have also been attracted to them when not prepared for sexual interactions. The fact that the incentive properties of the ultrasounds were dependent on ovarian hormones suggests that they function as a sexual rather than as a social incentive. Thus, while ultrasonic vocalizations fail to affect copulatory behavior, which is quite logical since they are not needed when the copulating individuals are already in proximity, they function as sexual incentives, which is also quite logical since they are most helpful for localizing a potential mate at a distance.

Knowing that ultrasonic vocalizations function as sexual incentives in mice, and that their impact on the perceiving female organism is dependent on ovarian hormones, it might be worthwhile to determine the stimuli controlling their production. Likewise, it would be interesting to know whether ultrasound production is dependent on gonadal hormones just as sensitivity to them is. Before looking into these issues, it is convenient to recall that only males produce them. Thus, the perceiving organism is always a female, while the producing organism is always a male. For this reason, the following discussion of ultrasound production must be limited to the male. Male mice do not vocalize when alone. They do not waste their energy in singing for themselves like males of some other species do. The low baseline makes it easy to determine the stimuli that activate vocalizations. There is no background noise disturbing the experiments. These almost ideal circumstances have permitted scientists to show that chemicals released by a female are necessary and sufficient stimuli. They are necessary because when males with cuts in the vomeronasal tract are exposed to a female there are no vocalizations. Likewise, lesion of the olfactory bulbs suppressed responding to the female. Both these manipulations caused major disruption of the accessory olfactory system. Manipulations disrupting the main olfactory system failed to modify vocalizations (Bean, 1982). These data, then, show that there is no response to a female in the absence of a functioning accessory olfactory system. This system is, therefore, *necessary* for vocalizations. Another study confirmed these observations, but also reported contradictory data (Wysocki *et al.*, 1982). When using sexually inexperienced males, like those used in the Bean (1982) study, lesion of the accessory olfactory system reduced vocalizations in response to a female. However, when sexually experienced males were employed, the same lesion only marginally affected vocalizations. This observation challenges the conclusion that olfaction is *necessary* for female-induced vocalizations. In contrast to the results obtained with normal females as stimuli, responses to female mouse urine or to an anesthetized female were also strongly reduced in experienced animals. These observations were extended many years later (Sipos *et al.*, 1995). Lesions in the main olfactory system, the accessory olfactory system or both failed to modify the vocalizations emitted by sexually experienced males in response to a female mouse. When the female was replaced by fresh female urine, it was found that main or accessory olfactory system lesion reduced responding to a similar degree. Lesions of both systems eliminated responses to female urine.

The results reported by Wysocki *et al.* (1982) and by Sipos *et al.* (1995) have important implications. One is that chemical stimuli are not at all required for a sexually experienced male's vocalizations when exposed to a living female mouse. If the male has no sexual experience, then chemicals seems to be of importance. Learning, then, must be an important determinant of the stimuli that activate ultrasonic vocalizations. It would seem that some chemicals function as unconditioned stimuli and that these stimuli become associated with other stimuli emitted by the female or with contextual stimuli during or immediately following the execution of copulatory behavior. There was, at some point, much confusion as to whether olfactory stimuli really are unconditioned, or if sexual experience is always necessary. Several older papers suggested that experience in fact was a requisite (see Nyby *et al.*, 1983, for a discussion). This opinion was soon abandoned (Nyby *et al.*, 1985) and it is now generally accepted that female odors are unconditioned stimuli for the activation of male mouse ultrasonic vocalization. Nevertheless, some texts will still express the now obsolete idea of the opposite. Such texts should preferably be allowed to enter the oblivion of history.

The role of learning in sexual behavior will be extensively discussed in Chapter 7 and there is no need to start this discussion now. For the time being, it is sufficient to conclude that olfactory stimuli are the only ones that are known to be unconditioned activators of ultrasonic vocalizations in mice. As mentioned, the odor of a female will make a male vocalize whether he is sexually experienced or not and mice without sexual experience will not vocalize in response to a female unless they have an intact olfactory system. This latter observation is, indeed, the basis for the assertion that odor is a necessary stimulus in inexperienced males.

Having established that olfaction is *necessary* for the activation of ultrasonic vocalizations in males without sexual experience, we can now proceed with some speculations concerning the source of the olfactory stimuli and the conditions under which they are effective. Males vocalize readily in response to odors obtained from different parts of the female body, including the head and the vagina. As mentioned above, female urine is also effective regardless of sexual experience, while stimuli obtained from other males are ineffective (Nyby *et al.*, 1977b). These data show that olfactory stimuli are *sufficient* for inducing vocalizations in the male. A subsequent study confirmed that male mice vocalize in response to urine from an intact female, but rarely in response to urine from another male (Whitney and Nyby, 1979). However, the males do not distinguish between urine from females in proestrus, diestrus or metestrus. Not even ovariectomized females were inferior to proestrus subjects. Thus, ovarian hormones do not seem to be of much importance for the capacity of the donor's urine to evoke vocalizations in males. Urine from hypophysectomized females, however, failed to activate vocalizations. Nevertheless, the lack of a relationship between the presence or absence of ovarian hormones and the capacity of female urine to activate vocalizations in the male could make it doubtful that this behavior has any sexual meaning. If these calls indeed had sexual significance, then they could be expected to be specific to situations where sexual interaction is potentially possible. Otherwise it could be far more reasonable to imagine that they have some social significance, or no significance at all. I favor a different explanation, though. It is quite possible that the stimulus control of vocalization is rather lax, so that the males invite any member of the opposite sex to sexual interaction independently of whether that individual is willing to engage in sexual activities or not. Once proximity has been established, the male will sooner or later find out. The cost of the potential extra work may be so low that natural selection has not eliminated the habit of calling to any female. Despite the lack of precise stimulus control of the males' vocalizations, they do not respond with vocalizations to any conspecific. Male urine and odors from various parts of the male body are unable to activate vocalizations, as mentioned.

Visual or auditory cues are neither sufficient nor necessary for the activation of vocalizations (Whitney *et al.*, 1974). A female in total darkness is just as efficient as a female in light, showing that visual stimulation does not make any contribution to the male's response. A deafened male responds just as readily as a male with perfect hearing, showing that auditory stimuli are of no importance.

The studies summarized in the preceding paragraphs obliged us to conclude that olfactory stimuli from the female are necessary and sufficient for activating vocalizations in male mice. However, the odor of a female is less efficient than a living female for the promotion of vocalization (Nyby *et al.*, 1981). Moreover, while males habituate (in the sense of displaying a smaller response) to repeated presentations of olfactory stimuli, they do not habituate to a living female (Dizinno *et al.*, 1978). Thus, while odor is *sufficient* for activating vocalizations it is not *equivalent* to a living female. A living animal obviously emits a lot of stimuli in addition to odor, but it is not easy to identify those of importance here. We know that neither vision nor hearing is necessary or sufficient, but perhaps they can synergize with olfaction in some hitherto unknown way. Only more detailed studies of the stimulus control of male mouse ultrasonic vocalization could solve this most delicate mystery.

If male ultrasonic vocalizations have sexual incentive properties, in the way that they make the male more attractive to females, then we could consider the vocalizations as part of sexual behavior. In that case, we would expect them to be under the control of testicular hormones in exactly the same way as other aspects of male sexual behavior are. Thus, they should appear around puberty, when the testicles have initiated androgen production, but be absent before, they should be reduced by castration and they should be restored to precastrational levels by treatment with androgens. The main testicular androgen, testosterone, or some of its metabolites, dihydrotestosterone or estradiol, should be efficient. All these predictions have been confirmed in experimental studies. The female-induced vocalizations are rare or absent in prepubertal male mice but appear at the age of sexual maturity (Whitney *et al.*, 1973). Castration reduces vocalizations in response to an estrus female and this reduction can be reversed with testosterone treatment (Dizinno and Whitney, 1977). Further studies replicated this finding and revealed that dihydrotestosterone was far less efficient than testosterone in restoring vocalizations (Nunez *et al.*, 1978). Both hormones had similar peripheral effects, as evidenced by identical weight of the seminal vesicles in animals treated with testosterone and with dihydrotestosterone. Estradiol was moderately efficient in restoring vocalizations in this study. Other data show that the effects of dihydrotestosterone may vary between strains (Bean *et al.*, 1986). The effectiveness of testosterone appears to be beyond doubt, while dihydrotestosterone and estradiol sometimes are effective, sometimes not. Nevertheless, it is quite clear that female-induced ultrasonic vocalizations are androgen-dependent in the male mouse in the same way as other aspects of male sexual behavior. We even know the site of action of androgens within the brain. Female-induced vocalizations can be restored to precastrational levels in castrated males with testosterone implants into the preoptic area, but not by implants into the amygdala, septum, anterior or ventromedial hypothalamus or ventral tegmental area (Nyby *et al.*, 1992; Matochik *et al.*, 1994; Sipos and Nyby, 1996, 1998). These data show that androgens act to restore vocalizations in a brain structure that is also critical for male copulatory behavior.

The preceding discussion of female-induced ultrasonic vocalizations in male mice can be summarized in a few sentences. The vocalizations do not seem to modify copulatory interactions, but they may contribute to the male's sexual incentive properties. The crucial stimulus for the production of vocalizations is olfactory, but other unknown stimuli are also important. The male responds with vocalizations to females in all phases of the estrus cycle, and perhaps also to ovariectomized females. It appears, then, that the stimuli emitted by the female are not under the control of ovarian hormones. On the contrary, the male's production of vocalizations is dependent on androgens and the androgens act within the preoptic area. The female's response to these ultrasonic vocalizations is similar in estrus/proestrus and in diestrus. This shows that her reaction to vocalizations is different from her sexual responses to other stimuli emitted by the male. For example, a female mouse never responds with lordosis to a male's mount in diestrus and she will never display proceptive behaviors as a reaction to the male's approaches outside of the period of estrus. This means that the female's response to vocalizations is not dependent on ovarian hormones in the same strict way as copulatory behaviors are. However, ovariectomized females do not respond to the male's vocalizations, as I mentioned a couple of paragraphs ago (Pomerantz *et al.*, 1983). Treatment with estradiol + progesterone restores responding. Thus, ovarian hormones *are* important, but short-term variations in their concentration, like those occurring during the estrus cycle, do not manifest themselves as altered responsiveness. In this context, it may be worth the trouble to point out that vocalizations differ from copulatory behavior in other aspects too. For example, in adult, ovariectomized females, injections of testosterone activates vocalization to the same degree as that observed in castrated males given the same testosterone dose (Nyby *et al.*, 1977a). It was concluded that androgens or their metabolites have an activational effect on ultrasound production while not being necessary during the differentiation of the brain. This is certainly an original proposal, suggesting that the role of steroid hormones in vocalization is much different from their role in copulatory behavior.

I have now reviewed in some detail sex behavior-related ultrasonic vocalizations in mice. Many studies, some of them quite complicated, have been summarized. It would not surprise me if the reader had become somewhat overloaded after this. As a gesture of compassion, I will here provide a kind of summary of the summary. It should be helpful, particularly for confronting the ensuing analysis of ultrasonic vocalizations in rats.

We have seen that it is only the male who vocalizes, that his vocalizations have little or no impact on copulatory behavior, but that they may serve to increase his sexual incentive value. Furthermore, we have mentioned that olfactory stimuli from receptive females are efficient for activating vocalizations, but that living animals are still more efficient. Indeed, while some living conspecifics, for example a hypophysectomized female, is just as efficient as an intact, sexually receptive female, her odor does not activate vocalizations. Finally, it was shown that gonadal hormones determine the male's response, in the way that a castrated male does not vocalize in response to a receptive female. If such a male is treated with androgens, vocalizations are restored.

Ultrasonic vocalizations in rats

From a comparative perspective it could be most useful to examine ultrasonic vocalizations in closely related species. Many important questions could be asked and we have sufficient data to answer them. One such question is whether vocalizations are also a male prerogative in other species. Another is whether vocalizations influence copulatory behavior and/or the incentive properties of the

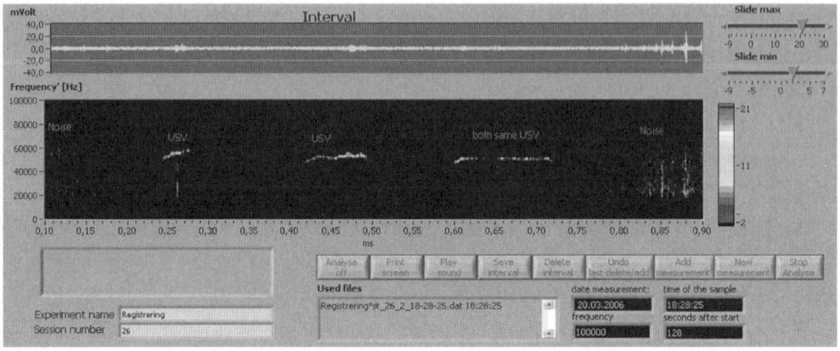

FIGURE 3.1 Ultrasonic vocalizations recorded from a sexually receptive female rat in response to exposure to an inaccessible male.

emitting animal. A third may be concerned with the stimuli required for activation of vocalizations. It could also be illuminative to determine their hormone dependency. If the answers to all these questions were similar to those obtained in studies on mice, we could conclude that the function and control of vocalizations have been conserved and that generalizations among species is meaningful. On the other hand, if there were important species differences, we would have to conclude that interspecies generalizations are meaningless or, still worse, they might be misleading. Let us begin with an analysis of vocalizations in rats. An example of a record of female rat precopulatory vocalizations are shown in Figure 3.1 and Figure 3.2 is a photograph of the experimental setup used for studying the emission of ultrasounds.

Abundant 22 kHz vocalizations following ejaculation in the male rat were the first high frequency sounds described in association with rat sexual behavior (Barfield and Geyer, 1972). The poetic spirit of Ron Barfield made him coin the expression 'post-ejaculatory song'. This song has been much studied and many functions have been attributed to it. The song is essentially limited to the absolute refractory period, the time following an ejaculation during which it is impossible to reactivate male sexual behavior. This interval constitutes about 75% of the post-ejaculatory interval. During the remaining 25% the male does not sing and it is possible to reactivate his behavior by stimuli heightening his arousal, like pinching his tail or applying electric shocks to his skin. Since the song coincides in time with a period where the male is unresponsive to the female, it was suggested that its function was to inhibit female sexual behavior and to maintain the female at some distance and consequently delay her sexual advances (Barfield and Geyer, 1975; Geyer and Barfield, 1980). Albeit interesting, this hypothesis turned out to be false. Devocalized males have post-ejaculatory intervals of the same length as vocalizing males and the female does not keep a longer distance between herself and a vocalizing male than between herself and a devocalized male (Thomas *et al.*, 1982a). Furthermore, if the female is given the opportunity to escape from the male following the ejaculation she stays away for a longer time if the male is devocalized, which is exactly the contrary to what would be predicted from the hypothesis

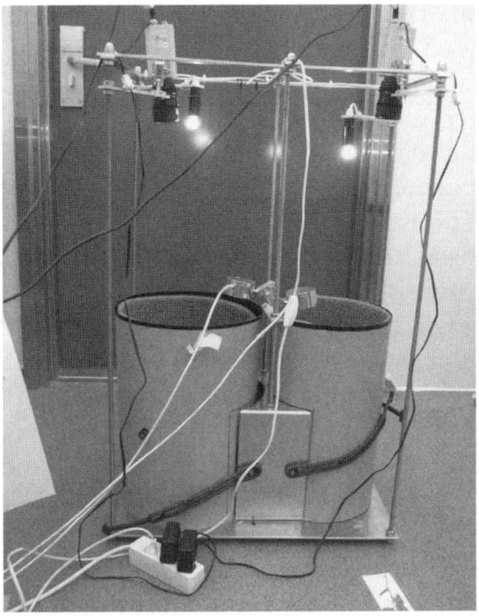

FIGURE 3.2 The experimental setup employed for the recording of the vocalizations shown in Figure 3.1. The female was housed in the small cylinder, while the male was introduced into the larger one. At the bottom of each cylinder, facing the other, is an opening covered with wire mesh. This allows the experimental subjects to smell, see and hear each other. No physical interaction is possible, not even snout to snout, because there is a 1 cm separation between the cylinders. Note the microphone located just above each cylinder and the sound absorbing cover on the inside cylinder walls. This arrangement makes it possible to record sounds from both animals simultaneously and in separate channels. The microphone over each cage is so oriented that only sounds originating within that cage are recorded. The camera above each cage is synchronized with the sound recording equipment, which allows an analysis of the behavior displayed at the moment of vocalization.

mentioned above (Thomas *et al.*, 1981). The fact that the female returns faster to a vocalizing male than to a devocalized male could actually suggest that the vocalizing male is more attractive. This is in line with other data showing that sexually receptive females display solicitations in response to castrated males when exposed to the 22 kHz vocalizations (McIntosh *et al.*, 1978). The results of this study as well as those of the Thomas *et al.* (1981, 1982a) studies suggest that these vocalizations enhance rather than inhibit female sexual behavior. We are now faced with a rather significant amount of contradictory data, making it impossible to arrive at a reasonably well-founded conclusion. However, a nicely designed study in a seminatural environment failed to detect any effect at all of the male's 22 kHz vocalizations on female behavior (Anisko *et al.*, 1978). It would seem, then, that the most likely conclusion is that the post-ejaculatory song has no consequences for social or sexual behaviors.

In addition to the post-ejaculatory 22 kHz vocalizations described above, the male rat occasionally emits pre-ejaculatory vocalizations of the same frequency. These pre-ejaculatory calls are of shorter duration than those emitted during the post-ejaculatory interval. The first description of these calls comes from a study where rats were allowed to copulate until exhaustion (Brown, 1979). When approaching the fifth ejaculation males began to emit these calls. The appearance of the 22 kHz calls coincided with an increase in the number of mounts without intromission. The males actually mounted more but intromitted less. An analysis of the behavior of the female partner showed that this was caused by a lessened tendency to display lordosis in response to male mounting. Furthermore, the females showed more aggressive responses (turning away, kicking and boxing) when pursued by the male. This is most interesting since aggressive encounters are another situation associated with emission of 22 kHz calls (Lore *et al.*, 1976). Thus, it is quite possible that the pre-ejaculatory calls are unrelated to copulatory behavior. Instead, they may be a simple side effect of aggressive behaviors on the part of the female. It can also be noted that these vocalizations are emitted in several other stressful situations like exposure to a cat (Blanchard *et al.*, 1991), handling (Brudzynski and Ociepa, 1992), defeat (Corrigan and Flannelly, 1979) or isolation (Francis, 1977). Whatever their function may be, it is probably related to something entirely different from copulatory behavior. In fact, playback of 22 kHz vocalizations has been reported to reduce locomotor activity for a short time immediately following the end of the playback period (Brudzynski and Chiu, 1995). During playback, however, there was no change in activity. If we try to relate this observation to the post-ejaculatory song and the initial proposal concerning its possible function, we face a considerable intellectual challenge. It was, as mentioned a few lines above, shown that the song lasted until the end of the relative refractory period. During that period, the listening female displayed low activity or even immobility (Geyer and Barfield, 1980). When the song ended, she resumed activity, which is exactly the opposite of what was found in the playback experiment. We seem to be forced to admit that the behavioral consequences of the 22 kHz vocalizations are not clear-cut. A possible explanation for these contradictory observations is that there are several types of 22 kHz calls, one specific for the post-ejaculatory interval, another for aggressive encounters, and so on. A small problem with this explanation is that there is no evidence for the existence of such differences. On the contrary, 22 kHz calls emitted under various circumstances seem to be quite similar both in duration and magnitude (Adler and Anisko, 1979). Sonograms and power spectra of calls emitted in response to stimuli as diverse as handling, footshock or intracerebral injections of carbachol are also very similar (Brudzynski *et al.*, 1991).

Males also emit vocalizations with a frequency around 50 kHz (Sales, 1972). Sexually experienced males start calling as soon as they are introduced into the mating test cage, before any female is present (Barfield *et al.*, 1979; Bialy *et al.*, 2000). These calls are absent in males lacking sexual experience, showing that learning is necessary for them to occur. Once copulation has been initiated, the male vocalizes before he mounts, intromits and ejaculates. The amount of vocalization is largest during the 10 seconds preceding ejaculation, somewhat lower in the period preceding an intromission and still lower before a mount (McIntosh and Barfield, 1980). These observations prompted the suggestion that the 50 kHz vocalizations somehow reflect the male's level of arousal. This may well be the case, but the crucial question is not what the vocalizations might reflect but what their effect on other animals, for example the copulation partner, might be. Since heterosexual copulation is an interaction between at least one male and one female, any functionally relevant effect of the male's vocalizations must be on this interaction. Because most calling occurs just before a sexual interaction, be it mount, intromission or ejaculation, it might be speculated that the calls' function would be to stop the female and facilitate the assumption of the lordosis posture. Chasing the female around the mating cage is one of the main tasks of the male in most copulation tests in rats and, as soon as the female interrupts a run and suddenly stops, the male frequently proceeds with a mount. This mount may or may not be associated with vaginal penetration (intromission) and intromission may or may not be associated with ejaculation. In any case, the female must stop running before the male can make a successful mount. It seems quite reasonable to assume that the male's calls could have such a stopping function and thereby facilitate the sexual interactions. This does not seem to be the case, though. A devocalized male mounts, intromits and ejaculates just as fast as a calling male, he makes the same number of mounts and intromissions and he has the same post-ejaculatory interval. If we focus on the female rather than on the male behaviors, it is found that the number of darts per unit time is lower in females copulating with devocalized males (Thomas *et al.*, 1981). However, this effect has no functional significance since no aspect of the sexual interaction is modified. We are forced to conclude that the male rat's 50 kHz vocalizations have no influence on copulatory behavior. That is exactly the conclusion we arrived at with regard to vocalizations in the male mouse. In contrast to the lack of a role in the control of copulatory behavior, I convinced you that the vocalizations enhanced the male mouse's incentive value. We will now see if this is the case also in rats.

It is not. A devocalized male rat is just as attractive as a vocalizing male. Furthermore, a devocalized male presented together with taped vocalizations is not more attractive than another male presented together with tape hiss (Thomas *et al.*, 1982b; White and Barfield, 1990). According to these studies, then, vocalizations do not contribute to a male rat's incentive value. However, results from an experiment in a Y maze seem to contradict this assertion (Barfield *et al.*, 1979). Female rats were allowed to choose between one arm in which male calls were played back and another without sound. The subjects preferred the arm with sound, an observation that could suggest that the calls have incentive properties. Unfortunately, this experiment does not provide conclusive evidence. It only shows that sound is preferred over no sound, but it is possible that any sound would have had the same effect as the male's tape-recorded vocalizations. We are forced to conclude that there is slight evidence for the hypothesis that male ultrasonic vocalizations have positive incentive properties. The negative evidence seems to be more convincing.

When discussing the mice data a couple of paragraphs ago, I mentioned that the female is silent during sexual encounters. That extraordinary quality is not found in rats. In fact, females vocalize as much as males and their vocalizations are indistinguishable from those of males, at least with regard to frequency, duration and amplitude (Thomas and Barfield, 1985). Studies of the possible behavioral consequences of female vocalizations have shown that females show more darting when they are devocalized than when they emit sounds. This effect may be caused by the fact that the male approaches and pursues a silent female more than a vocalizing female. Darting is a response activated by the male's approach or by direct tactile stimulation of the female's posterior. Thus, more frequent approaches and pursuits on the part of the male may lead to enhanced display of darting. The increased female darting as well as the enhanced male pursuit can be eliminated by playback of tape-recorded calls (White and Barfield, 1987, 1989). A reasonable interpretation of the fact that the female's vocalizations reduce the male's approaches to and pursuit of her would be that the sounds somehow keep the male away, i.e. they function as negative incentives. Furthermore, such a proposal is consistent with the data obtained in paced mating tests. As mentioned some pages ago, the female displays avoidance behaviors after intromissions and ejaculation and, when the female can pace sexual interaction, the intervals between successive intromissions is longer than when the male is in control. It may also be remembered that paced mating is associated with higher fertility than male-controlled mating (Coopersmith and Erskine, 1994). Ultrasonic vocalizations may contribute to slow down the copulatory interactions by keeping the male away for some time, thereby keeping the interintromission interval closer to that assuring maximal fertility. Instead of making this rather obvious proposal, White and Barfield (1989) suggest that 'female vocalizations may function proceptively, to stimulate male sexual responses' (p. 232). While the data show that *devocalization* enhances male sexual responses, the authors conclude the opposite. They continue by proposing that 'when the female is unable to vocalize, she increases her rate of other proceptive behaviors such as darting' (p. 232). Here we have a nice example of anthropomorphism blended with teleology. First, it is assumed that the female thinks that her silence makes her less attractive and that she consequently has to enhance her attractivity by some other means, for example by darting and running more. Just as a distasteful woman may enhance her attractiveness by buying haute couture clothes from Christian Dior and perfumes from Chanel. Second, the rat needs to be cognizant of the purpose of her behavior. Although the female rat may be flattered by being attributed such exquisite intellectual capacities, both assumptions are completely unfounded.

The conclusion of the devocalization study I presented in the preceding paragraph is nicely supported by a most interesting observation: deafening of the male rat has the same effect as devocalization of the female. Both female darts and male pursuits are increased (White and Barfield, 1987). Here it is difficult to imagine that the female enhances her darting because she wants to make herself more attractive. It is unlikely that the females know that their partner is deaf, however exquisite their intelligence may be. Thus, the change in female behavior must be secondary to deafness-induced changes in male behavior. This proposal also applies to the White and Barfield (1989) experiment. The devocalized female did not modify her behavior because she was silent. She did so because her silence had modified the male's behavior.

Before leaving the role of female vocalizations, it must be insisted upon the fact that their effects on male copulatory behavior are minor. Despite the effects on approach and pursuit, they do not modify any of the fundamental parameters of male behavior, such as the number of pre-ejaculatory intromissions or the ejaculation latency. I would even venture to conclude that female rat ultrasonic vocalizations are of no functional significance with regard to copulatory behavior.

Despite the slight evidence for a role of ultrasonic vocalizations in rat sexual behavior, I will briefly examine the stimuli required for their activation and their dependence on gonadal hormones. Such an analysis could shed some light on the question of whether they are sexual responses or not. It is also easily done, because data from rats are quite scarce.

As might be expected, olfactory stimuli seem to be important for the activation of ultrasonic vocalizations both in male and female rats. Bedding soiled by a member of the opposite sex is more efficient than clean bedding (Geyer and Barfield, 1978) and exposure to an invisible and untouchable, devocalized male is sufficient for enhancing vocalizations in females (White *et al.*, 1991). Moreover, odors from an intact male or a sexually receptive female are more effective than odors from a castrated male or an ovariectomized female for stimulating vocalizations. All these data clearly suggest that olfactory stimuli are sufficient for enhancing vocalizations in rats. Whether they are necessary or not is unknown. I have not been able to find any study determining whether anosmic rats vocalize or not, making it impossible to conclude anything at all concerning the necessity of olfactory stimuli. An observation of some importance in this context is that a living, sexually receptive female is far more efficient than her odor for stimulating vocalizations (Geyer and Barfield, 1978), at least in sexually experienced males. This would suggest that, although olfaction is sufficient, other stimuli contribute in a significant way. The other stimuli are not auditory, since adding tape-recorded vocalizations to the presence of a devocalized male does not change his capacity to stimulate vocalizations in female rats (White *et al.*, 1991). On the contrary, if male vocalizations are played back in the absence of any male, then female vocalizations are slightly stimulated (White *et al.*, 1993). This finding is important, because it shows that the female somehow responds to the male's vocalizations when no other stimulus emitted by the male is present. Such a situation could easily occur in the wild. A female rat may be walking around and a distant male may start to vocalize at some point. The female would vocalize in response, thereby establishing some kind of contact. She could easily localize the calling male and approach him and the male could also approach the calling female. What then might follow should be left in the obscure, for simple decency. The only problem with this nice story is that there are not much data supporting it. At the same time, there are no data contradicting it.

Compared to mice, relatively little is known about the stimulus control of ultrasonic vocalizations associated with sexual behavior in rats. We know that olfactory stimuli are sufficient both in rats and mice and we also know that they are necessary in mice. Whether this is also the case in rats is, as mentioned, unknown. Furthermore, we know that sexual experience is not necessary for vocalizations in mice. The role of experience in rats is not well known. All rat studies I have knowledge of, except one, have employed sexually experienced animals, probably because the possible importance of learning was of no concern for the scientists involved. The exception study (McGinnis and Vakulenko, 2003) employed an original procedure in which the male subject first was exposed to a female behind a wire net for 5 min. The female was then removed and vocalizations were recorded during the following 5 min. In this procedure, sexually inexperienced males vocalize equally after exposure to an ovariectomized or a sexually receptive female. However, they vocalize more following exposure to the receptive female than following exposure to a male. After the acquisition of sexual experience, exposure to the receptive female induces more vocalizations than exposure to the ovariectomized female. If we assume that the after-effects of exposure to a female are identical to exposure itself, then these data suggest that a female is a weak unconditioned stimulus for the activation of ultrasonic vocalizations in male rats. At the same time, they show that learning greatly enhances the response. Whether these conclusions also apply to vocalizations in the female is not known.

There is only one aspect of ultrasonic vocalizations in which rats and mice appear to be equally well known: their dependence on gonadal hormones. Castrated male rats vocalize a lot less than intact males (Geyer *et al.*, 1978) and ovariectomized females vocalize less than females rendered sexually receptive by treatment with estradiol and progesterone (Thomas and Barfield, 1985). Just as is the case with the induction of lordosis, sequential treatment with estradiol and progesterone is the most efficient procedure for stimulating vocalizations in ovariectomized females (Matochik *et al.*, 1992a). Further support for a role of ovarian hormones is provided by a study of variations in vocalizations in response to a male during the estrus cycle. The amount of vocalizations emitted during proestrus – early estrus is far above that observed at any other stage of the cycle (Matochik *et al.*, 1992b). There is, then, solid evidence for a role of ovarian hormones in the control of female rat vocalizations. The situation in the male is not as clear-cut. Although there is no doubt that castration reduces vocalizations, there is some confusion concerning the crucial hormone. An early study comparing the ability of testosterone, dihydrotestosterone and estradiol with regard to capacity to restore vocalizations in castrated males showed that only testosterone was effective (Matochik and Barfield, 1991). Since testosterone is aromatized to estradiol *and* reduced to dihydrotestosterone and since neither estradiol nor dihydrotestosterone was active by itself, these data suggest that estrogen and androgen receptors must be stimulated simultaneously. The ineffectiveness of estradiol was confirmed several years later (Cooke *et al.*, 2003), but the more recent study found that dihydrotestosterone re-established vocalizations. The combination of dihydrotestosterone and estradiol was no more effective than dihydrotestosterone alone. These results clearly contradict the Matochik and Barfield (1991) data. One replicable observation is that estradiol does not affect vocalizations in the male rat. The role of the androgen receptor is unclear with one study finding it necessary *but not* sufficient and another study finding it necessary *and* sufficient. A third study, showing that an androgen receptor antagonist, hydroxyflutamide, blocks testosterone-induced restoration of vocalizations in castrated animals when implanted into the ventromedial nucleus of the hypothalamus (Harding and McGinnis, 2004) confirms that the androgen receptor is necessary. However, whether it is sufficient or not for allowing the male rat to produce ultrasonic vocalization in response to a female is an unsettled issue. Nevertheless, there is no doubt that these vocalizations are not produced by males lacking gonadal hormones.

There are some curious differences between rats and mice with regard to the intracerebral localization of androgen effects on ultrasonic vocalizations. While several studies have established that androgens within the medial preoptic area activate vocalizations in male mice this does not seem to be the case in rats. Although there are no data on the effects of local administration of androgens to the preoptic area in castrated male rats, there is a study evaluating the effects of local administration of an androgen receptor antagonist (Harding and McGinnis, 2004). Intrapreoptic hydroxyflutamide did not reduce the capacity of testosterone to restore vocalizations in castrated males. This indicates that the preoptic area is not the site of action of androgens. As was mentioned in the preceding paragraph, the androgen antagonist was efficient when administered into the ventromedial nucleus, suggesting that this might be an important structure for vocalization in rats. This notion is supported by the observation that even rather small lesions of this nucleus reduce vocalizations (Harding and McGinnis, 2005). If these observations were to be confirmed, we are facing an intriguing species difference between male rats and mice.

Ultrasonic vocalizations have also been studied in hamsters, voles and other rodents. I do not consider it necessary to exhaust either myself or the reader with a review of any of these other species. The purpose of the analysis of vocalizations can be attained with the already extensive discussion of mice and rats.

2.3 Conclusions concerning auditory sexual incentive stimuli (ultrasonic vocalizations) in rodents

I decided to use vocalizations as an example of the ways in which potential sexual incentive stimuli need to be analyzed before any conclusion as to their incentive properties can be drawn. We should be able to learn a few lessons from this. Several of these lessons are applicable to any sexual incentive stimulus, while some are limited to those incentive stimuli that are emitted in response to the presence of a conspecific. First, however, we need to discuss the analysis of sexual incentive stimuli from an epistemological point of view. This discussion will appear in the following paragraph.

Odor, or body shape, are stimuli emitted whether there is any potential partner present or not. They are not, therefore, responses to the presence of a potential mate but context-independent characteristics of the subject. At difference, vocalizations are, according to abundant data mentioned above, emitted in response to the presence of an individual of the opposite sex. They are also emitted in other situations, like aggressive encounters, but there also they are a response to a conspecific. Vocalizations are, then, context-dependent at difference to odor and shape. The fact that they are emitted in quite specific situations easily leads to the notion that they communicate something in this situation. It may also provoke the idea that they are emitted with an intention, for example that the female rat vocalizes with the intention of attracting a male or to regulate some aspect of sexual interaction. In other words, she vocalizes with the purpose of influencing some aspect of the male's behavior. Here we are immediately back to a teleological explanation of vocalizations. Explanations in terms of intention are not immediately given to the emission of context-independent stimuli. We do not easily maintain that the female rat smells in a particular way because she has the intention of attracting a male. In case of a human female, however, we could say so, and sometimes with good reasons.

Although we do not immediately attribute an intention to the female rat's smell, we might almost spontaneously attribute it a purpose. The purpose of her particular smell is to attract a male, for example. As I did extensively in Chapter 1, I will continue to insist on the inadequacy of teleological explanations. Instead, we can maintain that the female vocalizes because some stimulus from the male, like his odor, excites a particular central motive state, and that particular central motive state somehow excites the motor neurons required for vocalization. This sequence of neural events can be conceived as automatic, as a part of the functional structure of the central nervous system, where intention or purpose neither is required nor useful. Likewise, there is no need to think that the male produces odors with the purpose of making the female vocalize. His smell just happens to have this effect.

We have now replaced teleological explanations for ultrasonic vocalizations with a mechanistic cause–effect explanation within incentive motivational theory. It is possible to go a step further in reductionism, as brilliantly shown some years ago (Blumberg, 1992). The following account is little more than a commented summary of the Blumberg (1992) paper. As explained early in this chapter, ultrasounds are produced when air is forced through the strongly constricted vocal cords. This means that they are emitted during expiration, the part of the respiratory cycle where the thoracic cage reduces its volume. It so happens that the respiratory cycle is coordinated with locomotion in quadruped species. This coordination is necessary because respiration and locomotion partially use the same parts of the body, for instance the thoracic cage and associated musculature. Expiration in a running quadruped starts at the moment the first forelimb hits the ground. Mechanical forces acting upon the foreleg lead to extra compression of the thorax, facilitating expiration. However, thoracic compression should not be too large or too fast. A way to avoid this is to constrict the larynx. The idea here is that constriction of the vocal cords in the larynx is a response to thoracic load caused by locomotion and that ultrasounds are a side effect of this physiological reaction. In support of this notion, it is reported that vocalizations in copulating animals are closely associated with intense locomotion, for example immediately before a mount or intromission.

In these moments the male normally runs behind the fast moving female, which suddenly stops just before a mount is made. Likewise, observation of females in slow motion shows that sounds are emitted shortly after the forepaws hit the ground, exactly as would be predicted from the locomotion–expiration hypothesis. The idea that the ultrasounds are side effects of a physiological process does not exclude that they may have effects on a perceiving organism. What it does exclude, though, is that the emitting subject has any intention.

This intriguing line of reasoning illustrates something extremely important and often overlooked: that animals may emit stimuli by accident or coincidence. I presented these arguments only to show that there are several ways to get rid of intentions or purposes as explanatory elements.

A lesson we can learn from our analysis of ultrasonic vocalizations has to do with their uncertain function in relation to sexual behavior. They may contribute to an animal's incentive value, but the data are scarce and mixed. Vocalizations do not appear to modulate copulatory behavior in any functionally relevant way. Despite this, almost every paper on the subject ends with a long list of the possible functions of ultrasonic vocalizations. This is natural, since most scientists certainly want their work to have some meaning. It might appear rather ridiculous to invest large efforts and much time in a long series of experiments just to conclude that the subject of study was, at the end, a non-subject. This applies, evidently, not only to studies of vocalizations and sex, but to all areas of scientific inquiry. It might be insinuated, though, that a more rational approach would have been to first establish the function of these vocalizations and then proceed with studies of their exact stimulus control, hormone-dependency, central nervous control, etc. The inverse procedure was used here as in so many other cases. The lesson is, then, that before performing detailed studies of a phenomenon we should assure that the phenomenon really is worth studying. Otherwise we make irresponsible use of our own and our students' time as well as of the tax-payers' money, provided we are supported by public funds. A suggestion as to how to proceed when judging the importance of a phenomenon will be outlined in the next lesson.

The second lesson is that behavioral research needs to be guided by coherent theoretical frameworks. If someone more than myself would take the trouble to go through the rather extensive literature on the subject of sexual behavior, this unlikely person would discover that most studies are descriptive and that at a rather low level of abstraction. This is legitimate, and even admirable. As we have seen, modern science is focused on careful description of the phenomena of nature. The problem in some of the literature on sex behavior is that this changes when we come to the conclusions or implications of the descriptive work. We are frequently exposed to far-fetched speculations in all directions, particularly concerning the function or adaptive value of the behaviors or behavioral mechanisms that were described. An explicit theoretical framework rejecting the use of teleological explanations at the same time as it provides criteria for the evaluation of any phenomenon with regard to its possible behavioral importance would be extremely helpful. It could assist us in the process of understanding the observations we have made and the behavior described through these observations, and it could provide guidelines for the generation of testable hypotheses about cause–effect relationships. I will try to illustrate the usefulness of an explicit theoretical framework by applying the incentive motivational model of sexual behavior to an analysis of the many descriptions and tests of hypotheses concerning ultrasonic vocalizations presented in many preceding paragraphs.

The first question must be whether ultrasonic vocalizations enter into the framework of incentive motivation or if they are more appropriately analyzed within the theoretical context of ethological/Hullian motivation theory. The question could be restated by asking whether they should be considered a part of copulatory behavior or of behaviors associated with approach and establishment of initial contact. We already know that copulatory behavior consists of a series of stereotyped, tactile reflexes in both males and females. Once initiated, copulatory behavior is not easy to modify (see Figure 3 in (Ågmo, 1999) for an illustration of this). I repeat once more that the ease of activation of these copulatory reflexes, be it mount with pelvic thrusting or lordosis, is certainly determined by motivation. I also repeat that the incentive motivational framework is not particularly useful here. Ethological/Hullian models are more appropriate. Sounds emitted by the partner could, in principle, be powerful activators of copulatory reflexes in both males and females. The male's calls could be the stimulus triggering lordosis, or ear wiggling. The female's calls could be the immediate cause of the male's mounts. Alternatively, it could be assumed that ultrasonic calls do not trigger any copulatory reflex but instead have a modulating action. We now know that all this is false, but we know that because these hypotheses have been empirically tested. Even though ultrasonic vocalizations do not affect the stereotyped copulatory reflexes, it is still possible that they have some role in the processes of finding a mate and establishing initial contact.

The events determining approach to a potential mate are preferentially analyzed in terms of incentive motivation. The clear-cut question here would be if vocalizations function as sexual incentives or not. Again, an additional/alternative question could be whether they modify the impact of other incentive stimuli or not. Considering that ultrasonic vocalizations may be perceived by individuals at some distance, they seem particularly suitable for having sexual incentive properties. This hypothesis could easily be tested, and it was tested many years ago in one mouse experiment and in a few rat experiments. While the single mouse study found that the male's vocalizations indeed have incentive properties for the female, two of the rat studies failed to detect an effect (Thomas *et al.*, 1982b; White and Barfield, 1990). The third maintained that male vocalizations had incentive properties (Barfield *et al.*, 1979). The latter study was not reported *in extenso* but briefly mentioned in a review. Thus, the rat results are, as we already know, mixed. Moreover, the few existing rat studies have evaluated the role of male vocalizations but ignored those of females. That female vocalizations were ignored in mice is quite understandable, since female mice do not vocalize. Nevertheless, despite the obvious importance of the question of vocalizations as sexual incentives, the number of studies evaluating their incentive properties is only a small fraction of the number of studies evaluating their effects on copulatory behavior. The reason for the almost obsessive concentration of research on copulatory behavior and the surprising neglect of incentive motivational aspects has been and is a lack of a coherent theoretical framework applicable to sexual behavior. Without such a framework there is nothing on which to focus our attention and hang up our concepts.

I hope that the attentive reader has noticed that it was possible to give a meaningful summary of a sizeable part of the long discussion of ultrasonic vocalizations in just a couple of lines. Instead of long-winded descriptions of many experiments, sometimes rather disparate ones, we could put existing knowledge together in an easily understandable system. It was not by accident that I dedicated so many pages to a presentation of parts of the research performed on ultrasonic vocalizations. I gave the facts, to the best of my knowledge. The end result, at least for myself, was quite confusing. A suffocating amount of detailed information prevented me from getting a productive idea about the role of these vocalizations in sexual behavior. The confusion was replaced by insight when the facts were put into a theoretical model. I do believe that one of the basic impediments for a faster scientific progress, particularly in the area of sexual behavior, is the overabundance of data and the paucity of sound theoretical analysis. An additional impediment comes from the tremendous amount of quasiscientific notions of sexual behavior circulating in popular magazines and invading some scientific texts, perverting the minds of the feeble.

For the duration of the third lesson, it is convenient to assume that ultrasonic vocalizations indeed have strong positive incentive properties in the way that they activate approach behaviors on the part of a listener. An individual may approach another individual for many reasons. In the case of a rat, approaching and establishing close physical contact with another rat could be a thermoregulatory response activated by low ambient temperature. Perhaps it could also be a response to a predator. A group of rats might be less vulnerable to attack than a single individual. Approach could also be activated by social incentive motivation. The simple fact of being close to another rat seems to be rewarding. There are much data showing that rats actively seek social contact (Latané and Glass, 1968; Eckman *et al.*, 1969; Latané, 1969; Latané *et al.*, 1972, 1973; Sloan and Latané, 1974) and there is no doubt that a rat is a positive social incentive for other rats. However, and what might be of particular importance in the present context, ultrasounds emitted by a rat do not seem to contribute to its social incentive properties (Borden *et al.*, 1977). Deafened rats approach and interact with other rats exactly as hearing rats do. We conclude that vocalizations do not contribute to a rat's social incentive properties.

Although nothing suggests that vocalizations are socially attractive among rats, we can for a while imagine that a rat approaches a vocalizing conspecific more than a non-vocalizing. This imaginary approach could be caused by any of the following central motive states: thermoregulatory, protection-seeking, social contact, sexual contact, and perhaps others. Under standard laboratory conditions there is no reason to believe that adult rats need to activate thermoregulatory responses. Likewise, there is no need for a rat to seek protection. There can be no perceived danger in an environment to which the rat has been thoroughly familiarized, as is the case with standard testing environments. Thus, there are two likely explanations for the rat's approach. One is a search for social contact and the other is search for sexual contact. In order to show that the approach is determined by sexual motivation rather than purely social motivation, we need to perform a couple of experiments. The most straightforward is probably to determine the effects of removal of the gonads from the subject. Neither males nor females show any sexual motivation in the absence of gonadal hormones. At the same time, it is known that the presence or absence of gonadal hormones do not affect social motivation, at least in some procedures (discussed in Ågmo *et al.*, 2004). If a subject without adequate amounts of gonadal hormones, for example a castrated male or an ovariectomized female, approach the vocalizing subject as much as an animal with gonadal hormones, for example an intact male or an estrus female, then we can conclude that the intensity of approach is determined by social motivation. If the approach is more intense in the presence of gonadal hormones than in their absence, it can be suggested that sexual motivation is important for approach behavior. Observing animals that have engaged in extensive sexual activity immediately before testing could further support this notion. There are many reasons to believe that such activity temporarily reduces sexual motivation (Ågmo *et al.*, 2004) while it does not affect social motivation (Sloan and Latané, 1974). If an effect of this manipulation is observed, then the idea of sexual rather than social motivation is reinforced. Although there certainly are several more ways to substantiate further sexual versus social motivation, the point should be clear: it is not self-evident that approaching another animal, even if it is of the opposite sex, is an indication of sexual motivation. Alternative explanations need to be ruled out by experimental tests. The message of the fourth lesson is, then, that the simple establishment of incentive properties of a stimulus is not enough. It must also be determined for what kind of motivation the stimulus is an incentive. In other words, which is the central motive state activated by the incentive.

The fifth, and fortunately last, lesson is very simple. We should always be humble in our conclusions and joyfully accept that we cannot answer all questions. In particular, we should avoid far-fetched speculations and never stretch our data beyond what they really show. We should also be aware of what the data do not show. Applying this to the ultrasonic vocalizations, a sufficiently humble conclusion would be that we do not know whether they have any function or not. In case they do have a function, we do not know which. We do not even know whether they are worth studying at all. Would the sound produced by a sneeze be worth studying? Another aspect of humility obliges me to accept that we cannot answer this question without having adequate data.

3 Olfactory sexual incentive stimuli

The rather dismaying conclusion ending the very long section on ultrasonic vocalizations should by no means discourage us. We will now turn to another kind of incentive. Our knowledge of these incentives is quite solid, even if not complete. I am talking about olfactory incentives. Before entering into any discussion of odors, though, we need to take a brief look at the structure and function of mammalian olfactory systems. Some knowledge of these systems is essential if we want to have an informed opinion about the possible role of odors as sexual incentives. This becomes particularly important when we turn to human sexual incentives, where there is a vast mythology but few facts. Some of the mythology is based on wanting knowledge about the basics of mammalian olfactory systems, and it will probably die by itself as soon as such knowledge is provided. The following introduction to the physiology of olfaction is a contribution to the death of some myths.

3.1 The accessory olfactory system

Most mammals are equipped with at least two olfactory systems, the main olfactory system and the accessory olfactory system. The latter is frequently called the vomeronasal system, because it originates in the vomeronasal organ, a mucosal pocket enclosed within a capsule partially formed by the vomer bone. The following brief introduction to the fascinating world of this organ is essentially based on the excellent reviews by Døving and Trotier (1998) and by Halpern and Martínez-Marcos (2003).

There is one vomeronasal organ on each side of the base of the nasal septum. In rodents, each organ consists of an elongated tube communicating with the floor of the nasal cavity through a narrow duct (Figure 3.3). This duct constitutes the only connection with the outside world. In some other mammals, like carnivores or ungulates, there is an additional connection between the vomeronasal organ and the outside, the nasopalatine canal (sometimes called the incisive canal) connecting the vomeronasal organ with the mouth. In cats, for example, the canal runs through

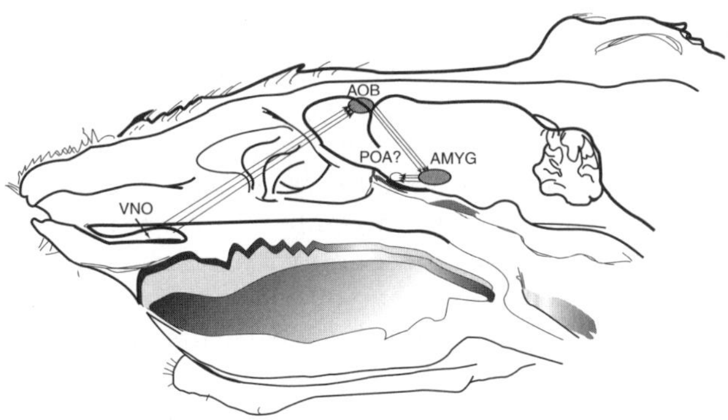

FIGURE 3.3 A drawing of a rat's head showing the position of the vomeronasal organ and its neural connections. VNO, vomeronasal organ; AOB, accessory olfactory bulb; AMYG, posterodorsal amygdala; POA, preoptic area. (Courtesy of Claudia Rodríguez.)

the incisival foramen and ends immediately behind the upper incisor teeth. The existence of a functional nasopalatine canal makes it possible for chemicals to enter the vomeronasal organ from the mouth as well as from the nose. Nevertheless, entry into the mucus filled cavity through any of the access ways is not unproblematic. The only means would be simple diffusion against an outward flow of mucus if no active transport mechanism existed. This would be an exceedingly inefficient process, making entry extremely slow and the concentration of any exogenous chemical in the vomeronasal organ would, consequently, be extraordinarily low. In view of this it is not surprising that a transport mechanism indeed exists.

Soluble chemicals may enter the vomeronasal organ with the aid of a pumping mechanism, creating a massive flow. In order to understand this mechanism, we need to take a look at the vascular system and the innervation of the vomeronasal organ. A branch of the sphenopalatine artery, the septal artery, provides the main blood supply. Blood leaves through one or two large veins surrounded by several layers of smooth muscle. Other important components of the pumping system may be the cavernous tissue and the exocrine, mucus-producing glands found in the vomeronasal organ. The blood vessels as well as the erectile (cavernous) tissue receive a dense adrenergic innervation through the nervus canalis pterygoidei (the Vidian nerve) originating in the pterygopalatine (sphenopalatine) ganglion. The Vidian nerve carries sympathetic information in fibers originating in the internal carotid plexus as well as parasympathetic information from the facial nerve (cranial nerve VII). The parasympathetic nerve terminals are mainly found on the exocrine glands and on blood vessels.

Enhanced blood flow to the cavernous tissue combined with reduced outflow would efficiently fill the vomeronasal organ with blood in a way similar to what occurs in the corpora cavernosa of the penis during erection. An increase in cavernous tissue volume would cause efflux of liquid from the organ. This efflux could be enhanced by increased secretion from the exocrine glands. Flow of liquid *out* from the vomeronasal organ is, thus, easy to explain. However, we were interested in how chemicals enter the organ, not in how they might be expelled. The problem is that the inward movement of liquid is much more difficult to explain than the outward. A popular hypothesis is that the large vein or veins draining blood from the organ may be constricted at their origin inside the vomeronasal organ. Constriction would reduce the veins' volume, creating a reduced pressure in the vomeronasal cavity. Liquid would thereby be sucked in from the outside. Although attractive, this hypothesis encounters at least one problem: if venous outflow were reduced, we could expect an increase in blood pressure within the vomeronasal organ and an ensuing filling of the cavernous tissue. Thus, although the vein/veins may reduce their volume by constriction, it is most likely that such a constriction would increase the volume of the cavernous tissue, making the end result difficult to predict. This notion is reinforced by the most interesting observation that the blood supply to the vomeronasal organ is organized similarly to that in other erectile tissues, namely the penis (Salazar *et al.*, 1997). The ambiguity inherent in any prediction concerning the consequences of venous constriction makes experimental test of these consequences quite urgent. A few such tests have indeed been performed. For example, Meredith and O'Connell (1979) blew an odor over the entrance to the vomeronasal organ without detecting any response in the neurons in the accessory olfactory bulb, but when the nasopalatine nerve was stimulated simultaneously with the odor blow, the neurons responded. This was interpreted as evidence for the existence of a pumping mechanism in the vomeronasal organ controlled by the nasopalatine nerve. The employment of an airborne odorant was rather peculiar, since the vomeronasal organ is supposed to be specialized in responding to non-volatile molecules. Furthermore, it ensures an intense stimulation of the main olfactory system. I assume that the reasoning was that neurons in the accessory olfactory bulb would not respond to stimulation of the main olfactory system.

A study in cats clearly shows that stimulation of sympathetic fibers, as well as adrenalin injection, draws fluid into the vomeronasal organ, while stimulation of the Vidian nerve, carrying parasympathetic fibers, enhanced the excretion of fluid (Eccles, 1982). This response was blocked by atropine, suggesting that the response was mediated by muscarinic acetylcholine receptors. Although the exact vascular and/or erectile events underlying the vomeronasal pump remain quite obscure, its existence seems established beyond reasonable doubt. Thus, we can conclude that soluble odorants indeed have access to the receptors lining the vomeronasal organ. Moreover, the fact that the flow can be reversed means that molecules having entered the organ with the aid of the pump also can be expelled by making the pump work in the opposite direction. Removing molecules is just as important as getting them in. Stimulation of the sensory cells should not last too long. A last but important question concerns the circumstances under which the pump sucks and under which it expels. Some data from hamsters suggest that suction is activated by any novel situation where the animal's attention is attracted (Meredith, 1994). This response would allow the subject to sample molecules in all kinds of novel situations. Whether it reacts to them or not is another question.

Quite a lot is known about the sensory receptors in the vomeronasal organ. I will discuss neither the families of genes coding for these receptors nor the transduction mechanisms. These subjects have been reviewed by real experts elsewhere (e.g. Mombaerts, 2004; Elsaesser and Paysan, 2005) and I have nothing to add. Suffice to say that the vomeronasal receptors are located on the apical dendritic knob of the bipolar vomeronasal neurons. Unlike the dendritic knob on sensory neurons in the main olfactory system, the vomeronasal knobs have no cilia. Instead, each knob contains up to 100 microvilli about 3–6 μm long and with a diameter of about 0.1 μm. It is generally believed that the chemotransduction occurs in the microvilli, but there is, for the moment, no direct evidence showing that this is the case. The axons from the vomeronasal sensory neurons form the vomeronasal nerves. These nerves follow the septum and enter the brain through the cribriform plate. They continue to glomeruli in the accessory olfactory bulb, located in the dorsocaudal part of the main olfactory bulb. There they make contact with the dendrites from the mitral cells. Thus, the first synapse in the vomeronasal system is located to the glomeruli (the name of these structures refers to their globose appearance). An interesting feature is that the mitral cells of the accessory olfactory bulb elaborate an extensive dendritic branching pattern such that each cell innervates multiple glomeruli. This never occurs in the main olfactory bulb, where each mitral cell innervates only one glomerulus. A consequence of this difference in nervous organization is most likely a different coding system for the sensory information.

The mitral cells project directly to layer 1 of the ipsilateral corticomedial nuclei of the amygdala. Here, interactions with the main olfactory systems are likely. In fact, stimulation of the main olfactory system can activate cells within the posteriomedial cortical amygdala that are also activated by stimulation of the accessory olfactory system (Licht and Meredith, 1987). Interestingly, there seems to be a one-way interaction between the olfactory systems, in the way that the main olfactory system may indirectly project to cells that are located in the accessory olfactory system's secondary projection area, while cells from this system never project to cells located in the main olfactory system's secondary projection areas. From the amygdaloid complex, fibers run in the stria terminalis to the ventromedial hypothalamus and the medial preoptic area as well as to the premammillary nucleus. Branches also arrive at the bed nucleus of stria terminalis. A substantial part of the tertiary projections of the accessory olfactory system reaches structures known to be crucial for male and female sexual behavior, and also to structures important for neuroendocrine regulation. However, unlike the main olfactory system, it does not appear to have access to the thalamus and in turn to neocortical regions. Some would interpret that in anthropomorphic terms, stating that information from the vomeronasal organ has no access to the conscience. Anyway, considering the projections to hypothalamic structures, it is not surprising that reproductive physiologists have attributed many functions to the central projections of the vomeronasal organ.

Electrophysiological responses to chemicals in vomeronasal and accessory olfactory tract neurons

I hope by now to have convinced most readers that many mammals have a vomeronasal organ, that sensory neurons in this organ project their axon to the glomeruli in the accessory olfactory bulb, where they connect with the dendrites of mitral cells, and that these mitral cells project their axons to the the ipsilateral corticomedial nuclei of the amygdala. It should also be clear that chemicals can enter the vomeronasal organ with the aid of a pumping system and that they can be expelled by reverse action of the same pump. We have also learned that the sensory neurons most likely have chemoreceptors that are located on the microvilli of the dendritic knob. All this is very exciting, but I have still not, even with a single word, concluded that this system indeed transmits sensory information. Before we can venture to draw any conclusion of this kind, we need to convince ourselves that the sensory neurons send action potentials when their receptors are stimulated with appropriate compounds, and that these action potentials are received in the glomeruli and transmitted to the mitral cells. We would also like to be sure that the mitral cells indeed modify the activity of neurons in their projection area, layer 1 of the corticomedial nuclei of the amygdala. Only when this has been unequivocally shown, can we conclude that the vomeronasal organ is a functional sense organ. It is never sufficient to know that all the ingredients of a particular system are present in a particular species. The ingredients may be vestigial, leftovers from an earlier period in evolution, or they may never have had any function. We need to remember that evolution is a random process and that functionless structures are not necessarily eliminated by natural selection. If they were neutral with regard to survival value, they could persist for millions of years. Thus, the simple presence of all the necessary machinery for something does not automatically imply that the machinery is used. Many misunderstandings in behavioral neuroscience and in biology could have been avoided by following this basic principle.

How about the electrophysiological evidence for a functional vomeronasal/accessory olfactory system? Some, but not much. Starting with the sensory neurons themselves: convincing data show that they indeed respond to some exogenous chemicals, for example compounds found in urine or excretions from the preputial gland. These compounds enhanced the firing rate of the sensory neurons in *ex vivo* studies on sensory epithelium from the mouse (Holy *et al.*, 2000; Leinders-Zufall *et al.*, 2000). We consider this as sufficient evidence for a functional activation of sensory neurons by exogenous chemicals. We can now turn our attention to the first synaptic relay in the accessory olfactory bulb. Mitral cells in slices of rat accessory olfactory bulb show excitatory post-synaptic potentials and, eventually, firing in response to stimulation of the vomeronasal nerve (Jia *et al.*, 1999). *In vivo* studies in mice have established that a drop of male urine on the nose of a female or exposure to bedding soiled by males leads to activation of neurons within the accessory olfactory bulb (Binns and Brennan, 2005). An analogous study in male mice reports similar results. Neurons, probably mitral cells, in the accessory olfactory bulb were activated when the experimental males were allowed to make physical snout-body contact with slightly anesthetized females. No response could be recorded before physical contact, although the males must have been exposed to volatile odors from the females immediately upon introduction into the test chamber. Furthermore, there was no response when the males were exposed to volatile odorants, such as amyl acetate, or putative pheromones like some farnesene derivatives, on a cotton swab held a short distance from the nostrils (Luo *et al.*, 2003).

We have now shown that second order neurons are activated by stimulation of vomeronasal sensory neurons. It now remains to show that the third order neurons, those in the posterodorsal medial amygdala, also respond to stimulation of the sensory neurons. In addition, we should also show that the projection areas of the amygdaloid neurons, such as the medial preoptic area or ventromedial nucleus, respond when vomeronasal receptors are appropriately stimulated. I have been unable to find electrophysiological data showing that this is the case. However, this may be entirely due to ignorance combined with incompetence and does not preclude that such data exist. One study, though, addressed the issue in a somewhat ambiguous way. Female mice received a small drop of urine on the tip of the nose and single unit activity was recorded in the medial amygdala. Exposure to male urine enhanced the firing rate of neurons in the medial amygdala (Binns and Brennan, 2005). So far everything is good, but the interpretation of these results is unclear. Urine contains several volatile odors, and stimulation of the main olfactory system likely occurred simultaneously with the vomeronasal system. Considering that the main olfactory system also projects to the amygdaloid neurons receiving vomeronasal input, the enhanced activity could have been caused by stimulation of that system rather than by the vomeronasal system. These data do not, therefore, convince us that stimulation of vomeronasal neurons indeed leads to changes in the firing pattern of neurons in the amygdala.

Other available data may perhaps help us solve the problem. For example, male hamsters were implanted with an electrode into the vomeronasal organ and electrical stimulation was applied to one side for 45 min. The animals were killed after another 45 min, and expression of the immediate early gene *fos* was quantified in several brain areas. A significant difference in *fos* expression between the stimulated and unstimulated side was obtained only in the accessory olfactory bulb and in the medial amygdala, that is in the primary and secondary projection areas (Meredith and Fewell, 2001). Perhaps these data are sufficient for asserting that stimulation of the vomeronasal organ can lead to changes in nervous activity in the medial amygdala. What is somewhat disturbing is that tertiary projection areas, that is areas receiving the axons of the amygdaloid cells innervated by axons from the mitral cells in the accessory olfactory bulb, such as the bed nucleus of the stria terminalis and medial preoptic area, did not show any *fos* response to vomeronasal stimulation. This coincides with an earlier study, again in male hamsters (Swann *et al.*, 2001). The sexually experienced subjects were exposed to a cotton swab soaked in vaginal secretions from sexually receptive female hamsters and it was ensured that the animals made physical contact with the swabs with the snout. One hour later they were killed and *fos* expression was analyzed in olfactory/vomeronasal projection areas. The areas where *fos* activation was found included the amygdala, the bed nucleus of the stria terminalis and the preoptic area. Some animals were tested after complete removal of the vomeronasal organ through a rather elegant surgical procedure. The elimination of vomeronasal input did not change *fos* expression in the structures just mentioned. It is not unreasonable to conclude, as the authors did, that *fos* activation in these areas is not dependent on stimulation of the accessory olfactory system. Some other animals had the olfactory receptors in the main olfactory system destroyed by intranasal administration of zinc sulfate. The vomeronasal system was left intact. These animals failed to show activation of *fos* in the amygdala, the bed nucleus of the stria terminalis and the preoptic area. This carefully done study clearly shows that the secondary and tertiary projection areas of the accessory olfactory system are activated by stimulation of the main olfactory system and *not* by activation of the vomeronasal system, at least in response to sexually relevant chemical stimuli.

A caveat needs to be introduced immediately. The quantification of *fos* expression is not a direct measure of neuronal firing and it could be argued that the meaning of activation or no activation of *fos* is open to discussion. There is, however, abundant evidence showing that *fos* is a reliable marker of nervous activity, but I accept that it would have been more convincing to have direct electrophysiological data. As mentioned, I have been unable to find reports of electrophysiological recordings in the posterodorsal medial amygdala, or anywhere in the amygdala for that matter, or the preoptic area in response to specific stimulation of vomeronasal sensory neurons. A reason for this may be that scientists have tried to detect such responses, but failed. Negative data, particularly negative data contradictory to dominant beliefs or prejudices, are neither funny nor easy to publish. Perhaps there are some data on the subject put away in a dusty drawer many years ago. We will probably never know.

The conclusion of our analysis of the functionality of the vomeronasal/accessory olfactory system is quite simple: the sensory neurons respond to chemicals, they transmit information to the mitral cells in the accessory olfactory bulb, and the mitral cells probably transmit this information to the posterodorsal medial amygdala. From there on, or perhaps therein, it appears that the main olfactory system and the vomeronasal/accessory olfactory system converge. The vomeronasal contribution to nervous activity flowing out from the amygdala appears to be modest at most. I will not try to make any more elaborate analysis than this. Rather, I will proceed with an analysis of the possible influences of stimulation of the vomeronasal organ on sexual behavior. I will entirely ignore the possible neuroendocrine roles of this organ and strictly concentrate on its role in sexual behavior.

Functions of the vomeronasal organ in relation to sexual behavior

Some believe that the vomeronasal organ is specialized for the detection of nonvolatile chemicals, that is chemicals that are not airborne but in water solution (Wysocki *et al.*, 1980). A consequence of this belief is that the sensing subject needs to be in direct physical contact with a liquid containing the chemical and literally suck it into the nose. The vomeronasal pump will then introduce the compound into the vomeronasal organ. These events may occur when a rat is sniffing urine or feces from other animals, or when a cat licks the genitals of another cat. However, as I have mentioned many times already, stimuli that have the capacity to enhance sexual incentive motivation and initiate approach behaviors need to act at a distance. Their main function is to attract the sensing subject to a potential mate. Nonvolatile compounds are entirely useless for that purpose, since the animals need to be in touch, so to say, before this kind of compound can reach the sense organ. On the other hand, there are some data suggesting that vomeronasal sensory neurons also can respond to volatile chemicals, and that could imply a potential role in sexual incentive motivation. Unfortunately, the results supporting the notion that the vomeronasal system responds to volatile chemicals do not establish whether these chemicals are airborne or in water solution. In one *ex vivo* study, volatile compounds in solution were applied to dissected vomeronasal neurons (Sam *et al.*, 2001), and in another, this one using living mice, the subjects were allowed to sniff cotton swabs impregnated with the odorant (Trinh and Storm, 2003). Although these observations clearly suggest that vomeronasal neurons can respond to volatile chemicals, they give no clue to whether these chemicals must be applied directly to the nose or not, in other words whether the neurons respond or not to airborne molecules. This crucial question has, fortunately for the present discussion, been addressed in a most elegant way. The study of responses of accessory olfactory bulb mitral neurons to chemicals, mentioned a few paragraphs ago (Luo *et al.*, 2003), convincingly showed that these neurons *do not* respond to volatile chemicals at a distance. Direct physical contact between the nose and the odorant source is required. However, contradictory results were obtained in a functional magnetic resonance imaging (fMRI) study in anesthetized female mice. Here, the test compounds were mixtures of several volatiles or urine from male mice. The experimental subjects were exposed to an air flow carrying the molecules to be tested. The experimental setup ensured that there was no direct contact between the nose and the liquid containing the test molecules. Any response must, therefore, be caused by airborne molecules. It turned out that the blood flow in both the main and accessory olfactory bulbs was changed after exposure to the volatile compounds as well as after exposure to urine (Xu *et al.*, 2005). These data may be interpreted as showing that the accessory olfactory system responds to airborne chemicals. In the face of these contradictory findings, any conclusion is risky. However, the Luo *et al.* (2003) study employed behaving mice, and they made direct recordings of nervous activity, while the Xu *et al.* (2005) experiment employed anesthetized animals and they inferred nervous activity from modifications of blood flow. If required to decide as to which study we should attribute the largest force of conviction, I would undoubtedly choose the study of behaving animals and direct recordings. Having done so, I would also venture to pose that the accessory olfactory system only responds to chemicals that enter the nose in water solution. The only occasion at which chemicals enter the nose in water solution is when the subject is sniffing an object. Chemicals on its surface will then be dissolved in the excretions emanating from the nostrils, and can thereby be pumped into the vomeronasal organ. In this context, we should remember that the pump appears to be activated by any situation associated with enhanced arousal (Meredith, 1994). An excellent example of an arousal-enhancing situation is sniffing an unknown object.

Most unfortunately, the fact that the sensing individual needs to make direct physical contact with the source of a chemical before the accessory olfactory system is activated precludes a participation in the process of distant mate detection and initial approach behaviors. It is only when approach is completed that this system may come into play. Therefore, it can have no important role in incentive motivational processes activated by distant stimuli. This assertion does not exclude the possibility that the vomeronasal organ contributes to the sexual incentive properties of proximate stimuli. A very neat experiment has evaluated this possibility. Female ferrets were allowed to sniff urine from either a male or a female for 5 min. Drops ($17\,\mu l$) of male and female urine had been placed on a glass slide 24 hours before use. During these 24 hours, the slide had been sitting in a fume hood. The purpose was to eliminate volatile odors before the urine was employed in testing, thereby minimizing stimulation of the main olfactory system and maximizing stimulation of the accessory olfactory system. The vomeronasal organ had been removed from some females. The time spent investigating the urine did not differ between females with or without a vomeronasal organ. Moreover, both groups of females investigated male urine for a longer time than female urine. Unexpectedly, the females without a vomeronasal organ also investigated male anal scents more than female anal scents, while no such difference appeared in the females with a vomeronasal organ (Woodley *et al.*, 2004). This experiment unequivocally shows that the vomeronasal organ does not contribute, in the slightest, to the incentive properties of stimuli from an individual of the opposite sex, even when the subject is free to interact physically with these stimuli and absorb them into the nose. Thus, the vomeronasal organ is not important for the incentive properties of either distant or of proximate sexual incentive stimuli.

Contrary data were obtained in an experiment where the vomeronasal organ was removed from a group of females tested in an open field. The females could approach either an intact or a castrated male behind a wire mesh. Controls spent more time in the vicinity of the intact male than in that of the castrated male, while subjects without a vomeronasal organ spent the same time in the vicinity of both males (Romero *et al.*, 1990). A problem with this experiment is that the wire mesh probably made direct physical contact between the incentive males and the experimental females difficult or impossible. It is, then, unclear how chemicals from the male entered the females' vomeronasal organ. In view of this, the interpretation of the data is unclear. One explanation is that the vomeronasal organ has different roles in ferrets and in rats. More data are required before we can determine if this is the case or not. Nevertheless, abundant data, some of them presented in preceding paragraphs, show that in the male the vomeronasal organ is *not* important for the sexual incentive properties of chemicals emitted by females. In females, the role of the vomeronasal organ for the sexual incentive properties of males is unclear.

Contrary to its uncertain role in incentive motivational mechanisms causing approach behaviors or behaviors maintaining contact after that approach is accomplished, the vomeronasal organ/accessory olfactory system could be important for the initiation of copulatory behavior in individuals already in close proximity. As pointed out on several occasions already, when approach is completed, another equally crucial process will start: the activation of copulatory reflexes. We will now turn to an examination of studies having determined the role of the vomeronasal organ in this process. To continue with ferrets for a while, it can be mentioned that the females used in the experiment described in the preceding paragraph were also tested for receptivity, anogenital investigation of the male or aggressive behaviors towards the male. Females with and without vomeronasal organ did not differ in any respect (Woodley *et al.*, 2004). Just as was the case with sexual incentive motivation, the vomeronasal organ does not appear to play any role in female ferret sexual behavior. The weight of these observations may perhaps be reduced by the fact that the size of the vomeronasal organ in ferrets, a carnivore, is inferior to that found in rodents, and its morphology is not identical (Wiler *et al.*, 1999). Perhaps the inferior size of the vomeronasal organ in ferrets could account for the difference between female rats and ferrets mentioned previously.

If size indeed were of importance, then we would expect the vomeronasal organ in a rodent to have profound influences on sexual behavior. Let us start with an examination of its role in the male rat. Removal of the organ has, in reality, very small effects. A rather radical intervention, consisting of complete deafferentation of the accessory as well as the main olfactory bulb, failed to reduce partner preference. Please note that the procedure left the bulbs themselves essentially undamaged. It was only the sensory input to them that was eliminated. Furthermore, measures of copulatory behavior, like the number of mounts and intromissions as well as the latency to mount, were unaffected, although the proportion of tests with ejaculation was somewhat reduced and the ejaculation latency somewhat enhanced. Anogenital investigation was reduced after deafferentation (Edwards *et al.*, 1996). It is impossible to determine if these effects were caused by the lack of the accessory or the main olfactory system, but the results strongly suggest that input from neither of these systems is crucial for copulatory behavior and is of no importance for the sexual incentive properties of a female. I would like to repeat and insist on the fact that the olfactory bulbs were not much damaged in the deafferentation procedure. This is most important to remember, because removal of the olfactory bulbs eliminates the sexual incentive value of the female (Edwards *et al.*, 1990). As shown a few lines above, these dramatic effects cannot be due to the absence of chemosensory input, but must be a consequence of the disruption of nervous processes internal to the olfactory bulb. The many reports of effects of lesion of the olfactory bulbs have often been erroneously interpreted as illustrating the role of chemosensory input. Much confusion, some of it quite severe, has its origin in this misinterpretation of otherwise excellent experimental data.

It must be noted that the consequences of combined elimination of sensory input to the accessory and main olfactory systems are quite variable. In some studies, the effects are minor, like in the Edwards *et al.* (1996) experiments described above. In another study, employing the same deafferentation procedure, effects were likewise minor. In fact, in tests performed one week after surgery, the experimental and control males showed identical behavior. One month after surgery, the deafferentated males showed impaired ejaculatory behavior (Bergvall *et al.*, 1991). This, however, may be due to secondary degenerations in the olfactory bulbs more than to lack of sensory stimulation. Unfortunately, no histological study of the olfactory bulbs was performed (at least it was not reported) making it impossible to determine whether this explanation is correct or not. In contrast to these data, other experiments have revealed more substantial deficits in copulatory behavior after deafferentation of the olfactory bulbs. Almost every aspect of male copulatory behavior showed reduction (Larsson, 1971; Edwards and Davis, 1997). Contradictory data are always annoying and they make it difficult to propose some founded conclusion. In addition, the fact that sensory input to both the main *and* accessory olfactory bulb was eliminated or much reduced makes it impossible to attribute the variable effects to any specific system. A potential solution to this last dilemma is offered in two papers where the vomeronasal organ was removed without damaging the main olfactory system, at least without causing extensive damage to it. In one study, it was found that sexually inexperienced male rats showed a quite substantial reduction of copulatory behavior following vomeronasal organ removal. The effect was less pronounced in experienced males, but the number of pre-ejaculatory mounts was increased and the number of intromissions was decreased, while the latencies to intromission and ejaculation were prolonged. The proportion of mounts ending in intromission, sometimes called the hit rate or the intromission ratio, was also reduced (Saito and Moltz, 1986a). These behavioral effects suggest that the animals without vomeronasal organ had some problems with erection or activity in the striated penile muscles. An erectile deficiency nicely explains why the animals were slower to intromit despite the fact that they mounted as fast as the control animals. It also explains the increase in the number of mounts. Increased number of mounts is a kind of standard effect of all experimental manipulations that reduce the males' capacity to intromit. A reduced capacity to achieve intromission also explains the enhanced ejaculation latency. A certain number of intromissions is required for triggering ejaculation and, if they are difficult to achieve, the rat not only mounts more, but he requires also a longer time to obtain the critical intromission number. Enhanced number of mounts and prolonged ejaculation latency were found in another study of copulatory behavior in sexually experienced rats after removal of the vomeronasal organ (Kondo *et al.*, 2003). Indeed, the number of mounts as well as the ejaculation latency was doubled in absence of the organ. These effects are remarkably similar to those reported in the first study, and allow us to propose the rather intriguing conclusion that removal of the vomeronasal organ causes problems with erection and/or activity in the penile striated muscles.

Results from other species could perhaps give us some additional clues as to effects of vomeronasal organ removal on male sex behavior. Unfortunately, this is not evident at all. For example, male hamsters without vomeronasal organ show large deficits in sexual behavior if the organ is removed before the subjects have acquired sexual experience (Powers and Winans, 1975; Meredith, 1986). Most annoyingly, neither of these studies reports complete behavioral data, so we do not know exactly what is meant by 'severe deficit'. A follow-up report does not offer any additional useful information (Winans and Powers, 1977). The second study (Meredith, 1986) reports only the number of intromissions per minute, making it impossible to determine how behavior actually was modified. One possible explanation is that nothing else was modified and the authors disliked reporting non-significant data, or they considered additional data unnecessary. In any case, incomplete behavioral data are a nuisance in the scientific literature and probably reflect the employment of sloppy or incompetent referees in too many journals. Tragically, the habit of reporting incomplete data has persisted for a long time, because the author persists in mentioning only the number of intromissions per minute when talking about inexperienced male hamsters without a vomeronasal organ in studies published 17 long years after the previous one (Westberry and

Meredith, 2003a, b). Effects of ablation of that organ on the intromission rate were similar to those described in the earlier study. Interestingly, if the males without vomeronasal organ were pre-exposed to vaginal fluids from sexually receptive female hamsters, their intromission rate was the same as that of intact animals. Likewise, if the animals were given an intracerebroventricular injection of GnRH, their intromission rate increased to the level of the control animals. This latter finding replicates an earlier study from the same laboratory (Meredith and Howard, 1992). In contrast to the data obtained in hamsters without sexual experience, removal of the vomeronasal organ from sexually experienced male hamsters does not affect copulatory behavior (Meredith, 1986). Needless to say, there was no sign of erectile deficiencies in that study. Finally, there is a rather funny study where sexually experienced males had their vomeronasal organ removed and then tested with an anesthetized male as partner. The peculiarity with this male was that his hindquarters were scented with estrous female vaginal fluids. It turned out that males with and without vomeronasal organ tried to copulate with this poor male with equal enthusiasm (Clancy *et al.*, 1984).

It is almost inevitable to make a small parenthesis here. The vaginal discharge from hamsters has attracted much attention during the last few years, and a large amount of knowledge has accumulated concerning its content. It so happens that hamster vaginal fluids contain a protein called aphrodisin (Briand *et al.*, 2004). The protein has been isolated and sequenced and the structure of the gene encoding it has been determined. Aphrodisin belongs to a family of proteins known as lipocalins, which are supposed to carry volatile odorants to olfactory receptors in the olfactory system. The aphrodisin gene is expressed in the hamster vagina, uterus and in the Bartholin's glands (Kruhøffer *et al.*, 1997). For those not familiar with the anatomy of the female genital tract, I can mention that Bartholin's glands are two small, round structures, one on either side of the vaginal opening. It has also been reported that the gene is expressed in the parotid glands in females (Mägert *et al.*, 1999). An interesting observation is that gene expression is already evident in prepubertal females. The protein was given its name (rather distasteful to some, for example me) because it supposedly stimulates male hamster copulatory behavior. In support for this hypothesis, the Clancy *et al.* (1984) study, mentioned at the end of the preceding paragraph, is frequently cited. The stimuli employed in that study were vaginal fluid or a high molecular weight fraction of this fluid. Whereas vaginal fluid stimulated copulatory behavior equally well in animals with and without vomeronasal organ, the high molecular weight fraction was effective only in controls. This fraction probably contained aphrodisin, and this observation supports the notion that the protein acts through receptors in the vomeronasal system. The capacity of aphrodisin to stimulate mounting of an anesthetized, scented male has been confirmed (Singer *et al.*, 1986). Interestingly, recombinant aphrodisin was found to be ineffective (Jang *et al.*, 2001). The usual explanation for this latter fact is that aphrodisin does not activate vomeronasal receptors by itself. Rather, it carries smaller hydrophobic molecules, volatile or not. Although aphrodisin has become kind of popular, it is far too early to assess its real importance for male copulatory behavior. What we already now can conclude, though, is that it can have no importance for sexual incentive motivation, simply because it acts in the vomeronasal organ. After this digression we immediately return to our main subject.

The male hamster data are, after all, rather clear. The vomeronasal organ is of slight or no importance for copulatory behavior in experienced males. It seems to be of some importance in males without sexual experience, but that importance should not be overestimated. Pre-exposure to estrous female odors can compensate for the absence of the organ. It is evident, then, that the activation or execution of copulatory behavior does not require a vomeronasal organ neither in inexperienced nor in experienced males. The fact that vomeronasal input is not necessary does not exclude some minor modulatory action, though. We do not know if erection in hamsters is affected by vomeronasal organ removal, as it seems to be in rats.

I will not spend hundreds of pages analyzing the effects of the vomeronasal organ on male sexual behavior. The subject has already started to be boring, so I will just mention that male guinea-pigs without a vomeronasal organ vocalize less than guinea-pigs with it in response to vaginal excretions (Eisthen *et al.*, 1987). After losing its vomeronasal organ, the gray mouse lemur (*Microcebus murinus*), a small primate originally from Madagascar, shows reduced intensity of mounting and anogenital investigations. This is difficult to interpret, though, since the animal displays a lower general activity as well. Indeed, the author concluded that the behavioral effects were not due to chemosensory deficit *per se* (Aujard, 1997). The deeper meaning of these observations is a bit unclear, but they do not really challenge the conclusion presented a few lines above.

We are now ready for a general conclusion concerning the role of external stimuli impinging on the vomeronasal organ for sexual incentive motivation and the activation of copulatory behavior. The discussion of incentive motivational effects can be very short: Since the central nervous projections of the vomeronasal organ are activated only by chemicals absorbed through direct physical contact with the stimulus object, it cannot participate in incentive motivational processes initiated by distant stimuli. The essence of these processes is that they are activated long before physical contact has been established. That leaves us with a potential role for the vomeronasal organ in the activation of copulatory behavior. In rats, it does not seem to affect activation at all, but rather it seems to play some role in mechanisms of erection and/or activity in striated penile muscles. This appears to be unexplainable, and we will be forced to return to that issue shortly. In other rodent species, like hamsters, it is has no effect on copulatory behavior in experienced individuals, but it may have some modulatory effect in hamsters without sexual experience. As I mentioned some pages ago, it is of limited importance in a carnivore, the ferret. The modest or non-existent effects of removal of the vomeronasal organ in ferrets were attributed to the organ's small size in this species. This explanation is at best partially true, since the effects of its removal in rodents are not dramatic despite its larger size. In other species having a vomeronasal organ, data are too scarce for drawing any conclusion whatsoever.

Before leaving the vomeronasal organ, I need to write something about its potential function in females. It would be most inconvenient to be accused of sexist attitudes for not doing that. Moreover, although females normally are vastly more complicated than males, frequently on the border of being entirely incomprehensible, we have here a most unusual exception. Females without a vomeronasal organ show a much reduced lordosis behavior, at least if they happen to belong to the rodent species of rats, mice or hamsters (Saito and Moltz, 1986b; Saito *et al.*, 1988; Keller *et al.*, 2006). The much reduced lordosis behavior approaches the levels of control subjects after injection of GnRH (Mackay-Sim and Rose, 1986; Saito and Moltz, 1986b). At least in mice, data show that the females without the organ distinguish an intact male from an estrous female and a castrated male from an intact male (Keller *et al.*, 2006). This is not surprising and coincides with data from males. The fact that lordosis is impaired is curious and the additional fact that it can be re-established by injections of GnRH needs comment. It was shown many years ago that GnRH enhances lordosis in estrogen-primed female rats, so it is difficult to see why it should not do it also in rats with vomeronasal damage. Both the estrogen-primed females in the classic studies and the females in the vomeronasal studies had low initial levels of lordosis, the ideal situation for the detection of stimulatory effect. There is absolutely no reason to start speculations about a role for altered GnRH release after vomeronasal removal simply because GnRH treatment restores behavior. To make a simple example: I could maintain that I often feel very tired after having lectured for several hours in a row without pause. The sensation of tiredness could be completely reversed by sniffing an appropriate dose of cocaine. That would probably make me return to the activity level I had before the lectures. Would it be reasonable to suggest that the effects of the lectures were caused by insufficient cocaine concentrations in my brain? If someone considers this an inadequate example because cocaine is a xenobiotic and GnRH is not, I could simply add that cocaine releases dopamine. The explanation would then be that my post-lecture fatigue was due to a lecture-induced reduction of dopamine release, an equally absurd conclusion if additional evidence, for example direct estimation of dopamine release, does not exist. The fact that 'something' can reverse the effects of an experimental manipulation in no way proves that this 'something' was affected by the manipulation.

A crucial question is whether there is any real evidence relating the behavioral effects of vomeronasal organ removal to deficient GnRH release or not. For example, if such a release indeed were important for lordosis in the females and intromission rate in hamsters we would expect a GnRH antagonist to reduce lordosis and intromission rate in the same way as vomeronasal organ removal does. Moreover, the effect should not be mediated by alterations in steroid hormone concentrations since these concentrations are not altered by vomeronasal organ removal. This probably means that we would need to limit ourselves to rapid effects of the antagonist. There are very few data on this subject, curiously enough. In fact, the only study I have been able to find was performed on ewes (Caraty *et al.*, 2002). The pentapeptide antagonist teverelix (Ac-D-Nal-D-Cpa-D-Pal-Ser-Tyr-D-Hci-Leu-Lys-(iPr)-Pro-D-Ala-NH_2) did not have any effect at all on sexual behavior in ewes carrying an estradiol implant sufficiently large to induce estrous behavior and an LH surge. On the contrary, GnRH had the typical stimulatory effect observed in rodents. The lack of an effect of the GnRH antagonist suggests that GnRH is not involved in the physiological control of estrus in the ewe. If that had been the case, the antagonist should have reduced some aspect of sexual behavior. The fact that GnRH nevertheless stimulated receptivity is a powerful indicator that the effect is pharmacological. With 'pharmacological' is understood an effect obtained after administration of a compound different from the effect that the endogenously released compound has. I will not now go into a discussion of how this is possible, but there are convincing explanations for how it can occur. It is impossible to know if the data from the ewe also apply to rodents, but the similarity in the effects of GnRH between this ovid and rodents is striking. Suffice to conclude that there is no evidence suggesting that GnRH release has any physiological role in sexual behavior, except indirectly through its important actions in the endocrine system. More studies evaluating the acute effects of GnRH antagonists are urgently needed. It must also be remembered that data from males are entirely lacking. Nevertheless, so far we have not obtained any clear explanation as to why GnRH can reduce the effects of vomeronasal organ removal in females and in sexually experienced hamsters except the suggestion that it is a purely pharmacological effect, unrelated to any consequence of ablation of the vomeronasal organ.

There is another possible reason for the notion that alterations in GnRH are causally related to vomeronasal organ removal. Some structural evidence shows that GnRH containing cells in the hypothalamus are a third order projection area of neurons from the vomeronasal organ. However, structural coincidence does not by itself suggest a cause–effect relationship between the vomeronasal organ and GnRH release, and still less between GnRH release and copulatory behavior.

I have discussed the accessory olfactory system quite extensively. The reasons, as outlined many pages ago, were that it is a relatively little known system and that it has been ascribed important functions in sexual behavior. It seemed, at least to me, important to examine carefully each part of the system with the purpose of determining if there are functional sensory neurons, where these neurons project in the brain and, above all, if stimulation of the sensory neurons leads to changes in nervous activity in the primary, secondary and tertiary projection areas. Related to this was the question of whether activation of vomeronasal sensory neurons leads to any nervous activity that cannot be explained by events in the main olfactory system. The conclusion was that the sensory neurons are equipped with functional chemoreceptors and that stimulation of these receptors causes the neurons to send action potentials to the mitral cells in the accessory olfactory bulb, and the mitral cells in turn send action potentials to the posterodorsal amygdala. There, available data suggest, the accessory and main olfactory systems converge. It appears that activation of the vomeronasal receptors alone is not sufficient for modifying nervous activity in the tertiary projection areas like the medial preoptic area or the ventromedial nucleus of the hypothalamus. The main olfactory system needs to be stimulated in addition. Whether stimulation of the main olfactory system alone is sufficient for obtaining these effects is unclear at present. Furthermore, the importance of the accessory olfactory system for sexual behaviors should not be overestimated. It has, at most, some modulatory role. One exception to this may be found in animals without sexual experience in some species.

3.2 The main olfactory system

The role of olfaction, that is to say of the main olfactory system, in sexual behavior is not much better known than that of the vomeronasal system. If, to make things easy, we start with its role in copulatory behavior, a review of existing data tells us that the importance of the main olfactory system is modest. We have already insinuated that the elimination of olfactory input has variable effects. Early studies (Stone, 1922) showed that anosmia did not modify the copulatory behavior of male rats, not even in their first encounter with a receptive female. This observation was confirmed about 50 years later (Cain and Paxinos, 1974). However, later studies have not always replicated this observation. For example, Bergvall *et al.* (1991) reported that peripheral anosmia reduced the intensity of male rat copulatory behavior when the subjects had been anosmic for a month. After one week, though, their behavior was unaffected. Anosmia-induced degeneration of neurons in the olfactory bulb may be the cause of this delayed effect. In fact, it is well known that removal of the olfactory bulbs frequently, but not always, has deleterious effects on male copulatory behavior. It appears that peripheral anosmia normally has rather limited effects, while lesions of the olfactory bulbs frequently have more pronounced effects (reviewed in Hull *et al.*, 2002). There is no doubt, though, that even these effects are modulatory in the way that the copulatory behavior may be reduced or absent in some individuals while others may copulate normally. This shows that neither olfactory input nor the olfactory bulbs are *necessary* for male sexual behavior.

An important part of male rat copulatory behavior is erection. Without appropriate erection it is unlikely that the male will achieve intromission and treatments interfering with erectile capacity always have dramatic effects on copulatory behavior. Curiously enough, just like humans the rat may have erections without any apparent cause. These 'spontaneous' erections are frequently associated with seminal emission (Orbach, 1961). Their frequency is low and they have a maximum during the day (Kihlström, 1966). We should remember that rats are nocturnal, meaning that they sleep a substantial portion of the light period. Humans, at least most of us, are diurnal and we sleep during the dark period. By some unknown coincidence, spontaneous erections in humans are most common during this latter period. Everyone is familiar with the concept of nocturnal or morning erections. When rats and humans display low activity, as they do during the period where sleep is the dominating behavior pattern, spontaneous erections display high activity. Perhaps because neither rats nor humans have anything better to do.

The rather low frequency of spontaneous erection can be enhanced by exposing a male rat to airborne chemicals from a sexually receptive female. If the female is located downwind, no effect of her presence is observed, but as soon as she is located upwind the male starts to display erections (Sachs, 1997). These erections have been termed non-contact erections, but I would prefer to call them female enhanced spontaneous erections. Since airborne chemicals are involved, it is reasonable to suppose that the main olfactory system is responsible for their detection and the ensuing physiological effects. This supposition has been substantiated. Olfactory bulbectomy eliminated female enhanced spontaneous erections and destruction of the nasal mucosa with $ZnSO_4$ produced a marked reduction. Removal of the vomeronasal organ had no effect (Kondo *et al.*, 1999). The authors then proceeded to test the efficiency of devocalized females in activating female enhanced spontaneous erections. No difference was found between vocalizing and silent females. Finally, visual contact between the male and the female was obstructed by an opaque screen, without affecting the frequency of erection. This unusually well done series of studies convincingly show that the main olfactory system, and nothing else, mediates the female enhanced spontaneous erections.

Another elegant study, this time performed in dogs, recorded intracavernous pressure changes in response to female odors. Simultaneously, the electromyograms from the ischiocavernosus and bulbospongiosus muscles were obtained (Shafik, 1997). As mentioned long ago, these muscles are important for penile insertion. Results showed that vaginal odors from donors in all stages of the estrous cycle were equally effective in enhancing intracavernous pressure and all were equally ineffective in modifying electromyographic activity in the striated penile muscles. Interestingly, the pressure response was abolished after anesthesia of the nasal mucosa, showing that olfactory input is essential. These data together with those from rats mentioned in the preceding paragraph nicely demonstrate that odors can enhance sexual arousal, understood as enhanced genital blood flow. As always, it is important to remember that the term arousal has many different meanings depending on the context in which it is used. In the context of sexual behavior, however, it has only one acceptable meaning, which obviously is the one employed here. It is also important to note that, although activation of the olfactory system is crucial for female enhanced spontaneous erections, it is not for erection *in copula*. The alterations in copulatory behavior sometimes reported in anosmic rats are not indicative of any important problems with erection. Hence, during actual sexual interactions there must be other stimuli in addition to odor that efficiently activate erectile mechanisms.

As we will see in a few lines, odors from receptive female rats are powerful sexual incentives for male rats. They will approach such odors. In the description of the incentive motivational model I made many pages ago, it was mentioned that positive incentives activate approach behaviors, that is organized activity in skeletal muscles. In addition, it was mentioned that they may activate visceral responses. A typical example are the effects of the sight or smell of food. Even before starting to eat, salivation is frequently enhanced and the release of some hormones, like insulin, and gastric secretions are stimulated. These responses constitute the cephalic phase of digestion, according to those working with ingestive behaviors (Giduck *et al.*, 1987). They are very similar to the physiological reactions produced by the presence of food within the digestive system. In parallel to this, some of the visceral reactions associated with actual copulatory behavior are activated by the incentive stimulus before copulation has been initiated. The female enhanced spontaneous erections are a beautiful example of one of the properties of sexual incentives. We will see several similar examples when we arrive at the discussion of sexual incentives in the human.

In the female, the effects of anosmia are slight. The receptivity of mice is somewhat reduced after destruction of the olfactory epithelium (Edwards and Burge, 1973) while that of female rats is unaffected (Edwards and Warner, 1972). When the main olfactory bulb is removed (probably associated with removal of the accessory olfactory bulb), receptivity is much reduced in mice (Edwards and Burge, 1973; Edwards, 1974) but enhanced in rats (Edwards and Warner, 1972; al Satli and Aron, 1977; Williams *et al.*, 1992). These observations are remarkable for at least two reasons. First, they are consistent. The inhibitory effect observed in mice is similar to what was reported after disruptions of the accessory olfactory system (see above) and can perhaps be attributed to damage to that system also in the studies mentioned here. In rats, all available data show facilitated receptivity after removal of the olfactory bulbs. There is no contradictory report. Second, an important species difference appears to exist between rats and mice. The opposing role of the olfactory bulbs in rats and mice is not easy to explain. In fact, I cannot even imagine any convincing explanation. Although I cannot propose any explanation, I can insist on the fact that we need to be cautious when generalizing from one species to another, even closely related species. This basic truth is not always remembered.

The sensory neurons of the main olfactory system respond to airborne chemicals. We call chemicals activating olfactory receptors odors and we all know that odors are airborne, so the preceding statement should not be surprising. One consequence of this is that the odor source may be distant. We need not be in direct physical contact with another individual to feel its odor. This makes it possible for the main olfactory system to be involved in mechanisms of sexual incentive motivation and odors may function as sexual incentive stimuli. The fact that odors seem to be of slight or no importance for the execution of copulatory behavior in no way excludes them from being important sexual incentives. It may be convenient to recall the reasoning I presented in relation to ultrasonic vocalizations for making this argument more compelling. When animals are already close to each other, as they always are when copulating, there is no need for stimuli that act from a distance. Especially not since we know that copulatory behavior is heavily dependent on tactile stimulation, something that indeed requires close proximity if not outright intimacy. On the contrary, the process of finding a mate requires the employment of distant sources of stimulation and these distant sources must activate approach behaviors if they are going to lead to something. They must be positive incentives. After having concluded that, I will now proceed with an analysis of odors as sexual incentives.

We know that a sexually inexperienced male rat will approach even an inaccessible female rat in estrus (e.g. Hetta and Meyerson, 1978; Meyerson *et al.*, 1979; Vega-Matuszczyk and Larsson, 1993; Vega-Matuszczyk *et al.*, 1994; López *et al.*, 1999; Ågmo, 2003). We also know that a sexually inexperienced male will approach the odor of a female rat in estrus (Le Magnen, 1952; Ågmo, 2003). Many earlier studies maintained that only sexually experienced males approached odors from receptive females (e.g. Carr *et al.*, 1965, 1970; Stern, 1970), but that conclusion was based on results from procedures presenting a complex odor mixture rather than pure female odors. Thus, the assumption that sexually receptive females or the odor of such females are unconditioned incentives for a male rat is entirely justified. Furthermore, we can conclude that olfactory stimuli alone act as incentives, showing that they are *sufficient*. The next question is whether olfactory stimuli are *necessary* for a female rat's incentive properties. Indeed they are. Anosmic males do not approach a sexually receptive female (Bergvall *et al.*, 1991). The procedure employed in this study was such that the males were not allowed to copulate with the females. The female sexual incentive was separated from the experimental male by a wire mesh. Although the animals could see and hear each other through the mesh, no direct sexual interaction was possible. This means that the procedure evaluates the intensity of approach to the female and nothing else. It is, then, an exquisite measure of incentive motivation. Moreover, it affords the female the possibility to provide three kinds of stimulation to the male: olfactory, visual and auditory. The anosmic males could see and hear the female, yet they did not approach her. The rather clear conclusion is, then, that olfactory stimuli are *necessary*. In addition, the results tell us that neither sound nor sight is *sufficient*. This elegant and simple experiment offered a lot of information.

Now let us look at another that did not offer any useful information at all. If male rats are made anosmic by deafferentation and then allowed to choose between copulating with a receptive female and arguing with a non-receptive one, they prefer the receptive female in the sense that they spend more time in a compartment containing a tethered, receptive female than in a compartment containing a tethered, non-receptive female (Edwards *et al.*, 1996). The conclusion was that partner preference was not reduced by absence of olfactory input. Since the deafferentated males indeed copulated with the receptive female, it is not surprising that they preferred her. Most unfortunately, this study has sometimes been interpreted as if olfactory stimuli are unnecessary for the receptive female's incentive properties. This erroneous interpretation is possible only if we forget that a sexually accessible female offers many kinds of stimulation in addition to odor, for example mechanical stimulation of the intromitting penis. That should be enough for keeping a male in proximity for a while. Thus, this experiment does not give us any meaningful information whatsoever concerning a female rat's incentive properties. The males were deprived of one source of stimulation, but were given plenty of alternatives already known to be strong incentives. Here we have a beautiful example of how sloppy design of an experiment may pass unnoticed even by most respectable colleagues and how erroneous conclusions may originate because of deficient procedures.

Not only has the stimulus modality responsible for a female rat's sexual incentive properties been identified as olfaction, but also the source of the olfactory stimulus is known. Several studies have reported that the preputial glands (in females sometimes called clitoral glands) are important and perhaps even crucial, at least in rats and mice (Bronson and Caroom, 1971; Orsulak and Gawienowski, 1972; Gawienowski *et al.*, 1975, 1976). Interestingly, the production of the odorant in the preputial glands is dependent on estrogens in females and on androgens in males (Marois and Marois, 1974; Thody and Dijkstra, 1978; Donohoe *et al.*, 1981; Pietras, 1981; Alves *et al.*, 1986). Some efforts have been made to identify the compound in the preputial secretions that carries the incentive properties. At least one compound from female rat preputial gland that has incentive properties for males but not for females has been isolated and characterized by mass spectrometry (Kannan and Archunan, 2001). Further studies of that compound are undoubtedly required before any firm conclusion as to its biological importance can be drawn. The data reported so far are very exciting, though.

The role of olfaction should not be overestimated despite the fact that olfactory stimuli are both necessary and sufficient for female rodents' incentive properties. It is not certain that a living female's incentive properties are carried exclusively by her odor. In fact, there are subtle differences between the male's responses to a female and his responses to the odor of a female. Males habituate rapidly to the odor of urine from receptive females, but they never habituate to a living female (Ågmo, 2003; Ågmo *et al.*, 2004). Males continue to approach even an inaccessible female for a seemingly unlimited number of tests. We found no reduction in the intensity of the males' approach behaviors over 10 tests performed every 48 hours or so, notwithstanding the absence of direct physical interaction with the females. This is quite different from what happens when males are exposed to odors of females instead of females. Olfactory preference tests have shown that odors from receptive females are attractive at the first exposure, but at the third exposure they are not more attractive than the odor of a non-receptive female or of water (e.g. Pankevich *et al.*, 2004). These data clearly show that female odor is not equivalent to a female. Odor is necessary and sufficient for the immediate response to a female, but it needs to be supplemented with something in order to maintain the female's incentive properties under conditions of repeated exposure. With what it needs to be supplemented is presently unknown. One possibility is that odor has to be combined with vocalizations in order to avoid habituation. This possibility is currently being evaluated in my laboratory and, by the time this book is off press, there might well be an answer.

We have talked a lot about the importance of odors emitted by females and their effects on males. On the contrary, the effects of male odors on females have not been mentioned. This unpleasant discrimination is entirely unacceptable and I will immediately try to correct it. Sexually experienced, ovariectomized female rats were reported to spend more time investigating odors from an intact male than odors from a castrated male. This was the case regardless of whether the females had been primed with estradiol + progesterone or not. In contrast, sexually naïve females spent more time investigating odors from the intact male only when made sexually receptive by sequential treatment with estradiol + progesterone (Carr et al., 1965). It is important to note that the females were exclusively exposed to airborne odors. A follow-up study confirmed these results (Carr *et al.*, 1970). Another elegant experiment revealed that intact female rats in proestrus approach an intact male more than a castrated male. The same result was obtained in ovariectomized females given estradiol + progesterone. Likewise, both kinds of females approach the *odor* of an intact male more than the *odor* of a castrated male. After transection of the olfactory nerves, disrupting sensory input to both the main and accessory olfactory systems, the females did not distinguish between intact and castrated males. Finally, an effort was made to distinguish the contributions of the main olfactory system from those of the accessory olfactory system. This was done by removing the vomeronasal organ. Females without a vomeronasal organ failed to distinguish between an intact and a castrated male (Romero *et al.*, 1990). Considering that the vomeronasal organ does not respond to airborne chemicals, these data contradict the observations by Carr *et al.* (1965, 1970) where such odors were found reliably to produce approach to an intact male. Furthermore, in the test procedure employed by Romero *et al.* (1990), the male incentives were separated from the experimental females by a wire mesh. Because of that it is doubtful whether the females were exposed to anything but airborne odors. The results of this experiment are difficult to interpret. In any case, the data show that some chemical stimulus is *necessary* to give a male sexual incentive properties.

Female mice also prefer odors from intact males over those of castrated males (Ninomiya and Kimura, 1988; Drickamer, 1989; Drickamer and Brown, 1998), but this preference is unrelated to the phase of the estrus cycle (Scott and Pfaff, 1970; Hayashi and Kimura, 1978). This fact suggests that odors of male mice are sexual incentives for females independently of the presence or absence of ovarian hormones. A more recent study confirms this supposition. Four groups of sexually inexperienced, ovariectomized mice were treated either with oil, estradiol, progesterone or estradiol + progesterone and the time they spent investigating odors from intact males or non-receptive females was recorded. None of the hormone treatments affected the females' behavior. All groups spent much more time investigating the odor of intact males than that of non-receptive females (Moncho-Bogani *et al.*, 2004). These data, then, confirm that female mice are attracted to male odors by a mechanism independent of ovarian hormones. This surprising feature of female mice could suggest that odors of male mice are equally efficient as social and sexual incentives and always superior to odors of a non-receptive female. As pointed out before, a male cannot be a sexual incentive for an ovariectomized, untreated female since such a female never engages in sexual interaction. It is also important to keep in mind that the females used in the studies mentioned above had direct access to the odor source, meaning that both the main and accessory olfactory systems should have been stimulated.

The role of odor for the sexual incentive properties of non-rodent species is somewhat less clear. I will not try to make an extensive comparative review, but some words on the role of olfactory stimuli in non-human primates are unavoidable. An old study revealed that male rhesus monkeys failed to perform a learned operant with access to an attractive female as reward when olfaction was temporarily impeded. With a functional olfactory system, the males performed the operant with high frequency (Michael and Keverne, 1968). It must be noted that the females were made attractive with an intravaginal, estrogen-containing sponge. If estrogen were injected systemically, then the males responded equally well with and without a functional olfactory system. This study revealed that olfactory stimuli might be important for the female rhesus monkey's incentive properties under particular circumstances. Furthermore, it suggested that the source of the olfactory stimulus was the vagina, since intravaginal estrogens were more efficient than systemic. In further studies, the active components of vaginal secretions were identified and repeatedly shown to enhance mounting in males when applied to the sexual skin of untreated, ovariectomized females (Michael and Keverne, 1970; Curtis *et al.*, 1971; Michael *et al.*, 1971). The compounds responsible for the incentive properties of the secretions were found to be rather short, volatile, aliphatic fatty acids (acetic, propanoic, methylpropanoic, butanoic, methylbutanoic and methylpentanoic acid). The ensemble was given the poetic name of copulin. As the name suggests, copulin does not only enhance the female's sexual incentive properties in the way that males approach scented females more than unscented, but also stimulates the execution of copulatory behaviors like masturbation, mounting, intromission and ejaculation (Michael *et al.*, 1977).

In all the above-mentioned, promising experiments, a crucial control was missing. The ovariectomized, untreated females routinely used for testing the effect of copulin were unscented in control conditions but copulin-scented in the experimental condition. Thus, the odor of copulin constituted a new stimulus in the testing situation. It is known that novelty may enhance sexual behavior, probably as a side effect of enhanced arousal, in primates and other mammals. An adequate control procedure in the copulin studies would have been to compare the effects of a novel but meaningless odor to those of copulin. In fact, this was done in an ingenious experiment. Male rhesus monkeys were exposed either to a galbazene and grisalva mixture or to copulin. Both stimuli were equally effective in stimulating copulatory behavior (Goldfoot, 1981). It would seem, then, that novelty was the critical factor, not the chemical composition of the olfactory stimulus. In another experiment, it was found that sexually experienced rhesus males sniffed wood blocks scented with mid-cycle vaginal secretions more than any other odorous stimulus, including copulin and the galbazene/grisalva mixture. Inexperienced males did not show any differential interest in the several odors presented. These data suggest that some component of vaginal secretions other than copulin is critical for the incentive properties of these secretions. Furthermore, they show that the incentive properties are learned. Otherwise, experienced and inexperienced subjects should have responded equally. These interesting observations require confirmation before any concluding statement can be made. Nevertheless, it appears that the interest in copulin was short lived. It is almost never mentioned nowadays. Perhaps the same fate awaits the aphrodisin from hamsters mentioned a couple of pages ago. Only the future will tell us.

In parenthesis I might mention that all effects of copulins and the other chemicals mentioned in the preceding paragraph must be mediated by the main olfactory system since the rhesus monkey does not possess a functional vomeronasal organ.

Besides the experiments described above, where male monkeys were allowed to interact physically with females, there are some data suggesting that stimuli emitted by female rhesus monkeys at a distance are efficient sexual incentives. The habitual effect of distant sexual incentives is to activate responses in skeletal muscles leading to a reduction of the distance to the source of the incentive stimuli. In addition, incentives may activate visceral responses as outlined in the brief presentation of incentive motivation theory in Chapter 2.

In an interesting experiment, male rhesus monkeys were exposed to an inaccessible, estrus female. The males displayed erection and some also masturbated during exposure to the female. In absence of the female, the frequency of erection was only about 15 % of that observed with the female present. It must be noted, though, that most of the erections displayed were incomplete (the glans protruding from the penile sheath with or without significant penile extension). Full erection (penis fully erect and oriented at an angle less than 90° from the male's trunk) was rare in response to the inaccessible female and they were virtually non-existent in her absence. Likewise, no male masturbated during the observation period with the female absent (Pomerantz, 1990, 1991, 1995). In this experiment the males could smell, see and hear the female. Therefore, it is impossible to determine whether the males responded to an olfactory, visual or auditory stimulus. This ambiguity was avoided in an experiment using male marmosets as subjects. They responded with erection to the odor of an estrus female in an experiment where visual and auditory stimuli were absent (Ziegler *et al.*, 2005). These data show that the capacity of distant sexual incentive stimuli to induce sexual arousal in primates is not limited to the rhesus monkey. There is a lot of anecdotal data concerning erection evoked by the presence of females from many other species of mammals, but sound experimental analysis is limited to the studies just mentioned.

Just for the sake of completeness, I have to mention that there is one visceral reaction in addition to erection that is associated with exposure to distant sexual incentives, which has been studied in a few mammalian species: the release of gonadal hormones. Such release occurs in male rats and in males of some other species not only during and after sexual activity but also when exposed to an inaccessible female or to the odor of a receptive female (Ågmo, 1976; Kamel *et al.*, 1977; Bonilla-Jaime *et al.*, 2006). In the same way, female rats respond with LH release when exposed to male odors (Beltramino and Taleisnik, 1983; Tsai *et al.*, 1994). These observations are an eloquent example of a non-vascular visceral response activated by an incentive stimulus before and even in the absence of direct interaction with the individual emitting the stimulus in non-primate mammals. Similar endocrine responses have been recorded in non-human primates. The male common marmoset responds with substantial testosterone release to the odor of an estrus female (Ziegler *et al.*, 2005).

3.3 Other potentially chemosensitive structures in the nose

Besides the main and accessory olfactory systems, two other supposedly chemosensitive systems have been described. One is the septal organ of Masera, frequently called only the septal organ, a cluster of cells located on the nasal septum ventral to the main olfactory epithelium and caudal to the vomeronasal organ. This organ has been described in many mammalian species, including rodents and lagomorphs. The sensory neurons express several olfactory receptor proteins, similar to those in the main olfactory neurons. A single neuron expresses only one receptor (Tian and Ma, 2004). Their axons project to a subset of glomeruli in the ventromedial part of the posterior main olfactory bulb. Some of these glomeruli are exclusively innervated by axons from the septal organ of Masera. Other sensory neurons project to glomeruli receiving their main input from the main olfactory epithelium (Lèvai and Strotmann, 2003). The possible behavioral role of the septal organ is presently unknown, but a popular hypothesis has been that it alerts a resting animal of potentially important odors. Due to its position in the nasal cavity, it is automatically exposed to the air flow during normal respiration in a more efficient way than the main olfactory epithelium. Although this hypothesis was launched in 1943 (Rodolfo-Masera, 1943), it was not subjected to an experimental test until 1995. Lesion of the septal organ failed to modify male rats' reaction to novel odors (Gianetti *et al.*, 1995). If these data had not existed, I would have suggested that the septal organ could contribute to the detection of a mate and thereby be of importance for sexual incentive motivation. Most unfortunately, this suggestion seems entirely unwarranted.

Another olfactory structure, this one located in the opposite part of the nose, is the septal organ of Grüneberg (Grüneberg, 1973). This is a group of cells in the rostrodorsal part of the nasal septum. The Grüneberg organ has so far only been described in mice, but it seems to have attracted some recent attention. It is known that the sensory neurons within the organ express the olfactory marker protein, a protein found in all main olfactory and vomeronasal sensory neurons. About 500 sensory neurons form the Grüneberg organ and they are arranged in clusters and project their single axons in discrete bundles to a subset of glomeruli in the caudal main olfactory bulb (Fuss *et al.*, 2005; Fleischer *et al.*, 2006). This subset of glomeruli appears to be part of what have been termed 'atypical' or 'necklace' glomeruli. They are characteristically innervated by centrifugal acetylcholine esterase containing axons. The projection to these peculiar glomeruli, together with the location of the receptors, right at the entrance to the nose, prompted the suggestion that they may be involved in early detection of biologically relevant olfactory stimuli. This is certainly an exciting possibility and it is not impossible that the septal organ of

Grüneberg and its central projections are important for the sexual incentive properties of olfactory stimuli. Only experimental evidence can transform this speculation into some form of founded hypothesis. Nevertheless, it may be interesting to observe that there are still essentially unexplored olfactory sensory systems.

No discussion of nasal contributions to sexual behavior would be complete without mentioning the *nervus terminalis*. This nerve, sometimes considered a cranial nerve, has been described in almost all classes of vertebrates, including mammals and among those also the human (Fuller and Burger, 1990; Demski, 1993). The cell bodies of the nervus terminalis are strung out along the ventral telencephalon in an area extending caudally from the entry of the olfactory nerve. It closely follows the ventromedial surface of the olfactory bulbs and projects to the olfactory tubercle, the septum and the preoptic area (Bojsen-Møller, 1975). Indeed, the nervus terminalis constitutes the only direct projection from the nasal cavity to brain sites important for sexual behavior, notably the preoptic area. This has led to the suspicion that this nerve may be important for sex behavior. Furthermore, many studies have shown that GnRH containing neurons originate in the olfactory placode and then migrate along the vomeronasal and terminal nerve projections to their final destination (Schwanzel-Fukuda, 1999). Failure of migration may be the cause of a rare disease, hypogonadotrophic hypogonadism or Kallmann's syndrome (Schwanzel-Fukuda *et al.*, 1989). A fascinating detail worth mentioning is that Kallmann's syndrome is associated with anosmia. However, more recent research has challenged the notion that GnRH neurons originate in the olfactory placodes. It appears that their embryonic origins can be traced to the anterior pituitary placode and the cranial neural crest. From these sites they migrate to their final destinations. During this migration, they transiently associate with the developing olfactory system (Whitlock, 2005). This issue is not crucial to the present discussion since we, for the moment, are not interested in problems of development. In contrast, we need to find out what the nervus terminalis can do to sex behavior in adult individuals.

Perhaps it can do a lot. At least in hamsters, the nervus terminalis is colocalized with the vomeronasal nerve to the midbulb level. This means that most studies of vomeronasal organ removal in hamsters have accidentally also lesioned the nervus terminalis. The importance of this was evaluated in a study employing sexually experienced male hamsters in which the nervus terminalis had been transected without causing much damage to the vomeronasal nerve (Wirsig and Leonard, 1987). The effects on copulatory behavior were quite specific and, more interestingly, they did not show any recovery. A somewhat reduced proportion of the animals copulated after lesion, but those who did had the same latencies to mount and intromission as the controls. However, the lesioned animals displayed an enhanced number of preejaculatory intromissions without any accompanying increase in ejaculation latency. This suggests that something had happened to ejaculatory mechanisms after the lesion. Besides tests for copulatory behavior, the hamsters were subjected to an odor detection test. The task was to find a chocolate cookie buried under one inch of wood shavings. The lesioned animals found the cookie as fast as controls. They were also exposed to a vaginal smear from a receptive female, and the time spent sniffing or licking the smear was recorded. Lesioned animals spent less time investigating the smear than controls. However, animals that had suffered damage to the olfactory bulbs, rather than to the nervus terminalis during the lesioning procedure, also showed reduced investigation. This is important because lesion of the nervus terminalis is not possible without inflicting some damage to the olfactory bulbs. That damage rather than the nervus terminalis lesion appears to be the cause of the deficits in investigational behavior. On the contrary, the effects on mating behavior were specific to the subjects with bilateral damage to the nervus terminalis. Finally, serum concentration of testosterone was determined. The lesion had no effect. This carefully controlled study shows that the nervus terminalis may have some importance for male hamster copulatory behavior. Most unfortunately, this single study has not been replicated and we have no data from any other species. The reason for this apparent lack of interest in the role of the nervus terminalis is probably that the surgical technique required for complete and selective lesion of it is extremely difficult.

We cannot leave the Wirsig and Leonard (1987) study without making one additional comment. To give that comment some sense, we need to remember that sexually experienced hamsters were employed. Not long ago we learned that removal of the vomeronasal organ in sexually experienced male hamsters had slight or no effect on copulatory behavior (Meredith, 1986). Likewise, we learned that when removing the vomeronasal organ it is almost impossible not to disrupt the nervus terminalis. It is most likely, then, that the subjects without vomeronasal organ also were without a functioning nervus terminalis. Yet their copulatory behavior was essentially unaffected. A possible reason for the discrepancy between the data reported by Wirsig and Leonard (1987) and those by Meredith (1986) is that the effect of the nervus terminalis lesion on copulatory behavior found in the former experiment was limited to an increased number of pre-ejaculatory intromissions. Furthermore, there was a slight reduction of the proportion of animals that copulated at all. The net effect of an enhanced number of pre-ejaculatory intromissions in those animals that copulated plus a small reduction of the number of copulating animals could mean that the total number of intromissions displayed in the lesioned group was equal to that in the control group. We also know that ejaculation latency was not changed by the lesion, meaning that the mean number of intromissions per unit of time could have been the same in both groups. Now we should remember that Meredith and colleagues have the bad habit of reporting only the mean number of intromissions displayed per minute in their papers. It is quite likely that they do not analyze, or perhaps they do not even care about, any other behavioral measure. Perhaps the effects of the lesion were identical in the two contradictory studies. However, since the behavior was quantified in different ways, they happen to report different results. This is pure speculation, but we cannot come any farther *because one of the studies reported incomplete behavioral data.* I invested so many lines and so much effort in the presentation of the nervus terminalis experiment only because I wanted to come to this conclusion. Perhaps it is a consequence of a fundamental default in my character or intellectual maturity, or both, but I cannot insist enough on the importance of a careful and complete description of the behavior we are studying. Someone else may see things in our data that we are unable to see ourselves and we should be charitable enough to give our neighbors this opportunity. I can see no convincing reason why we should not. A more spiteful thought may also come to mind. Namely that incomplete behavioral data are reported simply because those not reported are too lousy to be mentioned or they contradict what we want to conclude, or simply because they are confusing to us and we do not understand them. Whatever the motives may be for not reporting complete behavioral data, I disagree with this habit.

Before leaving the nervus terminalis, I just want to mention a study in female hamsters (Wirsig-Wiechmann, 1993). Again, selective lesions of the nerve were performed and, as controls, both sham operated animals and animals with damage inflicted to the forebrain were employed. This damage was of the same magnitude as the forebrain damage caused by lesioning the terminal nerve. A rather unusual procedure was employed for evaluating receptivity. First, a male induced a lordosis. Then the male was removed and lordosis interrupted. After varying intervals tactile stimulation was applied to the lumbosacral area with a small paint brush. The latency from stimulation until the display of lordosis was recorded. It turned out that subjects with lesion of the nervus terminalis had a shorter latency to display lordosis compared to prelesion tests, while none of the control groups showed this reduction. However, the effect was small and depended on the interval between removal of the male and application of the tactile stimulus. A reasonable conclusion would be that this small and unreliable effect is of no functional (as opposed to statistical) significance. The author herself concluded that, 'the nervus terminalis plays a minor role in mediating sensory processing during reproductive encounters' (Wirsig-Wiechmann, 1993, p. 867). To my taste, this is an overstatement. This is all we know about the nervus terminalis and female sexual behavior.

Finally, we are at the point where we can end the discussion of olfactory incentive stimuli. None of the lesser known nasal systems, namely, the septal organ of Masera, the organ of Grüneberg or the nervus terminalis has been shown to contribute, in the slightest, to the central nervous or peripheral processing of sexual incentive stimuli. This conclusion is based on very few data and it is not impossible that it will have to be modified at some future date. However, that would surprise me.

4 Visual incentive stimuli

The widespread conviction that odors are basic for non-human copulatory behaviors has led to a neglect of vision. Furthermore, blinding animals is not a standard procedure in any animal behavior laboratory and it would probably be difficult to convince an ethics committee about the urgency of studying copulatory behavior in blinded animals. It is particularly so, since the limited amount of available data suggest that visual stimuli are unimportant for copulatory behavior in the species where they have been studied. Blind rats show a completely normal behavior (Stone, 1922, 1923; Beach, 1942) and so do blind humans as far as I know.

Visual stimuli are also unimportant for sexual incentive motivation, at least in rats and ferrets. A series of experiments in which male rats were tested in total darkness (infrared lighting together with an infrared sensitive video camera was used for observation) showed that their approach behavior to a receptive female was indistinguishable from that seen when rats are tested in light (Hetta and Meyerson, 1978; Ågmo, 2003). In male and female ferrets, approach behavior was evaluated in a situation where the experimental subjects were exposed to either an active individual of the opposite sex behind Plexiglas doors allowing the passage of air (sight, sound and odor condition) or to an anesthetized individual of the opposite sex behind cardboard screens, again allowing the passage of air (odor only condition). The intensity of approach behaviors was similar in the two conditions in both males and females (Kelliher and Baum, 2002). Thus, at least in one rodent and one carnivore species we know that vision is of little or no importance.

In primates, the situation may be quite different. We know that visual sexual incentive stimuli are of paramount importance in the human and there is a small amount of data suggesting that visual stimuli may enhance sexual arousal also in a non-human primate. The human data will be discussed at length in Chapter 4. Here I will briefly mention a couple of studies made in the chacma baboon, *Papio ursinos*. This species attracted scientists' attention because the females show an unusually large swelling of the genitals and the perineum during estrus. Genital swelling around estrus is known to occur in females of many primate species, but zoologists have suggested that the swelling in the chacma baboon is far larger than that of any other primate species. Because of that, it was quite natural to imagine that this conspicuous visual stimulus might have some effects on sexual arousal in males. In a very creative study, individually caged males were exposed either to estradiol-treated females with full-blown sexual swelling or to ovariectomized females without hormone replacement, and consequently no sexual swelling. The females were kept in cages at some distance from the male and no sexual interaction was allowed. Obviously, the males could see, hear and smell the females at all times. In order to evaluate male sexual arousal, in this study as well as in many others defined as erection and masturbation to ejaculation, the number of seminal spots found below the males' cages was counted every morning. It so happens that the semen from the male chacma baboon coagulates very rapidly after ejaculation. Incidentally, we have already seen this phenomenon in rats. The male chacmas try to collect their own semen with the hands upon ejaculation and then eat it, but a small amount always escapes between the fingers and falls down on the floor underneath the cage. There it forms a visible, characteristic spot. Exposure to a female with swollen genitalia increased the median number of spots from 0, the value obtained during exposure to non-swollen females, to 6. These data are suggestive, but they do not really show that visual stimuli are of any particular importance. The estradiol-treated females could have displayed proceptive behaviors (visual stimuli), they might have vocalized (auditory stimuli) and they certainly smelled (olfactory stimuli) differently from the females not given hormones. In order to determine the critical stimulus modality, the same experiment was repeated, but now a screen was placed between the cages housing the females and the males' cages. The screen prevented the males from receiving visual stimulation from the females, but auditory and olfactory stimuli were not prevented from reaching them. It turned out that the males' response to the estradiol-treated females was no different from that to the untreated females (Bielert, 1982). The rather clear conclusion from this experiment was that visual stimuli are critical for the distant female chacma baboon's capacity to enhance male sexual arousal.

The study described in the preceding paragraph established that visual stimuli are important for distant female-induced sexual arousal in male chacma baboons. The next problem to solve was, logically, the identification of the exact visual stimulus. Female chacmas display proceptive behaviors when exposed to males in periods with swollen genitals. Both phenomena are under the control of estrogens. However, the genital swelling continues to increase with increasing doses of estradiol even after the frequency of proceptive behaviors has arrived at its maximum. This makes it possible to vary the amount of genital swelling while keeping proceptive behaviors at a constant level. Interestingly, results show that the males' response intensity is determined by the degree of genital swelling and unrelated to the display of proceptive behaviors (Bielert and Anderson, 1985). This conclusion was further supported in a very nice study in which ovariectomized females were fitted with a plastic model of fully swollen genitalia. These females efficiently enhanced male sexual arousal, although they did not display any proceptive behaviors at all and had no smell of an estradiol-treated female (Girolami and Bielert, 1987). This splendid series of studies convincingly show that distant visual stimuli can enhance sexual arousal in a primate species. I know of no data from other non-human primates which clearly demonstrate that visual stimuli are efficient sexual incentives. Likewise, I know of no data concerning the role of visual stimuli for the enhancement of sexual arousal in female non-human primates. Nevertheless, the results from the chacma baboon suggest that the human male is not the only primate responding sexually to distant visual stimuli.

5 General conclusion concerning sexual incentive stimuli in non-human animals

Auditory stimuli seem to be of limited importance. Most research effort has been focused on the elucidation of the role of high frequency vocalizations in rodents in the control of copulatory behavior. It appears that their function, if any, is limited to modulate the frequency of some proceptive behaviors in female rats. Concerning sexual incentive motivation, we have too few experimental data for daring to propose any function at all. At the same time, it would be premature to conclude that high frequency vocalizations have no function. Nevertheless, for the moment it seems appropriate to suggest that the role of auditory stimuli, and consequently hearing, in copulatory behavior and in sexual incentive motivation is limited. Turning to olfactory systems, it seems warranted to pose that the vomeronasal system is unimportant for sexual incentive motivation in both sexes, while it may influence some aspects of copulatory behavior. The most consistent sequel of vomeronasal organ removal is undoubtedly reduced expression of lordosis in females. This is a very intriguing observation, since we know very well that lordosis is a tactile reflex. Chemical stimuli cannot be particularly important, because lordosis is easily activated by manual stimulation of the flanks and perineal region. Some scientists have routinely replaced a mounting male with their own hand with excellent results. How removal of the vomeronasal organ can modify a tactile reflex is mysterious. Furthermore, the deleterious effect on lordosis of vomeronasal organ removal can reliably be blocked by the administration of GnRH and it can also be reduced by increasing the dose of estradiol given to ovariectomized females. None of these effects is easy to explain.

Elimination of sensory input to the main olfactory system does not have any noticeable effect on copulatory behavior, neither in males nor females. On the contrary, sexual incentive motivation appears to be severely disrupted. It is probably not controversial to propose that airborne molecules are important for activating approach behaviors in male and female mammals while not being important for the execution of sexual reflexes, i.e. copulation. The olfactory bulbs are, however, important for female copulatory behavior. Lesion of them consistently inhibits lordosis in mice and facilitates it in rats. We are here faced with a double mystery. First, it is not evident how to account for the rather dramatic species difference. For all we know, the neurobiology of lordosis is similar in rats and mice. Second, how do the olfactory bulbs modulate a tactile reflex? This, in fact, is the same problem we faced a few lines ago when trying to explain the effects of vomeronasal organ removal. Despite a lot of experimental efforts by first class scientists over decades, we are still not in a position to understand how the olfactory bulbs modify receptivity. With regard to males, we have seen that the effects of destruction of the olfactory bulbs are most variable, ranging from none to severe disruption of copulatory behavior. Again, we do not know how to explain this inconsistency between studies.

Vision does not seem to be of much importance either for sexual incentive motivation or for regulation of copulatory behavior in the few non-human species in which it has been studied. Nevertheless, some data suggest that visual stimuli are effective in enhancing at least some aspects of sexual arousal in non-human primates. The fundamental importance of cutaneous stimulation for copulatory reflexes is beyond doubt. Likewise, its lack of importance for sexual incentive motivation is obvious. It would appear that we have to agree with the proposal made by Frank Beach more than 50 years ago: copulatory behavior is a multisensory process. Lack of one sensory modality can be compensated for by other modalities. An excellent review of the multisensory hypothesis was published several years ago (Stern, 1990). Despite its age it is still the most eloquent exposition of Beach's hypothesis I have ever read. Turning our attention from copulation to incentive motivation responsible for approach to a potential mate, we have to accept that we know far less. What we can deduce without any experimental data, though, is that the critical stimuli must be acting at some distance from their source. We can exclude tactile stimuli and all chemicals that are not airborne. This leaves us with olfactory, visual and auditory stimuli. In the human, it is not impossible that all these kinds of stimuli are of importance. In non-human mammals, the main group of stimuli with sexual incentive properties is probably that of airborne odorants. Perhaps sounds may be important.

All these conclusions are essentially based on data from rodents. I have mentioned the rodent focus of this chapter many times and I have also explained the reasons for that focus. For the question of clarity, I will do it once more. My main concern throughout this long chapter has been to try to produce a systematic presentation of experimental facts. If we suspect that a certain stimulus or group of stimuli may have sexual incentive properties, there is a certain number of things we need to know. Among those are, obviously, if we have functional sensory receptors that react to the stimulus. We also need to know how the stimulus reaches the receptors. As might be remembered, this was an important issue with regard to the vomeronasal organ. If the stimulus has access to the receptors, and if the receptors respond to the stimulus, then we need to show that these receptors have connections with the central nervous system. If they have, then we have to know where they project. Moreover, in addition to determining the structural basis, we need to know that the projections are functional. This can, for example, be shown by making electrophysiological recordings of action potentials throughout the projection areas in response to stimulation of the receptors or of the first order neurons. The fact that we can visualize connections within the brain with one tracing technique or another does not automatically assure that these connections are functional. The vomeronasal organ is an excellent example. We can trace connections from the nose through the amygdala and stria terminalis to the preoptic area. However, when stimulating the first order sensory neurons we never get a response beyond the amygdala. In order to get that, we need also activation of the main olfactory system. This extremely important fact imposes some limits on what we can expect the vomeronasal organ to do.

After having established the functionality of the receptor and determined to where in the brain we can follow action potentials when the receptor is stimulated, we need to proceed with behavioral studies. What happens to behavior when we eliminate the receptor, or its connection with the brain, or the propagation of action potentials from one point to another within the brain? All these are classical questions in neuroscience and all need an answer if we want to understand the neurobiology of a behavior. This is not enough, though. If we want a complete understanding, it is also obligatory to know how and when the stimulus is produced by the individual emitting it. Ultrasonic vocalizations can be a good example. How, where and under which circumstances are these sounds produced? Are they produced in response to a potential mate, in response to any surprise, after exercise, before dinner and so on. In the case of sexual incentives, we can also legitimately ask whether the production of the stimulus is under the control of gonadal hormones, as sexual functions in general are. I will not persist in making the list of questions longer, but I confirm that it can be done. Independently of this, this chapter has been a humble introduction to the kind of questions we need to ask and the kinds of answer we require. If these questions were asked more frequently and more systematically, then many of the myths surrounding sexual behavior would never have arisen and we would perhaps have less prejudice, be more curious about facts and, above all, more cognizant about our ignorance.

Summary

1 Non-human mammals may respond to stimuli or to qualities of stimuli that human sensory systems are unable to detect. Some of these stimuli or stimulus qualities may have sexual incentive properties.
2 Among the most studied of the stimuli beyond the capacities of the human are high frequency sounds produced by rodents. The mechanisms of sound production have been described and the impact of high frequency sounds on listening individuals has been evaluated. There is some evidence suggesting that these sounds activate sexual approach behaviors in female mice. In rats, data are unclear. The elimination of sound production, by cutting laryngeal nerves, has small and inconsistent effects on copulatory behavior in rats and mice. Likewise, deafening has marginal effects on this behavior.
3 Mammals, with the exception of some primates like the human, have two olfactory systems: the accessory olfactory system, sometimes called the vomeronasal system, and the main olfactory system. The vomeronasal system is sensitive to chemicals in solution; the main olfactory system is sensitive to airborne chemicals.
4 The role of chemicals acting in the vomeronasal system on sexual incentive motivation and on copulatory behaviors is unclear. Results from experimental studies are frequently contradictory. The most reasonable proposal is that the vomeronasal system exerts no influence on incentive motivational processes but it may have some marginal role in the control of copulatory behaviors.
5 The sexual incentive properties of several odors have been convincingly established in several mammalian species. Odors are important for initiating approach to a potential mate, but less so for the execution of copulatory reflexes. Airborne chemicals may also produce erection, i.e. sexual arousal, in rats and monkeys. Likewise, they enhance the release of pituitary hormones in some species, at least.
6 Visual stimuli seem to be unimportant among all mammals except primates. Studies of the chacma baboon have nicely shown that males of this species respond to the sight of the swollen sex skin of distant estrus females.
7 Knowledge of sexual incentives in non-human mammals is still fragmentary. Suggestions for focusing the scientific effort on some fundamental issues include the use of a comprehensive theoretical framework, the abandonment of the prejudice that all phenomena of nature have a function as well as the careful, detailed and exact description of behavior.

References

Adler, N. and Anisko, J. (1979). The behavior of communicating: an analysis of the 22 kHz call of rats (*Rattus norvegicus*). *Am Zoologist* **19**, 493–508

Ågmo, A. (1976). Serum luteinizing hormone and testosterone after sexual stimulation in male rabbits. *Acta Physiol Scand* **96**, 140–142

Ågmo, A. (1999). Sexual motivation. An inquiry into events determining the occurrence of sexual behavior. *Behav Brain Res* **105**, 129–150

Ågmo, A. (2003). Unconditioned sexual incentive motivation in the male Norway rat (*Rattus norvegicus*). *J Comp Psychol* **117**, 3–14

Ågmo, A., Turi, A.L., Ellingsen, E. and Kaspersen, H. (2004). Preclinical models of sexual desire: Conceptual and behavioral analyses. *Pharmacol Biochem Behav* **78**, 379–404

al Satli, M. and Aron, C. (1977). Influence of olfactory bulb removal on sexual receptivity in the rat. *Psychoneuroendocrinology* **2**, 399–407

Alves, A.M., Thody, A.J., Fisher, C. and Shuster, S. (1986). Measurement of lipogenesis in isolated preputial gland cells of the rat and the effect of oestrogen. *J Endocr* **109**, 1–7

Anderson, J.W. (1954). The production of ultrasounds by laboratory rats and other mammals. *Science* **119**, 808–809

Anisko, J.J., Suer, S.F., McClintock, M.K. and Adler, N.T. (1978). Relation between 22-kHz ultrasonic signals and sociosexual behavior in rats. *J Comp Physiol Psychol* **92**, 821–829

Aujard, F. (1997). Effects of vomeronasal organ removal on male socio-sexual responses to female in a prosimian primate (*Microcebus murinus*). *Physiol Behav* **62**, 1003–1008

Barfield, R.J., Auerbach, P., Geyer, L. A. and McIntosh, T.K. (1979). Ultrasonic vocalizations in rat sexual behavior. *Am Zoologist* **19**, 469–480

Barfield, R.J. and Geyer, L.A. (1972). Sexual behavior: ultrasonic postejaculatory song of the male rat. *Science* **176**, 1349–1350

Barfield, R.J. and Geyer, L.A. (1975). The ultrasonic postejaculatory vocalization and postejaculatory refractory period of the male rat. *J Comp Physiol Psychol* **88**, 723–724

Beach, F.A. (1942). Analysis of the stimuli adequate to elicit mating behavior in the sexually inexperienced male rat. *J Comp Psychol* **33**, 163–207

Bean, N.J. (1982). Olfactory and vomeronasal mediation of ultrasonic vocalizations in male mice. *Physiol Behav* **28**, 31–37

Bean, N.J., Nyby, J., Kerchner, M. and Dahinden, Z. (1986). Hormonal regulation of chemosignal-stimulated precopulatory behaviors in male housemice (*Mus musculus*). *Horm Behav* **20**, 390–404

Bell, R.A., Noble, M.E. and Daves, W.F. (1971). Echolocation in the blinded rat. *Percept Psychophys* **10**, 112–114

Beltramino, C. and Taleisnik, S. (1983). Release of LH in the female rat by olfactory stimuli. Effects of the removal of the vomeronasal organs or lesioning of the accessory olfactory bulbs. *Neuroendocrinology* **36**, 53–58

Bergvall, Å.H., Vega-Matuszczyk, J., Dahlöf, L.G. and Hansen, S. (1991). Peripheral anosmia attenuates female-enhanced aggression in male rats. *Physiol Behav* **50**, 33–40

Bialy, M., Rydz, M. and Kaczmarek, L. (2000). Precontact 50-kHz vocalizations in male rats during acquisition of sexual experience. *Behav Neurosci* **114**, 983–990

Bielert, C. (1982). Experimental examinations of baboon (*Papio ursinus*) sex stimuli. In *Primate communication* (C.T. Snowdon, C.H. Brown and M.R. Petersen, eds), pp. 373–395. Cambridge University Press: Cambridge

Bielert, C. and Anderson, C.M. (1985). Baboon sexual swellings and male response: a possible operational mammalian supernormal stimulus and response interaction. *Int J Primatol* **6**, 377–393

Binns, K.E. and Brennan, P.A. (2005). Changes in electrophysiological activity in the accessory olfactory bulb and medial amygdala associated with mate recognition in mice. *Eur J Neurosci* **21**, 2529–2537

Blanchard, R.J., Blanchard, D.C., Agullana, R. and Weiss, S.M. (1991). Twenty-two kHz alarm cries to presentation of a predator, by laboratory rats living in a visible burrow system. *Physiol Behav* **50**, 967–972

Blumberg, M.S. (1992). Rodent ultrasonic short calls: locomotion, biomechanics, and communication. *J Comp Psychol* **106**, 360–365

Blumberg, M.S. and Alberts, J.R. (1990). Ultrasonic vocalizations by rat pups in the cold: an acoustic by-product of laryngeal braking? *Behav Neurosci* **104**, 808–817

Blumberg, M.S. and Sokoloff, G. (2001). Do infant rats cry? *Psychol Rev* **108**, 83–95

Bojsen-Møller, F. (1975). Demonstration of the terminalis, olfactory, trigeminal and perivascular nerves in the rat nasal septum. *J Comp Neurol* **159**, 245–256

Bonilla-Jaime, H., Vázquez-Palacios, G., Arteaga-Silva, M. and Retana-Márquez, S. (2006). Hormonal responses to different sexually related conditions in male rats. *Horm Behav* **49**, 376–382

Borden, R., Walker, M.R. and Latané, B. (1977). Ultrasounds and social attraction in rats: concomitants or determinants? *Bull Psychonom Soc* **10**, 89–91

Briand, L., Trotier, D., Pernollet, J.C. (2004). Aphrodisin, an aphrodisiac lipocalin secreted in hamster vaginal secretions. *Peptides* **25**, 1545–1552

Bronson, F.H. and Caroom, D. (1971). Preputial gland of the male mouse: attractant function. *J Reprod Fert* **25**, 279–282

Brown, R.E. (1979). The 22-kHz pre-ejaculatory vocalizations of the male rat. *Physiol Behav* **22**, 483–489

Brudzynski, S.M. and Chiu, E.M.C. (1995). Behavioural responses of laboratory rats to playback of 22 kHz ultrasonic calls. *Physiol Behav* **57**, 1039–1044

Brudzynski, S.M. and Ociepa, D. (1992). Ultrasonic vocalization of laboratory rats in response to handling and touch. *Physiol Behav* **52**, 655–660

Brudzynski, S.M., Ociepa, D. and Bihari, F. (1991). Comparison between cholinergically and naturally induced ultrasonic vocalization in the rat. *J Psychiatr Neurosci* **16**, 221–226

Cain, D.P. and Paxinos, G. (1974). Olfactory bulbectomy and mucosal damage: Effects on copulation, irritability, and interspecific aggression in male rats. *J Comp Physiol Psychol* **86**, 202–212

Caraty, A., Delaleu, B., Chesneau, D. and Fabre-Nys, C. (2002). Sequential role of E2 and GnRH for the expression of estrous behavior in ewes. *Endocrinology* **143**, 139–145

Carr, W.J., Loeb, L.S. and Dissinger, M.L. (1965). Responses of rats to sex odors. *J Comp Physiol Psychol* **59**, 370–377

Carr, W.J., Wylie, N.R. and Loeb, L.S. (1970). Responses of adult and immature rats to sex odors. *J Comp Physiol Psychol* **72**, 51–59

Clancy, A.N., Macrides, F., Singer, A.G. and Agosta, W.C. (1984). Male hamster copulatory responses to a high molecular weight fraction of vaginal discharge: Effects of vomeronasal organ removal. *Physiol Behav* **33**, 653–660

Cooke, B.M., Breedlove, S.M. and Jordan, C.L. (2003). Both estrogen and androgen receptors contribute to testosterone-induced changes in the morphology of the medial amygdala and sexual arousal in male rats. *Horm Behav* **43**, 336–346

Coopersmith, C. and Erskine, M.S. (1994). Influence of paced mating and number of intromissions on fertility in the laboratory rat. *J Reprod Fert* **102**, 451–458

Corrigan, J.C. and Flannelly, K.J. (1979). Ultrasonic vocalizations of defeated male rats. *J Comp Physiol Psychol* **931**, 115

Curtis, R.F., Ballantine, J.A., Keverne, E.B., Bonsall, R.W. and Michael, R.P. (1971). Identification of primate sexual pheromones and the properties of synthetic attractants. *Nature* **232**, 396–398

Demski, S. (1993). Terminal nerve complex. *Acta Anat* **148**, 81–95

Dizinno, G. and Whitney, G. (1977). Androgen influence on male mouse ultrasounds during courtship. *Horm Behav* **8**, 188–192

Dizinno, G., Whitney, G. and Nyby, J. (1978). Ultrasonic vocalizations by male mice (*Mus musculus*) to female sex pheromone: experiential determinants. *Behav Neural Biol* **22**, 104–113

Donohoe, S.M., Thody, A.J. and Shuster, S. (1981). Effect of α-melanocyte-stimulating hormone and ovarian steroids on preputial gland function in the female rat. *J Endocr* **90**, 53–58

Døving, K.B. and Trotier, D. (1998). Structure and function of the vomeronasal organ. *J Exp Biol* **201**, 2913–2925

Drickamer, L.C. (1989). Odor preferences of wild stock female house mice (*Mus domesticus*) tested at three ages using urine and other cues from conspecific males and females. *J Chem Ecol* **15**, 1971–1987

Drickamer, L.C. and Brown, P.L. (1998). Age-related changes in odor preferences by house mice living in seminatural enclosures. *J Chem Ecol* **24**, 1745–1756

Eccles, R. (1982). Autonomic innervation of the vomeronasal organ of the cat. *Physiol Behav* **28**, 1011–1015

Eckman, J., Meltzer, J.D. and Latané, B. (1969). Gregariousness in rats as a function of familiarity of environment. *J Personal Soc Psychol* **11**, 107–114

Edwards, D.A. (1974). Non-sensory involvement of the olfactory bulbs in the mediation of social behaviors. *Behav Biol* **11**, 287–302

Edwards, D.A. and Burge, K.G. (1973). Olfactory control of the sexual behavior of male and female mice. *Physiol Behav* **11**, 867–872

Edwards, D.A. and Davis, A.B. (1997). Deafferentation of the olfactory bulbs of male rats reduces erection to remote cues from females. *Physiol Behav* **62**, 145–149

Edwards, D.A., Griffis, K.T. and Tardivel, C. (1990). Olfactory bulb removal: effects on sexual behavior and partner-preference in male rats. *Physiol Behav* **48**, 447–450

Edwards, D.A., Walter, B. and Liang, P. (1996). Hypothalamic and olfactory control of sexual behavior and partner preference in male rats. *Physiol Behav* **60**, 1347–1354

Edwards, D.A. and Warner, P. (1972). Olfactory bulb removal facilitates the hormonal induction of sexual receptivity in the female rat. *Horm Behav* **3**, 321–332

Eisthen, H.E., Wysocki, C.J. and Beauchamp, G.K. (1987). Behavioral responses of male guinea pigs to conspecific chemical signals following neonatal vomeronasal organ removal. *Physiol Behav* **41**, 445

Elsaesser, R. and Paysan, A. (2005). Morituri te salutant? Olfactory signal transduction and the role of phosphoinositides. *J Neurocytol* **34**, 97–116

Fleischer, J., Hass, N., Schwarzenbacher, K., Besser, S. and Breer, H. (2006). A novel population of neuronal cells expressing the olfactory marker protein (OMP) in the anterior/dorsal region of the nasal cavity. *Histochem Cell Biol* **125**, 337–349

Francis, R.L. (1977). 22-kHz calls by isolated rats. *Nature* **265**, 236–238

Fuller, G.N. and Burger, P.C. (1990). Nervus terminalis (cranial nerve 0) in the adult human. *Clin Neuropathol* **9**, 279–283

Fuss, S.H., Omura, M. and Mombaerts, P. (2005). The Grueneberg ganglion of the mouse projects axons to glomeruli in the olfactory bulb. *Eur J Neurosci* **22**, 2649–2654

Gawienowski, A.M., Orsulak, P.J., Stacewicz-Sapuntzakis, M. and Joseph, B.M. (1975). Presence of sex pheromone in preputial glands of male rats. *J Endocr* **67**, 283–288

Gawienowski, A.M., Orsulak, P.J., Stacewicz-Sapuntzakis, M. and Pratt, J.J. Jr (1976). Attractant effect of female preputial gland extracts on the male rat. *Psychoneuroendocrinology* **1**, 411–418

Geyer, L.A. and Barfield, R.J. (1978). Influence of gonadal hormones and sexual behavior on ultrasonic vocalization in rats: I. Treatment of females. *J Comp Physiol Psychol* **92**, 438–446

Geyer, L.A. and Barfield, R.J. (1980). Regulation of social contact by the female rat during the postejaculatory interval. *Anim Learn Behav* **8**, 679–685

Geyer, L.A., Barfield, R.J. and McIntosh, T.K. (1978). Influence of gonadal hormones and sexual behavior on ultrasonic vocalization in rats: II. Treatment of males. *J Comp Physiol Psychol* **92**, 447–456

Gianetti, N., Saucier, D. and Astic, L. (1995). Analysis of the possible alerting function of the septal organ in rats: a lesional and behavioral study. *Physiol Behav* **58**, 837–845

Giduck, S.A., Threatte, R.M. and Kare, M.R. (1987). Cephalic reflexes – their role in digestion and possible roles in absorption and metabolism. *J Nutr* **117**, 1191–1196

Girolami, L. and Bielert, C. (1987). Female perineal swelling and its effects on male sexual arousal: an apparent sexual releaser in the chacma baboon (*Papio ursinus*). *Int J Primatol* **8**, 651–666

Goldfoot, D.A. (1981). Olfaction, sexual behavior, and the pheromone hypothesis in rhesus monkeys: a critique. *Am Zoologist* **21**, 153–164

Grüneberg, H. (1973). A ganglion probably belonging to the N terminalis system in the nasal mucosa of the mouse. *Zeitschr Anat Entwicklungsgesch* **140**, 39–52

Halpern, M. and Martínez-Marcos, A. (2003). Structure and function of the vomeronasal system: an update. *Progr Neurobiol* **70**, 245–318

Harding, S.M. and McGinnis, M.Y. (2004). Androgen receptor blockade in the MPOA or VMN: effects on male sociosexual behaviors. *Physiol Behav* **81**, 671–680

Harding, S.M. and McGinnis, M.Y. (2005). Microlesions of the ventromedial nucleus of the hypothalamus: effects on sociosexual behaviors in male rats. *Behav Neurosci* **119**, 1227–1234

Hayashi, S. and Kimura, T. (1978). Effects of exposure to males on sexual preference in female mice. *Anim Behav* **26**, 290–295

Hetta, J. and Meyerson, B. J. (1978). Sexual motivation in the male rat. A methodological study of sex-specific orientation and the effects of gonadal hormones. *Acta Physiol Scand* **Suppl. 453**, 1–67

Holy, T.E., Dulac, C. and Meister, M. (2000). Responses of vomeronasal neurons to natural stimuli. *Science* **289**, 1569–1572

Hull, E.M., Meisel, R.L. and Sachs, B.D. (2002). Male sexual behavior. In *Hormones, brain and behavior*, vol. 1 (D.W. Pfaff, A.P. Arnold, A.M. Etgen, S.E. Fahrbach and R.T. Rubin, eds), pp. 3–137. Academic Press: New York

Jacobs, G.H. (1993). The distribution and nature of color vision among the mammals. *Biol Rev* **68**, 413–471

Jang, T., Singer, A.G. and O'Connell, R.J. (2001). Induction of *c-fos* in hamster accessory olfactory bulbs by natural and cloned aphrodisin. *Neuroreport* **12**, 449–452

Jia, C., Chen, W.R. and Shepherd, G.M. (1999). Synaptic organization and neurotransmitters in the rat accessory olfactory bulb. *J Neurophysiol* **81**, 345–355

Jourdan, D., Ardid, D., Chapuy, E., LeBars, D. and Eschalier, A. (1997). Audible and ultrasonic vocalizations elicited by a nociceptive stimulus in rat: relationship with respiration. *J Pharmacol Toxicol Meth* **38**, 109–116

Jürgens, U. (2002). Neural pathways underlying vocal control. *Neurosci Biobehav Rev* **26**, 235–258

Kamel, F., Wright, W.W., Mock, E.J. and Frankel, A.I. (1977). The influence of mating and related stimuli on plasma levels of luteinizing hormone, follicle stimulating hormone, prolactin and testosterone in the male rat. *Endocrinology* **101**, 421–429

Kannan, S. and Archunan, G. (2001). Chemistry of clitoral gland secretion of the laboratory rat: assessment of behavioural responses to identified compounds. *J Biosci* **26**, 247–252

Keller, M., Pierman, S., Douhard, Q., Baum, M.J. and Bakker, J. (2006). The vomeronasal organ is required for the expression of lordosis behaviour, but not sex discrimination in female mice. *Eur J Neurosci* **23**, 521–530

Kelliher, K.R. and Baum, M.J. (2002). Effect of sex steroids and coital experience on ferrets' preference for the smell, sight and sound of conspecifics. *Physiol Behav* **76**, 1–7

Kihlström, J.E. (1966). Diurnal variations in the spontaneous ejaculations of the male albino rat. *Nature* **209**, 513–514

Kirky, R.F. and Blumberg, M.S. (1998). Maintenance of arterial pressure in infant rats during moderate and extreme thermal challenge. *Dev Psychobiol* **32**, 169–176

Kondo, Y., Sudo, T., Tomihara, K. and Sakuma, Y. (2003). Activation of accessory olfactory bulb neurons during copulatory behavior after deprivation of vomeronasal inputs in male rats. *Brain Res* **962**, 232–236

Kondo, Y., Tomihara, K and Sakuma, Y. (1999). Sensory requirements for noncontact penile erection in the rat. *Behav Neurosci* **113**, 1062–1070

Kruhøffer, M., Bub, A., Cieslak, A. *et al.* (1997). Gene expression of aphrodisin in female hamster genital tract segments. *Cell Tissue Res* **287**, 153–160

Larsson, K. (1971). Impaired mating performances in male rats after anosmia induced peripherally or centrally. *Brain Behav Evol* **4**, 463–471

Latané, B. (1969). Gregariousness and fear in laboratory rats. *J Exp Soc Psychol* **5**, 61–69

Latané, B., Edwards, J., Steele, C. and Walton, D. (1973). Social attraction among and between albino and hooded rats. *Bull Psychonom Soc* **2**, 20–22

Latané, B. and Glass, D.C. (1968). Social and nonsocial attraction in rats. *J Personal Soc Psychol* **9**, 142–146

Latané, B., Joy, V., Meltzer, J., Lubell, B. and Cappell, H. (1972). Stimulus determinants of social attraction in rats. *J Comp Physiol Psychol* **79**, 13–21

Le Magnen, J. (1952). Les phenomènes olfacto-sexuels chez le rat blanc. *Arch Sci Physiol* **6**, 295–331

Leinders-Zufall, T., Lane, A.P., Puche, A.C. *et al.* (2000). Ultrasensitive pheromone detection by mammalian vomeronasal neurons. *Nature* **405**, 792–796

Lèvai, O. and Strotmann, J. (2003). Projection pattern of nerve fibers from the septal organ: DiI-tracing studies with trangenis OMP mice. *Histochem Cell Biol* **120**, 483–492

Licht, G. and Meredith, M. (1987). Convergence of main and accessory olfactory pathways onto single neurons in the hamster amygdala. *Exp Brain Res* **69**, 7–18

López, H.H., Olster, D.H. and Ettenberg, A. (1999). Sexual motivation in the male rat: the role of primary incentives and copulatory experience. *Horm Behav* **38**, 176–185

Lore, R., Flannelly, K. and Farina, P. (1976). Ultrasounds produced by rats accompany decreases in intraspecific fighting. *Aggr Behav* **2**, 181

Luo, M., Fee, M.S. and Katz, L.C. (2003). Encoding pheromonal signals in the accessory olfactory bulb of behaving mice. *Science* **299**, 1196–1201

Mackay-Sim, A. and Rose, J.D. (1986). Removal of the vomeronasal organ impairs lordosis in female hamsters: effect is reversed by luteinizing hormone releasing hormone. *Neuroendocrinology* **42**, 489–493

Mägert, H.J., Cieslak, A., Alkan, O., Lüscher, B., Kauffels, W. and Forssmann, W.G. (1999). The golden hamster aphrodisin gene. *J Biol Chem* **274**, 444–450

Marois, M. and Marois, G. (1974). Étude du contrôle par les hormones sexuelles mâles et femelles des glandes préputiales du rat. *J Gynécol Obstét Biol Reprod* **3**, 1151–1167

Matochik, J.A. and Barfield, R.J. (1991). Hormonal control of precopulatory sebaceous scent marking and ultrasonic mating vocalizations in male rats. *Horm Behav* **25**, 445–460

Matochik, J.A., Barfield, R.J. and Nyby, J. (1992a). Regulation of sociosexual communication in female Long-Evans rats by ovarian hormones. *Horm Behav* **26**, 545–555

Matochik, J.A., Sipos, M.L., Nyby, J.G. and Barfield, R.J. (1994). Intracranial androgenic activation of male-typical behaviors in house mice: motivation versus performance. *Behav Brain Res* **60**, 141–149

Matochik, J.A., White, N.M. and Barfield, R.J. (1992b). Variations in scent marking and ultrasonic vocalizations by long-Evans rats across the estrous cycle. *Physiol Behav* **51**, 783–786

McGinnis, M.Y. and Vakulenko, M. (2003). Characterization of 50-kHz ultrasonic vocalizations in male and female rats. *Physiol Behav* **80**, 81–88

McIntosh, T.K. and Barfield, R.J. (1980). The temporal patterning of 40–60 kHz ultrasonic vocalizations and copulation in the rat (*Rattus norvegicus*). *Behav Neural Biol* **29**, 349–358

McIntosh, T.K., Barfield, R.J. and Geyer, L.A. (1978). Ultrasonic vocalizations facilitate sexual behavior of female rats. *Nature* **272**, 163–164

Meredith, M. (1986). Vomeronasal organ removal before sexual experience impairs male hamster mating behavior. *Physiol Behav* **36**, 737–743

Meredith, M. (1994). Chronic recording of vomeronasal pump activation in awake behaving hamsters. *Physiol Behav* **56**, 345–354

Meredith, M. and Fewell, G. (2001). Vomeronasal organ: electrical stimulation activates Fos in mating pathways and in GnRH neurons. *Brain Res* **922**, 87–94

Meredith, M. and Howard, G. (1992). Intracerebroventricular LHRH relieves behavioral deficits due to vomeronasal organ removal. *Brain Res Bull* **29**, 79

Meredith, M. and O'Connell, R.J. (1979). Efferent control of stimulus access to the hamster vomeronasal organ. *J Physiol* **286**, 301–316

Meyerson, B.J., Eliasson, M. and Hetta, J. (1979). Sex-specific orientation in female and male rats: development and effects of early endocrine manipulation. *Adv Biosci* **25**, 451–460

Michael, R.P. and Keverne, E.B. (1968). Pheromones and the communication of sexual status in primates. *Nature* **218**, 746–749

Michael, R.P. and Keverne, E.B. (1970). Primate sex pheromones of vaginal origin. *Nature* **225**, 84–85

Michael, R.P., Keverne, E.B. and Bonsall, R.W. (1971). Pheromones: isolation of male sex attractants from a female primate. *Science* **172**, 964–966

Michael, R.P., Zumpe, D., Richter, M. and Bonsall, R.W. (1977). Behavioral effects of a synthetic mixture of aliphatic acids in rhesus monkeys (*Macaca mulatta*). *Horm Behav* **9**, 296–308

Mombaerts, P. (2004). Genes and ligands for odorant, vomeronasal and taste receptors. *Nature Rev Neurosci* **5**, 263–278

Moncho-Bogani, J., Lanuza, E., Lorente, M. J. and Martinez-Garcia, F. (2004). Attraction to male pheromones and sexual behaviour show different regulatory mechanisms in female mice. *Physiol Behav* **81**, 427–434

Ninomiya, K. and Kimura, T. (1988). Male odors that influence the preference of female mice: roles of urinary and preputial factors. *Physiol Behav* **44**, 791–795

Nunez, A.A., Nyby, J. and Whitney, G. (1978). The effects of testosterone, estradiol, and dihydrotestosterone on male mouse (*Mus musculus*) ultrasonic vocalizations. *Horm Behav* **11**, 264–272

Nunez, A.A., Pomerantz, S.M., Bean, N.J. and Youngstrom, T.G. (1985). Effects of laryngeal denervation on ultrasound production and male sexual behavior in rodents. *Physiol Behav* **34**, 901–905

Nyby, J., Bigelow, J., Kerchner, M. and Barbehenn, F. (1983). Male mouse (*Mus musculus*) ultrasonic vocalizations to female urine: Why is heterosexual experience necessary? *Behav Neural Biol* **38**, 32–46

Nyby, J., Dizinno, G. and Whitney, G. (1977a). Sexual dimorphism in ultrasonic vocalizations of mice (*Mus musculus*): gonadal hormone regulation. *J Comp Physiol Psychol* **91**, 1424–1431

Nyby, J., Kay, E., Bean, N.J., Dahinden, Z. and Kerchner, M. (1985). Male mouse (*Mus musculus*) attraction to airborne urinary odors of conspecifics and to food odors: effects of food deprivation. *J Comp Psychol* **99**, 479–490

Nyby, J., Matochik, J.A. and Barfield, R.J. (1992). Intracranial androgenic and estrogenic stimulation of male-typical behaviors in house mice (*Mus domesticus*). *Horm Behav* **26**, 24–45

Nyby, J., Wysocki, C.J., Whitney, G. and Dizinno, G. (1977b). Pheromonal regulation of male mouse ultrasonic courtship (*Mus musculus*). *Anim Behav* **25**, 333–341

Nyby, J., Wysocki, C.J., Whitney, G., Dizinno, G., Schneider, J. and Nunez, A.A. (1981). Stimuli for male mouse (*Mus musculus*) ultrasonic courtship vocalizations: presence of female chemosignals and/or absence of male chemosignals. *J Comp Physiol Psychol* **95**, 623–629

Orbach, J. (1961). Spontaneous ejaculation in the rat. *Science* **134**, 1072–1073

Orsulak, P.J. and Gawienowski, A.M. (1972). Olfactory preferences for the rat preputial gland. *Biol Reprod* **6**, 219–223

Pankevich, D.E., Baum, M.J. and Cherry, J.A. (2004). Olfactory sex discrimination persists, whereas the preference for urinary odorants from estrous females disappears in male mice after vomeronasal organ removal. *J Neurosci* **24**, 9451–9457

Peichl, L. (2005). Diversity of mammalian photoreceptors properties: adaptations to habitat and lifestyle? *Anat Rec Part A* **287A**, 1001–1012

Pietras, R.J. (1981). Sex-pheromone production by preputial gland: the regulatory role of estrogen. *Chem Senses* **6**, 391–408

Pomerantz, S.M. (1990). Apomorphine facilitates male sexual behavior of rhesus monkeys. *Pharmacol Biochem Behav* **35**, 659–664

Pomerantz, S.M. (1991). Quinelorane (LY163502), a D2 dopamine receptor agonist, acts centrally to facilitate penile erections of male monkeys. *Pharmacol Biochem Behav* **39**, 123–128

Pomerantz, S.M. (1995). Monoamine influences on male sexual behavior of nonhuman primates. In *The pharmacology of sexual function and dysfunction* (J. Bancroft, ed.), pp. 201–211. Elsevier: Amsterdam

Pomerantz, S.M., Nunez, A.A. and Bean, N.J. (1983). Female behavior is affected by male ultrasonic vocalizations in house mice. *Physiol Behav* **31**, 91–96

Powers, J.B. and Winans, S.S. (1975). Vomeronasal organ: critical role in mediating sexual behavior of the male hamster. *Science* **187**, 961–963

Riley, D.A. and Rosenzweig, M.R. (1957). Echolocation in rats. *J Comp Physiol Psychol* **50**, 323–328

Roberts, L.H. (1972). Correlation of the respiration and ultrasound production in rodents and bats. *J Zool* **168**, 439–449

Roberts, L.H. (1975a). Evidence for the laryngeal source of ultrasonic and audible cries in rodents. *J Zool* **175**, 243–257

Roberts, L.H. (1975b). The rodent ultrasound production mechanism. *Ultrasonics* **13**, 88

Rodolfo-Masera, T. (1943). Su l'esistenza di un particulare organo olfattivo nel setto nasale della cavia e di altri roditori. *Arch Ital Anat Embriol* **48**, 157–212

Romero, P.R., Beltramino, C.A. and Carrer, H.F. (1990). Participation of the olfactory system in the control of approach behavior of the female rat to the male. *Physiol Behav* **47**, 685–690

Rosenzweig, M.R., Riley, D.A. and Krech, D. (1955). Evidence for echolocation in rats. *Science* **121**, 600

Rossier, J., Häberli, C. and Schenk, F. (2000). Auditory cues support place navigation in rats when associated with a visual cue. *Behav Brain Res* **117**, 209–214

Sachs, B.D. (1997). Erection evoked in male rats by airborne scent from estrous females. *Physiol Behav* **62**, 921–924

Saito, T.R., Hokao, R., Imamichi, T. (1988). Incidence of lordosis behavior in the female rat following removal of the vomeronasal organ. *Exp Anim* **37**, 93–95

Saito, T.R. and Moltz, H. (1986a). Copulatory behavior of sexually naive and sexually experienced male rats following removal of the vomeronasal organ. *Physiol Behav* **37**, 507–510

Saito, T.R. and Moltz, H. (1986b). Sexual behavior in the female rat following removal of the vomeronasal organ. *Physiol Behav* **38**, 81–87

Salazar, I., Sanchez Quinteiro, P., Cifuentes, J.M., Fernández, P. and Lombardero, M. (1997). Distribution of the arterial supply to the vomeronasal organ in the cat. *Anat Rec* **247**, 129–136

Sales, G.D. (1972). Ultrasound and mating behaviour in rodents with some observations on other behavioural situations. *J Zool* **168**, 149–164

Sam, M., Vora, S., Malnic, B., Ma, W., Novotny, M.V. and Buck, L.B. (2001). Odorants may arouse instinctive behaviours. *Nature* **412**, 142

Sanders, I., Weisz, D.J., Yang, B.Y., Fung, K. and Amirali, A. (2001). The mechanism of ultrasonic vocalization in the rat. *Soc Neurosci Abstr* **88**, 19

Schwanzel-Fukuda, M. (1999). Origin and migration of luteinizing hormone-releasing hormone neurons in mammals. *Microsc Res Tech* **44**, 2–10

Schwanzel-Fukuda, M., Bick, D. and Pfaff, D.W. (1989). Luteinizing hormone-releasing hormone (LHRH)-expressing cells do not migrate normally in an inherited hypogonadal (Kallmann) syndrome. *Mol Brain Res* **6**, 311–326

Scott, J.W. and Pfaff, D.W. (1970). Behavioral and electrophysiological responses of female mice to male urine odors. *Physiol Behav* **5**, 407–411

Shafik, A. (1997). Olfactory-corporeal reflex: description of a new reflex and its role in erectile process. *Eur Urol* **32**, 105–109

Singer, A.G., Macrides, F., Clancy, A.N. and Agosta, W.C. (1986). Purification and analysis of a proteinaceous aphrodisiac pheromone from hamster vaginal discharges. *J Biol Chem* **261**, 13323–13326

Sipos, M.L. and Nyby, J.G. (1996). Concurrent androgenic stimulation of the ventral tegmental area and medial preoptic area: synergistic effects on male-typical reproductive behaviors in house mice. *Brain Res* **729**, 29–44

Sipos, M.L. and Nyby, J.G. (1998). Intracranial androgenic activation of male-typical behaviours in house mice: Concurrent stimulation of the medial preoptic area and medial nucleus of the amygdala. *J Neuroendocr* **10**, 577–586

Sipos, M.L., Wysocki, C.J., Nyby, J.G., Wysocki, L. and Nemura, T.A. (1995). An ephemeral pheromone of female house mice: perception via the main and accessory olfactory systems. *Physiol Behav* **58**, 529–534

Sloan, L. and Latané, B. (1974). Sex and sociability in rats. *J Exp Soc Psychol* **10**, 147–158

Smith, J.C. (1979). Factors affecting the transmission of rodent ultrasounds in natural environments. *Am Zoologist* **19**, 431–441

Stern, J.J. (1970). Responses of male rats to sex odors. *Physiol Behav* **5**, 519–524

Stern, J.M. (1990). Multisensory regulation of maternal and masculine sexual behavior: a revised view. *Neurosci Biobehav Rev* **14**, 183–200

Stone, C.P. (1922). The congenital sexual behavior of the young male albino rat. *J Comp Psychol* **2**, 95–153

Stone, C.P. (1923). Further study of sensory functions in the activation of sexual behavior in the young male albino rat. *J Comp Psychol* **3**, 469–473

Swann, J., Rahamann, F., Bijak, T. and Fiber, J. (2001). The main olfactory system mediates pheromone-induced fos expression in the extended amygdala and preoptic area of the male syrian hamster. *Neuroscience* **105**, 695–706

Thody, A.J. and Dijkstra, H. (1978). Effect of ovarian steroids on preputial gland odours in the female rat. *J Endocr* **77**, 397–403

Thomas, D.A. and Barfield, R.J. (1985). Ultrasonic vocalization of the female rat (Rattus norvegicus) during mating. *Anim Behav* **33**, 720–725

Thomas, D.A., Howard, S.B. and Barfield, R.J. (1982a). Male-produced postejaculatory 22-kHz vocalizations and the mating behavior of estrous female rats. *Behav Neural Biol* **36**, 403–410

Thomas, D.A., Howard, S.B. and Barfield, R.J. (1982b). Male-produced ultrasonic vocalizations and mating patterns in female rats. *J Comp Physiol Psychol* **96**, 807–815

Thomas, D.A., Talalas, L. and Barfield, R.J. (1981). Effect of devocalization of the male on mating behavior in rats. *J Comp Physiol Psychol* **95**, 630–637

Tian, K. and Ma, M.H. (2004). Molecular organization of the olfactory septal organ. *J Neurosci* **24**, 8383–8390

Trinh, K. and Storm, D.R. (2003). Vomeronasal organ detects odorants in absence of signaling through the main olfactory epithelium. *Nature Neurosci* **6**, 519–525

Tsai, Y.F., Wang, H.J., Pi, W.P. and Tai, M. Y. (1994). Age-related changes in the pattern of luteinizing hormone release induced in female rats by male rat urine. *Neurosci Lett* **180**, 179–182

Vega-Matuszczyk, J., Appa, R.S. and Larsson, K. (1994). Age-dependent variations in the sexual preference of male rats. *Physiol Behav* **55**, 827–830

Vega-Matuszczyk, J. and Larsson, K. (1993). Sexual orientation and sexual motivation of the adult male rat. *Physiol Behav* **53**, 747–750

von der Emde, G. (2004). Distance and shape: perception of the 3-dimensional world by weakly electric fish. *J Physiol (Paris)* **98**, 67–80

Westberry, J.M. and Meredith, M. (2003a). Pre-exposure to female chemosignals or intracerebral GnRH restores mating behavior in naive male hamsters with vomeronasal organ lesions. *Chem Senses* **28**, 191–196

Westberry, J.M. and Meredith, M. (2003b). The influence of chemosensory input and gonadotropin releasing hormone on mating behavior circuits in male hamsters. *Brain Res* **974**, 1–16

White, N.R. and Barfield, R.J. (1987). Role of the ultrasonic vocalization of the female rat (*Rattus norvegicus*) in sexual behavior. *J Comp Psychol* **101**, 73–81

White, N.R. and Barfield, R.J. (1989). Playback of female rat ultrasonic vocalizations during sexual behavior. *Physiol Behav* **45**, 229–233

White, N.R. and Barfield, R.J. (1990). Effects of male pre-ejaculatory vocalizations on female receptive behavior in the rat (*Rattus norvegicus*). *J Comp Psychol* **104**, 140–146

White, N.R., Colona, L.C. and Barfield, R.J. (1991). Sensory cues that elicit ultrasonic vocalizations in female rats (*Rattus norvegicus*). *Behav Neural Biol* **55**, 154–165

White, N.R., Gonzalez, R.N. and Barfield, R.J. (1993). Do vocalizations of the male rat elicit calling from the female? *Behav Neural Biol* **59**, 76–78

White, N.R., Prasad, M., Barfield, R.J. and Nyby, J.G. (1998). 40- and 70-kHz vocalizations of mice (*Mus musculus*) during copulation. *Physiol Behav* **63**, 467–473

Whitlock, K.E. (2005). Origin and development of GnRH neurons. *Trends Endocr Metab* **16**, 145–151

Whitney, G., Alpern, M., Dizinno, G. and Horowitz, G. (1974). Female odors evoke ultrasounds from male mice. *Anim Learn Behav* **2**, 13–18

Whitney, G., Coble, J.R., Stockton, M.D. and Tilson, E.F. (1973). Ultrasonic emissions: Do they facilitate courtship of mice? *J Comp Physiol Psychol* **84**, 445–452

Whitney, G. and Nyby, J. (1979). Cues that elicit ultrasounds from adult male mice. *Am Zoologist* **19**, 457–463

Wiler, E., Apfelbach, R. and Farbman, A.I. (1999). The vomeronasal organ in the male ferret. *Chem Senses* **24**, 127–136

Williams, G.W., McGinnis, M.Y. and Lumia, A.R. (1992). The effects of olfactory bulbectomy and chronic psychosocial stress on serum glucocorticoids and sexual behavior in female rats. *Physiol Behav* **52**, 755–760

Winans, S.S. and Powers, J.B. (1977). Olfactory and vomeronasal deafferentation of male hamsters: histological and behavioral analyses. *Brain Res* **126**, 325–344

Winslow, J.T. and Insel, T.R. (1991). Infant rat separation is a sensitive test for novel anxiolytics. *Progr Neuro-Psychopharmacol Biol Psychiatr* **15**, 745–757

Wirsig-Wiechmann, C.R. (1993). Nervus terminalis lesions: II. Enhancement of lordosis induced by tactile stimulation in the hamster. *Physiol Behav* **61**, 867–871

Wirsig, C.R. and Leonard, C.M. (1987). Terminal nerve damage impairs the mating behavior of male hamsters. *Brain Res* **417**, 293–303

Woodley, S.K., Cloe, A.L., Waters, P. and Baum, M.J. (2004). Effects of vomeronasal organ removal on olfactory sex discrimination and odor preferences of female ferrets. *Chem Senses* **29**, 659–669

Wysocki, C.J., Nyby, J., Whitney, G., Beauchamp, G.K. and Katz, Y. (1982). The vomeronasal organ: primary role in mouse gender recognition. *Physiol Behav* **29**, 315–327

Wysocki, C.J., Wellington, J.L. and Beauchamp, G.K. (1980). Access of urinary nonvolatile odors to the mammalian vomeronasal organ. *Science* **207**, 781–783

Xu, F., Schaefer, M., Kida, I. *et al.* (2005). Simultaneous activation of mouse main and accessory olfactory bulbs by odors or pheromones. *J Comp Neurol* **489**, 491–500

Yajima, Y., Hayashi, Y. and Yoshii, N. (1982). Ambiguus motoneurons discharging closely associated with ultrasonic vocalization in rats. *Brain Res* **238**, 445–450

Ziegler, T.E., Schulz-Darken, N.J., Scott, J.J., Snowdon, C.T. and Ferris, C.F. (2005). Neuroendocrine responses to female ovulatory odors depends upon social condition in male common marmosets, *Callithrix jacchus*. *Horm Behav* **47**, 56–64

CHAPTER

4

The stimulus control of sexual incentive motivation: the human

1 Introduction

In human society, we frequently meet individuals of the sex which we prefer for sexual encounters. In isolated rural areas these encounters may be limited to one or two or even fewer every day, while in urban areas they may be counted in thousands. Just a short walk on boulevard Hausmann in Paris, on Kurfürstendamm in Berlin or on 7th avenue on Manhattan on a Saturday around noon will put many hundreds of individuals of the preferred sex within a distance of one meter, or perhaps less if the Saturday happens to be in December. On the streets, at work, in the post office, at dinner parties, during coffee breaks, on the subway, almost everywhere we meet potential mates. Occasionally, some of these potential mates attract us. The question to be examined here is how this attraction is initiated and how it eventually can lead to sexual interaction. This is a very dangerous subject, because we are now entering into something that has been illustrated, discussed and analyzed in thousands of works of fiction and almost as many works of popular science. Still worse, most of us have some personal experience in this kind of affairs, providing us with first hand knowledge. There are also those considering themselves real experts, because of an unusually high number of exploits. Almost all of us have some ideas about how to attract a member of the preferred sex and the most audacious have clear notions on how to proceed to the successful establishment of sexual commerce. It is not impossible that most or all of these ideas are seriously lacking in scientific support. In this chapter, I will present current knowledge about sexual incentives in the human. The preceding analysis of sexual incentives in non-human animals provides a valuable, or perhaps even indispensable, background for the arguments to be advanced on the following pages. At the same time, we need to accept that humans are different from all other animals in some respects. This issue belongs, I fear, to philosophy and I will not make any profound exploration of the uniqueness of humans. Rather I will point out some specific differences and keep the discussion of them at a low level of abstraction.

Human sexual behavior is fundamentally different from that of most other mammals in many aspects. One difference is that the human may engage in sexual activity at any time during the adult life span. This is not the case in rats, for example. When a female rat comes into proestrus/estrus, she becomes a sexual incentive. The estrus female emits some kind of stimulus, for example odor, which has sexual incentive properties for all or at least most males. In the wild, she is followed by a pack of males, all of which try to copulate with her, and many succeed. Female copulatory behavior continues until the period of estrus ends (Calhoun, 1962; Robitaille and Bouvet, 1976). Furthermore, the female will copulate with many males during this period. When estrus ends, the female is much less attractive and she will not display any copulatory behavior. Moreover, if a male tries to copulate with a non-estrus female it is extremely unlikely that an intromission will be achieved. In the absence of lordosis, the vaginal opening is not accessible.

Males of many mammalian species are capable of displaying copulatory behavior anytime from puberty until death. Others, like seasonal breeders show copulatory behavior only during a particular season. Animals belonging to such species have sexual incentive properties only during the breeding season.

In the adult human, both males and females are constantly ready for sexual interaction. There are no physical constraints on sexual activity as there are in many other species. A woman may permit access to her vagina at any moment during the menstrual cycle. This, in fact, is also the case for females of some other primate species. While copulatory behavior of females of most mammalian species is under an absolute control of ovarian hormones, making it possible for them to display this behavior only for a short period around ovulation, copulatory behavior of women and some other female primates do not show this absolute dependency on ovarian hormones. A major consequence of this is that women and some other female primates are very similar to men and males of those other primate species, respectively, in the way that they can display copulatory behavior at any moment of their adult life. There is, then, a fundamental difference between women and most other female mammals with regard to their capacity to display copulatory behavior. We could express this in a slightly more poetic way by saying that while most female mammals have a copulatory behavior controlled by cyclic events in their ovaries, women have been liberated from the ovarian dictatorship and are free to copulate when they may wish to. We will return to the endocrine control of sexual behavior in Chapter 5.

Despite being able to display sexual activity at any moment, a woman does normally not copulate with any man and few men try to copulate with any woman. A female rat, on the contrary, is completely promiscuous. Whether the wild female rat shows some selectivity or not is unknown, but laboratory data suggest that mate choice is random in each copulatory encounter (Ferreira-Nuño *et al.*, 2005). In contrast to rats, in primates like the human the presence of a sexually receptive female is not enough for activating sexual incentive motivation in all available males. Likewise, a man does not activate sexual incentive motivation in all women. Most men and women are attracted only to some individuals of their preferred sex.

This means that there is, somehow, a process of choice involved. Of all available potential sexual partners, only a few have the capacity to activate sexual incentive motivation in a particular individual. We can reformulate this statement and propose that one individual may activate sexual incentive motivation in some other individuals but not in all. Whatever the stimuli emitted by that individual are, their impact is variable between one individual and another. This is something we do not see in rats. There, all males, or at least an overwhelming majority of males, are attracted by the odor of a receptive female, and all, or an overwhelming majority of, receptive females are attracted by the odor of an intact male. While sexual incentive stimuli have similar effects in most individuals when we are talking about rats, there is a huge interindividual variation in the human. The task of identifying these stimuli is thereby tremendously complicated.

The interindividual variability is undoubtedly largest with regard to the incentive stimuli activating approach behaviors. Some may find men with a long neck extremely attractive, others find women with a deep, vibrant voice irresistible and still others may be attracted by an acid sense of humor in a woman with blonde, short hair. The activation of approach behaviors by particular sexual incentive stimuli determines with whom we will seek to establish sexual contact. The selective activation of approach by a specific incentive stimulus or a set of incentive stimuli constitutes the basic mechanism of mate choice. The choice of mate does not need to have major consequences if the initial encounter happens to be unique or if it leads to a short relationship with the individual emitting these incentive stimuli. But humans have a tendency to form lasting relationships sooner or later and then the choice of partner becomes quite important. The fact that humans, in contrast to rats and most other mammals, frequently establish lasting relationships further complicates any analysis of sexual incentives. Moreover, although sexual incentive properties probably are involved in the immediate attraction to a partner and in the establishment of short-term relationships, there may be many other factors determining the choice of long-term partners. That issue is far beyond my capacities to analyze, but it is important to keep in mind that most humans will be engaged in one or more long-term relationships and the choice of the partner for such relationships may be entirely independent, or only partially dependent, on sexual incentive properties.

The factors directing choice of long-term partners have been analyzed by social scientists of all kinds and I will not venture into any detailed discussion of them. It appears, though, that choice of partner was ruled by practical considerations for centuries. Marriage was, for the wealthy, a convenient procedure for distribution or accumulation of wealth and, for less wealthy women, it was the only legitimate way to obtain an income. As eloquently expressed by a late 19th century Swedish novelist, marriage is the only lawful form of prostitution. In the social context, where marriage was a requisite for socially acceptable sexual activity, and where the choice of marriage partner was based on practical reasons, it would probably have been outrageous to mention sexual incentive properties as a cause for marriage. This state of affairs may have changed slightly with the decline in the status of marriage and as a consequence of the enhanced economic independence of women, but the change has certainly not been of such a magnitude that long-term partners are chosen entirely according to sexual incentive properties. Consequently, there will be no further discussion of the mechanisms involved in the choice of long-term partners.

The enduring habit of establishing long-term relationships based on factors other than sexual incentive properties may have several consequences. One that immediately comes to mind is that of future sexual functioning within the relationship. If the selected partner has no or weak sexual incentive properties for the other member of the couple, it is not impossible that sexual dysfunction will follow, at least in the long run. When we arrive at the discussion of sexual dysfunctions we will see that the partner indeed may be important, both for etiology and treatment. Infidelity and even break up of the relationship are other possible results of deficient incentive properties on the part of one or both members of the couple.

One of the popular myths, or truths depending on our personal inclination, of contemporary society is that human couples are united by an emotional bond called romantic love. It appears that this concept refers to some emotional state, essentially irrational, bringing two individuals together. Although romantic love is not equivalent to sexual attraction, it has been maintained that at least the possibility of sexual interaction is a requisite for its occurrence (Fisher, 2004). It is also assumed that romantic love is independent of trivial considerations, such as social status or intellectual capacity of the partner. It is also independent of traditional moral rules, since the object of love may be already married, in love with someone else, or even a close relative such as a sister or brother. All these 'immoral' romantic loves seem to be the favorites in many successful television series or best-selling popular novels. The interesting thing here is that the factors underlying the romantic love, these mysterious, irresistible, irrational forces directing approach behaviors, seem to be processes very similar to what I would call sexual incentive motivation. As repeatedly mentioned, this motivation directs approach behaviors toward the stimulus originating it, the incentive, in an automatic way without any intervention of the will. This latter is not my invention. I just remind you about the opinion of Saint Augustine regarding the lack of voluntary control over sexual reactions. The conclusion here is that the notion of romantic love makes exactly the same predictions as to the determinants of approach behaviors as sexual incentive motivation theory. We will return to this issue in a couple of pages.

Another important difference between human and non-human sexual incentives is that most or perhaps all human sexual incentives are learned. Some may be learned by association with copulatory behaviors or other forms of genital stimulation. For example, stimuli present in the context where copulatory behavior is executed can acquire sexual incentive properties through classical conditioning. This will be amply analyzed in Chapter 7. Furthermore, any stimulus may acquire sexual incentive properties through social learning, including formal and informal instruction. Some have gone so far as to maintain that all human sexuality is socially constructed. Whether this is true or not is irrelevant for our present discussion, but we need to be aware of the fact that any stimulus can, at least in principle, acquire sexual incentive properties through one or other form of learning. This means that the identification of human sexual incentive stimuli is an essentially hopeless enterprise. Despite this rather negative appraisal, I dare to add that there are some stimuli that have sexual incentive properties for many, perhaps even for most, humans. Whether they are learned incentives or not has no importance here. It seems that the naked body of someone belonging to our preferred sex has sexual incentive properties for most individuals, at least those belonging to the European culture. Likewise, it seems that watching other people engaged in precopulatory or copulatory behaviors has sexual incentive properties for most people. I will not pursue this discussion now, I just want to suggest that *there are* at least some stimuli that have sexual incentive properties for most humans. This makes a discussion of sexual incentives in the human meaningful. If there were no interindividual consistency in incentive stimuli, any discussion of them would have had interest only for me.

During the ensuing discussion it is useful to keep some of the arguments expressed in the preceding paragraphs in mind. The most important is that humans are unlike other animals. An explicit recognition of this fact has some quite important consequences for the behavioral concepts we use to analyze human sexual behavior. Another, perhaps more obvious consequence of the different status of humans relative to other mammals, is that the techniques and procedures used for studying sexual incentive phenomena and copulatory behavior in non-human mammals are distinct from those that have been employed in humans. Experiment, in the classical sense, is the basis for non-human research, but is rarely employed in research on humans. The availability of gonadal hormones is easily manipulated for experimental purposes in non-human mammals while it is only done for therapeutic reasons in the human. Random samples of humans cannot be subjected to experimental modifications of hormones in the same way as random samples of rats can be. We do not put electrodes into people's brains to record nervous activity in response to sexual incentives, neither do we intentionally lesion interesting brain structures. While we can analyze nervous activation at a cellular level by determinations of expression of immediate early genes, like *fos*, in non-human mammals, we are limited to comparisons of cerebral blood flow before and after sexual events in humans through imaging techniques. The spatial resolution of these techniques is modest at the very best and vastly inferior to analyses of *fos* expression. Furthermore, whereas an overwhelming amount of data shows that *fos* expression is indeed induced by enhanced nervous activity, imaging techniques are making inferences about nervous activity from hemodynamic responses in brains kept in an intense magnetic field.

There are also many limitations on the behavioral procedures judged appropriate for human use. We do not experimentally determine how many ejaculations a man can achieve before sexual exhaustion and we do not investigate whether the replacement of the partner with a fresh one can reactivate copulatory behavior in exhausted men as it does in exhausted rats. In studies of sexual reward and reinforcement, we routinely allow a rat access to copulation as reinforcement, while a human is reinforced by the presentation of a photo of a naked individual of the preferred sex rather than with sexual activity. Taken together, the limitations applying both to neurobiological and behavioral procedures acceptable for human use mean that our knowledge of the mechanisms involved in human sexuality remains far inferior to that we have concerning some other mammals. Please note that I do not argue for unrestricted experimentation on human subjects. The limitations I have mentioned, as well as many others, are there for the protection of human beings and it is not in the interest of the scientific community to have them removed. Nevertheless, it is useful to keep in mind that many biological and behavioral mechanisms of human sexuality need to be inferred from data obtained in other mammals or from less than perfect data obtained from humans under special and sometimes even embarrassing circumstances. This does not exclude, though, that we also have some good data from humans, particularly at a descriptive level.

2 Human sexual incentive stimuli: visceral reactions

2.1 General comments

After having made explicit some of the peculiarities with the human, we can now turn to what should be the subject of this chapter. For the moment we will focus upon sexual excitement induced by external stimuli. Environmental stimuli having the capacity to produce sexual excitement can be included under the definition of sexual incentives. Strangely enough, the term incentive incites me immediately to a short recapitulation of some fundamentals of incentive motivation theory. Any positive incentive should, by definition, activate approach behaviors if the central motive state is in an appropriate condition. Approach behaviors are, even in the human, nothing more than activity in skeletal muscles. In addition to the activation of approach behaviors, incentives can also turn on visceral reactions, and this can occur long before actual commerce with the incentive begins. We can, then, simultaneously approach an incentive through the use of skeletal muscles and manifest visceral reactions. When we analyzed sexual incentives in non-human mammals, basically in rats, we concentrated on the activities of skeletal muscles initiated by the incentive. In contrast to visceral reactions, which are difficult to observe, skeletal muscle activity is evident in behavior. This simple fact has certainly contributed to a focus on skeletal muscle activity in most non-human studies of sexual incentives. In the human, however, it has turned out to be much easier to study visceral than somatic responses. A visceral response, like enhanced genital blood flow, is rather easy to observe and quantify, and such responses are usually not considered as offensive or immoral if not made too evident in public. Human somatic responses to sexual incentives would require the creation of a situation where sexual interaction eventually would be possible. Otherwise, no approach responses involving skeletal muscles could be observed. Copulatory behavior has occasionally been studied in the laboratory, as we know by now, but then the partners have been pre-established rather than selected from a range of alternatives by the experimental subjects. I will, because of simple decency, not enter into a detailed description of what a study of human somatic responses would look like. In any case, I can guarantee that it would be judged as immoral and unethical. Rather than taking all the risks involved in offending the moralists, scientists have preferred to stick to visceral responses when evaluating the effects of sexual incentives in the human. We will now look at two of these visceral responses.

2.2 Endocrine responses to sexual incentives

In Chapter 3, we saw that distant sexual incentives, even conditioned, could enhance pituitary release of luteinizing hormone (LH) and increase in blood androgen concentrations. In fact, the release of androgens in response to distant sexual stimuli appears to be widespread among male mammals. Curiously, it does not seem to be a consistent response in men. Several studies have exposed men to pornographic movies and failed to find any change in plasma androgen concentrations, while others have reported an increase (see Krüger *et al.*, 1998 for a discussion and references). Even copulation or masturbation to orgasm has unreliable effects on androgen concentrations in men. It seems that male non-human mammals of many species reliably respond with enhanced testosterone release when exposed to sexual incentives and during and after copulatory activity, while men display a much more variable response. The conclusion of this short discussion of visceral responses is that erection is more reliable than androgen release in men exposed to sexual incentive stimuli. Other hormones seem to be affected in an equally unreliable way by sexual incentives. Pornographic movies or pictures fail to modify plasma concentrations of gonadotropins, adrenocorticotropic hormone (ACTH), prolactin, cortisol, adrenaline and noradrenaline in young men of good health (Rowland *et al.*, 1987; Carani *et al.*, 1990; Krüger *et al.*, 1998; Exton *et al.*, 2000). Curiously enough, erotic still images were found to enhance plasma concentrations of ACTH, prolactin, noradrenaline and cortisol in another study, also conducted with young men as subjects (Codispotti *et al.*, 2003). It is very difficult to speculate about possible reasons for this discrepancy, but the notion that erection is the only reliable visceral response to sexual incentives is not weakened by this study.

Women's endocrine responses to sexual incentives have been little studied. Only one study provides some information. Nine young women were exposed to a pornographic movie and blood was sampled, through an indwelling catheter, every 5 min before, during and after exposure. Although subjective sexual arousal was reported to be much elevated by the movie, there was no effect on plasma concentrations of adrenaline, cortisol or prolactin. Contrary to this, noradrenaline concentration was elevated (Exton *et al.*, 2000). This may have some relation to the increase in systolic blood pressure produced by the movie. A single study is obviously not enough for concluding that the endocrine system in women does not respond to sexual incentives, but the similarity to the results obtained in men is striking. Since the several studies in men coincide, with one unexplainable exception, in showing that there is no endocrine effect of sexual incentives, I would venture to suggest that this is also the case in women. I would, nevertheless, feel more comfortable with additional data in support of that conclusion. The lack of popularity of studies in women makes a hope for more data slightly unrealistic, nonetheless. This lack is not necessarily due to some kind of sexism on the part of scientists. It can perhaps be explained by the difficulties associated with the cyclic variation of hormone release in females of all mammalian species. Despite this apparent lack of interest in women, any physiologist knows that females of several non-human species present substantial post-coital changes in hormone release. Species with copulation-induced ovulation constitute the most eloquent example.

2.3 Enhanced genital blood flow

During the discussion of non-human sexual incentives, I mentioned an example of a visceral reaction provoked by an incentive, penile erection in the rat and monkey. We learned that rats enhance the frequency of spontaneous erection in response to odors from a distant female. However, this visceral reaction is not a constant response to an incentive stimulus, because we also learned that during actual copulatory activity, erection in rats is not continuous but associated with mounting. The female odor, however, can be expected to be present with almost constant intensity throughout the copulatory interaction. Female odor cannot, therefore, be the stimulus controlling erection during copulatory encounters. Furthermore, the visceral response of erection does not occur in anticipation of a mount, for example during pursuit of the female, but during the initial stages of mounting when physical contact with the female has already been established. It appears, then, that erection in response to distant incentive stimuli emanating from an inaccessible female is controlled by stimuli different from those operating during copulatory interaction, at least in rats.

In the human, the situation may be similar. Particularly young men, and sometimes men not so young any more, get an erection in response to an incentive even in situations where sexual interaction is excluded from the outset. Reading a pornographic magazine could be such a situation. It is extremely unlikely, if not outright impossible, that the man reading it expects the lady on the photograph to materialize before him and propose sexual intercourse. He can have fantasies of that sort, but that is a question we will not address now. Just as the rat, he responds with erection to stimuli emitted by an inaccessible female. It should not be controversial to propose that men frequently display visceral reactions to sexual incentives long before actual sexual interaction takes place as well as in the absence of the possibility of such interactions. There are much data showing that this kind of response is not limited to men. In healthy women, enhanced genital blood flow is an extremely consistent reaction to sexual incentives in the form of pornographic movies or pictures with erotic content (see Rellini *et al.*, 2005, for several references). It appears to be just as unfailing as erection in healthy men.

As pointed out in the preceding paragraph, the main visceral response to distant sexual incentives in the human is enhanced genital blood flow. This response is called sexual arousal, as mentioned long ago. The distinction between this visceral reaction and any mental state produced by sexual incentive stimuli should always be clearly maintained. Nevertheless, terms like 'subjective arousal' are sometimes used to refer to the conscious perception of genital arousal. It appears, though, that subjective arousal is only partially determined by genital blood flow. Studies in men and women have revealed modest correlations between questionnaire or other reports of arousal and objective measurements of genital response, viz. penile tumescence and vaginal blood flow (Laan *et al.*, 1995). It is generally believed that the correlation is higher in men than in women (Heiman, 1977; Steinman *et al.*, 1981). One probable reason for this is that erection is a more notorious event than vaginal lubrication. Sensory feedback from a fully or partially erect penis is probably more intense than that from a more or less lubricated vagina. In consequence, it is easier for men to be conscious of their degree of erection than it is for women to be conscious of their degree of vaginal lubrication. However, some recent data suggest that the coincidence between subjective and genital arousal in women may have been underestimated (Rellini *et al.*, 2005).

I will now turn to an analysis of the stimulus control of the sexual arousal response in the human. The evidence for a role of olfaction in human sexual incentive processes is not overpowering. In fact, there are no data concerning the capacity of odorous stimuli to promote erection. Likewise, there are no data concerning the effects of odorous stimuli on vaginal blood flow. There is, however, a curious little experiment in which 28 young women either were exposed to a pornographic movie or were asked to produce sexual fantasies. While doing that, the women were exposed to one of three odors. A neutral odor (distilled water), a masculine odor (a commercial perfume belonging to the 'fresh fougere category', whatever that means) or a feminine odor (another commercial perfume belonging to the aldehydic floral category) were employed. Vaginal blood flow was estimated with a photoplethysmograph. Results showed that the odor failed to modify the response to the pornographic movie, while the 'masculine' odor enhanced vaginal blood flow during the production of sexual fantasies. The effect was seen in the follicular phase but not in the periovulatory phase of the menstrual cycle (Graham *et al.*, 2000). This study suggests that synthetic odors may have some effect on the capacity of mental representations to stimulate vaginal blood flow. Combined with visual stimuli, though, the contribution of this kind of odor is marginal. A firm conclusion as to the role of olfactory incentive stimuli in the activation of sexual visceral responses is impossible in view of the limited data available.

The role of hearing is equally unclear. An accomplished study evaluated the response to erotic/pornographic and romantic audiotapes in a sample of undergraduates (Heiman, 1977). Both questionnaire and physiological estimations of the intensity of response to the stories told on the tapes were performed. Physiological measurements included penile tumescence in men and vaginal blood volume in women. These genital responses were, not surprisingly, used as physiological indicators of sexual arousal. The pornographic stories stimulated erection and enhanced vaginal blood flow while the romantic stories were ineffective. Questionnaire responses coincided with the physiological indicators of sexual arousal. As always, the correlation between genital and questionnaire responses was higher in men than in women. This beautiful study seems to suggest that auditory stimulation enhances sexual arousal in both men and women. However, a problem with the interpretation of the data is that we cannot know whether the pornographic audiotapes evoked visual mental representations that were sexually arousing. It is possible to argue that the function of hearing was to evoke such mental imagery and that the imagery rather than the verbal content was the direct cause of arousal. The capacity of mental representations to induce sexual arousal has already been mentioned several times and the possibility that the verbal message was only triggering mental images needs to be seriously considered. In fact, sexually explicit visual imagery is an important part of physically arousing sexual fantasies (Leitenberg and Henning, 1995). Considering this it must be accepted that the Heiman (1977) study does not provide unequivocal support for a role of auditory stimulation in sexual arousal.

Another way to evaluate the importance of hearing for sexual arousal is to add auditory stimulation to visual stimuli. If sound plus image were more efficient than image alone, we might conclude that auditory stimulation indeed contributes to arousal. The experimenter-provided visual stimulation could be expected to neutralize any possible effects of mental imagery evoked by the sound. For example, a pornographic movie could be shown with and without the accompanying sound. The movie images would mask or at least substantially reduce the impact of any mental imagery occurring while watching the movie. One carefully controlled study employing a group of 133 young men, all of them undergraduate students, revealed that the sound accompanying the sexual activities depicted in pornographic movies indeed enhanced physiological arousal above that displayed after watching the same movies without sound. Physiological arousal was quantified by recording penile tumescence with a strain gauge. In addition, the students were asked to rate their level of arousal on a Likert scale. These ratings coincided with the physiological recordings in showing that movies presented with sound were more arousing than the same movies without sound (Gaither and Plaud, 1997). Pornographic movies illustrating different kinds of sexual activities were also compared in the same study. The activities included cunnilingus, fellatio, mutual oral sex, intercourse with the man superior, intercourse with the woman superior and intercourse with rear entry. In other words, many of the sexual activities making up any pornographic movie worth watching were shown. The results showed that fellatio was inferior to mutual oral sex and intercourse with female superior. This last finding does not appear to be particularly important in the present context, but it is a nice illustration of the precision of the analyses of the properties of a stimulus determining its impact on arousal that are carried out in some laboratories. Returning to the role of sound, it can be concluded that auditory stimulation contributes to sexual arousal even in a situation where visual imagery should be of reduced importance. Whether the sounds stimulated other kinds of imagery not brought to mind by the visual stimuli or not is unknown. There are other studies (e.g. Golde *et al.*, 2000) confirming that auditory stimulation, although producing sexual arousal by itself, is inferior to the combination of visual and auditory stimulation.

It could also be argued that the possible confound produced by mental representations of sexually arousing stimuli is exaggerated. Although mental representation, fantasies, can be arousing to the point of orgasm, there are large interindividual differences both in the occurrence of fantasies and in their impact on sexual arousal. A small proportion of individuals does not seem prone to fantasizing at all, and others do it infrequently (Leitenberg and Henning, 1995). Several studies have determined that mental representation, even in situations where subjects are explicitly instructed to produce fantasies with sexual content, are inferior to external stimuli with sexual connotations. In one study, subjects were exposed to either a pornographic movie segment, pornographic slides, audiotaped descriptions of the activities displayed in the movie segments, written descriptions of them and, finally, the subjects were asked to produce mental representations of exactly the same activities as were exposed in the movie segment. It may also be important to note that the slides actually were frames taken from the pornographic movie segment, meaning that the factual content of all stimuli were identical. Data show that the pornographic movie enhances physiological arousal (evaluated as penile tumescence) more than all other stimuli. Furthermore, looking at slides, listening to an audiotape, or reading were equally effective, and superior to mental representation but inferior to the movie. Although mental representations did produce some arousal, their low effectiveness compared to the other stimuli suggests that they do not mediate the effects of these other stimuli on sexual arousal.

Personally, I think it is premature to conclude that mental representations, fantasies, are unimportant for sexual arousal produced by external stimuli. The vividness of the mental representations may be heavily influenced by external stimuli, and that may be of some importance for the arousal response. At the same time, the position that external stimuli act by activating mental representations of a sexual nature and that these representations then activate some physiological processes leading to enhanced genital blood flow introduces an unnecessary complication. It is more parsimonious to pose that the external stimuli directly activate some physiological mechanisms leading to enhanced genital blood flow without the detour over representations. However, it is quite unlikely that visual stimuli in the form of printed or handwritten letters directly lead to sexual arousal. Likewise, an auditory stimulus consisting of someone reading a pornographic story is unlikely directly to produce sexual arousal in a listener. In both these cases, it is almost certain that the verbal message leads to mental representations, perhaps in the form of imagery, which then produces arousal. Otherwise we are forced to assume that the visual stimulus properties of a text with pornographic content differs, in some perceptible way, from the visual stimulus properties of, for example, a recipe for making a pumpkin pie.

Regarding an auditory stimulus, we would have to assume that the sonoric qualities of a pornographic story are perceptibly different from those of a travel story. This possibility appears extremely unlikely, forcing us to accept that sexually arousing verbal messages acquire their arousing properties via the activation of mental representations. A possible compromise between the notion that all stimuli producing sexual arousal need to activate mental representations with a sexual content in order to do so and the contrary notion of a direct connection between stimulus and arousal is to assume that some stimuli, like verbal accounts with arousing effects, act via mental representations while other stimuli, like a silent pornographic movie, directly stimulate sexual arousal.

The capacity of certain stimuli to produce sexual arousal in the absence of mental representations is probably most evident in the case of tactile stimuli. For example, mechanical stimulation of the clitoris is a most efficient stimulus for enhancing vaginal blood flow. Reflex connections between the sensory receptors in the clitoris and the clitoral/vaginal vascular system (Shafik, 1995) as well as between sensory receptors in the penis and its vascular supply have been described (see Giuliano and Rampin, 2004, for a review). Women with spinal cord transection at levels above the sacrum will respond to clitoral stimulation with tumescence of that structure and vaginal lubrication (Sipski *et al.*, 2001). Men in a similar condition also experience erection in response to tactile stimulation of the penis (Riddoch, 1917; Kuhn, 1950; Courtois *et al.*, 1993). Women and men with spinal transection do not consciously register the stimulation, meaning that they do not feel it. This makes it extremely unlikely that the mechanical stimulus activates mental representations with a sexual content. Indeed, it is just as unlikely that this is the case as it is unlikely that the visual stimulus characteristics of a pornographic text directly modify clitoral/vaginal blood flow. The inevitable conclusion must be that the compromise hypothesis outlined above is correct or at least far more likely than any of the alternative hypotheses.

A summary of current knowledge of the stimulus control of human sexual arousal is now in place. Humans, women and men, respond to certain kinds of distant stimuli with enhanced genital blood flow. These stimuli may be visual and/or auditory. Olfactory stimuli seem to be of slight importance. The exact characteristics of the visual stimuli that carry their sexually arousing properties have not been determined in any detail, but it is clear that color is of little importance. Black and white movies are efficient, but no direct comparison between such and color movies has been done. Nevertheless, I allow myself to conclude that color is not essential. We also know that moving pictures are more efficient than slides. The reason for the superiority of motion is not known, but it may be speculated that moving pictures look more realistic than slides. It is also indisputable that the theme illustrated in the pictures and movies must be sexual, preferably depicting an explicit copulatory activity such as vaginal intercourse in varying positions, fellatio, cunnilingus or anal intercourse. Geometric figures or movies depicting landscapes or flower gardens are inefficient. Concerning verbal messages, either read or listened to, the content needs to be sexual if arousal is to be produced. This is probably a triviality and I will not insist upon the point. Non-verbal auditory stimuli are probably also efficient. Humans produce a lot of sounds in many contexts, and at least some humans vocalize intensely while displaying copulatory activities. During copulation, there are also sounds coming from other parts of the body than the mouth or nose. I do not think of vulgarities such as farting, but the movement of the genitals in and out of body orifices also produce sounds of varying intensity and quality. Paraphernalia, such as the bed or dining table, may also make rhythmic noises during copulatory activity, to much amusement for the neighbors and to the dismay of the sound producers. All these non-verbal sounds associated with sexual activity might have arousing properties. This possibility has not been unequivocally evaluated, but the fact that the addition of non-verbal sounds to a pornographic movie enhances its arousing effect suggests that these kinds of sounds indeed are arousing.

Some proximate stimuli, here meaning touch stimuli, are sexually arousing. There are innumerable descriptions of the use of tactile stimuli for producing sexual arousal in the popular literature and many in more serious works. The arousing properties of tactile clitoral or penile stimulation have probably been known since the dawn of mankind. In the 1920s, this old knowledge entered into some marriage manuals, which recommended sexually inexperienced young men to stimulate the clitoris of their equally inexperienced wife before penetrating her vagina. The purpose was to enhance her enjoyment of the marital duties. Curiously enough, it was even recommended that the clitoral stimulation should be performed with the mouth and tongue. In some countries, the application of this advice would have been a criminal offense. More socially acceptable forms of arousing tactile stimulation are non-genital mouth-to-body contact and mouth-to-mouth or tongue-to-tongue contacts. All these behaviors frequently go under the label 'kissing'. I have never seen any laboratory study of the genital response to kissing, but anecdotal evidence suggests that this activity in fact may be arousing. I will not analyze non-genital tactile stimulation, but it is probably safest to conclude that it may be sexually arousing, at least in some cultures.

If we were to try to establish the order of arousing potency of different kinds of sexual incentive stimuli, we would probably find that genital tactile stimulation is the most efficient. The order of arousing potency of other kinds of tactile stimulation and of distant stimuli is unclear.

We have now briefly summarized current knowledge concerning the stimulus control of sexual arousal in the human. I have retained the old term 'arousal' throughout this chapter, although we learned in Chapter 2 that it has been replaced with the term 'excitation', at least when talking about the sequence of human sexual behavior. I have done so for a very simple reason: in the clinical literature, the term 'arousal' is consistently employed when talking about enhanced genital blood flow in women. Lack of genital response to sexual stimuli is actually named 'female sexual arousal disorder'. In men, the equivalent condition used to be called impotence, but the name erectile disorder has become increasingly popular. I do not know why the term 'arousal disorder' is not employed when talking about men, although the problem referred to is a problem of arousal. Furthermore, the cause of impotence as well as of lack of clitoral/vaginal response to sexual stimuli is lack of enhanced genital blood flow both in women and men. Thus, it might seem reasonable to use the same term for the same disorder in both sexes. Why this is not the case could perhaps be explained by sociologists or feminists. I have an additional reason for having preserved the term 'arousal'. The phase of the human sexual response cycle now called 'excitation' encompasses events immediately preceding actual copulation *and* copulation. In the Masters and Johnson (1966) model, the 'arousal' phase seems to *precede* copulation. In the preceding pages, we have mentioned many studies of 'arousal' where copulation never took place. Listening to pornographic stories, or watching an adult movie (using a very stupid American euphemism) was never followed by copulation in any of the studies mentioned. 'Arousal' was studied as an isolated phenomenon, not as part of a sequence of events. 'Excitation', as it is understood in the discourse on human sexual behavior would, therefore, have been inadequate.

We will now turn to an analysis of the stimulus control of events preceding arousal. The arousal response was described as a visceral response to sexual incentive stimuli. Such stimuli also activate somatic processes. Salient among those are the display of approach behaviors, leading to establishment of initial contact with a potential mate, as mentioned many times already. Very little is known about these behaviors and their control by external stimuli or mental representations. Whereas our knowledge of sexual arousal is sufficient for presenting some informed speculations and a few firm conclusions, we can do nothing of the sort with regard to approach to a potential mate. The following section is, consequently, essentially a work of fiction. I am deeply embarrassed over this sad fact and I decided to put that section after the discussion of sexual arousal with the hope that most readers will be exhausted by now and prefer to immediately jump to the next chapter, which is far more exciting.

3 Human sexual incentive stimuli: approach behaviors

3.1 Visual incentive stimuli

For years I have tried to find some kind of reasonably accurate data on what are or may be the sexual incentive stimuli activating approach in the human. Some sociologists and others have ventured to speculate about the factors determining the attractiveness of humans, but they have not focused on the subject at hand here. Rather they have studied the factors that may make a human attractive for marriage and that is something entirely different from the factors that determine sexual incentive value. There are, however, a couple of studies that have evaluated the role of visual stimuli in the form of body shape for attractiveness. I seriously doubt that the attractiveness of a certain body shape is a main determinant of the sexual incentive properties of the individual having that shape, but I will briefly mention those studies anyway. What I will not mention here are the speculations emanating from evolutionary psychology, since they probably are totally irrelevant for the issue at hand and for any other issue for that. They will be formally dismissed in a later section.

A few rather old studies tried to match men's preferences for certain female body types with personality characteristics. They did not directly address the issue of sexual incentive properties of the body shapes, but the aspects of body shape that were evaluated seem to be those considered sexual incentives at the time and perhaps also at present. These aspects were breast size, buttock size and leg thickness. Furthermore, it was maintained that contemporary society, that is American society in the 1960s, classified men as Breast Men, Buttock Men and Leg Men, according to which part of the female body they gave most importance. In an amusing study, men were asked to make pairwise comparisons of all possible combinations of silhouettes of nude women with varying breast sizes, buttock sizes and leg thickness. The results were subjected to a most complex statistical analysis, which allowed for the identification of five groups of men. The first group is those preferring women with ample breasts, small buttocks and moderate legs. These men were frequently of working class origin with definite career plans. They are affiliative and friendly and read many magazines. The second group liked everything moderate, breasts, buttocks and legs. Members of this group are helpful and free of minor vices and they do not read *Playboy*. The third group likes women with large breasts and legs, but with moderate buttocks. Men in this group are disorganized, feel guilty and inferior and are distractible and non-persistent in their work. A college career for these men is unlikely to be associated with success. The fourth group likes women with moderate breasts, small buttocks and sturdy legs. Women with that body shape do not appear to be particularly attractive to me and the men belonging to this group are unfriendly and non-affiliative. They were breast fed as infants. The fifth and last group preferred large breasts, moderate buttocks and long, slender legs. Men in this group smoke and tend to heavy drinking, are non-conformists and do what they like. They prefer their fathers to their mothers (Wiggins *et al.*, 1968; Wiggins and Wiggins, 1969; Wiggins, 1971). It is evident that these kinds of data are essentially irrelevant to the issue of how potential sexual incentives activate approach behaviors and to the issue of what these incentives are.

We may find some of the comments taken from the Wiggins papers ludicrous. However, they probably reflect the dominant social prejudices of the time in the same way as contemporary research reflects dominant prejudice of our time. We do not notice this until time has passed, because humans have the strange tendency to interpret the world according to the prejudices with which we live, believing that they represent unchanging facts of life.

Some more recent studies have also exposed men to drawings of nude females and determined the attractiveness of additional body features, like a thin waist or the waist to hip ratio in addition to the traditional breasts and buttocks (Horvath, 1979, 1981; Singh and Young, 1995; Furnham *et al.*, 1998; Maisey *et al.*, 1999). None of these efforts offer any information more useful than what we can extract from the Wiggins studies. However, they illustrate that our knowledge of the sexual incentive properties of stimuli stems entirely from studies of the viscerosomatic responses to them. The stimuli responsible for the initial activation of the desire to engage in sexual activities, the phase preceding excitation or arousal in the human sexual response sexual cycle, remain unknown.

3.2 Chemicals as human sexual incentives

General comments

In old times, psychologists excelled in creating extravagant theories of human behavior based on a series of hypothetical, usually unverifiable, constructs. Among those theories we find William McDougall's analysis of human motivation (McDougall, 1914). The concept of inherited instincts, innate tendencies to direct attention to particular stimuli and to respond to them for the purpose of attaining some goal, was basic in the theory. The concept of instinct became popular and psychologists elaborated endless lists of instincts thought to determine human behavior. McDougall's theorizing showed some characteristics typical of German and Japanese fascism and soon fell in disrepute. So did the instinct idea.

Tragically enough, the concept of instinct has experienced a renaissance in later years. It will be discussed in the last section of this chapter. However, some experimental studies of potential sexual incentives implicitly based on the notion of instinct will be mentioned here, since we are presently discussing experimental data as opposed to speculation.

In fact, among the many rather exotic hypotheses concerning innate reaction tendencies to stimuli are those suggesting that a human being's sexual motivation can be enhanced by a class of chemicals known as pheromones. In order to appreciate this hypothesis, we need to put it in the appropriate context. As always, we begin with a definition. The original definition of pheromone, 'substances which are secreted to the outside by an individual and received by a second individual of the same species, in which they release a specific reaction' (Karlsson and Lüscher, 1959) is probably unsurpassed. More recent definitions have added some vagueness and that is particularly evident in the non-scientific literature. For example, according to *Wikipedia*, the free encyclopedia, 'a pheromone is any chemical produced by a living organism that transmits a message to other members of the same species' (http://en.wikipedia.org/wiki/Pheromone, retrieved March 18, 2006). This definition is seriously lacking in at least one aspect. It is unclear how we possibly could know what the message being transmitted is unless it has some physiological or behavioral manifestation in the receiving organism. If there is no overt manifestation, we cannot even know whether any message at all has been transferred or not. In the scientific literature, the pheromone concept has also become more vague (see e.g. Wysocki and Preti, 2004 for a good review). The original definition currently applies only to two of four classes of pheromones. One of these is releaser pheromones, chemicals modifying the behavior of the receiver. The other is priming pheromones, chemicals that primarily affect neuroendocrine functions. The endocrine effects may eventually translate into modifications of behavior. Then we have signaler pheromones, chemicals signaling kinship, place in a hierarchy, degree of parasite infection and many other things. Finally, there are modulator pheromones, chemicals having actions on mood or emotions. The experimental evidence for these latter two classes of pheromones is of variable strength and not necessarily sufficiently strong for convincing a critical mind that it is meaningful to make these additions to the pheromone notion. We will, then, stick to the original definition.

Pheromones have been described in many invertebrates and vertebrates. The molecules acting on the vomeronasal organ of hamsters and rats, as well as the airborne chemicals attracting a female to a male, are examples of mammalian pheromones. Their sexual incentive properties have been established in many convincing ways and I have presented data showing that these properties are not dependent on learning. We must, therefore, accept that some mammals show innate, or instinctive if we like that term, reactions to some chemicals. Personally, I prefer to say that some mammals present unconditioned responses to some chemicals. These chemicals are, consequently, unconditioned stimuli. The advantage of these terms compared to the term 'instinct' or 'innate reaction tendency' is that we can place both the stimulus and the response within the context of behavior theory or learning theory if we want to. By doing so, we gain access to a substantial theoretical and empirical framework within which we can understand, and put to experimental test, the functioning of these stimuli and responses. There is no usable theoretical or empirical framework for 'instincts' or 'innate reaction tendencies' allowing us to test hypotheses concerning their function or even to prove their existence.

Pheromones are expected to act on the vomeronasal organ. This expectation is completely unfounded, but it is, nevertheless, held by many scientists and by almost all those considering themselves knowledgeable without being scientists. In order to understand this common misconception, we have to take a short look at recent history. An influential review (Wysocki, 1979) made a strong case for the accessory olfactory system as a hard-wired arrangement ensuring that the receiver of a pheromone would react in a specific way, independent of momentaneous variations in external or internal states. Furthermore, this system was considered as projecting to central nervous areas concerned with reproductive events. A role for the main olfactory system in responses to pheromones was considered unlikely. Since the review included substantial experimental support for this notion, it received wide acceptance, and soon came to be regarded as an undisputable fact. Indeed, in a splendid paper published by one of the prominent scientists in the field it was stated that pheromones activate the vomeronasal system, thereby eliciting a characteristic array of innate reproductive and social behaviors (Dulac, 1997).

Only recently, has this notion been challenged. It is now evident that the main olfactory system also responds to airborne chemicals satisfying the criteria for being called pheromones (Baxi *et al.*, 2006).

Because of the fact that pheromones can be airborne molecules acting at receptors in the main olfactory system, we should abandon the association between actions in the vomeronasal system and pheromonal activity. Nevertheless, the notion that pheromones *must* act on the vomeronasal system has become deeply rooted and it is apparently not easy to abandon. Those attracted by the idea that chemicals with pheromonal activity are important sexual incentive stimuli in the human have gone to great lengths in trying to show that humans are equipped with a functional vomeronasal system. Some of this research will be reviewed in the following section.

In addition to the question of the existence of a vomeronasal organ in the human, there is another issue that has been ignored by the proponents of the pheromone–vomeronasal system as a mediator of sexual incentive properties of chemical stimuli. We know from Chapter 2 that airborne chemicals do not activate sensory neurons in the vomeronasal organ. Direct physical contact between the nose and the chemical in solution seems to be necessary. This poses no problem in many non-human species where sniffing of the partner, including the genitals, is a common part of the prelude to sexual interaction. However, sniffing of the partner's genitals is a most irregular behavior in the human. Even sniffing of other parts of the human body is not a regular feature of human social interactions. In fact, it is not unreasonable to propose that humans sniffing each other have already established some intimate contact. Considering this, it is extremely unlikely that a possible pheromone acting on the vomeronasal system has any role in approach behaviors. It could have some role in arousal processes occurring late in sexual interactions. If so, however, their importance must be marginal if any, since we know that other kinds of stimuli efficiently promote arousal in the absence of olfactory stimuli. This does not exclude, in any way, that odors may become conditioned sexual incentives. However, this is completely independent of any pheromonal properties, since such properties are regarded as unconditioned, inborn or innate.

Speculations about human pheromones have influenced some research but, in particular, they have given rise to many myths concerning the importance of certain chemicals as sex attractants (a popular expression for sexual incentive). It would be premature to enter a discussion of sexual incentive functions of chemicals and of the vomeronasal organ without some mention of the possible existence of it and of an accessory olfactory system in the human.

Is there a functional accessory olfactory system in the human?

Human vomeronasal cavities were described in 1703 by the Dutch anatomist Frederic Ruysch. A reproduction of a drawing of his can be found in Trotier *et al.* (2000). More recent studies have shown that about 70% of human corpses have a vomeronasal pit at least on one side of the nasal septum. Endoscopic examination of a large number of living humans revealed a similar proportion of individuals with at least one vomeronasal pit (Trotier *et al.*, 2000). This is certainly an interesting observation, but it would be still more interesting if it could be shown that these pits contained sensory neurons and that these neurons sent axons into the brain as occurs in many other mammals. It would also be interesting to know if molecules from the exterior had any chance of entering the mucus-filled pit, in other words if the pumping mechanism described in other mammalian species also exists in humans. The evidence is negative with regard to all this. Several studies have failed to detect neurons in adult vomeronasal epithelia and there is no trace of axons leaving the vomeronasal epithelium (reviewed in Meredith, 2001). Even if there were, these axons would have nowhere to go, since the human does not have an accessory olfactory bulb (see Meisami and Bhatnagar, 1998). Finally, there is no indication of the existence of mechanisms that could pump liquids in and out of the pit (Trotier *et al.*, 2000). The inevitable conclusion seems to be that there is no functional vomeronasal organ or accessory olfactory system in humans.

This notion is reinforced by a study comparing responses to androstenone in humans with and without a detectable vomeronasal organ. Furthermore, in those with a detectable organ the opening was sometimes covered and sometimes not. No difference in response was observed between subjects with and without a vomeronasal organ and there was no difference between tests with uncovered and with covered opening in those having a vomeronasal organ (Knecht *et al.*, 2003). I cannot resist mentioning, albeit parenthetically, that androstenone is a steroid with putative pheromone function, at least in pigs, suggested to be helpful in the context of artificial insemination of sows (Reed *et al.*, 1974). It has been the favorite test compound in studies of pheromone communication in humans. It is, at least to me, somewhat unclear why scientists interested in human pheromones have decided that humans and pigs should respond to the same kind of chemical. No insinuations are appropriate here. Nevertheless, the absence of a functional vomeronasal organ in humans seems to have been established beyond reasonable doubt. This is not surprising, since it is also absent in many other primates including most old world monkeys and hominoids (Meisami and Bhatnagar, 1998; Smith *et al.*, 2001; Zhang and Webb, 2003).

The consequences of this are not particularly serious, since chemicals can act in the main olfactory system and have pheromonal activity anyway. Nevertheless, the absence of a functional vomeronasal organ in the human is a catastrophe for those insisting on an obligatory association between that organ and pheromonal activity.

I have dedicated many lines to the discussion of the human vomeronasal organ. My intention with that was not only to add a page or two to this chapter, but also to draw attention to some of the many unfounded myths circulating in popular writings.

Sexual incentive properties of odors

While chemicals certainly possess unconditioned sexual incentive properties in non-human mammals, the existence of unconditioned, chemical sexual incentives in the human is uncertain. Similarly, the existence of conditioned olfactory stimuli with sexual incentive properties in the human is unknown. While we know that visual or auditory stimuli with sexual content enhance genital blood flow, there is no published study showing that an olfactory stimulus can have this effect. The closest approximation is a study where young women, as always undergraduates, were asked to read either an explicit account of a sexual encounter between a man and a woman on a train or a story about a French journalist. There was no sexual content in the story about the journalist. Half of the subjects were invited to wear a mask impregnated with 300 μg of androstenol and the other half wore a control mask. The mask covered the mouth and the nose, assuring that the steroid reached the olfactory epithelium. After having finished reading, the subjects answered a questionnaire evaluating sexual arousal and some other mental events. Results showed that androstenol had no effect on self-rated sexual arousal independently of the reading material provided. Reading the pornographic text produced, as expected, a higher level of arousal than the neutral text. The conclusion was that androstenol did not influence sexual feelings (Benton and Wastell, 1986). An interesting aspect of this study was that the chemical may have come in contact with the vomeronasal organ, if it had existed.

In another study, 39 young men and 39 young women, all undergraduates at a Canadian university, watched a 15 min slide show together with an opposite sex confederate. The confederate wore either a few drops of an androstenol solution applied on each wrist and behind each ear, or a synthetic musk applied on the same places, or no odor. After the slide show the experimental subjects were asked to answer a questionnaire and to rate the attractiveness of the partner. Finally, the subjects were asked about the purpose of the study and then subjected to an odor test. None of the subjects was aware of the real purpose and all subjects included in the analyses were able to detect the odor of androstenol. Data showed that androstenol as well as musk failed to affect the ratings of the partner's attractiveness. The conclusion of this interesting study was clear-cut: 'The failure to find positive effects under these circumstances suggests that it is premature to classify androstenol as a human pheromone, the enthusiasm of the perfume industry notwithstanding' (Black and Biron, 1982, p. 329).

The lack of effect of androstenol was confirmed in another quite ingenious approximation to a test of sexual incentive motivation. Groups of homo- and heterosexual young men and women were asked either to arouse themselves with self-produced sexual fantasies or to enhance their appetite with self-chosen alimentary fantasies. The efficiency of the fantasies was determined by recording heart rate rather than genital blood flow, but heart rate is well known to increase during sexual arousal. Indeed, the sexual fantasies enhanced heart rate while the alimentary fantasies failed to do so. After the induction of sexual or alimentary arousal, the subjects were asked to wait for a few minutes in a waiting room with several chairs. One of the chairs had been discretely scented with androstenone. The chair chosen for sitting on during the waiting period was recorded. It was reasoned that if androstenone had sexual incentive properties, then the subjects would chose the scented chair more than the others, particularly since their sexual arousal previously had been enhanced by the fantasies. The data revealed that the androstenone chair was not chosen more than the other chairs regardless of whether the subjects were sexually aroused or not (Pause, 2004). Again, the conclusion must be that the steroid does not have any sexual incentive properties. Most humans are able to smell androstenol and androstenone (Gower and Ruparelia, 1993), so the failure to find an effect cannot be due to sensory incapacity.

The three studies described above are the only ones I know of that have tested the hypothesis that an olfactory stimulus has sexual incentive properties in the human employing procedures that afford some scientific credibility. For the sake of completeness, I will mention a fourth study, although it does not offer much useful information. However, it can be employed as an example of the convoluted ways in which some enthusiasts design their experiments. As usual, undergraduate students in psychology or biology were used as subjects. There were 76 of them, 38 men and 38 women. They were asked to wear a perforated necklace containing a cotton swab soaked in an androstenol solution from the afternoon of one day until 9.30 in the morning of the next day. When showing up at 9.30 to return the necklace, the students were invited to answer a questionnaire and to fill out a diagram consisting of three concentric circles. The central circle depicted the subjects and the second one depicted others. The space between this second and the third circles was divided into segments and each segmented was labeled with some category of relationship to others, like relatives, friends, fellow students etc. This appears to have had some decorative function, because these relationship categories were never mentioned again. The subjects were asked to recall all verbal interactions they had had since waking up that morning and mark them with an arrow, the head pointing to the one that initiated the interaction. It turned out that women wearing the androstenol necklace initiated more verbal exchanges with men than women wearing the control necklace. There was no necklace-effect in men. The results were interpreted as showing that women exposed to the androstenol solution were more responsive to men than women in the control group (Cowley and Brooksbank, 1991). Whether the number of verbal exchanges early in the morning has some sexual significance or not is impossible to determine. Just to stimulate curiosity, I mention that the rationale for the authors' conclusion will become explicit in a few paragraphs.

There are many additional curious studies of the influence of chemicals on human behavior and even on sexual functions, but none of them is of any relevance to the problem of olfactory sexual incentives, or the role of pheromones for human sexuality. Nevertheless, I will mention one of these studies as a kind of entertainment. The scientists spent five weeks observing the choice of stalls in a rest-room for men and in another for women. Some weeks half of the stalls were treated with a solution containing androstenol and the other half, the control stalls, were treated with androsterone, a steroid without known pheromonal activity. Compared to baseline weeks, men avoided the androstenol-treated stalls during treatment weeks. Women did not show any change in choice between control and treatment weeks. These data were interpreted as showing that androstenol acts as a spacing pheromone in men (Gustavson *et al.*, 1987). Since the subjects evidently avoided the androstenol treated stalls, androstenol must be a negative incentive, at least according to the concepts of incentive motivational theory. Although this fascinating rest-room study is of no direct relevance for sexual behavior, its results make it difficult to imagine that androstenol would be a sexual incentive.

Some comments must be made on the choices of olfactory stimuli employed in the studies mentioned and in the many not worth mentioning. We have already seen that the androgenic steroid androstenol and its derivatives are popular. The reason for this is that odors produced by boars are known to enhance the proceptive behaviors of sows. The abundant saliva excreted by sexually excited boars was suspected to contain the relevant odor and it was soon found out that two steroids, produced in the testes and excreted by the submaxillary glands, androstenol and androstenone, were the critical molecules (Melrose *et al.*, 1971; Signoret, 1976). The capacity of these compounds to stimulate pig sex has been confirmed on many occasions and they are even commercially available in the form of an aerosol spray named *Boar Mate* (Knight, 1985). So far, neither androstenol nor androstenone has been shown to have similar effects in any other species. The fact that some scientists believe that humans are more similar to pigs than to other mammals is not easy to explain. Nevertheless, the belief in some communities between human and porcine behavior is nicely illustrated in the Cowley and Brooksbank (1991) androstenol study. After having found that androstenol enhanced women's verbal exchanges with men, they most reasonably concluded that this effect was caused by an increased responsiveness to males. In the following paragraph, they remark that androstenol is known to rouse estrus sows to be sexually receptive and they continue by suggesting that there is much to be gained from a comparative approach. In fact, their study was based on the notion that data from sows could be immediately generalized to women. Whether this is to be taken as a compliment to the fair sex is uncertain.

A favorite odor source in human pheromone studies is sweat collected from the armpits. We can probably all agree that the odor produced by non-deodorized armpits is strong and salient, but everyday experience insinuates that it is not particularly attractive to potential sex partners. Nevertheless, there are those who have tried to show that armpit sweat is a positive sexual incentive. A concoction with a composition purportedly similar to underarm secretions was provided to 36 women. They were instructed to apply the concoction under the nose, on the cheeks and behind the ears at least once every other day for 6 weeks. Subjects recorded their sexual activities in a diary and they were also interviewed a few times during the study. The concoction was found to increase the frequency of sexual intercourse and formal dating. The authors concluded that the concoction acted as a sex attractant, enhancing the women's sexual attractiveness (McCoy and Pitino, 2002). In the language of incentive motivational theory, this means that the concoction had sexual incentive properties. A similar study, this time using postmenopausal women (mean age was 57 years) as subjects, reported that the same concoction enhanced the frequency of petting and kissing. No effect on the frequency of intercourse was found in these older women. Again it was concluded that the concoction was a sex-attractant (Rako and Friebely, 2004). Another study, performed in young men, showed that another concoction prepared specially for men, this time employed as aftershave, enhanced the frequency of sexual intercourse and of 'sleeping next to a romantic partner'. It was concluded that the concoction increased the sexual attractiveness of men (Cutler *et al.*, 1998). This means that it functioned as a sexual incentive. It might seem, then, that chemicals similar to those found in the armpits are sexual incentives both for men and women. These data are not unproblematic, though. First, a re-examination of the data from the McCoy and Pitino (2002) and the Cutler *et al.* (1998) studies revealed that there was no increase in any sociosexual behavior in the groups carrying the concoction in either of the studies. The significances reported in the original reports were due to changes in the behavior of the control groups. Thus, 'the claim in both studies that the action of pheromones is through an increase in attractiveness to the opposite sex depends on unconvincing assumptions and is at best premature and receives no support in data' (Winman, 2004, p. 701). Second, the composition of the underarm secretions produced by men and women has been shown to be similar (Zeng *et al.*, 1996). If these secretions are chemically similar, then their perceived odor should also be similar. Men and women should, then, produce exactly the same 'sex attractant' and men should be attracted to other men just as much as to women and women should be attracted to other women just as much as they are attracted to men. A third problem comes from the fact that humans invest considerable resources in getting rid of armpit odors. In the late 1980s, Americans alone spent 1.6 billion dollars per year on products to eliminate or mask such odors (Ramirez, 1990) and that figure has probably not been reduced since then. If axillary odors indeed enhanced our attractiveness, there should be no reason for getting rid of them. Particularly not considering that we also invest huge sums in perfumes and other liquids with the purpose of enhancing our odorous attractiveness.

Almost absent from human studies of chemicals as sexual incentives are the only group of compounds that could be predicted to have that function. A long time ago it was found that vaginal secretions from women contain a number of aliphatic fatty acids such as propanoic, methylpropanoic, butanoic, methylbutanoic and methylpentanoic acid (Michael *et al.*, 1974). Most interestingly, the concentrations of these volatile fatty acids had their maximum during the late follicular phase (Michael *et al.*, 1975). Furthermore, the concentrations of them were rather similar to what had earlier been reported for several primate species (Michael *et al.*, 1972). The attentive reader may also remember that these acids have been found to enhance sexual arousal and copulatory behavior in nonhuman primates (see Chapter 3). In my humble judgment, it is surprising that scientists have tested chemicals working in boars but in no other mammal and stinking armpit products that most humans find detestable instead of carefully evaluating the volatile fatty acids for which some data from primates do exist. One possible reason for this is that the two studies performed in humans offered negative results. In the necklace study mentioned above (Cowley and Brooksbank, 1991), a group wearing a necklace containing the fatty acid mixture typical of the human vagina was included. Neither men nor women reported any change in verbal exchanges as a result of walking around with the necklace. This is not surprising, however. First, there is no reason to believe that heterosexual females should be aroused by typically female odors. Second, there is no reason to believe that the presence of these volatile fatty acids ever was perceived by the men interacting with the women in the study. Verbal exchanges, particularly in the morning, occur with some distance between the participants (the exception being those sharing a bed, but they were probably not many in the study), making it extremely unlikely that the concentration of the airborne fatty acids was above detection limit in the men's nose. Third, we do not know if the number of verbal exchanges with women is the most sensitive indicator of sexual incentive motivation in men. Thus, I conclude that the Cowley and Brooksbank (1991) study is inconclusive.

More conclusive data were obtained in a direct evaluation of the effects of vaginal odors on self-reported sexual motivation and frequency of intercourse. Sixty-two young couples were paid $1.00 per day for participating in the study. Perhaps that sum had some significance back in 1978. The women's task was to apply a mixture of aliphatic fatty acids similar to the one found in the vagina of women and apes to the skin of the chest at bedtime. Control odors were alcohol (the solvent used for the fatty acid preparation), water and a perfume called Heather. The women did not know which scent they applied. The participants were asked to keep a record of the time of scent application and the possible occurrence of intercourse. Thirty-four couples also answered a questionnaire purportedly evaluating sexual motivation, sex play and the occurrence of orgasm. There was no effect of vaginal odors or of any other scent on any parameter (Morris and Udry, 1978). This apparently well-controlled study strongly suggests that vaginal secretions do not function as sexual incentives in the human.

I mentioned in Chapter 3 that a biologically irrelevant novel odor stimulated sexual arousal and copulatory behaviors in male monkeys as much as vaginal secretions did. An entertaining experiment evaluated the possible effects of this kind of odor on the attractiveness of women. Male psychology undergraduates were told that they participated in a study of factors determining the 'first impression of others'. Each male was paired with a female confederate and subjected to an interview during which the subject was sitting side by side with the confederate. Completely non-controversial questions like, 'What are your favorite leisure time activities?' were asked and the confederate was instructed to give equally non-controversial answers according to a memorized script. After the interview, the subject (and the confederate) was asked to answer a questionnaire evaluating the attractiveness of the partner. Some confederates wore a perfume with the poetic name of Jungle Gardenia and others wore no scent. Other psychology students had rated the perfume as pleasant and olfactory potent. Some of the confederates were neatly dressed in blouse, skirt and hose while others were dressed in the typical student uniform, jeans and a sweatshirt. Results show that the scent made the informally dressed women appear more 'romantic' than the unscented. The opposite was true for the neatly dressed women. Regarding attractiveness, the only effect obtained was an interaction between scent and dress in the way that informally dressed women were more attractive if smelling of perfume, while the neatly dressed women were less attractive when perfumed. My conclusion is that you had better smell good if you are lousily dressed and better avoid it if your dress is cute. The author prefers a more conservative conclusion, namely:

> while pleasant aromas sometimes exert positive effects on social behavior, they do not always do so. Thus, contrary to what advertisements often suggest, unquestioned faith in the benefits of perfume, cologne, and similar products does not seem justified (Baron, 1981, p. 616).

I find no reason to disagree. This study illustrates quite neatly that non-natural odors may enhance attractiveness, or they may reduce attractiveness, depending on circumstances. Results from this study are just as unconvincing as those of the studies mentioned in the preceding paragraphs.

The preceding brief presentation of some data pertaining to the question of olfactory sexual incentives in the human should have made it clear that there is no evidence for any functionally important role for olfaction neither in sexual arousal nor in processes related to approach to a potential mate, or copulation for that matter. If we were extremely open minded and assumed, without having any good reason for it, that attractiveness is equivalent to sexual incentive properties, then we would have to conclude that either the role of olfactory stimuli for these properties in the human is so complex that it escapes our comprehension or that such stimuli are unimportant. My personal inclination for parsimony makes me stick to the latter alternative.

4 Human sexual incentives outside of the laboratory

4.1 General

What I have presented in the preceding sections is, evidently, a series of studies that were performed in the laboratory or in other situations where actual sexual interaction with the stimulus object or objects was out of the question. This situation is slightly similar to that found in people reading pornographic magazines or watching pornographic movies. Sexual interaction with the individuals described or depicted is impossible. A certain body shape can be judged attractive for many reasons, in the same way as a painting or a piece of furniture may be found attractive for a host of different reasons, and perhaps also for no conceivable reason at all. An important question is if the features of the human body found attractive in the laboratory also are authentic sexual incentives in the way that they activate sexual motivation in a setting where sexual contact would be possible. We know nothing of this. Speculations about human sexual incentives based on laboratory judgments of attractiveness are nothing more than risky assumptions of unknown veracity.

4.2 Dorothy Tennov's concept of limerence as an approximation to sexual incentive motivation

One of the few scientists who can provide some useful information as to how human sexual incentives may act in the initial stages of approach behaviors is Dorothy Tennov (Tennov, 1979). At the outset it must be made clear that her account is based on interviews rather than on direct observations of behavior. She interviewed a large number of persons (more than a 1000) over a period of several years and used her skills as an experienced psychologist for extracting meaningful information from these interviews. However, descriptions of past events or emotional experiences given in an interview do not necessarily coincide with reality. The memories of past events or emotional reactions may have been transformed in several ways unknown both to the interviewee her/himself and to the interviewing scientist. In a situation where we have almost no usable knowledge at all about how human sexual incentive motivation works outside the laboratory, we are forced to accept that even preliminary, imperfect notions may be helpful. In the best of cases, these kinds of uncertain notions can help us in formulating testable research hypotheses. The formulation of testable hypotheses is a requisite for progress in the field and, despite its shortcomings, Tennov's work may constitute a basis for informed hypothesis formulation. I have found no more appropriate account of the processes by which humans are attracted to each other in the first place, how they continue to be so and how attraction may end.

Although Tennov never used the expression 'sexual incentive', her analysis of a mental state called limerence can be of great interest in the present context. Limerence is defined in the following way:

> When a person, A, is in a state of receptivity and an attractive member of A's sexually preferred category, person B (hereafter referred to as 'LO' for limerence object), is presumed by A to have exhibited, or can be expected to exhibit, amorous interest in A, then A enters a state of limerence with LO as its object' (Tennov, 2001, p. 112).

We can see from this definition that the notion of expectancy is basic. The limerence object must emit some sign or signs that A interprets as indicative of a possibility for successful approach, i.e. a sign or signs giving A an *expectancy* of successful approach. What this sign is (or these signs are) is not mentioned, not even speculated about. Without having any data in support, I pose that the kinds of signs emitted by the limerence object, and their interpretation by the limerent person, are learned through the process of social learning. The existence of some kind of culturally independent, universal signs expressing acceptance or rejection of approach seems very unlikely. Concerning the processes determining the choice of limerence object, Tennov's data do not offer much information. It appears that the limerence object may be 'any attractive member of A's sexually preferred category'. The factors determining attractiveness were of no concern to Tennov. In fact, she considers that the initial contact between A and B basically is a random process.

Equally important as the origin of limerence is the end of it. There are at least three circumstances that efficiently can reduce or eliminate the state. One is the loss of hope of reciprocation by the limerence object. When success seems impossible, the state disappears. Another circumstance leading to a reduction of the intensity of limerence is that the limerence object enters into a committed and monogamous relationship with the limerent person. The effect is reinforced if the limerent person gets convinced that the reciprocation will be sustained. A third condition ending or reducing limerence toward a particular individual is the replacement of this individual as limerence object with another. This process appears to be facilitated if one or both of the preceding conditions are satisfied.

I would like to propose that the state of limerence is the conscious experience of sexual incentive motivation. There is indeed some support for this notion. The limerence object must be a member of the sex which the limerent person prefers for sexual interaction. In our discussion of non-human sexual incentive motivation, I repeatedly made clear that the sexual incentive stimulus must be emitted by an individual with whom sexual interaction is potentially possible. Otherwise the stimulus may activate social motivation rather than sexual. A female rat can be a sexual incentive only when receptive, a castrated male cannot be a sexual incentive, and so on. Likewise, limerence can only be activated by an individual with whom sexual interaction eventually may occur. We must remember that humans can have mental representations of events that have not yet occurred but that may occur in some future. This means that the limerent person's expectancy of future sexual interaction is enough for satisfying the criterion that only a potential mate can activate sexual incentive motivation. Moreover, it is likely that the limerence object activates visceral reactions indicative of sexual arousal, like erection (full or partial) in men and vaginal lubrication in women. Tennov does not provide us with data on this delicate issue, but mentions that sexual desire increases as the intensity of limerence increases. I do not think it is unreasonable to suppose that limerence, as she describes it, is associated with visceral reactions like erection and vaginal lubrication. Introspective data might support this contention, although such data are of no scientific value. Nevertheless, if correct, this point constitutes rather direct evidence for limerence being the conscious experience of sexual incentive motivation.

Additional arguments for considering limerence as a kind of sexual incentive motivation could come from an examination of the factors that end or reduce it. In non-human mammals, the execution of copulatory behavior is known to reduce the intensity of motivation, at least temporally. However, this does not appear to be the case with limerence. On the contrary, a majority of subjects report enhanced limerence following the execution of copulatory behaviors. This contradiction can perhaps be solved taking into account that the experience of limerence, or of human sexual incentive motivation, is partially or entirely determined by social learning. The emotional experiences activated by intimate human relationships are strongly affected by the beliefs about such relationships commonly held by the members of the culture. One such belief is that sexual activity reinforces interindividual emotional bonds. If the limerent person establishes a lasting and committed relationship with the limerence object, the limerent state will disappear, according to Tennov. It is not impossible that a lasting and committed relationship includes sexual activities. In case it does, and in case that sexual reward is obtained when executing these activities, no reduction in the sexual incentive value of the limerence object would occur, as we will see in Chapter 7.

The point to which I have been aiming for a couple of sentences is that, although limerence may be the conscious experience of sexual incentive motivation, this experience is only partially similar to sexual incentive motivation as known in non-human mammals. Learning, particularly social learning, is probably a main determinant of the human experience of sexual incentive motivation. The notion of any innate experience, independent of the cultural context, of sexual incentive motivation in the human is untenable. This means that in contrast to the automatic connection between sensory input and approach responses found in non-human mammals, no similar connections can be expected to be found in the human. However, as we have already seen, there may well be some automatic connections between environmental stimuli and arousal responses.

Assuming that limerence indeed is the conscious experience of something similar to sexual incentive motivation, and knowing that limerence is based on expectancy of successful approach, we are obliged to conclude that human sexual incentive motivation is activated by expectancy of success. This is a prominently cognitive process, based on subjective interpretations of signs emitted by the limerence object. The attribution of meaning to signs such as words, facial expressions or body gestures is the subject of semiotics, and any serious discussion of that issue would be extremely pretentious on my part. What I might dare to suggest, though, is that the interpretation of signs is something we learn. It is also most likely that we learn which signs to emit when we want to attract someone, in the same way as we learn which signs to emit when we want to repel someone. Human ethologists have discussed the possible existence of signs with unconditioned mechanisms of expression and interpretation, like facial expressions of emotions. Their underlying assumption was that both the expression and the interpretation of it were innate, or instinctive, automatic processes established in the brain through evolutionary processes (Ekman *et al.*, 1972; Eibl-Eibesfeldt, 1973). Although rather popular in the 1970s, these ideas seem to have disappeared from the scientific discussion of sexual incentives.

4.3 Human sexual incentives and social learning

The importance of learning for the emission as well as the interpretation of signs with sexual meaning has been stressed by numerous scientists. Freud was a significant proponent of the notion that the expression of sexuality was socially determined. Arguments on this issue are found in many of his books, but a good introduction can be found in his most fundamental treatise on sexuality (Freud, 1905). Sociologists have pushed the analysis of the role of learning still farther, arguing that the sexual meaning given to some human expressions and behaviors is solely a result of learning. This rather radical point of view is part of what is known as social constructionism. An extremely elegant outline of the social constructionist view on sexuality was published in its first edition more than 30 years ago (Gagnon and Simon, 2002; reference is to the second edition) and its influence on social scientists interested in problems of sex has been enormous. The social constructionist approach has been reinforced by the brilliant analyses of sexuality presented by Michel Foucault (1976). He considers sexuality itself as a social construction. According to the social constructionists, the role of biology in the determination of human sexual behaviors is negligible. This point of view is offensive to some neuroscientists, to many American homosexuals and to a few psychologists. The majority in the neuroscience community is simply unaware of the existence of the approach, but those who are aware respond frequently with disdain. Some homosexuals respond to the social constructionist's claim that all of human sexuality, including homosexuality, is a result of social learning with disbelief or outright hostility. Some psychologists, normally without the slightest knowledge of the biological sciences, prefer to cultivate a romantic relationship with renamed instincts and look upon social constructionists as kinds of heretics.

Despite the rather cold reception of social constructionism in the field of sex research, its importance can hardly be overestimated. I will illustrate the force of the approach with the aid of an extremely simple example related to the similarity between sexual and ingestive behaviors as far as social construction is concerned. Ingestion is a part of human behavior that is far more important for survival than sexuality and ingestive behaviors are displayed more frequently than copulatory behaviors. In fact, many humans eat three times a day or even more, while few humans have sex three times a day and still fewer have sex more than that. In addition, we ingest liquids almost constantly. I have personally, in many European and American towns, seen people carrying water bottles attached to their belt, and I have even seen people drinking from this bottle in public. What is the role of biology versus social learning in the control of this behavior? We can first look at *when* food and drink is ingested. Most people seem to get hungry at about the same time in a given location. In Scandinavia, for example, everyone seems to get hungry around 11.30. This, at least, is the time when most Scandinavians have a meal called lunch. In Mexico, few people get hungry at 11.30. In fact, the normal lunch hour is around 14.30. Both in Scandinavia and in Mexico, breakfast is usually eaten sometime between 7 and 8, so the duration of food deprivation between breakfast and lunch is very different. Is there any biological discrepancy between Scandinavians and Mexicans determining that Scandinavians eat 3 hours earlier? Certainly not. I could make a long list of examples, but it should be clear to everyone that the time we eat is socially constructed. We learn to get hungry at a certain time, and what time it is turns out to be unrelated to any known biological phenomenon. We can also ask questions about *what* we eat. In Sweden, lunch consists of a decent meal, a salad or soup followed by an entrée and normally a light desert and a cup of coffee. In Norway, lunch consists of a piece of dry bread with a transparent slice of ham on top and a glass of skimmed-milk. Despite this difference in behavior, Swedes and Norwegians belong both to the Nordic race, they live in closely similar climates, and so on. What biological difference could be the cause of the different composition of lunch? I have no idea. We all know how abundant the British breakfast is and of what it is made up, and we all know that the French or Italians eat a miserable piece or two of sweet bread for breakfast. Are there any basic biological differences between British and French or Italians determining the different breakfast behaviors?

In northern Sweden, a fish fermented for several months is considered a delicacy. Its odor is putrescent and highly aversive to non-Swedes but very attractive to flies. Every fly in the neighborhood immediately approaches the fish as soon as it is exposed. Some humans (notably Swedes from northern Sweden) consume this stinking food voraciously, while an overwhelming majority rejects it as unsuitable for human consumption. The biological factors behind these behaviors remain unknown. I suggest that most, if not all alimentary incentives are learned, independently of whether the food has positive or negative incentive properties. Finally, we need to ask how we eat, i.e. the alimentary behaviors displayed. In many Scandinavian homes, a more formal dinner party has a rather fixed ritual, exactly as it has in most other places of the world. We can start with the table. The tablecloth should preferably be of linen and the porcelain should be from Rörstrand. There should be at least four glasses for each guest; one for water, one for beer, one for aquavit, and one for wine. The glasses should, if the host can afford it, be from Orrefors. The cutlery should either be of silver or from stainless steel produced by Gense. The napkins, of cotton, should be tightly rolled and the roll should be kept tight with a silver ring and placed to the right of the plate. There should be lighted candles on the table and a vase with low flowers. You should always be able to see the guest in front of you over the flowers. I will stop here, but it would be ridiculous to maintain that anything of what I have described here has a biological basis. Likewise, during the good old times when smoking was not ferociously repressed by the health hysterics, you could smoke at any better dinner in the UK. However, it was unthinkable to smoke before the port had been served and the queen had been toasted. Despite a considerable effort on my part, it has been completely impossible to figure out the biological determinants of that behavior. It is not entirely grotesque to propose that most ingestive behaviors performed by humans are socially constructed. What we might suggest is that biological mechanisms partly determine the timing impact of alimentary incentive stimuli, and somehow influence the amount eaten at each meal. Thus, while the alimentary incentive stimuli are socially learned, their effects on the organism are determined by activities in the central motive state, which is the name we have given to the biological mechanisms controlling motivated behavior.

Examples very similar to the one just presented could easily be made with regard to sexual behaviors. When, where, with whom, and how we copulate is certainly determined by social learning. The words, bodily expressions like gestures or facial activities, images and other kinds of external animate and inanimate stimuli having a sexual meaning have acquired that meaning through social learning rather than natural selection. I side completely with the social constructionists on these issues. Any other opinion is, according to my humble judgment, founded on a superstitious belief in non-existent brain mechanisms. This means that the stimuli controlling sexual incentive motivation are learned and also that the motor patterns executed during copulation are learned. It seems that the only unconditioned part of human sexual behavior is the connection between tactile stimulation of the penis or clitoris and enhanced genital blood flow. Notwithstanding the basic importance of learning, it must be remembered that the impact of sexual stimuli as well as the propensity to execute motor patterns associated with copulatory behaviors in a particular moment are determined by the central motive state, a biological mechanism. Exactly as I did with the motivation to ingest food, I maintain that biological phenomena in the brain partly determine if and how much we will act upon sexual incentives.

There is no contradiction between the notion that sexual incentives and sexual behaviors are socially constructed (learned) and the idea that the motivational mechanisms determining the likelihood to react to the learned stimuli are influenced by unlearned biological mechanisms. Please note that I write 'influenced by' and not 'determined by' biological mechanisms. A particular stimulus, for example an attractive girl taking off her T-shirt and exposing the naked breasts, would not lead to any sexual behavior on the part of a young man if the event occurred on the beaches of Normandy. The act of taking off a T-shirt and exposing the breasts has no sexual meaning on a public beach in contemporary Europe. If the same girl would take off her T-shirt and expose her breasts in her bedroom, to which she had invited a young man encountered on the beach, it is most likely that the young man would respond to the stimulus with some kind of sexual approach. The young man's central motive state has probably not changed much from the beach to the bedroom, yet a constant stimulus (the girl taking off her T-shirt and exposing the breasts) had different effects in the two situations. The cause for this differential reaction must reside in his interpretation of the incentive stimulus. On the beach, taking off the T-shirt has no sexual meaning, but in the bedroom it has. The interpretation of the stimulus is obviously a result of social learning and probably independent from the central motive state. This means that the impact of a constant stimulus is not only determined by the central motive state, but also by processes of assigning meaning to the stimulus. Psychologists would certainly call these processes 'cognitive'. The label put on them is of slight importance, but it should be clear that the central motive state *influences* the likelihood to respond with a sexual act to a stimulus, but it does not *determine* this likelihood.

5 An unkind note on sociobiology or evolutionary psychology

After having argued for the crucial importance of social learning for the stimulus control of sexual behavior as well as for the motor patterns displayed during its execution, I will now briefly turn to the opposite point of view. Some scientists maintain that at least some sexual incentives are unconditioned, while other scientists seem to mean that most of them are so. With an unconditioned stimulus is, as mentioned many times already, understood a stimulus to which humans have an innate tendency to react in a predetermined way. This innate tendency was called an instinct until the concept came into bad repute many years ago. As we know, the main reason for the decline of the concept was that it does not explain anything at all in a satisfactory way. The postulation of an instinct typically answers the question 'why?' and, as we have seen, this question is no longer a part of science. When an animal or a human did something we could not easily explain and someone asked the question 'why did she react in that way?', the answer 'because of an instinct' seemed almost perfect, except for the fact it did not contain any useful information.

When the concept of instinct had spent a sufficiently long time in oblivion for a new generation to be unaware of its disrepute, an equivalent concept appeared. If someone still persisted in asking 'why did she react in that way?' the contemporary answer would be 'evolution has determined that she should do so because it is adaptive'. As was the case with the instinct explanation, the explanation in terms of evolution does not contain any useful information. However, just as was the case with the instincts, the explanation can never be refuted and it appears convincing to some. A meaningless answer to a meaningless question is perhaps always convincing. The revival of the notion of instinct under the cover name of sociobiology or evolutionary psychology has prompted some speculations concerning innate sexual incentives in the form of particular visual stimuli. Within the tradition of beliefs in innate reactions to some external stimuli we also find the hypotheses concerning sexual incentive properties of some chemicals, a subject already discussed and dismissed. Although I keep any idea of innate reaction tendencies as explanations for human behavior in low esteem, I feel obliged to make some mention of such ideas for the sake of completeness of the analysis of sexual incentives in the human.

In 1964, Hamilton proposed an explanation for the strange fact that most females in some species of hymenoptera (an order of insects including bees and ants, among others) are sterile and grant their entire life to the task of caring for their sisters (Hamilton, 1964a, 1964b). This apparently inadaptive behavior had always been explained as caused by some notion of 'good for the species'. This rather primitive teleology did not satisfy biologists and, when Hamilton brought the attention to the fact that the ants had more genes in common with their sisters than with their mother, and that caring for their sisters was a better way to assure survival of their own genes than having progeny, he solved a quite challenging problem. Another ant expert got the idea that if a principle of evolution, inclusive fitness, could convincingly explain an apparently incomprehensible aspect of ant behavior, then it should also be able to explain mysterious features of human behavior (Wilson, 1975). Indeed, Wilson launched a series of explanations of human behaviors inspired by the Darwinian theory of evolution. Other sociobiologists focused their attention on aggression, sex, parental investment and male and female reproductive strategies (e.g. Trivers, 1972; see also Ruse, 1979, for a discussion). This latter focus has become particularly evident in the branch called evolutionary psychology (Buss, 2004, 2005).

Some of the central proposals in evolutionary psychology may be quite entertaining, as long as not taken seriously. Evolutionary psychologists suggest, for example, that since men make a modest investment in reproduction (a short copulation) compared to that of women (at least 9 months of pregnancy and some period of maternal care), it is most reasonable for a woman to take care of her child, while for a man it is more adaptive to impregnate as many women as possible. He will not care for the children anyway. Thus, a well-adapted woman should be a faithful housewife, caring for her children, while an equally well-adapted man should spend his time pursuing other women with the intention of having sex with them. Furthermore, he should not pursue just any woman, but preferably beautiful, young and healthy girls. Such women assure maximum health to their offspring and thereby maximize the probability of survival of the man's genes. Since the behavior patterns described above are the most adaptive, those individuals displaying them deposit more genes in the population than those not displaying them and, after a sufficient number of generations, those genes have become common to the entire humankind. Thus, all women are faithful housewives and all men are rotten bastards not because they want to be, but because evolution has decided that it is the way it must be.

By some peculiar coincidence, the behaviors that evolution has found most adaptive correspond to the stereotypes of womanly and manly that are, at least tacitly, dominant among the American white, protestant middleclass (see e.g. Garcia, 1983; Sprecher *et al.*, 1987; Gagnon and Simon, 2002, for a discussion of social prejudice or stereotypes). There is no reason to make a sociological or cultural analysis of this fact here. This is a task for social scientists. Nevertheless, it may be convenient to remember that explanations in terms of final cause or purpose, teleological explanations, such as those behind the sex differences mentioned above, are always arbitrary. There exists a myriad of possible answers to the questions of 'why?' and 'what for?' are men more promiscuous than women. The answer we choose is inevitably determined by our personal beliefs. If we believe that male promiscuity indeed is adaptive, we can always create an evolutionary argument for that. Evolutionary arguments are, anyway, untestable, so we do not risk much. We may even invent an algorithm and perform complex calculations showing that the argument works. On the contrary, if we believe that male promiscuity is a result of social learning, we can find excellent arguments for that, and perhaps also some supporting data. The reason for the ease of finding multiple and contradictory explanations is that the initial question is meaningless. 'Why?' and 'what for?' have no place in science. It would be far more useful to find out how sexual motivation and copulation are activated, which constraints society has imposed on their expression, and so on. Then, if someone finds sex differences in promiscuity worth studying, men and women could be compared on relevant variables.

Sociobiology, under the name of evolutionary psychology, has become a frequent topic in popular magazines and in some psychology courses. This unfortunate state of affairs obliges me to dwell on the subject a little more. In addition to the criticisms of the analysis of alleged male promiscuity versus female purity already made, I will mention another popular notion in sociobiology that is equally entertaining and equally illustrative of the role of social stereotypes: the ideal of female beauty or attractiveness. The most attractive female body has a waist to hip ratio of 0.70 (Singh, 1993) according to sociobiology. The body shape associated with that ratio is assumed to be indicative of good health, which in turn is a result of good genes. The crucial importance of that figure for female attractiveness has been confirmed in a couple of empirical studies (e.g. Furnham *et al.*, 1998).

The obvious question is whether a waist to hip ratio of 0.70 is attractive because of evolutionary mechanisms or because of social stereotypes. An examination of one of the most important American periodicals, *Playboy*, could perhaps shed some light on this issue. The young ladies who expose themselves in that publication with an impressive monthly regularity frequently have a waist to hip ratio of 0.70. It has not changed between 1978 and 1998 according to one study (Katzmarzyk and Davis, 2001) and not between 1953 and 2003 according to another (Seifert, 2005). This latter study has also quantified the girls' bust to hip ratio during the same period, without finding any change. Interestingly, although basic body shape had not evolved there was a tendency for the playmates to be more slender in recent years. Slightly different results were obtained from another study of the same photos (Voracek and Fisher, 2002). Body mass index and bust to hip ratio decreased while waist to hip as well as waist to bust ratio increased. The conclusion was that the playmates body mass index has descended below corresponding population levels, whereas the waist to hip ratio has changed in the opposite direction, approaching population levels.

The studies mentioned so far have limited themselves to an analysis of the girls' body shapes without placing them in any wider social context. A different approach to the physical characteristics of playmates has been taken in social psychology. Here, an effort was made to relate the body shape of the Playmate of the Year to social and economic factors that same year. Among the factors included were unemployment rate, disposable personal income, consumer price index, birth rate and marriage rate. It turned out that Playmates of the Year in difficult years were older and taller and had a larger waist to hip ratio than those selected in good years (Pettijohn and Jungeberg, 2004). This might suggest that the ideals of beauty are influenced by socioeconomic factors rather than evolutionary pressures.

There may be many explanations for the real or imaginary changes in *Playboy* girls' appearance from the 1950s to the 2000s. Perhaps Hugh Hefner's choice of playmates has always been determined by changing evolutionary pressures of which he himself probably is unaware. Another alternative is that he and his staff have a sense of what sells, an intuitive knowledge of social stereotypes and their evolution over time. Whatever the answer to that question may be, Hugh Hefner should be quite flattered by the attention his magazine has attracted from so many scientists. Please note that I have only cited a fraction of the many studies of *Playboy*'s centerfold girls. While reading some of those papers, I almost started to consider the possibility of covering the walls of my office with the kind of photographic materials employed in them, for purely scientific purposes. Social stereotypes would associate such pictures with mechanical workshops or the men's section of a luxury public convenience, but by claiming that they are research materials they would nicely fit into any university office.

Playmates and porn stars usually have large breasts, either naturally or with the help of appropriate surgery and plastic materials, and large breasts are more attractive than small ones (Gitter *et al.*, 1983; Singh and Young, 1995; Furnham *et al.*, 1998). Sociobiology, just as the German philosopher Schopenhauer did more than 150 years ago (Schopenhauer, 1844), tells us that it is because large breasts assure abundant food for the infant. Schopenhauer also discussed the role of other characteristics of the female body contributing to its attractiveness, including what we now would call the body mass index. Within sociobiology, the body mass index is an indicator of good genes competing in importance with the waist to hip ratio (Tovée *et al.*, 1999, 2002; Furnham *et al.*, 2005). If nothing else, the similarity between the ideals of beauty or attractiveness put forth by an 19th century philosopher and present day sociobiologists shows that they are firmly rooted in Western culture. Moreover, the fact that Schopenhauer proposed that these characteristics were associated with inclusive fitness, although he did not use that expression, shows that the purported association between physical appearance and favorable qualities is by no means a derivative of evolutionary theory. Darwin's *On the origin of species* appeared 15 years after Schopenhauer had published *Die Welt als Wille und Vorstellung*. Social stereotypes or popular beliefs probably exerted a more profound influence than evolutionary theory ever did on the philosopher's thinking. It might also be illuminating to think about the title Schopenhauer gave to the chapter where he analyzes female attractiveness, *The Metaphysics of Sexual Life*. As a philosopher, he was aware of the fact that his arguments were purely teleological and he put them into the category where such arguments belong, metaphysics. Not all sociobiologists show this exemplary modesty.

The preceding paragraphs should have made it clear that there is a coincidence between social ideals of beauty or attractiveness and the sociobiological notion of favorable human characteristics that evolution has made us like. This coincidence may have many causes. One is that evolution, in fact, has determined what individuals in Western societies are attracted to. Another is that beauty or attractiveness are socially determined, in other words that social learning determines what we find attractive or beautiful. This is just one more example of the eternal and certainly meaningless question of whether our behavior is determined by our genes or by our environment. There is, as always, abundant evidence for each of these positions, and what each one of us believes is undoubtedly determined by personal inclination rather than by some process of reason. Fortunately, there is no need and no way to answer the question in the way it is posed. It is in itself based on many assumptions about the causation of behavior, and we do not know if any of these assumptions is supported by data or not. One of the basic assumptions behind the idea that evolution has made a certain shape of the female body more attractive than other shapes is that the attractive shape is associated with some biological qualities making that body a better mate than other bodies. With better mate is either understood a mate that can transmit genes carrying resistance to disease and all kinds of favorable characteristics or a mate whose maternal behavior will ensure larger survival of the offspring than other mates would ensure or, in the best of cases, a mate that possesses both qualities at the same time.

If we believe that the quest for reproductive success has determined how the individual with whom we prefer to reproduce should look, we need to make some assumptions:

1 Potential mates must provide reliable signs indicating their degree of fertility and the likelihood that they will produce multiple and healthy offspring.
2 These signs must vary in intensity or quality between and within individuals according to the individual's degree of fertility and likelihood of producing multiple and healthy offspring.
3 Potential mates must provide reliable signs indicating their capacity to assure survival of the offspring, ideally until they reach sexual maturity. These signs must have predictive value since the care of offspring will become of importance days, weeks, months or years after the moment of fertilization, depending on the length of gestation and obligatory maternal care.
4 The signs must function as unconditioned incentives, insofar as they must be able to activate approach behaviors on the part of individuals from the opposite sex, and they must do so without any previous experience with them.

This series of assumptions, or conditions necessary for any choice of mate to be meaningful in the sociobiological sense, has the advantage of being, at least in principle, testable in contrast to the question as it was initially posed: is the attractiveness of a certain body shape socially or genically determined? (I use the adverb genically rather than genetically, since the latter word stems from the noun genetics, the science studying heredity, while we do not refer to this science but to genes.) However, as we soon shall see, although the assumptions are all testable in principle, some are not in practice. In the following discussion, I will use the waist to hip ratio as an example of attractive body shape. Exactly the same arguments that will be mentioned can be applied to any characteristic supposedly attractive, be it large feet, small eyes, prominent nose or a loud voice.

The first assumption is that a waist to hip ratio of 0.70 is associated with larger fertility than other waist to hip ratios. This is absolutely necessary, because if the genes carrying the characteristic of being attracted to that waist to hip ratio are to be established in the population, the men having them must have more offspring than men not having them. That, in turn, means that the women having the best ratio also must have more offspring than other women. This is an empirical question. We could, for example, determine if the *Playboy* playmates have more children than women depicted at the before side in advertisements of dieting products. Or we could measure the waist and hip in thousands of young women and determine how many children they have 25 years later. However, such fascinating studies would not answer the question. Contemporary women have the habit of using contraceptives, or they may ask their men to do so. Evidently, the women may also use contraceptives because their man has asked them to. The use of contraceptives could easily bias the results. We know that contraceptive use is influenced by education, income, place of residence, racial belonging, and so on.

Many of the factors determining the frequency of use of contraceptives may also affect the waist to hip ratio. One way to avoid the bias would be to forbid the participants in our study to use contraceptives, but such an approach appears slightly unrealistic. In fact, the worldwide use of contraceptives has made impossible an unbiased study comparing the fertility of women with a waist to hip ratio of 0.70 to that of women with larger or smaller waist to hip ratios. A solution to this could be to use historical data from a time before contraceptives were introduced. Many countries have kept excellent birth registers for centuries. The only problem is that these birth registers do not normally include the waist to hip ratio. It is doubtful whether any do it. Perhaps some old photographs could be used for estimating this crucial ratio, but it is uncertain. In former times people had the unexplainable habit of being dressed when they went to the photographer and on the occasions where photographs were taken, like weddings or birthdays. Clothing makes any estimate of waist to hip ratio uncertain. Fortunately, this problem was not encountered by those using *Playboy* photographs. It is a tragedy for science that Hugh Hefner's grandfather didn't launch the magazine. In conclusion, there is no easy way and probably no way at all to correlate the fertility of long since dead women with their waist to hip ratio. Since it is also impossible to do it with contemporary women, we are faced with the conclusion that it is not possible directly to verify the hypothesis. Nevertheless, those favoring it have tried to obtain some indirect evidence in its favor, but again, this indirect evidence rests on some assumptions of unknown veracity.

One of these assumptions is that determinations of fertility could be replaced by determinations of health. The reasoning is that good health may be associated with a long and fertile life, thereby offering the opportunity to have many children and to raise them properly. Furthermore, good health must be dependent on good genes, and healthy women must have better genes than unhealthy women, at least according to sociobiological reasoning. The results from studies of the association between indices of health and the waist to hip ratio are mixed (Ho *et al.*, 2003; Martin *et al.*, 2003; Bigaard *et al.*, 2004) and there is no evidence that a ratio of 0.70 is a reliable indicator of better health than other ratios. When a sociobiological prediction exceptionally is confronted with data, it is not confirmed. After all, why should it be? Social prejudices are rarely based on reality. One example of this appears, in fact, in a study of the waist to hip ratio and incidence of ovarian cancer (Andersson *et al.*, 2004). Whereas the waist to hip ratio was a useless predictor, intense physical activity was associated with an increase in the risk. Social prejudice would have it that intense physical activity is good for your health.

The conclusion must inevitably be that the waist to hip ratio's value as an indicator of fertility is unknown and even impossible to get to know. Approximations, like considering the ratio as an indicator of good health rather than high fertility, have been tested and have failed to confirm the predictive value of it. Consequently, are we forced to conclude that the basic notion of the role of this ratio is not only unfounded but openly false. I could end the discussion of the waist to hip ratio here, since no more arguments are necessary to lay it dead. However, this analysis is not limited to that particular hypothesis, although I systematically use it as an example. We proceed, therefore, with a discussion of assumption number 2 on the list.

Women are fertile only a day or so around the moment of ovulation. There is no reason to believe that the waist to hip ratio substantially or even noticeably changes according to phase of the menstrual cycle. Therefore, it is not a reliable indicator of current fertility. This default can be made less important imagining that women with a convenient waist to hip ratio will form a stable couple with the man attracted to them. For how long have humans formed stable couples? And how stable are they? This last is an important point and it is associated with the question of the predictive value of the waist to hip ratio. Although the ratio does not change during the menstrual cycle, it certainly changes during a lifetime. A favorable ratio at the time a couple is established should predict the fertility of the woman for the entire duration of the couple. If we assume that male–female couples have been long lasting for the last thousands of years and, if we assume that primitive man formed a couple when very young, then the waist to hip ratio at an early age should predict the woman's fertility for a long time ahead. We do not know if this is the case and I cannot imagine any way it could possibly be known with certainty.

The problems caused by the lack of association between the waist to hip ratio and instant fertility can be summarized in the following way: any characteristic needs a long time to become the norm in a species through natural selection. Therefore, we must assume that a waist to hip ratio of 0.70 has been attractive for quite some time. Since it does not vary with instant fertility, its effectiveness must be based on the existence of stable couples. We have, and cannot have, any idea of whether primitive man formed stable couples or not. Furthermore, if stable couples existed for a substantial proportion of the fertile lifetime, the waist to hip ratio must have had predictive validity. This is not known and it cannot become known.

With regard to the potential relationship between waist to hip ratio and the capacity to care for the young we are in a state of ignorance of about the same kind as with regard to the relationship between this ratio and fertility. I do not know of any study showing that mothers' waist to hip ratio of 0.70 is associated with lower infant or child mortality than any other waist to hip ratio. Likewise, I do not know of any study showing that infant or child morbidity is lower in children of women with a waist to hip ratio of 0.70 than in other women. Extensive searches of many databases have failed to illuminate the issue and there is apparently no way to escape from the conclusion that we do not know anything about this.

The last assumption behind the idea that a certain waist to hip ratio is attractive because of some evolutionary mechanism is that its attractiveness does not depend on learning. If it did, then it would be socially and not evolutionary determined. This means that a certain waist to hip ratio must function as an unconditioned sexual incentive. It has never been shown that *any* waist to hip ratio has unconditioned sexual incentive properties. This excludes, evidently, any possibility for maintaining that a ratio of 0.70 is superior to other ratios. There are few scientists who have speculated about unconditioned sexual incentives in the human. The few that have taken the trouble to do so maintain that there is none (Hardy, 1964; Baldwin and Baldwin, 1997). All sexual incentives are, according to that point of view, conditioned. Mechanical stimulation of the genitalia is intrinsically rewarding and reinforcing, they argue. Any neutral stimulus that is present when genital stimulation occurs can become associated with the pleasure provoked by that stimulation, thereby acquiring the capacity to function as conditioned incentives. The notion of no unconditioned sexual incentive is most parsimonious and scientifically unobjectionable, but most problematic for the hypothesis of a waist to hip ratio of 0.70 as being not only an unconditioned incentive but also superior to other waist to hip ratios.

As we now have seen, none of the basic assumptions behind the idea that a certain waist to hip ratio is attractive because of an evolutionary process is supported by any experimental data. On the contrary, the limited amount of data available indicates that they are unfounded and simple logical reasoning shows that most of them, if not all, are beyond the reach of scientific evaluation. Their justification is entirely teleological and this at a very low level of abstraction.

An explanation of human behavior in terms of purpose or goal is entirely legitimate, since the human may have a representation of the goal and this representation functions as a cause for behavior. In this case, a waist to hip ratio of 0.70 would need to evoke a representation of a child being born and raised to puberty by an able and healthy mother. This representation may in fact exist in the head of a man in some overly romantic movies or best selling novels by Danielle Steele, but it is probably not a major force in most sexual encounters. Nevertheless, it nicely illustrates one of the possible sources of sociobiological teleology: social prejudice, or the conservative middle-class dreams of the ideal life.

Many notions of sociobiology, like the attractiveness of a waist to hip ratio of 0.70, are rather similar to numerous notions of psychoanalysis. Not in content, but in the way that they provide an explanation of the world around us that intuitively makes sense. This, combined with the employment of pseudoscientific arguments and apparent logic, makes such notions very attractive for informed laymen and the popular scientific magazines. It even attracts some behavioral scientists, particularly from the cognitive sciences. They feel probably at home with loose terminology, speculative arguments and sweeping generalizations.

My concern for understanding behavior in terms of cause–effect rather than in terms of hypothetical function as well as a predilection for experiments rather than speculations makes me unfit for appreciating the contributions of sociobiology. To persist in teleological explanations certainly assures continuity from antiquity and on but it can hardly point to the future. Molecular biology has revealed how our genes direct the synthesis of proteins and biochemistry has told us something about what these proteins do. The spectacular progress in these areas is based on experimental studies of cause–effect relationships, not on vain conjectures about final causes.

Summary

1 Verbal messages containing descriptions of sexual activities as well as visual stimuli, either photographs or moving pictures, illustrating copulatory behaviors enhance genital blood flow in women and men.
2 In the case of verbal messages, it is likely that they provoke generation of mental representations of sexual activities and these mental representations rather than the verbal message itself produce enhanced genital blood flow.
3 Whether the effects of visual stimuli are also mediated by mental representations of sexual activities is unknown.
4 Tactile stimulation of the genitals is an efficient stimulus for enhancing genital blood flow. This may be achieved through simple reflex connections, eliminating the need for mental representations as mediators.
5 The sexual incentive stimuli activating approach behaviors and initial establishment of contact between potential mates are almost completely unknown.
6 Studies evaluating the role of olfactory stimuli for the activation of sexual incentive motivation and approach behaviors have given consistent negative results.
7 The state of limerence may be a kind of subjective experience of sexual incentive motivation.
8 There are many reasons to believe that human sexual incentive stimuli are learned in some way or another. They can be regarded as socially constructed.
9 Sociobiology is the most disturbing example of teleological reasoning still surviving within biology. Many aspects of sexual behavior can be explained by reference to final causes, as in the time of Aristotle, according to this doctrine.
10 Some examples are presented that show how basic sociobiological notions are based on a series of untestable or false assumptions. Consequently, these notions have no place in science.

References

Andersson, J.P., Ross, J.A. and Folsom, A.R. (2004). Anthropometric variables, physical activity, and incidence of ovarian cancer: the Iowa women's health study. *Cancer* **100**, 1515–1521

Baldwin, J.D. and Baldwin, J.I. (1997). Sexual behavior. In *Encyclopedia of human biology*, vol. 7 (R. Dulbacco, ed.), pp. 829–838. Academic Press: San Diego

Baron, R.A. (1981). Olfaction and human social behavior: effects of a pleasant scent on attraction and social perception. *Pers Soc Psychol Bull* **7**, 611–616

Baxi, K.N., Dorries, K.M. and Eisthen, H.L. (2006). Is the vomeronasal organ really specialized for detecting pheromones? *TINS* **29**, 1–7

Benton, D. and Wastell, V. (1986). Effects of androstenol on human sexual arousal. *Biol Psychol* **22**, 141–147

Bigaard, J., Frederiksen, K., Tjønneland, A. *et al.* (2004). Waist and hip circumference and all-cause mortality: usefulness of the waist-to-hip ratio? *Int J Obes* **28**, 741–747

Black, S.L. and Biron, C. (1982). Androstenol as a human pheromone: No effect on perceived physical attractiveness. *Behav Neural Biol* **34**, 326–330

Buss, D.M. (2004). *Evolutionary psychology: the new science of the mind*, 2nd edn. Pearson: Boston

Buss, D.M. (ed.) (2005). *The handbook of evolutionary psychology*. John Wiley: Hoboken

Calhoun, J.B. (1962). *The ecology and sociology of the Norway rat*. US Government Printing Office: Washington, DC

Carani, C., Bancroft, J., Del Rio, G., Granata, A.R.M., Facchinetti, F. and Marrama, P. (1990). The endocrine effects of visual erotic stimuli in normal men. *Psychoneuroendocrinology* **15**, 207–216

Codispotti, M., Gerra, G., Montebarocci, O., Zaimovic, A., Raggi, M.A. and Baldaro, B. (2003). Emotional perception and neuroendocrine changes. *Psychophysiology* **40**, 863–868

Courtois, F.J., Charvier, K.F., Leriche, A. and Raymond, D.P. (1993). Sexual function in spinal cord injury men. I. Assessing sexual capability. *Paraplegia* **31**, 771–784

Cowley, J.J. and Brooksbank, B.W.L. (1991). Human exposure to putative pheromones and changes in aspects of social behaviour. *J Steroid Biochem Mol Biol* **39**, 647–659

Cutler, W.B., Friedmann, E. and McCoy, N.L. (1998). Pheromonal influences on sociosexual behavior in men. *Arch Sex Behav* **27**, 1–13

Dulac, C. (1997). Molecular biology of pheromone perception in mammals. *Semin Cell Dev Biol* **8**, 197–205

Eibl-Eibesfeldt, I. (1973). *Der vorprogrammierte Mensch: das Ererbte als bestimmender Faktor im menschlichen Verhalten*. Fritz Molden: Vienna

Ekman, P., Friesen, W.V. and Ellsworth, P. (1972). *Emotion in the human face: guidelines for research and an integration of findings*. Pergamon: New York

Exton, N.G., Truong, T.C., Exton, M.S. *et al.* (2000). Neuroendocrine response to film-induced sexual arousal in men and women. *Psychoneuroendocrinology* **25**, 187–199

Ferreira-Nuño, A., Morales-Otal, A., Paredes, R.G. and Velázquez-Moctezuma, J. (2005). Sexual behavior of female rats in a multiple-partner preference test. *Horm Behav* **47**, 290–296

Fisher, H.E. (2004). *Why we love: The nature and chemistry of romantic love*. H. Holt: New York

Foucault, M. (1976). *Histoire de la sexualité*. Gallimard: Paris

Freud, S. (1905). *Drei Abhandlungen zur Sexualtheorie*. F. Deuticke: Leipzig

Furnham, A., Dias, M. and McClelland, A. (1998). The role of body weight, waist-to-hip ratio, and breast size in judgments of female attractiveness. *Sex Roles* **39**, 311–326

Furnham, A., Petrides, K.V. and Constantinides, A. (2005). The effects of body mass index and waist-to-hip ratio on ratings of female attractiveness, fecundity, and health. *Pers Individ Differ* **38**, 1823–1834

Gagnon, J.H. and Simon, W. (2002). *Sexual conduct: the social sources of human sexuality*, 2nd edn. AldineTransaction: New Brunswick

Gaither, G.A. and Plaud, J.J. (1997). The effects of secondary stimulus characteristics on men's sexual arousal. *J Sex Res* **34**, 231–236

Garcia, L.T. (1983). Sexual stereotypes and attributions about sexual arousal. *J Sex Res* **19**, 366–375

Gitter, G.A., Lombranz, J., Saxe, L. and Bar-Tal, Y. (1983). Perceptions of female physique characteristics by American and Israeli students. *J Soc Psychol* **121**, 7–13

Giuliano, F. and Rampin, O. (2004). Neural control of erection. *Physiol Behav* **83**, 189–201

Golde, J.A., Strassberg, D.S. and Turner, C.M. (2000). Psychophysiologic assessment of erectile response and its suppression as a function of stimulus media and previous experience with plethysmography. *J Sex Res* **37**, 53–59

Gower, D.B. and Ruparelia, B.A. (1993). Olfaction in humans with special reference to odorous 16-androstenes: their occurrence, perception and possible social, psychological and sexual impact. *J Endocr* **137**, 167–187

Graham, C.A., Janssen, E. and Sanders, S.A. (2000). Effects of fragrance on female sexual arousal and mood across the menstrual cycle. *Psychophysiology* **37**, 76–84

Gustavson, A.R., Dawson, M.E. and Bonett, D.G. (1987). Androstenol, a putative pheromone, affects human (*Homo sapiens*) male choice performance. *J Comp Psychol* **101**, 210–212

Hamilton, W.D. (1964a). The genetical evolution of social behaviour. I. *J Theor Biol* **7**, 1–16

Hamilton, W. D. (1964b). The genetical evolution of social behaviour. II. *J Theor Biol* **7**, 17–52

Hardy, K.R. (1964). An appetitional theory of sexual motivation. *Psychol Rev* **71**, 1–18

Heiman, J.R. (1977). A psychophysiological exploration of sexual arousal patterns in females and males. *Psychophysiology* **14**, 266–274

Ho, S.Y., Lam, T.H. and Janus, E.D. (2003). Waist and stature is more strongly associated with cardiovascular risk factors than other simple anthropometric indices. *Ann Epidemiol* **13**, 683–691

Horvath, T. (1979). Correlates of physical beauty in men and women. *Soc Behav Pers* **7**, 145–151

Horvath, T. (1981). Physical attractiveness: the influence of selected torso parameters. *Arch Sex Behav* **10**, 21–24

Karlsson, P. and Lüscher, M. (1959). 'Pheromones': A new term for a class of biologically active substances. *Nature* **183**, 55–56

Katzmarzyk, P.T. and Davis, C. (2001). Thinness and body shape of Playboy centerfolds from 1978 to 1998. *Int J Obes* **25**, 590–592

Knecht, M., Lundtsröm, J.N., Witt, M., Hüttenbrink, K.B., Heilmann, S. and Hummel, T. (2003). Assessment of olfactory function and androstenone odor thresholds in humans with or without functional occlusion of the vomeronasal duct. *Behav Neurosci* **117**, 1135–1141

Knight, T.W. (1985). Pheromones in farm animals. *TIPS* **6**, 171–173

Krüger, T., Exton, M.S., Pawlak, C., von zur Mühlen, A., Hartmann, U. and Schedlowski, M. (1998). Neuroendocrine and cardiovascular response to sexual arousal and orgasm in men. *Psychoneuroendocrinology* **23**, 401–411

Kuhn, R.A. (1950). Functional capacity of the isolated human spinal cord. *Brain* **73**, 1–51

Laan, E., Everaerd, W., van der Velde, A. and Geer, J.H. (1995). Determinants of subjective experience of sexual arousal in women: Feedback from genital arousal and erotic stimulus content. *Psychophysiology* **32**, 444–451

Leitenberg, H. and Henning, K. (1995). Sexual fantasy. *Psychol Bull* **117**, 469–496

Maisey, D.S., Vale, E.L.E., Cornelissen, P.L. and Tovée, M.J. (1999). Characteristics of male attractiveness for women. *Lancet* **353**, 1500

Martin, A.D., Daniel, M., Clarus, J.P. and Marfell-Jones, M.J. (2003). Cadaver-assessed validity of anthropometric indicators of adipose tissue distribution. *Int J Obes* **27**, 1052–1058
Masters, W.H. and Johnson, V.E. (1966). *Human sexual response*. Little, Brown & Co.: Boston
McCoy, N.L. and Pitino, L. (2002). Pheromonal influences on sociosexual behavior in young women. *Physiol Behav* **75**, 367–375
McDougall, W. (1914). *An introduction to social psychology containing supplementary chapters on theories of action and on the sex instinct*. Methuen: London
Meisami, E. and Bhatnagar, K.P. (1998). Structure and diversity in mammalian accessory olfactory bulb. *Microsc Res Tech* **43**, 476–499
Melrose, D.R., Reed, H.C. and Patterson, R.L. (1971). Androgen steroids associated with boar odour as an aid to the detection of oestrus in pig artificial insemination. *Br Vet J* **127**, 497–502
Meredith, M. (2001). Human vomeronasal organ function: a critical review of best and worst cases. *Chem Senses* **26**, 433–445
Michael, R.P., Bonsall, R.W. and Kutner, M. (1975). Volatile fatty acids, 'copulins', in human vaginal secretions. *Psychoneuroendocrinology* **1**, 153–163
Michael, R.P., Bonsall, R.W. and Warner, P. (1974). Human vaginal secretions: volatile fatty acid content. *Science* **186**, 1217–1219
Michael, R.P., Zumpe, D., Keverne, E.B. and Bonsall, R.W. (1972). Neuroendocrine factors in the control of primate behavior. *Rec Progr Horm Res* **28**, 665–706
Morris, N.M. and Udry, J.R. (1978). Pheromonal influences on human sexual behaviour: an experimental search. *J Biosoc Sci* **10**, 147–157
Pause, B.M. (2004). Are androgen steroids acting as pheromones in humans? *Physiol Behav* **83**, 21–29
Pettijohn, T.F. II and Jungeberg, B.J. (2004). Playboy playmate curves: changes in facial and body feature preferences across social and economic conditions. *Pers Soc Psychol Bull* **30**, 1186–1197
Rako, S. and Friebely, J. (2004). Pheromonal influences on sociosexual behavior in postmenopausal women. *J Sex Res* **41**, 372–380
Ramirez, A. (1990). All about deodorants; the success of sweet smell. *New York Times* [Section 3], 10, column 1. 12-8-1990. New York
Reed, H.C., Melrose, D.R. and Patterson, R.L. (1974). Androgen steroids as an aid to the detection of oestrus in pig artificial insemination. *Br Vet J* **130**, 61–67
Rellini, A.H., McCall, K.M., Randall, P.K. and Meston, C.M. (2005). The relationship between women's subjective and physiological arousal. *Psychophysiology* **42**, 116–124
Riddoch, G. (1917). The reflex functions of the completely divided spinal cord in man, compared with those associated with less severe lesions. *Brain* **40**, 264–402
Robitaille, J.A. and Bouvet, J. (1976). Field observations on the social behaviour of the Norway rat, *Rattus norvegicus* (Berkenhout). *Biol Behav* **1**, 289–308
Rowland, D.L., Heiman, J.R., Glaude, B.A., Hatch, J.P., Doering, C.H. and Weiler, S.J. (1987). Endocrine, psychological and genital response to sexual arousal in men. *Psychoneuroendocrinology* **12**, 149–158
Ruse, M. (1979). *Sociobiology: sense or nonsense?*Reidel: Dordrecht
Schopenhauer, A. (1844). *Die Welt als Wille und Vorstellung, zweiter Band*. F.A. Brockhaus: Leipzig
Seifert, T. (2005). Anthropomorphic characteristics of centerfold models: trends toward slender figures over time. *Int J Eating Disord* **37**, 271–274
Shafik, A. (1995). The 'clitorimotor' reflex. *Int Urogynecol J Pelvic Floor Dysf* **6**, 329–336

Signoret, J.P. (1976). Chemical communication and reproduction in domestic animals. In *Mammalian olfaction, reproductive processes and behaviour* (R.L. Doty, ed.), pp. 243–256. Academic Press: London

Singh, D. (1993). Body shape and women's attractiveness. The critical role of waist-to-hip ratio. *Hum Nature* **4**, 297–321

Singh, D. and Young, R.K. (1995). Body weight, waist-to-hip ratio, breasts, and hips: role in judgments of female attractiveness and desirability for relationships. *Ethol Sociobiol* **16**, 483–507

Sipski, M.L., Alexander, C.J. and Rosen, R. (2001). Sexual arousal and orgasm in women: Effects of spinal cord injury. *Ann Neurol* **49**, 35–44

Smith, T.D., Siegel, M.I. and Bhatnagar, K.P. (2001). Reappraisal of the vomeronasal system of catarrhine primates: Ontogeny, morphology, functionality, and persisting questions. *Anat Rec* **265**, 176–192

Sprecher, S., McKinney, K. and Orbuch, T.L. (1987). Has the double standard disappeared? An experimental test. *Soc Psychol Quart* **50**, 24–31

Steinman, D.L., Wincze, J.P., Sakheim, D.K., Balrow, D.H. and Mavissakalian, M. (1981). A comparison of male and female patterns of sexual arousal. *Arch Sex Behav* **10**, 477–492

Tennov, D. (1979). *Love and limerence: the experience of being in love*. Stein and Day: New York

Tennov, D. (2001). Conceptions of limerence. In *Sexual appetite, desire and motivation: energetics of the sexual system* (W. Everaerd, E. Laan and S. Both, eds), pp. 111–116. Koninklijke Nederlandske Akademie van Wetenschappen: Amsterdam

Tovée, M.J., Hancock, P.J.B., Mahmoodi, S., Singleton, B.R.R. and Cornelissen, P.L. (2002). Human female attractiveness: waveform analysis of body shape. *Proc Royal Soc London Series B Biol Sci* **269**, 2205–2213

Tovée, M.J., Maisey, D.S., Emery, J.L. and Cornelissen, P.L. (1999). Visual cues to female physical attractiveness. *Proc Royal Soc London Series B Biol Sci* **266**, 211–218

Trivers, R.L. (1972). Parental investment and sexual selection. In *Sexual selection and the descent of man, 1871–1971* (B. Campbell, ed.), pp. 136–179. Heinemann Educational: London

Trotier, D., Eliot, C., Wassef, M. *et al.* (2000). The vomeronasal cavity in adult humans. *Chem Senses* **25**, 369–380

Voracek, M. and Fisher, M.L. (2002). Shapely centerfolds? Temporal changes in body measures: trend analysis. *Br Med J* **325**, 1447–1448

Wiggins, J.S. (1971). Men's preferences for different types of female figures. *Med Aspects Hum Sexual* **5**, 100–116

Wiggins, J.S., Wiggins, N. and Conger, J.C. (1968). Correlates of heterosexual somatic preference. *J Pers Soc Psychol* **10**, 82–90

Wiggins, N. and Wiggins, J.S. (1969). A typological analysis of male preferences for female body types. *Multivariate Behav Res* **4**, 89–102

Wilson, E.O. (1975). *Sociobiology. The new synthesis*. Belknap Press: Cambridge

Winman, A. (2004). Do perfume additives termed human pheromones warrant being termed pheromones? *Physiol Behav* **82**, 697–701

Wysocki, C.J. (1979). Neurobehavioral evidence for the involvement of the vomeronasal system in mammalian reproduction. *Neurosci Biobehav Rev* **3**, 301–341

Wysocki, C.J. and Preti, G. (2004). Facts, fallacies, fears, and frustrations with human pheromones. *Anat Rec Part A* **281A**, 1201–1209

Zeng, X.N., Leyden, J.J., Spielman, A.I. and Preti, G. (1996). Analysis of characteristic human female axillary odors: qualitative comparison to males. *J Chem Ecol* **22**, 237–257

Zhang, J. and Webb, D.M. (2003). Evolutionary deterioration of the vomeronasal pheromone transduction pathway in catarrhine primates. *Proc Nat Acad Sci USA* **100**, 8337–8341

CHAPTER

5

Endocrine control of sexual behavior

In the preceding chapters, I have given a summary of current knowledge of the stimulus control of sexual incentive motivation and copulatory behavior in some mammalian species including the human. Although knowledge is still lacking in some respects, we have a reasonably clear picture of how external stimuli control sexual behaviors. Now we will enter the world under the skin, particularly the world inside the skull, and there we will try to discover the tangible bases of the central motive state. It is not that I will try to materialize an abstract concept, but I will try to show that the brain needs to be exposed to some hormones in order to react to external stimuli with sexual responses, and I will try to show that these hormones act in specific brain structures when enabling them to these reactions.

1 Males

1.1 The importance of testicular hormones in men

Some effects of the removal of testicular hormones have been implicitly known for thousands of years. Starting at an unknown period of remote antiquity, castrated men, eunuchs, were used as servants in harems and as chamberlains to kings (Humana, 1973). The fact that eunuchs were preferred for employment as guards of and servants to women may suggest that already in these ancient times it was known that castrated men had reduced or no sexual interest. There are many anecdotes about preserved sexual behaviors in some eunuchs, but these anecdotes do not contradict the proposal that most of them showed no or much reduced sexual activity. Although none of the stories concerning sexual exploits of eunuchs is based on actual observation, it is not inconceivable that some men castrated after puberty retained some sexual activity for considerable periods of time. Furthermore, some eunuchs were not castrated in the contemporary sense of the word, extirpation of the testicles, but by alternative procedures such as crushing the testicles. This may well have led to sterility, but it is not certain or even likely that hormone production was eliminated. Such cases may be those figuring in the anecdotes, at least in many of them.

It appears that some eunuchs made brilliant careers and, during the Chou period in China (1122–221 BC), many castrated men worked as advisers to the emperors. It is said that some of them were so influential that they in reality were the rulers of the Chinese empire during certain periods of the Han, T'ang, Ming and Sung dynasties (Carlson, 1972). Likewise, eunuchs were employed as public officials in the Persian empire and much later in the Roman and Ottoman empires. The impressive political success of some eunuchs suggests that castration may have many beneficial consequences. This was made still more evident during the early Christian period. A passage from the new testament, Matthew 19:12, reads:

> For there are some eunuchs, which were so born from their mother's womb: and there are some eunuchs, which were made eunuchs of men: and there be eunuchs, which have made themselves eunuchs for the kingdom of heaven's sake. He that is able to receive it, let him receive it.

These words inspired many Christian men to consider castration as a means of facilitating entry into Paradise. For example, one of the great thinkers of the Greek church, Origen (complete name Oregenes Adamantius, born in 185 in Alexandria, Egypt and died in 254 in Tyre, Phoenicia) is said to have castrated himself with the purpose of avoiding sexual temptations. There was even a eunuch sect, the Valesii, active in the 3rd century, in which the members not only castrated themselves to the greater glory of God, but they also castrated any visitor to their monastery (see Kuefler, 2001 and Scholten, 1995 for more extensive discussions of eunuchs in antiquity). Again, the purpose was to reduce or eliminate the sexual urges and the sinful acts that these urges might have incited men to engage in. This means that the reduction in sexual activity following castration must have been known for more than a thousand years. Some other effects of products emanating from the testicles have also been known for a long time. In Italy, young boys were castrated in order to avoid the deepening of the voice associated with puberty. As adults, these men had a beautiful soprano voice, much appreciated by many music lovers. The practice of castration for the purpose of preserving a childish voice was discontinued by Pope Leo XIII in 1878. Here we find an example showing that some of the peripheral effects of testicular products were as well known as their central nervous effects.

The stories about eunuchs and castrati singers are certainly entertaining and stimulate our fantasies, but their value as evidence for the hypothesis that human male sexual behavior is dependent on testicular hormones is slight. Fortunately, there are many scientifically acceptable studies of the consequences of castration in men. Most, if not all, of them can be criticized because of the employment of non-representative samples, but this criticism may not be devastating. For easily understandable reasons, it is not possible to castrate a random sample of men. The men that were castrated had almost always been convicted of sexual crimes and they were normally offered a penalty rebate if they accepted castration. This practice was considered inhumane in many societies and castration is almost never performed anymore. In Norway, the procedure may still be legally performed after demand by the individual and approval by a committee, but no demand for castration has been presented since 1974. The situation is probably similar in many other countries. During the first half of the last century, though, castration of sexual offenders was regularly performed in Europe and some scientists profited from this to evaluate sexual behavior in these long-term castrates. It should be noted that all these castrations were performed post-pubertally. Thus, we do not know what the effects of prepubertal castration might be. Data from studies of castrated men in several European countries are unequivocal in the way that they all show that the intensity of sexual behaviors, including sexual dreams and fantasies, was much reduced (e.g. Wantoch, 1935; Bremer, 1958; Stürup, 1972; Heim and Hursch, 1979; Ortmann, 1980; Heim, 1981; Wille and Beler, 1989). In addition to castrating sexual offenders, medical professionals have castrated men for other reasons. One such reason is prostate cancer. Sexual behaviors are reduced or eliminated as a consequence (Bergman *et al.*, 1984). A most interesting study in castrated cancer patients evaluated the genital response to sexual incentives in the form of a pornographic movie. Rather than relying on questionnaire or interview data like the vast majority of studies performed on castrated men, this exceptional study employed an objective measure of genital response. Penile circumference was monitored using strain gauges placed at the penile base and close to the corona. While, as mentioned in a previous chapter, pornographic movies reliably increase penile circumference in most men, only 25% of the castrated cancer patients displayed this response to the movie. Still more interesting was the observation that the men who did show erection, manifested as increased penile circumference, had serum free testosterone concentrations about double of those not displaying erection (Greenstein *et al.*, 1995). This very important result suggests that adrenal androgen secretion may be sufficiently high, in some men, for maintaining some level of sexual functioning even in the absence of testicular androgens. It may be remembered that the only androgen source in castrated men is the adrenal gland. These data also constitute one of the very few existing pieces of direct evidence showing that the impact of sexual incentives is androgen-dependent in the human male.

The preceding brief review of sexual behaviors in castrated men should have made it evident that androgens indeed are important for these behaviors. This conclusion is somewhat obscured by two complications. One is that sexual offenders and prostate cancer patients are not necessarily representative of the general population of healthy men. The other is that humans produce androgens in the adrenal cortex. As we saw in the excellent Greenstein *et al.* (1995) study, this androgen production may contribute to the maintenance of sexual behaviors. Castration of the human male, as of other primates, does not reduce androgen availability as much as castration of rats and perhaps other rodents does. In contrast to primates, the rat adrenal cortex does not produce androgens (van Weerden *et al.*, 1992) and blood androgen concentrations rapidly fall below detection limits in male rats, something that does not occur in men (Bélanger *et al.*, 1989). Thus, there is no *a priori* reason to predict that the effects of castration on male sexual behaviors should be identical in humans and in rats. In fact, considering the difference in androgen availability after castration, its effects on sexual behaviors are remarkably similar in rats and men.

Nowadays, castration has been replaced by more sophisticated means to eliminate androgen actions. The most interesting for our purpose is the use of androgen receptor antagonists. One, cyproterone acetate, has been given both to sexual offenders and to prostate cancer patients. Results of the evaluation of sexual functioning in men being treated with cyproterone coincide nicely with those obtained in studies on castrated men. A clear-cut reduction in the intensity of all indicators of sexual behaviors was observed (Tennent, 1984; Bradford and Pawlak, 1993; Bradford and Pawlaka, 1993; Reilly *et al.*, 2000). These data convincingly show that blockade of androgen receptors leads to serious impairment of sex behaviors in men. The significance of this observation will become evident when we analyze the role of androgen receptors in non-human males.

Another androgen receptor antagonist, flutamide, has been compared to cyproterone in a large clinical study in prostate cancer patients. After about 6 months of treatment, spontaneous erections had declined and after about 1 year the same had happened to sexual activity. After 2 years of treatment, sexual activity had disappeared altogether in more than 80% of the patients. These data are suggestive, but the patients were rather old (with a mean of 71 years) at the beginning of the study and a substantial decline in sexual activity is not uncommon at this age (Beutel *et al.*, 2006). The employment of a control group would have been extremely useful, but since the purpose of the study was to compare cyproterone acetate with flutamide it was not considered necessary. By the way, the antagonists did not differ with regard to their effects on sexual functioning.

Another approach that can be employed to reduce the actions of androgens is to diminish their availability by pharmacological means. While castration appears repugnant to many humans, non-surgical procedures with the same effect seem to be entirely acceptable. One efficient way to diminish blood androgen concentrations by chemical means is to administer drugs impeding gonadotrophin secretion. Strangely enough, both gonadotrophin releasing hormone (GnRH) agonists and antagonists can be used to achieve the same end. The antagonists block the actions of gonadotrophins in the testicles, while the agonists exhaust pituitary gonadotrophin reserves by a poorly understood mechanism. Both treatments have the same end result in the form of a much reduced blood concentration of androgens. The effects of such treatments are very similar to those of castration, both in sex offenders (Thibaut *et al.*, 1993; Briken *et al.*, 2000, 2003; Krueger and Kaplan, 2001) and in healthy young men (Bagatell *et al.*, 1994). I will not pursue this issue any further, because I consider the evidence in favor for a considerable if not complete hormone-dependency of human male sexual behaviors convincing beyond reasonable doubt.

Before turning to studies of non-human mammals, I cannot resist the temptation to make an irrelevant comment about semantics. Treatment with androgen receptor antagonists and with drugs impeding gonadotrophin actions is called chemical castration in the popular press and some less popular outlets for written materials. On the contrary, castration is called surgical castration. Traditionally, the noun castration refers to the removing of the testicles. Thus, the expression surgical castration has no sense since surgery, in one form or another, is the only possible procedure for the removal of the testicles. Except, of course, chemical castration. This expression evokes immediately the though of some intricate procedure by which a testicle-dissolving chemical is introduced into the scrotum, leaving an empty but uncut pouch behind.

1.2 The importance of testicular hormones in non-human males

Peasants have known that castration modifies the behavior of domestic animals for centuries. Bulls were turned into oxen for making them more manageable and males of some other species were castrated for improving the taste of their meat. This shows that both behavioral and physiological consequences of testicular products in non-human animals have been known for a long time. Despite the age-old popular knowledge, it was only about 100 years ago that the first experimental study of the behavioral consequences of testicular removal in a mammal was published. Male rats were castrated and their copulatory behavior was found to disappear after some time (Steinach, 1894). The disappearance of sexual behaviors after castration has been confirmed in many mammalian species, including the human as mentioned in the previous paragraphs. A question that immediately posed itself to the scientists performing these studies was what the testicles did in order to maintain sexual behavior. Those who had castrated human males for thousands of years had never asked that interesting question, curiously enough. In contrast, a professor at the Universität Göttingen in Germany had not only asked the question but had also obtained the answer. In a study performed in roosters it had been found that extirpation of the testicles eliminated mating behaviors. However, if the testicles simply were moved from their normal location to the abdominal cavity rather than being removed from the animal, there was no decline in sexual behaviors. All nervous and vascular connections to the testicles had been cut during the process. A post-mortem histological examination revealed that the moved testicles had revascularized, but there was no indication of a re-established nerve supply. The conclusion was, logically, that the testicles produced some blood-borne factor that stimulated sexual behavior (Berthold, 1849). This suggestion was confirmed in rats in a now classic study in which injections of a testicular extract were found to restore copulatory behavior in castrated males (Nissen, 1929). The blood-borne factor emanating from cells in the testicle was identified in 1935. It turned out to belong to a group of lipids called steroids and was given the name testosterone. Soon after, it was shown that injections of testosterone, either free or in the form of the propionate ester, activated copulatory behavior in castrated male rats (Shapiro, 1937; Moore and Price, 1938; Stone, 1939). The capacity of testosterone to restore copulatory behavior in castrated males was rapidly confirmed in many different species of mammals (Beach, 1948). A clear picture of the endocrine control of copulatory behavior in male mammals had been obtained. There were no data contradicting the notion that testicular hormones, more specifically testosterone, were necessary. Furthermore, several studies revealed that the hormone acted directly within the brain when reactivating copulatory behavior in castrated male rats. Implants in the medial preoptic area were able to restore behavior. These implants had no effect on peripheral target tissues, showing that the hormone did not leak out into the circulation in physiologically significant amounts (e.g. Davidson, 1966). Accordingly, its action must have been within the brain.

1.3 The role of testosterone metabolites in non-human males

In the 1960s, it was discovered that a testosterone metabolite, 5α-dihydrotestosterone, was formed from circulating testosterone in peripheral target organs like the prostate or seminal vesicle. The enzyme responsible for the transformation of testosterone into dihydrotestosterone, 5α-reductase, was found to be present in target tissues. Dihydrotestosterone has a higher affinity for the testosterone receptor than testosterone itself and it was soon suggested that testosterone was a prehormone and that it needed to be transformed into dihydrotestosterone before having any physiological effect (Wilson and Gloyna, 1970). It was quite logical to suppose that this would also be the case for the actions of testosterone within the central nervous system, particularly since 5α-reductase had been shown to be present in the brain (Jaffe, 1969). To the surprise of many scientists, this was not the case. Dihydrotestosterone turned out to be quite inefficient for the restoration of copulatory behavior in castrated male rats (McDonald *et al.*, 1970). In addition, some studies had shown that androgen receptor antagonists failed to block male rat copulatory behavior while blocking testosterone actions in peripheral tissues such as prostate and seminal vesicle (Beach and Westbrook, 1968; Whalen and Edwards, 1969). These two lines of evidence made it possible to propose that stimulation of the androgen receptor within the central nervous system was not necessary for sexual behavior. At about the same time, another enzyme had been described, aromatase. This enzyme transforms testosterone into estradiol. Interestingly, aromatase was found mainly in the brain, particularly in areas thought to be important for copulatory behavior (Ryan *et al.*, 1972). It was already known that estradiol had some stimulatory effects on this behavior in castrated animals. In fact, estradiol was known to efficiently stimulate mounting, but its effects on intromission were slight (Beach, 1942a; Davidson, 1969; Södersten, 1973). The hypothesis that testosterone acts in the form of dihydrotestosterone in the periphery, maintaining the integrity of the penis and penile muscles, and in the form of estradiol in the brain, maintaining the integrity of the central nervous system processes necessary for sexual behaviors, was born from these two lines of evidence. Experimental support was soon obtained. It turned out that minute doses of estradiol combined with dihydrotestosterone was just as efficient as testosterone for the restoration of copulatory behavior in castrated rats (Baum and Vreeburg, 1973; Larsson *et al.*, 1973). The pioneering observations were soon confirmed in an impressive amount of studies employing different techniques. For example, the effects of brain implants of testosterone, estradiol and dihydrotestosterone were evaluated in castrated male hamsters and it was found that testosterone was more efficient than estradiol, while dihydrotestosterone was inactive with regard to stimulation of copulatory behavior (Lisk and Bezier, 1980). Inhibitors of aromatase were shown to block the capacity of testosterone to stimulate this behavior in castrated male rats (Beyer *et al.*, 1976; Moralí *et al.*, 1977) and an estrogen receptor antagonist was found to have similar effects (Luttge, 1975). Thus, it appears that the stimulation of androgen receptors alone is not sufficient for the activation of male sexual behavior in castrated male rats. This appears to be the case also in some other rodents. These observations led some scientists to consider that aromatization to estradiol was a requisite for all central nervous effects of androgens, at least as far as sexual behaviors are concerned. Furthermore, the notion was extended from rodents to other species and it appears in some textbooks as a general principle applicable to all mammals and also to birds.

Both the conclusion that all androgen actions are mediated by estradiol, and hence by the estrogen receptor, and its extension to non-rodent species may have been premature. I showed, in the early 1970s, that an androgen receptor antagonist efficiently blocked copulatory behavior in male rabbits (Ågmo, 1975) and that dihydrotestosterone could activate that behavior in castrated males (Ågmo and Södersten, 1975). In addition, a non-aromatizable, synthetic androgen, fluoxymesterone, was as efficient as testosterone for activating copulatory behavior in castrated rabbits (Ågmo, 1977). Likewise, dihydrotestosterone was found efficiently to restore copulatory behaviors in castrated male guinea-pigs, either after peripheral administration (Alsum and Goy, 1974) or after implantation into the brain (Butera and Czaja, 1989). In mice, systemic dihydrotestosterone restores sexual behavior to precastrational levels in some strains while this androgen is inactive in others (Luttge and Hall, 1973). There are also rat strains that are responsive to dihydrotestosterone (Olsen and Whalen, 1984). This androgen also activates sexual behaviors in the rhesus monkey (Phoenix, 1974), while estradiol is ineffective in this primate (Michael *et al.*, 1990). It seems safe to conclude that aromatization of androgens and consequent stimulation of estrogen receptors is of no importance for male sexual behavior in the rhesus monkey. In the crab-eating macaque, also called the long-tailed macaque or the cynomolgus monkey (*Macaca fascicularis*), the role or lack of a role of aromatization is less clear than in the rhesus. There are data showing that dihydrotestosterone is unable to restore sexual behavior in castrated males (Michael *et al.*, 1986) and that the aromatase inhibitor fadrozole reduces this behavior in castrated males given testosterone replacement. Strangely enough, the behavioral deficits caused by fadrozole were not reliably reversed with concurrent treatment with estradiol (Zumpe *et al.*, 1993, 1996), making it very difficult to understand by which mechanism fadrozole reduced sex behavior. This and other observations not mentioned here justify the conclusion that 'aromatization does not appear to be obligatory for the activation of male copulatory behavior' in the cynomolgus monkey (Wallen, 2005, p. 21).

Non-aromatizable androgens can also activate copulatory behaviors in rats (Moralí *et al.*, 2002). Furthermore, more recent rat data have convincingly shown that androgen receptor antagonists may have deleterious effects on male rat sexual behaviors. This assertion is based on a series of extremely interesting studies from Marilyn McGinnis laboratory at the Mount Sinai School of Medicine in New York. First they showed that systemic administration of an androgen receptor antagonist, OH-flutamide, efficiently blocks testosterone-induced restoration of copulatory behaviors in castrated male rats and reduced the intensity of the behavior in intact males (McGinnis and Mirth, 1986). They then proceeded with intracerebral administration of the antagonist and found that it blocked testosterone-induced restoration of sex behavior when implanted into some hypothalamic sites. Implants into the septum were ineffective while medial amygdala implants produced a partial inhibition of testosterone's effects (McGinnis *et al.*, 1996). The importance of androgen receptors at hypothalamic sites has been confirmed in later studies (Harding and McGinnis, 2004). It is important to point out that OH-flutamide does not bind to the estrogen receptor, so a blockade of estrogen effects cannot contribute to the actions of OH-flutamide.

1.4 The role of aromatization in men

Data concerning the role of aromatization in the human male are not abundant, but a few studies have determined the consequences of inhibition of aromatase. In one of them, six young (20–26 years) university students were treated with testolactone for 6 weeks in a dose that reduced plasma estradiol concentrations to less than 50% of pretreatment concentration. No effect on any indicator of sexual activity, like number of sexual acts, frequency of sexual thoughts or frequency of sexual excitement, was found (Gooren, 1985). These data suggest that aromatization is not a requisite for androgen actions on human male sexual behavior. This idea receives support from results obtained in other groups included in the same study. One group of 12 men was given the estrogen receptor antagonist tamoxifen for 12 weeks. Sexual activity was not affected. Another group of six 'agonadal' (probably meaning castrated) men, that were maintained on a treatment with testosterone undecanoate, a long-acting testosterone ester, were given dihydrotestosterone instead of testosterone undecanoate for a period of 9 weeks. In this group, blood concentrations of testosterone and estradiol were below detection limits already after 4 weeks of treatment, and remained so at 9 weeks. Nevertheless, the men in this group did not report any reduction in sexual behaviors.

The lack of effect of an estrogen antagonist as well as of an aromatase inhibitor in the Gooren (1985) study clearly suggests that estrogen receptors are not involved in the control of sexual behaviors in men. Likewise, the fact that dihydrotestosterone maintained sexual behaviors just as well as testosterone in the 'agonadal' group, despite the fact that no estradiol could be detected in the blood of these men, is additional strong evidence against a role for estrogen receptors. The ineffectiveness of aromatase inhibition has been confirmed in later studies. In one (Bagatell *et al.*, 1994), healthy men between 20 and 40 years of age were treated with a GnRH antagonist for 6 weeks. Ten men received daily injections of testosterone enanthate during this period, while 10 others received testosterone enanthate + the aromatase inhibitor testolactone. The 10 men given only the GnRH antagonist had a dramatic reduction in blood testosterone concentration during the treatment period and sexual behaviors were much reduced from week 4 of treatment and on. The frequency of intercourse, of spontaneous erection, of sexual desire and of sexual fantasies was diminished. After 6 weeks, the frequency of masturbation was also reduced. All these behaviors were unchanged in the group given the aromatase inhibitor in addition to testosterone, in spite of a 75% reduction of blood estradiol concentrations. Similarly, the men given testosterone enanthate without the aromatase inhibitor did not show any change of behavior. This elegant study not only shows that human male sexual behaviors are dependent on adequate blood androgen concentrations, but also that aromatization is of slight, if any, importance. A similar conclusion had already been proposed in a study of androgen supplementation in patients with primary testicular failure, complemented with a group of eugonadal medical students receiving testolactone (Gooren, 1987).

Although the total number of men having participated in the studies outlined in the preceding paragraphs is rather small, the consistent results inspire some confidence. It does not seem too adventurous to propose that aromatization, and hence estrogen receptors, are not of any crucial importance for human male sexual behavior. In that way, humans may be different from rats and some strains of mice. At the same time, there is a curious report describing a man who had been castrated bilaterally because of severe post-vasectomy orchialgia. He maintained an adequate sexual functioning with a combined treatment of estrogens + progesterone. In fact, this regimen seemed more effective than treatment with testosterone enanthate (Davidson *et al.*, 1983). The significance of this single individual should not be overestimated, but some additional data suggest that estradiol may contribute to the maintenance of sexual functioning in men who have been castrated because of prostate cancer (Ellis and Grayhack, 1963; Bergman *et al.*, 1984). However, although the sexual activity of the men treated with estrogens was above that of untreated castrated men, it was far lower than that found in intact men. Thus, estrogens do not seem to be particularly efficient for the maintenance of sexual behavior. In addition, prostate cancer patients have frequently reduced sexual activity for various reasons and do not represent an ideal group for evaluating hypotheses concerning the role of gonadal hormones for sexual behaviors. Nevertheless, despite all their limitations and weaknesses, the studies just mentioned suggest that it might be premature to discard completely the possibility that estrogens somehow influence human male sexual behaviors. At present, though, this possibility lacks convincing empirical support and appears rather remote.

1.5 Conclusion concerning the role of aromatization

The ensemble of the data presented in the preceding paragraphs shows that the activation of estrogen receptors is not indispensable for the activation of male copulatory behavior and that dihydrotestosterone, and consequently the exclusive stimulation of androgen receptors, in fact may be sufficient for the complete restoration of these behaviors in castrated males of many species. In the case of the rat, it appears that complete restoration of sexual behavior in castrated males, as well as the initial activation of the behavior at puberty and its maintenance throughout adult life, requires simultaneous stimulation of androgen and estrogen receptors within the central nervous system (Vagell and McGinnis, 1997). In men, the majority of existing data suggest that stimulation of the androgen receptor alone is sufficient for the activation and maintenance of sexual behaviors.

1.6 Studies on knockout mice

The intricate issue of the role of aromatization for androgen actions on male sexual behaviors has been splendidly illustrated in studies of male mice with deletions of genes coding for either the estrogen α, the estrogen β or the estrogen α and β receptors as well as for the androgen receptor. Mice carrying a gene deletion are normally called knockout mice and they are frequently given a first name containing the name of the gene product that has been eliminated. For example, a mouse not expressing the estrogen receptor α is called 'estrogen receptor α knockout'. Before making a brief description of the sexual behaviors in mice carrying the different gene deletions, a short note on sexual differentiation is in order. The subject of hormonal influences on morphological and behavioral development is extremely complicated and it would be utterly pretentious on my part to make believe that I could give a reasonable summary of these influences. Nevertheless, from my vantage point of profound ignorance I can allow myself to ascertain that steroid hormones from the developing testicle affect the development of the internal and external sexual organs. In the absence of a functional testicle or of functional androgen receptors, the individual will develop female external genitalia and the internal sexual organs will be abnormal. It is also very likely that fetal or perinatal steroid hormones have profound influences on the brain. In the case of a mouse or a rat, the brain will develop with the wiring typical of a female if no gonadal steroids are present during development and with the wiring typical of a male if such hormones are present. The early effects on the brain of steroid hormones are called 'organizational', in the sense that they organize the central nervous system to be like that of a male. The absence of gonadal hormones allows the brain to remain disorganized, that is to be like that of a female. I admit that this statement may give the impression of being somewhat odd, but I see no other way to interpret the term 'organizational'. Once the brain has either been organized or left disorganized, the blood concentration of gonadal hormones will be very low until puberty. At the moment of puberty, these hormones will display their 'activational' effects. They will act on the organization prepared by the fetal or perinatal hormone secretions when stimulating sexual behaviors.

In mice with gene deletions, the organizational effects of gonadal hormones will be abnormal. The precise abnormality depends on the deleted receptor/receptors, but it is evident that the anatomical and functional substrate for gonadal hormone action in adulthood is different from that of a non-manipulated animal. An unavoidable consequence is that organizational and activational effects are confounded when studying steroid hormone actions in adult animals carrying steroid receptor gene deletions. Although extremely interesting, the results of studies employing knockout models must be interpreted with much caution because of this confound. An absent organizational effect may lead to compensatory changes in completely unrelated receptor systems, for example in neurotransmitter receptors, or changes in the connectivity of some brain areas, and other unknown changes that we cannot imagine. As a result, it is very risky to advance hypotheses concerning physiological roles of hormone receptors solely on the basis of data from knockout studies. Fortunately, in the case of sexual behaviors we have data from many other sources and, as we shall see, the data obtained in studies of knockout mice mostly coincide with earlier observations employing other techniques for blocking receptor actions.

In this context I cannot refrain from making a small comment on the popularity of knockout mice. In many situations, they are the only option for evaluating the possible functions of a gene product. In these situations, their use is beyond reproach. However, in the case of gonadal steroid receptors, the situation is entirely different. There are many well characterized pharmacological tools available if we want to block a particular steroid receptor. Eliminating the gene coding for the receptor is, in this case, a rather bad alternative. We should keep in mind that a good antagonist blocks receptor action in an efficient way and it does so at a specific time. Thus, a receptor antagonist administered to an adult animal will act in a completely normal brain and at the moment we want it to. We can even make the antagonist act at a specific site, for example by intracerebral administration of minute amounts of it. This temporal and spatial specificity of action can never be obtained by gene deletions. Thinking it over, it is difficult to understand why knockout mice have become so popular in studies of behavioral effects of gonadal steroids. They carry many disadvantages and I cannot see one single advantage with them compared to classical pharmacological procedures employing the administration of antagonists. One possible explanation is that the generation of knockout mice frequently represents an achievement of molecular biology. Because it has become a kind of fashion to profit from new achievements in that area of biology, I suspect that some scientists are irresistibly attracted to knockouts. Nevertheless, sooner or later the attraction to novelty will be replaced by a realization of the disadvantages associated with the use of these mice.

After this comment on research tactics, we can now return to studies of male mice lacking genes for one or more gonadal hormone receptors. Mice without an estrogen receptor α show a deficient sexual behavior. They may make some mounts, but they rarely perform intromissions and ejaculation is almost never seen (Rissman *et al.*, 1997; Wersinger *et al.*, 1997; Ogawa *et al.*, 1998). These mice have blood androgen concentrations above normal (Eddy *et al.*, 1996; Rissman *et al.*, 1997), so their deficient copulatory behavior cannot be a consequence of reduced availability of androgens. The differences in blood androgen concentrations between the knockout and the wild type was, in fact, eliminated in some of the behavioral studies by using castrated males given testosterone replacement. What the data show is that the estrogen α receptor is necessary either for the organization of brain circuits involved in sex behavior or for the activational effects of gonadal hormones in post-pubertal animals or for both the organizational and activational effects of these hormones. They do not provide us with any information that we did not already have from experiments with neonatal or adult castration or neonatal or adult administration of estrogen receptor antagonists.

After elimination of the estrogen receptor β we see no deficit in sexual behavior (Ogawa *et al.*, 1999). This may be a new finding, because there are no published studies of sexual behaviors in animals treated with antagonists for this receptor. In mice lacking both the α and the β estrogen receptors, sexual behavior is completely absent. The males do not even mount (Ogawa *et al.*, 2000). The fact that the combined deletion of the α and the β estrogen receptor genes has more dramatic effects than deletion of the α receptor gene alone suggests that the β receptor has some role at some time, although its absence does not by itself modify male mouse sexual behavior. Nevertheless, data from the different estrogen receptor knockout mice convincingly show that estrogen receptors are needed for the full expression of male sexual behavior. One serious limitation to these studies is that the knockout mice were all generated by gene deletions in the 129 strain and the descendants were backcrossed into the C57/BL6 strain. We know that stimulation of estrogen receptors is necessary for sexual behavior in some strains but not in others, and currently there is no way of knowing if the results obtained in the single knockout strain are applicable to other strains. Of course, it is also impossible to generalize to other species since there are so many data showing that the need for estrogen receptor stimulation in the activation of male sexual behavior is not a universal characteristic.

There are also a few studies that have looked upon sexual behaviors in mice lacking the gene coding for aromatase. This kind of knockout mouse would have much reduced estrogen concentrations in the brain and in other tissues where aromatase normally is present. Like the estrogen receptor knockouts, the aromatase knockouts were generated in the 129 strain and backcrossed into the C57/BL6 (Honda *et al.*, 1998). Not surprisingly, the knockout show severely disrupted copulatory behavior. In addition, the males do not show any approach behaviors to estrus females or their odors in several different test procedures (Bakker *et al.*, 2002). The impaired sexual behavior of mice lacking the aromatase gene is not surprising in view of the results from mice lacking the estrogen receptor genes previously mentioned. Although the aromatase mice may have normal amounts of estrogen receptors, these receptors should be inactive because there are no estrogens that can bind to them. Thus, it should be expected that the behavioral consequences of lack of estrogen receptors or of aromatase should be very similar, an expectation that was experimentally confirmed. While the aromatase knockout mice certainly are most interesting, the data concerning sexual behaviors obtained from them do not offer any new information. The use of a new technique has only confirmed what we already knew. As was the case with the estrogen receptor knockouts, these data are valid only for the C57/BL6 strain. Most unfortunately, they offer no evidence whatsoever as to the role of aromatization in other mouse strains or in other species.

A spontaneous mutation of the androgen receptor has been described in several mammalian species including mice and rats. It was first described in the rat in 1964, but it was not until 1973 that a complete description of it became available (Stanley *et al.*, 1973). The animals affected were called Tfm (testicular feminized male). A single transition mutation within exon E of the androgen receptor gene is normally the cause (Yarbrough *et al.*, 1990). An equivalent condition in mice has also been described (Lyon and Hawkes, 1970). Due to a lack of a nucleotide in exon 1 of the androgen receptor gene, the receptor does not bind to the DNA molecule and has no effect when activated by androgens (Gaspar *et al.*, 1991; He *et al.*, 1991). Tfm rats and mice are identical to the human in the way that genetic males have female external genitalia and functional, non-descended testes. Many endocrine and other kinds of physiological studies have been performed on these animals, but there are few studies of sexual behavior (Olsen, 1979, 1992). Nevertheless, the existing studies show that male sexual behavior is severely impaired in tfm rats and mice, independently of whether the animals are castrated or not, and independently of adult hormone treatments (Ohno *et al.*, 1974). The data from studies on this natural mutation are unequivocal in the way that they clearly establish that androgen receptors are necessary for masculine sexual behavior in rats and mice.

This idea is further strengthened by a study employing mice in which the androgen receptor gene was artificially altered. The appearance of these male mice was that of a female, exactly as is the case in the natural testicular feminization syndrome. Furthermore, they displayed no sexual behavior at all, neither when intact nor when castrated and treated with dihydrotestosterone. All the wild type littermates mounted, intromitted and ejaculated when intact and displayed enhanced copulatory behavior after treatment with dihydrotestosterone (Sato *et al.*, 2004). The androgen receptor knockout mice were generated from embryos of the CD-1 strain and floxed C57/BL6 mice were backcrossed with CMV-Cre transgenic mice, according to the methods description. This elegant procedure is certainly comprehensible to some molecular biologists, but I must admit that the only thing clear to me is that the 129 strain was not involved. I mention this only because strain may be, as we already know, of crucial importance in studies of mouse behavior. Independently of this, and also taking into account the data from animals with the androgen insensitivity syndrome, we can probably accept the conclusion that a functional androgen receptor is necessary for male sexual behavior in at least some strains of mice and rats.

1.7 Studies on men with spontaneous gene deletions

There is one report of a man lacking a functional estrogen receptor because of a premature stop codon in exon 2 of the estrogen receptor gene (Smith *et al.*, 1994). This man showed no response to administration of estradiol despite a large increase in blood estradiol concentration. The patient had normal male genitalia with bilateral descended testes, each with a volume of 20–25 ml, and a normal-sized prostate gland. Semen analysis showed a reduced sperm motility (viability of 18% against around 50% in normal men) with a normal sperm count of 25 million per millilitre. He indicated strong heterosexual interests and reported normal sexual functioning, including presence of morning erections and nocturnal emissions. From the clinical data reported, it is quite evident that the man lacked the estrogen receptor α. Although data from a single case study are far from sufficient for any firm conclusion, the evident lack of alterations of sexual behaviors in this man is remarkably different from the severe consequences of the absence of the estrogen receptor α in male mice. In addition to underlining possible crucial species differences, the data from the human αERKO, so to say, reinforce the notion that estrogens are not important for sexual behaviors in the human male.

Some men have an inactivating mutation in the gene coding for aromatase, CYP19. Up to 2005, seven cases have been identified and studied (Jones *et al.*, 2006). This number is very low and any conclusion obtained from the studies of these seven men must be regarded as very uncertain. Nevertheless, sexual behaviors appear to be of normal intensity in all seven, although fertility may be reduced by the congenital lack of estrogens (Simpson, 2004). Plasma hormone concentrations are variable, with some men showing elevated testosterone concentrations (Morishima *et al.*, 1995), while others have concentrations close to the lower end of the normal range (Carani *et al.*, 1997). Obviously, estrogen is not found in the blood of these men. One case study suggests that estrogen treatment enhances sexual functioning above pretreatment levels (Carani *et al.*, 1999), while another study fails to find this effect (Carani *et al.*, 2005). However, the combined treatment with estradiol + testosterone was reported to increase masturbation, morning erections, erotic fantasies and 'libido' in one man. Actual copulatory behaviors were non-existent, because the patient was a catholic priest taking his faith seriously. In any case, these studies do not offer sufficiently clear results for any conclusion at all. Considering the previously mentioned modest or absent effects on sexual behaviors of estrogen receptor blockade and the apparently normal sexual behaviors in men lacking functional aromatase, I would suggest that activation of estrogen receptors is not a requisite for sexual behaviors in men. Here we have a nice example of a rather drastic species difference. While males of some mouse strains, like the 129 backcrossed into the C57/BL6, show a severely impaired sexual behavior if estrogen receptors are not stimulated, sexual behavior in the human male appears essentially independent of estrogen receptor activation.

The proposal of a modest or absent role for estrogen receptors in the control of human male sexual behavior is supported by additional observations. Most interestingly, just as in rodents there exist natural mutations in the androgen receptor gene that lead to a receptor that is entirely non-functional or that has a much reduced function. These rare mutations and their consequences were actually described in humans before similar mutations were found in other mammals. The most frequent of the human variant appears to be a point mutation in the androgen receptor gene localized on the long arm of the X-chromosome (Xq11–13). In addition to point mutations, several other kinds of errors in the androgen receptor gene have been described. In the case of a total loss of function, the complete androgen insensitivity syndrome appears. The affected individuals are born with female external genitalia, while having functional, non-descended testicles, producing amounts of androgens similar to those in a normal male. Since the fetal testicles secreted Müllerian inhibiting hormone they lack female internal genitalia, i.e. there are no fallopian tubes, no uterus and no upper vagina. Nevertheless, the external appearance is that of a female and therefore these genetic males are usually raised as girls. When androgen binding affinity is reduced but not absent, a conditioned called partial androgen insensitivity syndrome, the phenotype is most variable. It ranges from virilized female external genitalia (in its mildest form clitorimegaly without other anomalies) to almost normal male genitals, albeit with penile hypotrophy and/or hypospadia. For excellent summaries of both the complete and incomplete androgen insensitivity syndrome, the reader is referred to French *et al.* (1965) or to Patterson *et al.* (1994). Because of the large variability within the partial androgen insensitivity syndrome, I will concentrate the following discussion on the complete form.

It may be important to observe that what is now called androgen insensitivity syndrome was called testicular feminization for many years. The latter is the name found in most of the older literature and also in reports from non-human animals, as mentioned in the preceding section. Now and then the name Morris' disease (after the physician who first described it) is used.

Very few studies have investigated the psychosexual development in individuals suffering from complete androgen insensitivity syndrome, i.e. genetic males raised as females. This is in sharp contrast to the many reports of physiopathological aspects of the syndrome. In fact, there is only one study of psychosexual development satisfying elementary methodological criteria (Wisniewski *et al.*, 2000). According to that study, individuals with complete androgen insensitivity show a normal female sexual behavior, have normal levels of 'libido' and normal levels of sexual satisfaction. Some cases required surgical intervention because of a shallow vagina and most individuals had had their testicles removed. Vaginoplasty as well as gonadectomy were usually performed during early adolescence. Some patients received estrogen treatment. Since compliance with estrogen therapy was low, this latter did not seem to be of major importance. These observations suggest that a functional androgen receptor is the sole source of masculinization. The androgen insensitive fetus must have been exposed to estradiol produced by aromatization from testosterone, so estrogen receptors do not seem to contribute to masculinization in the human. Interestingly, presence of a Y-chromosome, of testicles (at least until adolescence), of above normal body length etc. does not appear to affect psychosexual development as a female. It must be noted, though, that this study included only 13 subjects. As a curiosity it may also be mentioned that one of them displayed homosexual practices. Nevertheless, the results from this study coincide with the notion that the human brain differentiates in a female direction in the absence of testosterone actions.

It could, obviously, be argued that the assignment of a female gender at birth, and the subsequent up-bringing as a girl, is the cause of the female identity. However, there are cases where genetic males with functional androgen receptors and normal testicular androgen production have been assigned female gender and subjected to surgery to change the external genitalia to female shortly after birth. The cause for this has usually been insults causing irreparable damage to the penis. Many of these individuals show a male-like behavior during childhood, and at puberty they find females more attractive than males (reviewed in Diamond and Sigmundson, 1997). This is exactly the opposite to what was observed with men suffering from complete androgen insensitivity, and would suggest that the organizing and activating effects of testosterone in the human is dependent on the androgen receptor.

It should be observed that clinical studies of rare diseases are almost always limited to very few cases and results are influenced by a variety of uncontrolled and/or unknown factors. This makes the establishment of cause–effect relationships almost impossible and we should avoid the founding of vast hypotheses upon this kind of half-anecdotal observation.

1.8 Conclusions concerning the hormonal control of male sexual behavior

Testicular hormones are necessary for sexual behaviors in all mammals, including the human. Castration and other means of reducing testicular hormone concentrations always lead to reduced intensity of sexual behavior. This statement does not imply that reduction of blood androgen concentrations to a very low level *eliminates* sexual behavior in every individual. We have had a few castrated rats that copulated with exactly the same intensity as intact rats for more than 2 months after castration. At sacrifice, their accessory sexual glands were completely atrophied, as is always the case in castrated animals, showing that blood androgen concentration was as much reduced in these animals as it was in the castrated rats that showed no copulatory behavior at all. The proportion of rats maintaining copulatory behavior for a long time in the absence of testicular hormones is well under 1%, at least in my laboratory. This kind of most unusual individual does not in any way contradict the conclusion that sexual behavior is much reduced when androgen concentrations are reduced. They just illustrate a very fundamental principle in medicine, biology, psychology and many other sciences: interindividual variation is considerable and, if we want to come across an exception to a principle, there is always one to find. By the way, the existence of exceptional rats and humans should make us very sceptical to case studies.

Since the main testicular hormone, testosterone, may be reduced to dihydrotestosterone and aromatized to estradiol, and since both 5α-reductase and aromatase are present at brain sites important for sexual behaviors, it is quite logical to ask whether these products of testosterone metabolism rather than testosterone itself are responsible for the behavioral actions of testicular hormones. As we have seen, there is no unequivocal answer to this question. Among rodents, it appears that both estradiol and consequently estrogen receptors, and dihydrotestosterone, and consequently the androgen receptor, are of importance for normal masculine sexual behavior in some species. However, sexual behaviors can be induced by the exclusive stimulation of androgen receptors in other rodent species. In non-rodent species, the confusion is still more pronounced. Some, like rabbits and rhesus monkeys, do not respond at all to estrogen treatment but respond well to treatment with dihydrotestosterone, showing that androgen receptor stimulation alone is sufficient for male sexual behaviors. Others, like pigs (Levis and Ford, 1989) and rams (D'Occhio *et al.*, 1985), seem to require simultaneous stimulation of androgen and estrogen receptors for the expression of the complete copulatory behavior pattern. The cynomolgus monkey represents a case where no firm conclusion is possible. In the human, we do not have enough data for proposing a definitive hypothesis, but there is no doubt that relevant clinical studies suggest that the androgen receptor alone is involved.

2 Females

The ovaries are hidden within the abdominal cavity. This may be a reason why they have not attracted so much attention as the clearly visible testicles. Furthermore, it has not been possible surgically to eliminate the ovaries and assure survival of the victim until recently. Therefore, there are no stories of great deeds performed by spayed women in antiquity, like the ones we have about eunuchs. In the same way, we have no old anecdotes about the sexual consequences of removal of the ovaries in women. This lack of historical interest in the sexual glands of women as well as of females of other mammalian species forces us to focus our attention on rather recent contributions. The word recent is here employed in relation to a time scale of the same length as the one employed during my discussion of eunuchs.

2.1 The role of ovarian hormones in non-primate female mammals

All those interested in non-human sexual behavior and particularly those interested in the sexual behavior of rodents have noticed that females are not like males. While one of the main attributes of the male character is its constancy, females are highly unstable. This is evident in, for example, female rats. One day, a female may display intense hopping, darting and ear wiggling as soon as approached by a male, and his tender mounts activate the most beautiful of lordoses. Twelve hours later, the same female may abstain from all reactions to the male's approaches, and if he would dare to mount he would be kicked in his face with a back paw. If he didn't understand the message but continued to mount, it is most likely that the female would turn on her back and push the male away with all four paws. If that did not put him off, the female would not hesitate to bite. Four or five days later, the female will again display hopping, darting, ear wiggling and lordosis in response to the male while all rejective behaviors will be suppressed. Such rapid and extremely systematic variations in the female's propensity to engage in sexual interactions and the associated changes in response to the male's approaches contrast dramatically to the male rat's behavior. He will try to copulate with any female at any time from puberty until well into old age and he systematically responds to a female's invitation with cordiality and enthusiasm. One rare exception might occur if the male were already sexually exhausted. Then he would probably respond with indifference. Misbehaviors like biting are, however, extremely unlikely.

The causes of the systematic variations in female sexual behavior have intrigued scientists, and probably also laymen, for a very long time. However, it was not until the 1920s that experimental studies of the physiological underpinnings of female sexual behavior became popular. In a classic paper, Long and Evans (1922) described not only the regular variation in female rat sexual behavior, but also the equally regular changes in vaginal cytology. Both these events were soon found to depend on secretions from the ovaries (Allen and Doisy, 1923). The Danish endocrinologist, Axel Hemmingsen, published a splendid, detailed study of the behavioral and physiological effects of ovarian, more precisely follicular, hormones in 1933. When he began his research in rats and mice in 1924, the chemical structure of the follicular hormones was unknown. At that time the hormones were normally obtained from follicular extracts and the active material that was isolated from such extracts was called oestrin. In order to evaluate carefully the actions of oestrin on female sexual responses, Hemmingsen devised a set of behavioral criteria for determining the intensity of these responses. The lowest intensity was described in the following way:

> The female runs away from the male, defends herself violently with one of the hind legs, or rises breast to breast to the male, or rolls on her back, or rises her hind legs, making mounting impossible. Quivering of the ears is entirely absent (Hemmingsen, 1933, p. 119).

This is a very poetic description of what we now would consider behaviors typical of a non-receptive female. An intermediate intensity of sexual behavior was described in this way: 'When mounted, the female shows both lordosis and quivering of the ears'. A slightly more intense reaction to the male occurs when: 'quivering of the ears is provoked by the male smelling at the vagina, but not by the mere presence of a male'. The highest level of sexual excitability was expressed as:

> spontaneous quivering of the ears (perhaps also the darting movements?) is seen even if there are only other normal females present (perhaps also if the female is alone?) (Hemmingsen, 1933, p. 123).

These behavioral criteria for quantifying the intensity of sexual behavior, and the magnitude of the oestrin effect, were based on careful cinematographic analyses of female rat sexual behavior. More than 70 years after its publication in a rather obscure journal, it is a real intellectual treat to read this brilliant paper. I have made this rather extensive mention of it only to pay tribute to an almost forgotten pioneer in the field of behavioral endocrinology.

The crucial role of follicular hormones for female sexual behavior is not limited to rats and mice. I will not review the huge amount of data that has been accumulated over the years, but it appears to be quite obvious that these hormones are necessary for the display of sexual behavior in most mammals. Possible exceptions include some primates, like the human, as we will see in a couple of paragraphs. However, follicular hormones are not the only ovarian products that are involved in female sexual behavior. Although the follicular hormones, i.e. estradiol, estrone or estriol, can activate complete female rat sexual behavior if administered in doses large enough, the addition of a small dose of progesterone renders a small, otherwise ineffective, dose of estradiol capable of activating all aspects of sex behavior (Beach, 1942b). In rats, estrogens usually need to be administered at least 24 hours before progesterone in order for the latter hormone to be effective. It is now known that estrogens activate the gene coding for the progesterone receptor, something resulting in a huge increase in the number of cytosolic progesterone receptors (see, for example, Scott *et al.*, 2002 and references therein). I will not describe the molecular actions of estrogens, or of progesterone, but it can be interesting to remember that the sequential actions of estrogens and progesterone appear to be important for female sexual behavior in many rodents. In other mammals, for example the ewe, the most efficient order of hormone administration is the reverse. If progesterone is given for a few days before estrogen treatment, then the estrogen dose needed to induce receptivity is reduced (Robinson, 1954, 1955; Caraty *et al.*, 2002). The molecular mechanisms underlying this are not known. In other species, female sexual behaviors are activated exclusively by estradiol and progesterone may have an inhibitory rather than facilitatory action. This is the case in rabbits and cats, for example (Michael and Scott, 1964; Beyer *et al.*, 1969; Beyer and McDonald, 1973). Thus, it is quite important to keep in mind that the hormone treatment regimen most efficient in rats is not necessarily so in females of other mammalian species.

2.2 The role of ovarian and adrenal hormones in female primates

While the paramount role of estrogens, and sometimes also progesterone, in the control of female sexual behaviors is clearly established in most mammalian species, the situation is less clear in primates. While non-primate female mammals usually display sexual behavior only during a limited period of the estrus cycle, generally the period during which estrogen concentrations are high, female monkeys copulate at any time during the cycle (Zuckerman, 1930). The first study that established this fact was published only a few years after the first descriptions of the dramatic cyclic variations in female rat sexual behaviors. This temporal coincidence probably contributed to the now firmly established notion that the hormone dependency of female rodent sexual behavior is far superior to that found in primates. There is, undoubtedly, a substantial amount of support for this notion, but female primate sexual behavior is, as we will see, by no means independent of gonadal hormones. Nevertheless, the fact that female monkeys copulate throughout the menstrual cycle was soon confirmed in a carefully designed study (Ball and Hartman, 1935). Many later studies have described copulatory behavior throughout the primate menstrual cycle (e.g. Michael and Zumpe, 1970; Slob *et al.*, 1978; Dixson, 2001). Some researchers report increases in some aspects of female sexual behaviors in the periovulatory period (e.g. Czaja and Bielert, 1975), while others fail to find any systematic variation during the menstrual cycle (e.g. Nikitopoulos *et al.*, 2005).

A particularly interesting example of a primate female with almost constant sexual behaviors throughout the menstrual cycle is the capuchin monkey. At difference to the rhesus and many other primate species in which the females show evident exterior signs of the periovulatory phase, like substantial swelling of the sexual skin, the capuchin monkey has concealed ovulation. With that is understood that there is no morphological or known behavioral changes associated with ovulation. Nevertheless, the male partner displays enhanced affiliative and copulatory behaviors during the periovulatory phase (Carnegie *et al.*, 2005). Since neither the female's behavior nor her appearance is different at that period, the males must use some unknown sign or signs for discovering that ovulation is around. If females in a species with concealed ovulation are more attractive to males around ovulation, it is no wonder that females in species displaying exterior morphological and/or behavioral signs during that period also are more attractive than at other periods of the menstrual cycle. The often described enhanced periovulatory sexual activity in female primates with visible signs of imminent ovulation can, therefore, be entirely explained as a consequence of more intense behavior on the part of the male during this period, as suggested many years ago (Michael and Bonsall, 1977). It would appear, then, that primate female sexual behavior is little influenced by the cyclic variations in blood concentrations of ovarian hormones.

The absence of a clear-cut relationship between phase of the menstrual cycle and the female monkey's ability to display sexual behaviors raises immediately doubts as to the importance of ovarian hormones in the control of that behavior. In fact, sexual behavior persists after ovariectomy in several primate species, including the rhesus, stumptail macaque, talapoin and marmoset (Slob *et al.*, 1978; Dixson, 2001). Some behavioral changes are frequently observed, but most of these changes may be attributed to modifications of the male's responses to ovariectomized females (Herbert, 1970). Such females are considerably less attractive to males than intact females are and this leads to reduced intensity of sexual activity. At difference to the slight effects of ovariectomy, suppression of adrenal androgen secretion with dexamethasone (Everitt and Herbert, 1971) or adrenalectomy (Everitt *et al.*, 1972) causes a strong reduction of female sexual behavior, at least in the rhesus monkey. Treatment with estradiol benzoate fails to restore sexual behaviors in these animals, while androgens like testosterone or androstenedione readily do so. These observations suggest that androgens rather than estrogens are important for female primate sexual behavior, at least in the rhesus monkey.

In other primate species, like the stumptail macaque, it appears that female sexual behavior is entirely independent of hormones. I already mentioned that ovariectomy fails to affect sexual behaviors in females of this species and there are data showing that these behaviors persist undiminished after treatment of ovariectomized females with dexamethasone or after combined ovari- and adrenalectomy (Baum *et al.*, 1978). Treatment with either estradiol or testosterone had little effect, which is not particularly surprising since the absence of these hormones did not reduce sexual behaviors in the first place. This carefully performed study could suggest that there are important species differences among primates with regard to the importance of androgens. While they seem to be crucial in rhesus females, they are not necessary for stumptail macaques.

Rather than using species differences as an explanation for conflicting observations, it can also be proposed that experimental results are unreliable, varying from one experiment to another. This latter idea is somewhat strengthened by the results of an annoying study in the rhesus monkey. Instead of reducing androgen availability by suppression of adrenal hormone production through treatment with dexamethasone or adrenalectomy as was done in the Everitt and Herbert (1971) and Everitt *et al.* (1972) studies, this study employed an androgen receptor antagonist. It could reasonably be predicted that the functional consequences of blocking receptors should be quite similar to those of reducing androgen availability. Unfortunately, this did not turn out to be the case. The androgen antagonist flutamide failed to affect sexual behaviors when given to intact, cycling rhesus females (Johnson and Phoenix, 1978). I called this an annoying study not because it was badly performed but because it is difficult to understand why the results are so different from those obtained in earlier studies. Despite a substantial effort, it has been impossible to find a convincing explanation. The Johnson and Phoenix (1978) experiment did confirm one thing, though. In agreement with many other studies, it was found that stage of the menstrual cycle was unrelated to the intensity of sexual behaviors.

I will not try to review all the studies that have tried to determine the importance of hormones for female, non-human primate sexual behaviors. The examples given in the preceding paragraphs should be enough to illustrate the state of confusion in which that field is submerged. To strengthen that impression it might be appropriate to mention that one prominent scientist still insists upon the existence of systematic variations in female primate sexual behavior according to phase of the estrus cycle with a maximum around ovulation. Associated with this comes also an insistence on the role of estrogens and the lack of a role for androgens. The arguments supporting his points of view have been summarized in several reviews (e.g. Wallen, 1990; Wallen, 2004). I consider myself unable to dismiss all these arguments and I feel for the moment compelled to conclude that the endocrine control of non-human primate female sexual behaviors remains mysterious.

It is impossible to leave this issue without trying to offer some kind of explanation for the current sad state of affairs. There are, in fact, several possible reasons for the confusion. The most important, I believe, is the use of small numbers of animals. Many of the primate studies mentioned here were performed on very few subjects. A number of subjects as low as four pairs is not unusual and few studies employed more than 10 pairs. The low number of subjects may have many practical justifications, the main one probably being the prohibitive costs of maintaining large primate colonies. In addition, primate work is time- and personnel-consuming. None of these arguments have any scientific validity, though. I cannot see much reason to believe that interindividual variability between primates should be any smaller than that found in rats or mice. On the contrary, I would suspect that variability among monkeys is superior to that found between rodents. In addition to a badly known genic variability, it is probably much easier to assure equality of experiences among individuals in a standard rodent colony than it is in a standard monkey colony. The shorter life span of a rodent makes it simply impossible to acquire the same amount of experience as a monkey. It may consequently be suspected that pre-experimental experiences play a larger role in primate than in rodent studies. In view of these trivial considerations and many other unmentioned, I doubt that primate studies *a priori* require fewer subjects than rodent studies. Furthermore, I can see no reason why data obtained from a primate should be more valid than data obtained from a rodent and there is no reason to assume that statistical procedures are more trustworthy when they are employed to analyze primate data than they are when used for rodent data.

To make the arguments exposed in the preceding paragraph clearer, I will give an example from my own rat data. Some time ago we were interested in determining whether a new compound could activate male rat copulatory behavior. During these studies we employed dihydrotestosterone as a kind of control. We ran several experiments, some with rats with very little sexual experience, some with rats with substantial sexual experience. For purposes of illustration, I pick four rats treated with dihydrotestosterone from a group of 12 animals with little experience and another four from a group of 12 animals with more sexual experience. I also make a random sample of four controls (untreated, castrated males) of the 12 used in each experiment. As can be seen in Figure 5.1a, it turns out that the experienced animals are far superior to the controls while the inexperienced animals are not. The conclusion would be that the effects of dihydrotestosterone on male rat sexual behavior depend on experience. If I instead include all 12 animals from each group with their respective control groups of 12 animals, there is no effect at all of dihydrotestosterone and, consequently, none of experience (Figure 5.1b). This example is certainly not perfect, but it should illustrate how random variations between individuals can affect experimental results when the number of subjects is low. As soon as we approach a more reasonable group size, the influence of exceptional subjects becomes much reduced.

Random interindividual variations in response to hormones may be the cause of much of the conflicting primate data. A wise solution to this apparent problem is simply to follow the majority. A majority of studies have reported that female

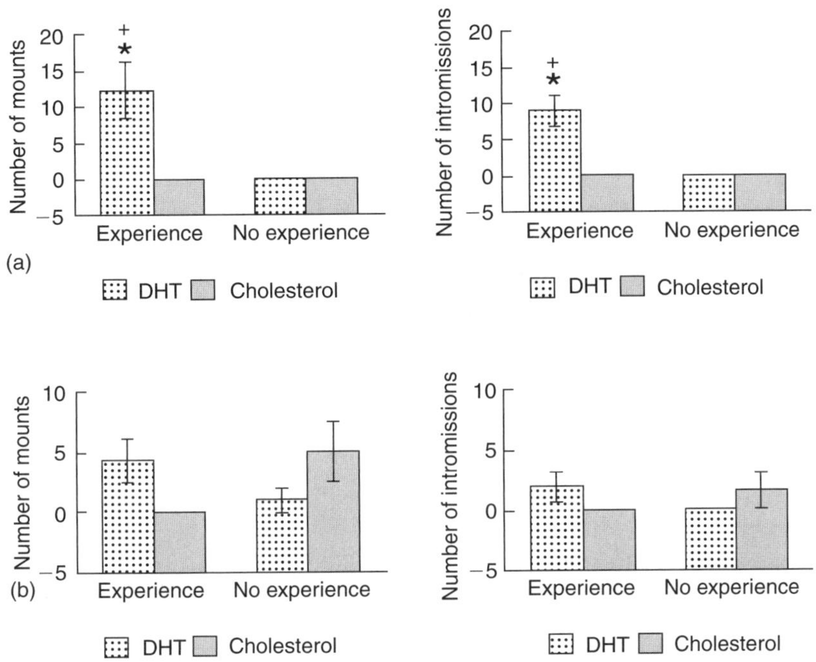

FIGURE 5.1 (a) Mean (± SE) number of mounts and intromissions in groups of four male rats each with extensive sexual experience (at least 15 tests terminated at the end of the first post-ejaculatory interval before the animals were castrated) and subcutaneously implanted with two 25 mm long Silastic® capsules containing dihydrotestosterone or cholesterol 35 days after castration (Experience) and from two other groups of four male rats each that had been exposed to sexually receptive females only twice before castration (No experience). The latter groups were also implanted with subcutaneous Silastic® capsules exactly as the experienced animals were. All four groups of four rats each were selected from treatment groups that contained 12 animals each. (b) Data from all 12 animals included in each group are illustrated in this figure. Note that the significant difference between the groups given dihydrotestosterone and cholesterol obtained with the small group sizes (illustrated in (a)) disappeared when data from more animals were added (as in (b)). Likewise, the significant difference between experienced and inexperienced rats treated with dihydrotestosterone obtained with the small group sizes (a) is no longer found when data from more animals are used (b). Data shown here are from a test performed 21 days after implantation of the Silastic® capsules. DHT, dihydrotestosterone. *, different from cholesterol according to the Mann-Whitney U-test, $P < 0.05$. +, different from animals with no experience (actually animals with little experience), $P < 0.05$. The experiment from which these data stem was performed by Xi Chu.

primate sexual behavior does not vary substantially during the menstrual cycle. We conclude, then, that it is independent of ovarian hormones. A majority of studies have found that androgens rather than estrogens are important. We conclude, then, that female primate sexual behaviors are androgen-dependent. Some may consider that this pragmatic way of reaching a conclusion is superficial. Perhaps it is, but it is at least based on some kind of rational reasoning. Furthermore, when we turn from non-human primates to women, we will find abundant support for both conclusions.

2.3 Hormones and sexual behavior in women

The endocrine control of sexual behavior in the human female has been the subject of a substantial amount of research. Undoubtedly, among the primates there is no species that has attracted as much attention as *Homo sapiens*. There are several good reasons for this. One is that disorders of the reproductive system are rather common, and one treatment for some of these disorders is ovariectomy. The considerable number of women who have been subjected to ovariectomy, from the early fertile age until well into menopause, has allowed researchers to evaluate the behavioral consequences in a rather large number of individuals. Another reason is that ovarian function ceases at a relatively young age in the human. Many women can expect to live 30 years or more after the ovaries have stopped producing the typical ovarian hormones, estrogens and progesterone. In many respects, the endocrine changes associated with menopause are similar to those occurring after ovariectomy. In fact, it has become customary to call the state related to ovariectomy in premenopausal women 'surgical menopause'. The fact that most women will spend a large part of their life without the cyclic variations in ovarian hormones typical of most other mammalian species poses some challenges for any hypothesis trying to relate sexual activity to blood hormone concentrations.

It seems convenient to begin the discussion of the human female with a summary of the hormonal changes occurring after puberty. We all know that women have a menstrual cycle, and any introductory physiology and even psychology textbook will provide us with an excellent description of the hormonal changes during the cycle. Suffice to mention that estradiol, the main follicular estrogen, is mainly secreted during the first half of the cycle with a peak around ovulation. Progesterone, produced by the corpus luteum, is elevated from shortly after ovulation until the beginning of menstruation, when it rapidly drops. This is the picture of the menstrual cycle generally divulged in textbooks and on Internet sites. However, in addition to producing estrogens and progesterone, the ovaries also produce androgens. In fact, human female blood contains several androgens, like dehydroepiandrosterone sulphate, dehydroepiandrosterone, androstenedione, testosterone and dihydrotestosterone. The physiologically most important of these is certainly testosterone. The blood concentration of this hormone varies according to the menstrual cycle, with a peak around ovulation and a minimum during the early follicular phase (Abraham, 1974; Longcope, 1986). About 25% of the testosterone is produced in the ovaries, while another 25% stems from the adrenals. The remaining 50% is produced in peripheral tissues by conversion of other androgens, mainly androstenedione (Burger, 2002).

The blood concentration of testosterone starts to decline in early adulthood, and it has been reported that the testosterone concentration at age 40–50 is about half of that found at age 20 (Zumoff *et al.*, 1995). Thereafter, the decline is slower. Of particular importance here is the fact that menopause fails to affect blood androgen concentrations (Davison *et al.*, 2005). A similar observation had been made a few years earlier (Burger *et al.*, 2000). In contrast to the lack of effect of menopause on blood androgen concentrations, it was found that the concentrations of sex hormone binding globulin was reduced at menopause. The reason for this was suggested to be the significant decline in estradiol occurring at menopause. Since part of the circulating androgens are bound to the sex hormone binding globulin, and thus not immediately bioavailable, a reduction of this globulin may in fact enhance androgen availability. Another study revealed that blood testosterone concentration increased from ages 50 to 89 when the values were adjusted for body mass index (Laughlin *et al.*, 2000). All these data convincingly show that women produce important amounts of androgens from the beginning of the fertile age until the end of life.

There are data suggesting that the ovarian stroma is one source of menopausal androgens. Indeed, ovariectomy in women between 55 and 75 years of age leads to a marked reduction of blood androgen concentrations (Laughlin *et al.*, 2000; Davison *et al.*, 2005). The impact of this reduction on sexual functions is, as we will see, not great but it is large enough to justify mentioning.

We can now place our discussion of the hormonal control of sexual behaviors in women within an appropriate endocrine context. From puberty until the climacteric, there is a large cyclic variation in the blood concentrations of estrogens and progesterone. During menopause, the concentration of these hormones declines sharply. From puberty until the end of life, androgens are also present in the circulation. In premenopausal women, there is a modest cyclic variation in the concentration of these hormones superimposed on a steady decline from age 20 to 45–50. Thereafter, there either is no change or a small increase. Any hypothesis concerning the endocrine control of women's sexual behavior needs to take the hormonal facts into consideration. If we propose that women are like rodents or sheep with estradiol + progesterone as the critical hormones, then we should expect large cyclic variations in sexual activity and a significant reduction during menopause, independently of whether it is surgical or natural. On the contrary, if we believe that women are like other primates with androgens as the crucial hormones, then we must expect a maximum of sexual activity during early fertile age followed by a decline until menopause. After that, sexual activity should remain at a stable level. We should expect ovariectomy, but not menopause, to produce a diminished sexual activity, since the ovaries contribute to the maintenance of blood androgen concentration both before and during menopause. Some slight variation during the menstrual cycle, with a maximum around ovulation when testosterone concentrations are at their highest may also be expected. All these predictions can be confronted to empirical data. Let us begin with the estradiol + progesterone hypothesis and the prediction that sexual activity varies according to the phase of the menstrual cycle, with a maximum around the estrogen peak associated with ovulation.

2.4 Variations in women's sexual behavior during the menstrual cycle

A large number of studies have addressed the issue of menstrual cycle-dependent variations in the intensity of sexual behavior in women. This is a well-liked issue that is discussed in popular magazines from one time to another, so I will not spend many lines on it. Data are conflicting and peaks of sexual intercourse frequency have been reported in the follicular phase just following menstruation (Spitz et al., 1975; Dobbins, 1980), in the periovulatory phase (Udry and Morris, 1968; Adams *et al.*, 1978) and in the luteal phase, just preceding the onset of menstruation (McCance *et al.*, 1937). To judge from the intercourse data, there is no systematic variation in sexual activity during the menstrual cycle. Instead of analyzing intercourse, some researchers have evaluated less tangible aspects of sexual behavior, such as sexual dreams and other mental representations of sexual activities (e.g. Schultz and Koulack, 1980). Frequency of masturbation and some more extravagant aspects of sexual behaviors have also been studied. Questionnaire evaluations of variations in sexual desire during the menstrual cycle have also been performed (Regan, 1996). Results from all these studies are confusing, with some reporting enhancement of dreams, desire or frequency of masturbation in one phase of the cycle, while other find increases in other phases and some find no effect. I see no need for providing a complete list of these studies. Nevertheless, the conclusion is that no conclusion as to an effect of menstrual cycle phase on sexual behaviors is possible. The cause of this regrettable state of affairs is that most of the published studies are based on self-reports of different kinds and introspective data do never inspire much confidence. By the way, the reports on variations in coital frequency are also based on self-reports and are certainly not more convincing than studies of other aspects of sexual behavior. Despite their appeal in the popular press and on some television channels, the kinds of data employed in the research summarized in this paragraph are useless for drawing any scientifically credible conclusion.

Fortunately, we have some more convincing data concerning variations in sexuality during the menstrual cycle. One study, for example, evaluated women's choice of visual entertainment during the estrus cycle. The available choices included a pornographic movie, a romantic movie, a comedy and an action movie. In addition to making a choice, the women were asked to rate the movies according to a 7-point scale ranging from 'a real bomb' to 'great fun and excitement'. It turned out that women in the period immediately following menstruation made most choices of pornographic films and rated such films more favorably than they did at other periods of the cycle. The lowest interest in pornographic movies was detected in the periovulatory period (Zillman *et al.*, 1994). A still more objective way to evaluate cyclic variations in sexual behaviors is to determine genital blood flow in response to sexual incentives in women in different phases of the menstrual cycle. This has been done in a few experiments. One study employed sexual fantasies in addition to pornographic movies as sexual incentives. No difference in vaginal response between phases of the menstrual cycle could be detected independently of whether fantasy or movie was used as incentive (Meuwissen and Over, 1992). The lack of effect of cycle phase on the vaginal response to a pornographic movie had been described earlier (Morrell *et al.*, 1984). Likewise, sexual arousal, as always quantified as vaginal blood flow, induced by audiotaped descriptions of sexual activities or fantasies failed to vary during the estrus cycle (Hoon *et al.*, 1982). Thus, three studies of vaginal response to sexual incentives coincide in concluding that the menstrual cycle does not modify that response. Another study, again evaluating vaginal blood flow in response to audiotapes with sexual content or fantasies with such content, found that the response to the sexual incentives was at a minimum during the periovulatory phase (Schreiner-Engel *et al.*, 1981). Thus, when estrogen concentrations are at their maximum, the vaginal response to a sexual incentive is at its minimum. This observation does certainly not support the notion that ovarian hormones control sexual behavior in the human female.

The four studies of sexual arousal mentioned so far estimated vaginal blood flow by photoplethysmography. An entirely different procedure for estimating vaginal response was used in a Dutch study. The temperature of a labium minus was recorded and it was supposed that an increase in temperature would reflect enhanced blood flow. Pornographic videos were used as sexual incentives. Although the videos turned out to increase temperature in the labia minora, there was no systematic variation in response magnitude during the menstrual cycle (Slob *et al.*, 1991). Thus, this study confirms the majority of those mentioned above despite the use of a different procedure. It cannot be too risky to conclude that as soon as an objective measure of sexual response is employed, the variation during the menstrual cycle is absent or at most marginal. To my critical mind, it must be concluded that any variation in sexual behaviors occurring during the menstrual cycle is of such a small magnitude that it can have no scientific or practical importance. Unlike rats and mice, women do not show any dramatic cyclic variation in the intensity of sexual behavior. Except, perhaps, an unusually low sexual activity during menstruation. This, however, can easily be explained by social conventions without any need for reference to endocrine mechanisms. Having mentioned social conventions, I cannot refrain from drawing the reader's attention to a nice little paper addressing the issue of research on the menstrual cycle.

> Knowledge claims by scientists and clinicians concerning the menstrual cycle are embedded in a history of misogyny: they are informed by popular beliefs which in turn reflect the culture of patriarchy rather than the experiences of women. Women, as subject to patriarchy, internalize misogynous myth and have beliefs about women as a group, from which they variously distance themselves (Nicolson, 1995, p. 783).

I was astonished myself when I discovered the amount of research that has been dedicated to a search for menstrual cycle effects on sexual behaviors in the human female. It should have been evident already several decades ago that if such effects indeed existed, they were too small to merit further research. In the absence of a rational explanation for the research effort invested in the issue, I must admit that the proposal made in the Nicolson paper is not directly outrageous.

2.5 Changes in women's sexual behavior at menopause

The preceding brief review of research on variations in sexual behaviors during the menstrual cycle did not offer much support for the hypothesis that ovarian hormones are closely involved in the control of sexual behaviors in women. The consequences of menopause, surgical or natural, for these behaviors do not seem to be more impressive than those of the menstrual cycle. The lack of effect of ovariectomy on sexual behavior in women was described many years ago (Filler and Drezner, 1944; Werner, 1946; Kinsey *et al.*, 1953) and has later been confirmed in a number of studies. Likewise, menopause does not seem to be associated with any particular decline in sexual activity. As always, questionnaire and self-report studies have given conflicting results and conclusions of such studies range from no effect (e.g. Kinsey *et al.*, 1953) to a substantial decline (Dennerstein *et al.*, 2002). Sexual activity is known to diminish with increasing age and this fact may explain the notion held by some that menopause is causally related to a reduced sexual activity. There is inevitably an age difference between pre- and menopausal women, and the effects attributed to menopause may just as well be attributed to the passage of time. Socially created expectations of a reduced sexual activity in menopausal women may be self-fulfilling. For the more cultivated reader I suggest a superb article written by two German psychoanalysts, published in German, where the rationale for our prejudices regarding appropriate climacteric female sexual behavior is given (Sies and Nestler, 1992). It is also useful to remember that social desirability is one of the main determinants of answers to questionnaire items (Hebert *et al.*, 1997; Plaud *et al.*, 1998). This important fact may account for the menopause-associated decline in sexual activities sometimes found. Furthermore, there are data suggesting that such a decline may be related to health and marital condition more than to menopause itself (Avis *et al.*, 2000). Rather than wasting time discussing the impressive amounts of paper employed in questionnaire studies, I will turn to some solid experimental data that will convince us that menopause has marginal effects on sex behavior functions. Obviously, I do not consider fertility as part of sexual behavior.

Vaginal blood flow in pre- and menopausal women was determined at baseline and during exposure to a pornographic movie. The premenopausal women had significantly higher baseline vaginal pulse amplitudes as determined by the photoplethysmographic procedure than the menopausal women. The authors explained this difference as a consequence of less estrogen actions on the vaginal epithelium in these women. I can see no reason to disagree. What is far more interesting, though, is that the response to the pornographic movie was similar in both groups, meaning that the menopausal women enhanced vaginal blood flow as much as the premenopausal women did (Laan and van Lunsen, 1997). A completely different procedure for evaluating genital blood flow responses to sexual incentives has been employed in several studies. Comparisons of MRI images of the female genitals obtained before, during and after viewing a pornographic movie show that exposure to sexual incentives enhanced the vestibular bulb, the labia minora and the clitoris to an equal extent in pre- and menopausal women (Suh *et al.*, 2004). Similar data had been published previously from the same group (Maravilla *et al.*, 2003b). Interestingly, clitoral size in the unaroused state does not differ between pre- and menopausal women (Maravilla *et al.*, 2003a). All these objective data point in the same direction, suggesting that menopause does not affect the genital response to sexual incentives. This being the case, it is possible to suggest that the effects of menopause on sexual functions observed in some questionnaire studies are indirect consequences of factors unrelated to sexual functions *per se*. Some of these factors have been mentioned above, and I will not repeat them here. Nevertheless, I conclude by stating that the considerable reduction in circulating estrogens typical of the climacteric and lasting for the rest of women's life has slight impact on sexual functions. This fact, in addition to the marginal or non-existent variations in sexual behaviors during the menstrual cycle despite the large variations in estrogen concentrations, inevitably suggests that neither short- nor long-term changes in estrogen availability is an important determinant of sexual behavior in the human female.

Before leaving the discussion of the effects of menopause, I need to make another comment on word usage. Menopause is, according to the online version of the Oxford Dictionary of the English Language, defined as 'permanent cessation of menstruation' (retrieved from http://dictionary.oed.com/cgi/entry/50173164?query_type=word&queryword=menopause&first=1&max_to_show=10&single=1&sort_type=alpha on July 5, 2006). In any woman, this period begins at the end of the climacteric and continues until the end of life. The word 'postmenopausal', then, acquires a peculiar meaning: after the menopause. Considering that menopause ends with the end of life, a postmenopausal woman must indeed have a low sexual activity, if any at all. Despite this quite obvious fact, some scientists report data on behavior in postmenopausal women. I wonder how these data were collected.

2.6 The role of androgens in women

If androgens are of importance for sexual behaviors in women, then we should expect no or minor variation during the menstrual cycle. As might be remembered, some studies have reported a slight increase in blood androgen concentrations around ovulation, but that increase is very small compared to what is seen with estrogens. The data reviewed a couple of paragraphs above show that variations in sexual functions during the menstrual cycle are small and unreliable, perhaps even absent. This coincides nicely with the androgen hypothesis. Ovariectomy has no effect on sexual behavior in women, as we have seen. We know that this intervention reduces androgen concentrations with as little as 25% (Judd *et al.*, 1974), or at most with 50% (Laughlin *et al.*, 2000). The concentration of estrogens is, not surprisingly, reduced far more. The modest reduction in androgen concentrations does not seem to affect sexual functions. This lack of effect is, nevertheless, easier to account for than the same lack of effect of the large reduction in estrogens. In men, it is known that the physiological concentration of androgens is above that needed for maintenance of sexual behaviors (Bhasin *et al.*, 2001). In women, it has been suggested that androgen insufficiency might be suspected as a cause for impaired sexual behaviors when the blood concentration is below 25% of the mean concentration for the age (Rivera-Woll *et al.*, 2004). If this is true, then the reduction in androgen concentrations following ovariectomy should not be large enough to affect sexual behaviors. This reasoning is supported by data mentioned above.

In order to observe a reduction in sexual behavior in women caused by androgen deficiency, a large decline in blood concentration appears to be necessary. There are indeed some women whose blood androgen concentration is extremely low. This is the case in women who have been subjected to adrenalectomy in addition to ovariectomy. Not only are the main sources of testosterone suppressed, but also most of the testosterone precursors that normally are transformed into testosterone in peripheral tissues are removed. The adrenal cortex is the main site of synthesis of these precursors, and extirpation of the adrenal will eliminate production of them. In adrenalectomized, ovariectomized women, sexual behaviors are much reduced (Waxenberg *et al.*, 1959) and they can be restored by treatment with androgens. Severe androgen deficiency following treatment with cytotoxic agents in ovariectomized women has also been reported to markedly decrease sexual responsiveness (Kaplan and Owett, 1993). These observations render strong support to the notion that androgens indeed are crucial for sexual behaviors in women. The deleterious effect of adrenalectomy + ovariectomy cannot be explained by a deteriorated health after these operations, because the women concerned reported improved health following surgery. The reason for removing the adrenals in addition to the ovaries was to eliminate all sources of estrogens in breast cancer patients, and getting rid of the tumor as well as the prospect of avoiding relapse had, not unexpectedly, beneficial health effects.

Since plasma androgen concentrations are not reduced during menopause, we should expect pre- and menopausal women to have the same level of sexual behaviors, provided that other factors that might affect sexual activity are excluded. In view of the lack of effect of menopause on the genital response to sexual incentives, we must conclude that this prediction is confirmed. The marginal importance of phase of the menstrual cycle, ovariectomy and menopause all suggest that androgens rather than estrogens are crucial for sexual behaviors in women. The large inhibitory effects of adrenalectomy combined with ovariectomy, and its reversal by androgen replacement, gives further support to the notion that androgens are important. I conclude by suggesting that women are more similar to other primates than to rodents with regard to the endocrine control of sexual behaviors, nothing else.

3 General conclusion

We have seen that that the hormonal control of sexual behaviors is far from elucidated. There is a substantial amount of confusing observations in non-human males and females. The importance of aromatization for male sexual behavior appears to vary between species, and also between strains within a species. Consequently, in several mammalian species it is unclear whether the androgen receptor or one or both of the estrogen receptors or all gonadal hormone receptors are necessary for the full expression of sexual behavior in males. Most fortunately, the terrible uncertainty reigning with regard to the hormonal control of non-primate male sexual behaviors disappears when we look at data from primates. In male monkeys and apes, including men, most evidence shows that aromatization is not necessary. Thus, the androgen receptor is sufficient and necessary for male primate sexual behavior. The primate data are quite consistent on this issue, making the preceding conclusion close to turn into an established truth.

In females, there is a considerable variation between mammalian species. In some, estrogen alone is enough for activation of full sexual behavior, and progesterone may be inhibitory in some of these species. In others, progesterone should precede estrogens for the full activation of receptivity. In rodents, estrogens synergize with progesterone in the induction of female sexual behaviors. It is impossible to find a general principle guiding the action of ovarian hormones on sexual behavior in non-primate mammals. The situation is not different in female non-human primates. We are again in the midst of confusion. Almost every hypothesis concerning the endocrine control of sexual behavior in these primates is debatable and the amount of contradictory observations is staggering. Fortunately, the situation is different for the human female. In women, all available data indicate that androgens are responsible for the activation of sexual behaviors while ovarian hormones are of marginal or no importance.

The endocrine control of human sexual behavior is better understood than that of any other mammal. This is a most surprising conclusion, since a huge amount of experimental work has been dedicated to non-human mammals. Rather than illuminating us, much of this research has increased the confusion. This is not the moment to pursue explanations for this state of affairs. It is amusing, though, that the most complex of all mammalian species turned out to be the one giving the most consistent data. How this is possible can be an excellent subject for speculation during periods of insomnia.

Summary

1 Some effects of castration of the human male have been known since antiquity. More recent data strongly suggest that stimulation of the androgen receptor is both necessary and sufficient for the full expression of sexual behavior in the human male.

2 Data from male non-human primates coincide with the human data in suggesting that the androgen receptor is crucial for sexual behaviors. Other gonadal hormone receptors do not seem to be involved in any significant way.
3 In non-primate male mammals the endocrine control of sexual behavior is unclear. In some species aromatization of testosterone to estradiol appears to be necessary, whereas this is not the case in other species. There may also be variations between strains within a species.
4 Female sexual behavior is dependent on ovarian hormones in non-primate mammals. There are frequently dramatic variations in sexual behavior between different phases of the estrus cycle or between breeding season and non-breeding season. In some species, estrogen alone is sufficient, in others estrogen synergizes with progesterone.
5 In non-human female primates, the role of ovarian hormones is unclear. However, most primate females are sexually receptive throughout the menstrual cycle. Androgens may be of importance, but data are contradictory.
6 In women, variations in sexual behavior during the menstrual cycle are minimal. Menopause does not seem to be a factor affecting sexual behavior more than marginally, at most.
7 Abundant evidence suggests that sexual behavior in the human female is dependent on androgens.

References

Abraham, G.E. (1974). Ovarian and adrenal contribution to peripheral androgens during the menstrual cycle. *J Clin Endocr Metab* **39**, 340–346

Adams, D., Gold, A. and Burt, A. (1978). Rise in female-initiated sexual activity at ovulation and its suppression by oral contraceptives. *New Engl J Med* **299**, 1145–1150

Ågmo, A. (1975). Cyproterone acetate diminishes sexual activity in male rabbits. *J Reprod Fert* **44**, 69–75

Ågmo, A. (1977). The comparative actions of fluoxymesterone and testosterone on sexual behavior and accessory sexual glands in castrated rabbits. *Horm Behav* **9**, 112–119

Ågmo, A. and Södersten, P. (1975). Sexual behavior in castrated rabbits treated with testosterone, oestradiol, dihydrotestosterone or estradiol in combination with dihydrotestosterone. *J Endocr* **67**, 327–332

Allen, E. and Doisy, E.A. (1923). An ovarian hormone: preliminary report of its localization, extraction and partial purification, and action in test animals. *J Am Med Assoc* **81**, 819–821

Alsum, P. and Goy, R.W. (1974). Action of esters of testosterone, dihydrotestosterone, or estradiol on sexual behavior in castrated male guinea pigs. *Horm Behav* **5**, 207–217

Avis, N.E., Stellato, R., Crawford, S., Johannes, C. and Longcope, C. (2000). Is there an association between menopause status and sexual functioning? *Menopause* **7**, 297–309

Bagatell, C.J., Heiman, J.R., Rivier, J.E. and Bremner, W.J. (1994). Effects of endogenous testosterone and estradiol on sexual behavior in normal young men. *J Clin Endocr Metab* **78**, 711–716

Bakker, J., Honda, S., Harada, N. and Balthazart, J. (2002). Sexual partner preference requires a functional aromatase (Cyp19) gene in male mice. *Horm Behav* **42**, 158–171

Ball, J. and Hartman, C.G. (1935). Sexual excitability as related to the menstrual cycle in the monkey. *Am J Obstet Gynecol* **9**, 117–199

Baum, M.J., Slob, A.K., de Jong, F.H. and Westbroek, D.L. (1978). Persistence of sexual behavior in ovariectomized stumptail macaques following dexamethasone treatment or adrenalectomy. *Horm Behav* **11**, 323–347

Baum, M.J. and Vreeburg, J.T.M. (1973). Copulation in castrated male rats following combined treatment with estradiol and dihydrotestosterone. *Science* **182**, 283–284

Beach, F.A. (1942a). Copulatory behavior in castrated male rats and its modification by estrogen administration. *Endocrinology* **31**, 679–683

Beach, F.A. (1942b). Importance of progesterone to induction of sexual receptivity in spayed female rats. *Proc Soc Exp Biol Med* **51**, 369–371

Beach, F.A. (1948). *Hormones and behavior; a survey of interrelationships between endocrine secretions and patterns of overt response*. Paul B. Hoeber: New York

Beach, F.A. and Westbrook, W.H. (1968). Morphological and behavioural effects of an 'antiandrogen' in male rats. *J Endocr* **42**, 379–382

Bélanger, B., Bélanger, A., Labrie, F., Dupont, A., Cusan, L. and Monfette, G. (1989). Comparison of residual C-19 steroids in plasma and prostatic tissue of human, rat and guinea-pig after castration: unique importance of extratesticular androgens in men. *J Steroid Biochem* **32**, 695–698

Bergman, B., Damber, J.E., Litttbrand, B., Sjogren, K. and Tomic, R. (1984). Sexual function in prostatic cancer patients treated with radiotherapy, orchiectomy or estrogens. *Br J Urol* **56**, 64–69

Berthold, A.A. (1849). Transplantation der Hoden. *Arch Anat Physiol Wissensch Med* **16**, 42–46

Beutel, M.E., Weidner, W. and Brahler, E. (2006). Epidemiology of sexual dysfunction in the male population. *Andrologia* **38**, 115–121

Beyer, C. and McDonald, P. (1973). Hormonal control of sexual behaviour in the female rabbit. *Adv Reprod Physiol* **6**, 185–219

Beyer, C., Moralí, G., Naftolin, F., Larsson, K. and Pérez-Palacios, G. (1976). Effect of some antiestrogens and aromatase inhibitors on androgen induced sexual behavior in castrated male rats. *Horm Behav* **7**, 353–363

Beyer, C., Vidal, N. and McDonald, P.G. (1969). Interaction of gonadal steroids and their effect on sexual behaviour in the rabbit. *J Endocr* **45**, 407–413

Bhasin, S., Woodhouse, L., Casaburi, R. *et al.* (2001). Testosterone dose-response relationships in healthy young men. *Am J Physiol Endocrinol Metab* **281**, E1172–E1181

Bradford, J.M. and Pawlak, M.A. (1993). Effects of cyproterone acetate on sexual arousal patterns of pedophiles. *Arch Sex Behav* **22**, 629–641

Bradford, J.M.W. and Pawlaka, A. (1993). Double-blind placebo crossover study of cyproterone-acetate in the treatment of the paraphilias. *Arch Sex Behav* **22**, 383–402

Bremer, J. (1958). *Asexualization: a follow-up study of 244 cases*. Oslo University Press: Oslo

Briken, P., Berner, W., Noldus, J., Nika, E. and Michl, U. (2000). Therapie mit dem LHRH-agonisten Leuprolinacetat bei Paraphilien und sexuell aggressiven Impulshandlungen. *Nervenarzt* **71**, 380–385

Briken, P., Hill, A. and Berner, W. (2003). Pharmacotherapy of paraphilias with long-acting agonists of luteinizing hormone-releasing hormone: A systematic review. *J Clin Psychiat* **64**, 890–897

Burger, H.G. (2002). Androgen production in women. *Fertil Steril* **77**, Suppl. 4, S3–S5
Burger, H.G., Dudley, E.C., Cui, J.S., Dennerstein, L. and Hopper, J.L. (2000). A prospective longitudinal study of serum testosterone, dehydroepiandrosterone sulfate, and sex hormone binding globulin levels through the menopause transition. *J Clin Endocr Metab* **85**, 2832–2838
Butera, P. and Czaja, J. (1989). Effects of intracranial implants of dihydrotestosterone on the reproductive physiology and behavior of male guinea pigs. *Horm Behav* **23**, 424–431
Carani, C., Granata, A.R.M., Rochira, V. *et al*. (2005). Sex steroids and sexual desire in a man with a novel mutation of aromatase gene and hypogonadism. *Psychoneuroendocrinology* **30**, 413–417
Carani, C., Qin, K., Simoni, M. *et al*. (1997). Effect of testosterone and estradiol in a man with aromatase deficiency. *New Engl J Med* **337**, 91–95
Carani, C., Rochira, V., Faustini-Fustini, M., Balestrieri, A. and Granata, A. R. M. (1999). Role of oestrogen in male sexual behaviour: insights from the natural model of aromatase deficiency. *Clin Endocrinol* **51**, 517–524
Caraty, A., Delaleu, B., Chesneau, D. and Fabre-Nys, C. (2002). Sequential role of E2 and GnRH for the expression of estrous behavior in ewes. *Endocrinology* **143**, 139–145
Carlson, M.L. (1972). *The rationale of eunuch power in the government of T'ang China, 618–805*, PhD thesis, University of Chicago.
Carnegie, S.D., Ferdigan, L.M. and Ziegler, T.E. (2005). Behavioral indicators of ovarian phase in white-faced capuchins (*Cebus capucinos*). *Am J Primatol* **67**, 51–68
Czaja, J.A. and Bielert, C. (1975). Female rhesus sexual behavior and distance to a male partner: relation to stage of the menstrual cycle. *Arch Sex Behav* **4**, 583–597
Davidson, J.M. (1966). Activation of the male rat's sexual behavior by intracerebral implantation of androgen. *Endocrinology* **79**, 783–794
Davidson, J.M. (1969). Effects of estrogen on the sexual behavior of male rats. *Endocrinology* **84**, 1365–1372
Davidson, J.M., Camargo, C., Smith, E.R. and Kwan, M. (1983). Maintenance of sexual function in a castrated man treated with ovarian steroids. *Arch Sex Behav* **12**, 263–274
Davison, S.L., Bell, R., Donath, S., Montalto, J.G. and Davis, S.R. (2005). Androgen levels in adult females: changes with age, menopause, and oophorectomy. *J Clin Endocr Metab* **90**, 3847–3853
Dennerstein, L., Randolph, J., Taffe, J., Dudley, E. and Burger, M.D. (2002). Hormones, mood, sexuality, and the menopausal transition. *Fertil Steril* **77**, Suppl. 4, S42–S48
Diamond, M. and Sigmundson, H.K. (1997). Sex reassignment at birth: long-term review and clinical implications. *Arch Ped Adolesc Med* **151**, 298–304
Dixson, A. (2001). The evolution of neuroendocrine mechanisms regulating sexual behaviour in female primates. *Reprod Fertil Dev* **13**, 599–607
Dobbins, J.G. (1980). Implications of a time-dependent model of sexual intercourse within the menstrual cycle. *J Biosocial Sci* **12**, 133–140
D'Occhio, M.J., Galil, K.A., Brooks, D.E. and Setchell, B.P. (1985). Differential effects of gonadectomy on sensitivity to testosterone of brain centers associated with gonadotrophin negative feedback and with mating behavior in rams. *J Endocr* **104**, 69–75
Eddy, E.M., Washburn, T.F., Bunch, D.O. *et al*. (1996). Targeted disruption of the estrogen receptor gene in male mice causes alterations of spermatogenesis and infertility. *Endocrinology* **137**, 4796–4805
Ellis, W.J. and Grayhack, J.T. (1963). Sexual function in aging males after orchiectomy and estrogen therapy. *J Urol* **89**, 895–899

Everitt, B.J. and Herbert, J. (1971). Effects of dexamethasone and androgens on sexual receptivity of female rhesus monkeys. *J Endocr* **51**, 575–588

Everitt, B.J., Herbert, J. and Hamer, J.D. (1972). Sexual receptivity of bilaterally adrenalectomized female rhesus monkeys. *Physiol Behav* **8**, 409–415

Filler, W. and Drezner, N. (1944). The results of surgical castration in women under forty. *Am J Obstet Gynecol* **47**, 122–124

French, F.S., van Wyk, J.J. and Baggett, B. (1965). Testicular feminization: clinical, morphological and biochemical studies. *J Clin Endocr Metab* **25**, 661–677

Gaspar, M.L., Bourgarel, P., Meo, T. and Tosi, M. (1991). Structure de l'ARN messager du recepteur des androgènes de souris et caracterisation moléculaire du mutant Tfm. *C R Soc Biol* **185**, 510–519

Gooren, L.J.G. (1985). Human male sexual functions do not require aromatization of testosterone: a study using tamoxifen, testolactone, and dihydrotestosterone. *Arch Sex Behav* **14**, 539–548

Gooren, L.J.G. (1987). Androgen levels and sex function in testosterone-treated hypogonadal men. *Arch Sex Behav* **16**, 463–473

Greenstein, A., Plymate, S.R. and Katz, P.G. (1995). Visually stimulated erection in castrated men. *J Urol* **153**, 650–652

Harding, S.M. and McGinnis, M.Y. (2004). Androgen receptor blockade in the MPOA or VMN: effects on male sociosexual behaviors. *Physiol Behav* **81**, 671–680

He, W.W., Kumar, M.V. and Tindall, D.J. (1991). A frame-shift mutation in the androgen receptor gene causes complete androgen insensitivity in the testicular feminized mouse. *Nucleic Acids Res* **19**, 2373–2378

Hebert, J.P., Ma, Y., Clemow, L. *et al.* (1997). Gender differences in social desirability and social approval bias in dietary self-report. *Am J Epidemiol* **146**, 1046–1055

Heim, N. (1981). Sexual behavior of castrated sex offenders. *Arch Sex Behav* **10**, 11–19

Heim, N. and Hursch, C.J. (11979). Castration of sex offenders: treatment or punishment? A review and critique of recent European literature. *Arch Sex Behav* **8**, 281–306

Hemmingsen, A.M. (1933). Studies on the oestrus-producing hormone (oestrin). *Skand Arch Physiol* **65**, 97–250

Herbert, J. (1970). Hormones and reproductive behaviour in rhesus and talapoin monkeys. *J Reprod Fert* **Suppl. 11**, 119–140

Honda, S., Harada, N., Ito, S., Takagi, Y. and Maeda, S. (1998). Disruption of sexual behavior in male aromatase-deficient mice lacking exons 1 and 2 of the *cyp19* gene. *Biochem Biophys Res Commun* **252**, 445–449

Hoon, P.W., Bruce, K.E. and Kinchloe, B. (1982). Does the menstrual cycle play a role in sexual arousal? *Psychophysiology* **19**, 21–27

Humana, C. (1973). *The keeper of the bed: the story of the eunuch.* Arlington Books: London

Jaffe, R.B. (1969). Testosterone metabolism in target tissues. Hypothalamic and pituitary tissued of the adult rat and human fetus, and the immature rat epiphysis. *Steroids* **14**, 483–498

Johnson, D.F. and Phoenix, C.H. (1978). Sexual behavior amd hormone levels during the menstrual cycles of rhesus monkeys. *Horm Behav* **11**, 160–174

Jones, M.E.E., Boon, W.C., Proietto, J. and Simpson, E.R. (2006). Of mice and men: the evolving phenotype of aromatase deficiency. *Trends Endocr Metab* **14**, 53–62

Judd, H.L., Lucas, W.E. and Yen, S.S. (1974). Effect of oophorectomy on circulating testosterone and androstenedione levels in patients with endometrial cancer. *Am J Obstet Gynecol* **118**, 793–798

Kaplan, H.S. and Owett, T. (1993). The female androgen deficiency syndrome. *J Sex Marital Ther* **19**, 3–24

Kinsey, A.C., Pomeroy, W.B., Martin, C.E. and Gebhard, P.H. (1953). *Sexual behavior in the human female*. Saunders: Philadelphia

Krueger, R.B. and Kaplan, M.S. (2001). Depot-leuprolide acetate for treatment of paraphilias: a report of twelve cases. *Arch Sex Behav* **30**, 409–422

Kuefler, M. (2001). *The manly enunuch: masculinity, gender ambiguity, and Christian ideology in late antiquity*. University of Chicago Press: Chicago

Laan, E. and van Lunsen, R. (1997). Hormones and sexuality in postmenopausal women: a psychophysiological study. *J Psychosom Obstet Gynecol* **18**, 126–133

Larsson, K., Södersten, P. and Beyer, C. (1973). Sexual behavior in male rats treated with estrogen in combination with dihydrotestosterone. *Horm Behav* **4**, 289–299

Laughlin, G.A., Barret-Connor, E., Kritz-Silverstein, D. and von Mühlen, D. (2000). Hysterectomy, oophorectomy, and endogenous sex hormone levels in older women: the Rancho Bernardo study. *J Clin Endocr Metab* **85**, 645–651

Levis, D.G. and Ford, J.J. (1989). The influence of androgenic and estrogenic hormones on sexual behavior in castrated adult male pigs. *Horm Behav* **23**, 393–411

Lisk, R.D. and Bezier, J.L. (1980). Intrahypothalamic hormone implantation and activation of sexual behavior in the male hamster. *Neuroendocrinology* **30**, 220–227

Long, A. and Evans, H.M. (1922). The oestrus cycle of the rat and its associated phenomena. *Mem Univ Calif* **6**, 1–148

Longcope, C. (1986). Adrenal and gonadal androgen secretion in normal females. *Clin Endocrinol Metab* **15**, 213–228

Luttge, W.G. (1975). Effects of anti-estrogens on testosterone stimulated male sexual behavior and peripheral target tissues in the castrate male rat. *Physiol Behav* **14**, 839–846

Luttge, W.G. and Hall, N.R. (1973). Differential effectiveness of testosterone and its metabolites in the induction of male sexual behavior in two strains of albino mice. *Horm Behav* **4**, 31–43

Lyon, M.F. and Hawkes, S.G. (1970). X-linked gene for testicular feminization. *Nature* **227**, 1217–1219

Maravilla, K.R., Cao, Y., Heiman, J. R. *et al.* (2003a). Imaging with MS-325 for evaluating female sexual arousal response: determination of intrasubject reproducibility. *J Magnetic Res Imag* **18**, 216–224

Maravilla, K.R., Heiman, J.R. and Garland, P.A. (2003b). Imaging of the sexual arousal response in women. *J Sex Marital Ther* **29**, Suppl. 1, 71–76

McCance, R.A., Luff, M.C. and Widdowson, E.E. (1937). Physical and emotional periodicity in women. *J Hyg* **37**, 571–611

McDonald, P., Beyer, C., Newton, F. *et al.* (1970). Failure of 5-alpha-dihydrotestosterone to initiate sexual behaviour in the castrated male rat. *Nature* **227**, 964–965

McGinnis, M.Y. and Mirth, M.C. (1986). Inhibition of cell nuclear androgen receptor binding and copulation in male rats by an antiandrogen, SCH16423. *Neuroendocrinology* **43**, 63–68

McGinnis, M.Y., Williams, G.W. and Lumia, A.R. (1996). Inhibition of male sexual behavior by androgen receptor blockade in preoptic area or hypothalamus, but not amygdala or septum. *Physiol Behav* **60**, 783–789

Meuwissen, I. and Over, R. (1992). Sexual arousal across phases of the human menstrual cycle. *Arch Sex Behav* **21**, 101–119

Michael, R.P. and Bonsall, R.W. (1977). Periovulatory synchronization of behaviour in male and female rhesus monkeys. *Nature* **265**, 463–465

Michael, R.P. and Scott, P.P. (1964). The activation of sexual behaviour in cats by subcutaneous administration of estrogen. *J Physiol* **171**, 254–274

Michael, R.P. and Zumpe, D. (1970). Rhythmic changes in the copulatory frequency of rhesus monkeys (*Macaca mulatta*) in relation to the menstrual cycle and a comparison with the human cycle. *J Reprod Fert* **21**, 199–201

Michael, R.P., Zumpe, D. and Bonsall, R.W. (1986). Comparison of the effects of testosterone and dihydrotestosterone on the behavior of male cynomolgus monkeys (*Macaca fascicularis*). *Physiol Behav* **36**, 349–355

Michael, R.P., Zumpe, D. and Bonsall, R.W. (1990). Estradiol administration and the sexual activity of castrated male rhesus monkeys (*Macaca mulatta*). *Horm Behav* **24**, 71–88

Moore, C.R. and Price, D. (1938). Some effects of testosterone and testosterone propionate in the rat. *Anat Rec* **71**, 59–78

Moralí, G., Larsson, K. and Beyer, C. (1977). Inhibition of testosterone-induced sexual behavior in the castrated male rat by aromatase blockers. *Horm Behav* **9**, 203–213

Moralí, G., Lemus, A.E., Munguía, R., García, G.A., Grillasca, I., and Pérez-Palacios, G. (2002). Hormone-like behavioral affects of levonorgestrel and its metabolites in the male rat. *Pharmacol Biochem Behav* **73**, 951–961

Morishima, A., Grumbach, M.M., Simpson, E.R., Fisher, C. and Qin, K. (1995). Aromatase deficiency in male and female siblings caused by a novel mutation and the physiological role of oestrogens. *J Clin Endocr Metab* **80**, 3689–3698

Morrell, M.J., Dixen, J.M., Carter, C.S. and Davidson, J.M. (1984). The influence of age and cycling status on sexual arousability in women. *Am J Obstet Gynecol* **148**, 66–71

Nicolson, P. (1995). The menstrual cycle, science and feminity: assumptions underlying menstrual cycle research. *Social Sci Med* **41**, 779–784

Nikitopoulos, E., Heistermann, M., de Vries, H., van Hooff, J.A.R.A.M. and Sterck, E.H.M. (2005). A pair choice test to identify female mating pattern relative to ovulation in long-tailed macaques, *Macaca fascicularis*. *Anim Behav* **70**, 1283–1296

Nissen, H.W. (1929). The effects of gonadectomy, vasotomy, and injections of placental and orchic extracts on the sex behavior of the white rat. *Genet Psychol Monogr* **5**, 455–547

Ogawa, S., Chan, J., Gustafsson, J.Å., Korach, K.S. and Pfaff, D W. (1999). Survival of reproductive behaviors in estrogen receptor β gene-deficient (βERKO) male and female mice. *Proc Nat Acad Sci USA* **96**, 12887–12892

Ogawa, S., Chester, A.E., Hewitt, S.C. *et al.* (2000). Abolition of male sexual behaviors in mice lacking estrogen receptors α and β (αβERKO). *Proc Nat Acad Sci USA* **97**, 14737–14741

Ogawa, S., Washburn, T.F., Taylor, J., Lubahn, D.B., Korach, K.S. and Pfaff, D.W. (1998). Modifications of testosterone-dependent behaviors by estrogen receptor-α gene disruption in male mice. *Proc Nat Acad Sci USA* **139**, 5058–5069

Ohno, S., Geller, L.N. and Lai, E.V.Y. (1974). *Tfm* mutation and masculinization versus feminization of the mouse central nervous system. *Cell* **3**, 235–242

Olsen, K.L. (1979). Induction of male mating behavior in androgen-insensitive (tfm) and normal (King-Holtzman) male rats: effects of testosterone propionate, estradiol benzoate and dihydrotestosterone. *Horm Behav* **13**, 66–84

Olsen, K.L. (1992). Genetic influences on sexual behavior differentiation. In *Handbook of behavioral neurobiology: sexual differentiation*, vol. 11 (A.A. Gerall, H. Moltz and I.L. Ward, eds), pp. 1–40. Plenum: New York

Olsen, K.L. and Whalen, R.E. (1984). Dihydrotestosterone activates male mating behavior in castrated King-Holtzman rats. *Horm Behav* **18**, 380–392

Ortmann, J. (1980). The treatment of sex offenders. Castration and antihormone therapy. *Int J Law Psychiatr* **3**, 443–451

Patterson, M.N., McPhaul, M.J. and Hughes, I.A. (1994). Androgen insensivity syndrome. *Baillière's Clin Endocrinol Metabol* **8**, 379–404

Phoenix, C.H. (1974). Effects of dihydrotestosterone on sexual behavior of castrated male rhesus monkeys. *Physiol Behav* **12**, 1045–1055

Plaud, J.J., Gaither, G.A. and Weller, L.A. (1998). Gender differences in the sexual rating of words. *J Sex Marital Ther* **24**, 13–19

Regan, P.C. (1996). Rhythms of desire: the association between menstrual cycle phases and female sexual desire. *Can J Human Sexual* **5**, 145–156

Reilly, D.R., Delva, N.J. and Hudson, R.W. (2000). Protocols for the use of cyproterone, medroxyprogesterone, and leuprolide in the treatment of paraphilia. *Can J Psychiat* **45**, 559–563

Rissman, E.F., Wersinger, S.R., Taylor, J.A. and Lubahn, D.B. (1997). Estrogen receptor function as revealed by knockout studies: neuroendocrine and behavioral aspects. *Horm Behav* **31**, 232–243

Rivera-Woll, L.M., Papalia, M., Davis, S.R. and Burger, H. G. (2004). Androgen insufficiency in women: diagnostic and therapeutic implications. *Hum Reprod Update* **10**, 421–432

Robinson, T.J. (1954). The necessity of progesterone with estrogen for the induction of recurrent estrus in the ovariectomized ewe. *Endocrinology* **55**, 403–408

Robinson, T.J. (1955). Quantitative studies on the hormonal induction of oestrus in spayed ewes. *J Endocr* **12**, 163–173

Ryan, K.J., Naftolin, F., Reddy, V., Flores, F. and Petro, Z. (1972). Estrogen formation in the brain. *Am J Obstet Gynecol* **114**, 454–460

Sato, T., Matsumoto, T., Kawano, H. *et al.* (2004). Brain masculinization requires androgen receptor function. *Proc Nat Acad Sci USA* **101**, 1673–1678

Scholten, H. (1995). *Der Eunuch in Kaisernähe: zur politischen und sozialen Bedeutung des Præpositus sacri cubiculi im 4. und 5. Jahrhundert n. Chr.* P. Lang: Frankfurt am Main

Schreiner-Engel, P., Schiavi, R.C., Smith, H. and White, D. (1981). Sexual arousability and the menstrual cycle. *Psychosom Med* **43**, 199–214

Schultz, K.J. and Koulack, D. (1980). Dream affect and the menstrual cycle. *J Nerv Mental Dis* **168**, 436–438

Scott, R.E.M., Wu-Peng, X.S. and Pfaff, D.W. (2002). Regulation and expression of progesterone receptor mRNA isoforms A and B in the male and female rat hypothalamus and pituitary following oestrogen treatment. *J Neuroendocr* **14**, 175–183

Shapiro, H.A. (1937). Effect of testosterone propionate upon mating. *Nature* **139**, 588–589

Sies, C. and Nestler, V. (1992). Soll und Haben. Die Wechseljährige zwischen Illusion und Wirklichkeit. *Psychoanalyse* **46**, 366–387

Simpson, E.R. (2004). Models of aromatase insufficiency. *Sem Reprod Med* **22**, 25–30

Slob, A.K., Ernste, M. and van der Werff ten Bosch, J.J. (1991). Menstrual cycle phase and sexual arousability in women. *Arch Sex Behav* **20**, 567–577

Slob, A.K., Wiegand, S.J., Goy, R.W. and Robinson, J.A. (1978). Heterosexual interactions in laboratory-housed stumptail macaques (Macaca arctoides): observations during the menstrual cycle and after ovariectomy. *Horm Behav* **10**, 193–211

Smith, E.P., Boyd, J., Frank, G.R. *et al.* (1994). Estrogen resistance caused by a mutation in the estrogen-receptor gene in a man. *New Engl J Med* **331**, 1056–1061

Södersten, P. (1973). Estrogen-activated sexual behavior in male rats. *Horm Behav* **4**, 247–256

Spitz, J.C., Gold, A.R. and Adams, D.B. (1975). Cognitive and hormonal factors affecting coital frequency. *Arch Sex Behav* **4**, 249–263

Stanley, A.J., Gumbreck, L.G., Allison, J.E. and Easley, R.B. (1973). Male pseudohermaphroditism in the laboratory Norway rat. *Rec Progr Horm Res* **29**, 43–64

Steinach, E. (1894). Untersuchungen zur vergleichenden Physiologie der männlichen Geschlechtsorgane insbesondere der accessorischen Geschlechtsdrüsen. *Pflügers Arch ges Physiol* **56**, 304–338

Stone, C.P. (1939). Copulatory activity in male rats following castration and injections of testosterone propionate. *Endocrinology* **24**, 165–174

Stürup, G.K. (1972). Castration: The total treatment. In *Sexual behaviors: social, clinical, and legal aspects* (H.L.P. Resnik and M.E. Wolfgang, eds), pp. 361–382. Little, Brown and Co.: Boston

Suh, D.D., Yang, C.C., Cao, Y., Heiman, J.R., Garland, P.A. and Maravilla, K.R. (2004). MRI of female genital arousal and pelvic organs during sexual arousal. *J Psychosom Obstet Gynecol* **25**, 153–162

Tennent, G. (1984). Review of research into the use of drugs and the treatment of sexual deviations with special reference to the use of cyproterone acetate (Androcur). In *Guidelines for the use of psychotropic drugs. A clinical handbook* (H.C. Stancer, P.E. Garfinkel and V.M. Rakoff, eds), pp. 411–426. Spectrum publications: New York

Thibaut, F., Cordier, B. and Kuhn, J.M. (1993). Effect of a long-lasting gonadotrophin hormone-releasing hormone agonist in six cases of severe male paraphilia. *Acta Psychiatr Scand* **87**, 445–450

Udry, J.R. and Morris, N.M. (1968). The distribution of coitus in the menstrual cycle. *Nature* **220**, 593–596

Vagell, M.E. and McGinnis, M.Y. (1997). The role of aromatization in the restoration of male rat reproductive behavior. *J Neuroendocr* **9**, 415–421

van Weerden, W.M., Bierings, H.G., van Steenbrugge, G.J., de Jong, F.H. and Schröder, F.H. (1992). Adrenal glands of mouse and rat do not synthesize androgens. *Life Sci* **50**, 857–861

Wallen, K. (1990). Desire and ability: hormones and the regulation of female sexual behavior. *Neurosci Biobehav Rev* **14**, 233–241

Wallen, K. (2004). Hormones and history: the evolution and development of primate female sexuality. *J Sex Res* **41**, 101–112

Wallen, K. (2005). Hormonal influences on sexually differentiated behavior in nonhuman primates. *Front Neuroendocrinol* **26**, 7–26

Wantoch, H. (1935). Die bisher in Schweiz gesammelten Erfahrungen über die Kastration von Sexualverbrechern. *Arch Kriminol* **96**, 78–80

Waxenberg, S.E., Drellich, M.G. and Sutherland, A.M. (1959). Role of hormones in human behavior. I. Changes in female sexuality after adrenalectomy. *J Clin Endocr Metab* **19**, 193–202

Werner, A.A. (1946). Syndrome characteristics for menopause, ovarian hypofunction and castration in the human female. *J Missouri State Med Assoc* **43**, 311–315

Wersinger, S.R., Sannen, K., Villalba, C., Lubahn, D.B., Rissman, E.F. and DeVries, G.J. (1997). Masculine sexual behavior is disrupted in male and female mice lacking a functional estrogen receptor α gene. *Horm Behav* **32**, 176–183

Whalen, R.E. and Edwards, D.A. (1969). Effects of the antiandrogen cyproterone acetate on mating behavior and seminal vesicle tissue in male rats. *Endocrinology* **84**, 155–156

Wille, R. and Beler, K.M. (1989). Castration in Germany. *Ann Sex Res* **2**, 103–133

Wilson, J.D. and Gloyna, R.E. (1970). The intranuclear metabolism of testosterone in the accessory organs of reproduction. *Rec Progr Horm Res* **26**, 306–339

Wisniewski, A.B., Migeon, C.J., Meyer-Bahlburg, F.L. *et al.* (2000). Complete androgen insensitivity syndrome: Long-term medical, surgical, and psychosexual outcome. *J Clin Endocr Metab* **85**, 2664–2669

Yarbrough, W.G., Quarmby, V.E., Simental, J.A. *et al.* (1990). A single base mutation in the androgen receptor gene causes androgen insensitivity in the testicular feminized rat. *J Biol Chem* **265**, 8893–8900

Zillman, D., Schweitzer, K.J. and Mundorf, N. (1994). Menstrual cycle variation of women's interest in erotica. *Arch Sex Behav* **23**, 579–597

Zuckerman, S. (1930). The menstrual cycle of the primates. Part 1. General nature and homology. *Proc Zool Soc London* 691–754

Zumoff, B., Strain, G.W., Miller, L.K. and Rosner, W. (1995). Twenty-four hour mean plasma testosterone concentration declines with age in normal premenopausal women. *J Clin Endocr Metab* **80**, 1429–1430

Zumpe, D., Bonsall, R.W. and Michael, R.P. (1993). Effects of nonsteroidal aromatase inhibitor, fadrozole, on the sexual behavior of male cynomolgus monkeys (*Macaca fascicularis*). *Horm Behav* **27**, 200–215

Zumpe, D., Clancy, A.N., Bonsall, R.W. and Michael, R.P. (1996). Behavioral responses to depo-provera, fadrozole and estradiol in castrated, testosterone-treated cynomolgus monkeys (*Macaca fascicularis*): the involvement of progestin receptors. *Physiol Behav* **60**, 531–540

CHAPTER

6

Neural control of sexual behavior

1 The male

1.1 Introduction

In the preceding chapter, we discussed the hormonal control of the reactivity of the central motive state. As we know, this state determines the likelihood of the display of sexual behavior in response to sexual incentives and the intensity of the behavior displayed if activated. It is quite evident that the hormones need to act on specific neurons somewhere in the brain or in the peripheral nervous system in order to have their stimulatory effect on the central motive state. The potential importance of the latter should not be underestimated. We all have a tendency to believe that the brain is omnipotent in the control of behaviors and that it is beyond discussion that hormones act in the brain when controlling behavior. This is not necessarily the case, though. Androgens, for example, have important effects on the penile muscles required for intromission in male rats and there are androgen receptors on the neurons innervating the genitals (Keast and Gleeson, 1998; Keast and Saunders, 1998; Keast, 1999). I will not examine the peripheral effects of hormones, but we should remember that the behaving organism is more than a brain.

Hormones have no effect unless bound to the appropriate receptor. One simple way to localize hormone action in the brain is, consequently, to localize the corresponding receptor. With regard to the gonadal hormones, the central nervous receptor distribution was determined many years ago and is now presented in most textbooks of behavioral endocrinology. Among the sites expressing androgen receptors are the medial preoptic area, the medial amygdala, the ventromedial nucleus of the hypothalamus and the bed nucleus of the stria terminalis. Estrogen receptors are found in the medial preoptic area, the ventromedial nucleus of the hypothalamus, the olfactory bulbs and at many other sites. In fact, the gonadal hormone receptors are widely distributed in the brain. So widely that the presence of receptors within a particular structure is not enough evidence for a functionally important role for that structure in the control of sexual behavior. On the contrary, it could be imagined that absence of receptors for gonadal hormones excludes a structure from being of potential importance. This is not entirely true, though. The activation of sexual behavior requires sensory information. Any incentive stimulus must be perceived and correctly identified before it can act on the central motive state. Consequently, at least some sensory pathway is absolutely necessary. Likewise, all behavior is activity in skeletal muscles, meaning that at least some motor systems must be functional if sexual behavior is to occur. It is not at all necessary that the sensory or motor structures involved are equipped with gonadal hormone receptors. It can be quite sufficient that some nervous structure integrating appropriate sensory input with suitable motor output is hormone sensitive. This, indeed, appears to be the case.

1.2 The medial preoptic area: motivation or execution

In the male, the medial preoptic area is essential for male sexual behavior. This has been shown in several vertebrate species (Hillarp *et al.*, 1954; Heimer and Larsson, 1966; Slimp *et al.*, 1978; Koyama *et al.*, 1984; Lloyd and Dixson, 1988) and it is most likely that equivalent structures in men are involved (see Paredes, 2003 and Paredes and Baum, 1997 for excellent reviews). Lesion of the medial preoptic area eliminates all sexual behavior irreversibly, provided the lesion includes most of the area. The dramatic deleterious effects are not due to damage to passing axons. Axon-sparing lesions of the medial preoptic area have the same effects as electrolytic or radiofrequency lesions (Hansen *et al.*, 1982). Thus, cell bodies located within the medial preoptic area are necessary. Electrical stimulation of this structure produces intense sexual behavior that lasts as long as stimulation continues (Schmidt, 1968; Merari and Ginton, 1975; Satou, 1984). This effect has consistently been observed in several vertebrate classes. Other means for stimulating neuronal firing within the preoptic area are also effective for enhancing the intensity of sexual behavior. Blockade of $GABA_A$ (γ-amino butyric acid) receptors with bicuculline leads to a dramatic reduction of the post-ejaculatory interval and an increase in the number of ejaculations per test (Fernández-Guasti *et al.*, 1986). Similar effects are seen when neurons are excited by depositing ferrous ions (Paredes and Ågmo, 1992). Furthermore, electrophysiological studies have shown that neurons within the medial preoptic area enhance firing rate during copulatory behavior. Sixteen per cent of the neurons did so during the male's pursuit of the female, another 35% during pelvic thrusting and 43% enhanced firing during the rapid withdrawal immediately following intromission. After an ejaculation, most neurons reduced firing (Shimura *et al.*, 1994). This study suggests that preoptic neurons are involved in the entire sequence of events forming male copulatory behavior. Finally, sexual behavior in inactive, castrated animals can be activated by implants of minute amounts of testosterone into the preoptic area (Davidson, 1966; Morgantaler and Crews, 1978; Balthazart and Surlemont, 1990), as always in several classes of vertebrates. The evidence for a role of the medial preoptic area in male sexual behavior throughout the vertebrates is, as we must accept, overwhelming. The conservation of function throughout a long evolutionary history from fish to primates strongly suggests that an equivalent area is also crucial in the human. We have, understandably, no experimental data confirming this, but I can see no good reason for doubts.

Over the years, there has been some discussion with regard to the exact role of the medial preoptic area. Some have proposed that it is involved in the executional but not the motivational aspects of sexual behavior. The main argument for that point of view comes from studies in which rats had been trained to perform operant responses for access to a receptive female, or where they had been trained to find a female in a specific location. After preoptic lesion, the males soon stopped copulating with the females but they still performed operant responses to get access to them (Everitt and Stacey, 1987; Everitt, 1990). Other studies have found that unconditioned responses, like pursuing a receptive female or approaching an inaccessible female, are abolished by preoptic lesions (Paredes *et al.*, 1993; Hurtazo *et al.*, 2003). Similarly, interest in odors from receptive females disappears after preoptic lesions in rats and ferrets (Paredes and Baum, 1995; Paredes *et al.*, 1998). These data show that the reactions to sexual incentives are reduced or abolished by preoptic lesions and such an effect is normally interpreted as motivational. It is possible that responses that have been conditioned with sexual reward are more resistant to disappearance following castration than copulatory behavior is. Some experimental data support this notion (Edwards and Einhorn, 1986), and I and others (Hull *et al.*, 1999; Paredes, 2003) are convinced that the medial preoptic area is crucial for sexual motivation. Whether it is crucial also for the execution of copulatory reflexes is not known. If an animal is not motivated to copulate, it has no reason to display any copulatory behavior. A Solomonic hypothesis, which I favor with insistence, is that the current excitability of the preoptic area determines whether remote sexual incentives will activate approach behaviors or not. It also determines whether other stimuli, probably tactile, will activate the copulatory reflexes of mount and contraction of the striated penile muscles or not. In other words, both motivation and likelihood of execution of copulatory reflexes are determined by the excitability of the medial preoptic area. The observation that long-term enhancement of neuronal excitability within that area by kindling can activate sexual behavior in otherwise inactive rats (Paredes *et al.*, 1990) may offer some support for the notion, at least concerning a role for the preoptic area in the control of sexual motivation.

1.3 A note on lordosis in male rats and the influence of the medial preoptic area and of the ventromedial nucleus of the hypothalamus

Male rats do not only display copulatory behaviors typical of their own sex. In addition, they readily display lordosis if given adequate tactile stimulation of the perineal region and the flanks, either by a mounting male or by the experimenter's hand. Since male rats do not frequently mount other males and since most males do not like to be mounted by other males, the latter procedure is the one most commonly used. In some strains, about 50% of intact, untreated males display lordosis, while the proportion of males showing this behavior may be 0 in other strains (Södersten *et al.*, 1974). The lordosis behavior is reduced following castration and can be restored by treatment with estradiol or testosterone (Södersten and Larsson, 1975). The effects of estradiol, as well as those of testosterone, are blocked by an estrogen receptor antagonist (Södersten and Larsson, 1974), suggesting that aromatization of testosterone and subsequent stimulation of estrogen receptors is crucial. In intact rats, there is no relationship between blood androgen or estrogen concentrations and the propensity to display lordosis (Södersten *et al.*, 1974), showing that males displaying female behavior have no endocrine alteration. Thus, the display of lordosis in some strains of male rats seems to be a normal element of the behavioral repertoire. What is of interest in the present context is that lesions of the medial preoptic area enhance lordosis behavior in male rats independently of whether the lesion is electrolytic or produced by an axon-sparing neurotoxin (Hennessey *et al.*, 1986; Olster, 1993). These data show that the preoptic area is not only essential for the display of male sexual behavior, but also that it inhibits the display of female sexual behavior in males. As we soon will see, the preoptic area also inhibits the display of lordosis in female rats.

The ventromedial nucleus of the hypothalamus is best known for its role in lordosis behavior in females of many rodent species as we will see in the next section. However, it also plays a role in lordosis behavior in the male rat. Indeed, lesion of the ventromedial nucleus completely eliminates this behavior (Chateau *et al.*, 1987). It would appear, then, that the neural control of male rat lordosis is identical to that of female lordosis. A question that immediately comes to mind regards the possible effects of ventromedial nucleus lesions on male behavior in male rats. If preoptic lesions facilitate female behaviors, then it could be expected that ventromedial nucleus lesions would facilitate male behaviors. This is exactly what happens. Bilateral lesion of this nucleus stimulates most aspects of copulatory behavior. The latencies to mount and intromit were reduced while the number of mounts, intromissions and ejaculations displayed during tests was enhanced (Christensen *et al.*, 1977). These observations confirm those of an earlier study (Dörner *et al.*, 1969), but contradict those of a more recent one (Harding and McGinnis, 2005). However, the lesions in the latter study generally left parts of the ventromedial nucleus intact and the remaining parts could well be enough for maintained function. The importance of lesion size will be discussed shortly. For the moment, I conclude that lesions of the ventromedial nucleus of the hypothalamus in male rats facilitate male sexual behavior. There seems to exist, then, a reciprocal inhibition between male and female behavior in the way that the site responsible for male behavior patterns inhibits the site responsible for female behavior patterns and vice versa.

1.4 The potential role of the prefrontal cortex

The role of the cerebral cortex in the control of male rat sexual behavior has been a subject of discussion for several decades. Early data suggested that the quantity of cortical tissue removed was the factor determining the effect on sexual behavior (Beach, 1940). This observation suggested that all cortical areas were equally important, or perhaps unimportant, for this behavior. This hypothesis was tested in a few experiments (Larsson, 1962, 1964). Results showed that the effects of localized lesions were modest, but suggested that frontal ablations might have larger effects than dorsolateral or posterior ablations. An interesting experiment in hamsters revealed that ablation of the cingulated and retrosplenial cortices enhanced the threshold for sexual arousal (Bunnell *et al.*, 1966). This observation was followed by a study of the effects of large ablations of the medial and dorsomedial prefrontal cortex (Soulairac and Soulairac, 1972). This kind of lesion seemed to have a reliable and quite peculiar effect on male rat sexual behavior. We confirmed these preliminary results in a quite extensive study of the consequences of extensive destruction of the medial and dorsal prefrontal cortex in male rats. Subjects with such a lesion frequently started to copulate after two hours or more of exposure to a receptive female, while controls began copulation within a minute or so. As soon as copulation had begun, the behavior of the lesioned animals was indistinguishable from that of intact controls. The number of pre-ejaculatory mounts and intromissions was not different from controls and the ejaculation latency, as well as the post-ejaculatory interval, were completely unaffected by the lesion (Ågmo *et al.*, 1995). It appears that the males identify the female as a sexual incentive with large difficulty, but once they have done that their behavior is like that of intact males. Interestingly, an injection of caffeine before the test ameliorates these males (Ågmo and Villalpando, 1995). Our conclusion was that coffee sharpens the mind so that correct identification of the incentive stimulus becomes faster. A more parsimonious explanation can be that caffeine enhances general arousal. By the way, another stimulant, amphetamine, had the same effect.

1.5 Effects of other brain lesions on male sexual behavior

A lesion with quite interesting effects has been performed in a small cluster of lumbar spinothalamic cells. These cells are specifically activated with ejaculation but not with other components of male sexual behavior (Truitt *et al.*, 2003). Destruction of these cells eliminates ejaculation without affecting any other aspect of male rat sexual behavior (Truitt and Coolen, 2002). Only further studies can determine the exact function of this structure in the activation of ejaculatory mechanisms.

While lesions of the medial preoptic area always eliminate male sexual behavior, under the condition that they involve most of the area, lesions of most other structures have variable and always minor effects. The exceptions may be the medial and dorsomedial prefrontal cortices and the lumbar spinothalamic cell cluster as illustrated above. However, more data on these structures are needed before any firm conclusion concerning their real role can be drawn. Structures where lesions produce minor and inconsistent effects include the different parts of the amygdala, the nucleus accumbens, the dorsolateral tegmentum, the ventromedial nucleus of the hypothalamus and the main olfactory bulb, as we have already seen. Some are effective only in animals without sexual experience, some affect one aspect of sexual function but not others, some are significant in one study but not in another, and so on. Results from these kinds of studies show that male sexual behavior is sensitive to events in the brain, but they do not provide any solid fundament for useful hypotheses concerning the central nervous control of sex behavior. If someone should be interested in the huge lesion literature, there is a very extensive review that I can recommend (Hull *et al.*, 2002).

1.6 A comment on the importance of lesion size

One important comment is in place now. I have repeatedly mentioned that lesions of the medial preoptic area need to be large if sexual behavior is to be permanently eliminated. To some, the insistence on large provokes suspicion as to the real importance of the lesion. Such an opinion reflects an immature idealism or outright ignorance. Most lesions need to be large if they are to be effective, probably because of the substantial redundancy typical of the brain. For example, hippocampus lesions start to affect learning and memory mechanisms when about 80% of the hippocampus has been destroyed (Moser *et al.*, 1995; Broadbent *et al.*, 2004), and respiration starts to become irregular when more than 80% of the cells in the preBötzinger complex has been eliminated (Gray *et al.*, 2001). For those unfamiliar with respiratory physiology, I can mention that the preBötzinger complex is the medullary region controlling the respiratory rhythm. Clinicians maintain that the first symptoms of Parkinson's disease appear when more than 80% of the substantia nigra dopaminergic neurons have disappeared. The insistence on lesions destroying most of the target structure is nothing peculiar to sexual behavior, as we now have seen, but a common feature of all lesions. Perhaps the use of larger lesions would have revealed reliable involvement of other structures in male sexual behavior. Personally, I feel more comfortable criticizing lesions for being too small rather than for being too large.

1.7 Indirect estimations of nervous activity in association with sexual behavior

Rather than lesioning or electrically stimulating parts of the brain or implanting gonadal hormones with the purpose of determining the involvement of these parts in the control of sexual behavior, more indirect procedures can be employed. Some years ago it became a kind of fashion to study the expression of the *c-fos* gene after exposing animals to sexually relevant events, such as the odor or the sight of a conspecific of the opposite sex, or allowing the subjects different amounts of sexual activity, ranging from performing or receiving one mount to several ejaculations. There is, as I mentioned somewhere, abundant evidence showing that the immediate early gene *c-fos* is activated whenever neural activity is heightened.

Enhanced expression of the *c-fos* gene product is, then, an indicator of neural activity. Results from the large amount of *c-fos* studies have essentially confirmed what we already knew, perhaps adding some detail here and there. Among the areas activated by sexual activity or sexually relevant stimuli are the olfactory bulbs, different parts of the amygdala, the bed nucleus of the stria terminalis, the medial preoptic area and many other sites (see e.g. Dudley *et al.*, 1992; Wersinger *et al.*, 1993; Bressler and Baum, 1996; Erskine and Hanrahan, 1997; Greco *et al.*, 1998). As we will see when we turn to the female, sex-induced *fos* expression in a structure does not necessarily mean that the structure is important for sexual behavior, so the functional as opposed to statistical significance of the *fos* data needs to be established by other methods, such as lesion or stimulation.

We have seen many examples of the usefulness of studying changes in genital blood flow in response to sexual incentives. Instead of that, some researchers have preferred to study changes in blood flow in the brain during exposure to such incentives. Changes in blood flow or in the oxygenation levels of hemoglobin are supposed to depend on altered nervous activity. In the marmoset, exposure to the odor of periovulatory females increased the signal from the preoptic area and the anterior hypothalamus more than exposure to the odor of ovariectomized females did. Furthermore, while the response to the periovulatory female odor outlasted the presence of the odor, the small response to the odor of ovariectomized female disappeared as soon as the odor was withdrawn (Ferris *et al.*, 2001). This study confirms that the preoptic area is activated by sexually relevant stimuli. Data from the human, strangely enough, do not systematically do so. Young men watching a pornographic movie while inside a magnetic resonance imaging machine showed enhanced activation in the inferior frontal lobe, cingulate and insular gyri, corpus callosum, thalamus, caudate nucleus, globus pallidus and the inferior temporal lobe. No such activation was found while watching a neutral movie. Not unexpectedly, subjects displayed activation of the occipital lobe when watching the pornographic as well as the neutral movie (Park *et al.*, 2001b). An interesting detail in this study is that penile erection was quantified during exposure to the movies in order to assure that the pornographic movie indeed succeeded in enhancing sexual arousal. It did. However, two hypogonadal men failed to respond with erection to that movie. They also showed less activation of the relevant brain sites than the responders did. Treatment with testosterone brought their level of brain activation as well as of erection in response to the pornographic movie up to that of the other men. Although data from two men are far from sufficient for a conclusion, it might be tentatively suggested that the pattern of brain activation observed is dependent on gonadal hormones. The fact that testosterone treatment of these hypogonadal men enhanced activation indicates certainly that it is androgen dependent, exactly as any human male sexual response should be. Again, this proposal goes far beyond what can be ascertained by data from two men, but it should at least be considered as worthwhile of further investigation.

A somewhat more recent study went even further and correlated the degree of erection with the hemodynamic brain response. The anterior cingulate and insular cortices, amygdala and secondary somatosensory cortices showed activation associated with sustained erection. The data from the hypothalamus were unclear. There was some activation at the onset of erection, but there was no difference in signal between no erection and sustained erection (Ferretti *et al.*, 2005). Erection was also monitored in another imaging study employing pornographic movies. The neutral movie used as control depicted sports. An overall comparison of activation between the pornographic and sports movies failed to detect any difference, but when erection was included as a regressor in the analysis, activation was found in the insular cortex and in the middle occipital and temporal gyri as well as in the cingulum, caudate-putamen and hypothalamus. All these data suggest that hemodynamic changes in response to visual sexual incentives are reliably obtained in the cingular and insular cortices. The hypothalamus, and by extension the preoptic area, is sometimes included among the sites responding to visual sexual stimulation. However, a study employing mechanical stimulation of the penis (masturbation) rather than pornographic movies failed to detect activity changes in the hypothalamus/preoptic area. The only structures showing enhanced blood flow were the insular cortex and the secondary somatosensory cortex (Georgiadis and Holstege, 2005).

While experimental studies in non-human animals uniformly have demonstrated the fundamental role of the preoptic area for male sexual behavior, the human imaging studies have only rarely found activation of that area during sexual arousal. In contrast, the only non-human primate imaging study I know of, performed in the marmoset, did, as I mentioned above, find activation of the preoptic area in response to sexually relevant odors. How can we explain the annoying discrepancy between non-human and human studies? One possibility that occurs to me is that the preoptic area, and its human equivalent, is a rather small structure compared to the cortical areas usually found lit up in imaging studies. It is likely that any signal generated in the small preoptic area is smaller than signals generated in the large cortical areas. This small signal may be lost when artifacts caused by minute head movements or other factors are eliminated. The marmoset study, showing preoptic area activation, employed a head-restraining device and a very strong magnetic field (9.5 tesla compared to fields of 1.5–2 tesla used in the human studies). A consequence of these differences may be that movement artifacts were smaller and resolution higher. This may have allowed for the detection of preoptic area activation. I recognize that this explanation is pure speculation, but I find it most unlikely that the human should be fundamentally different from other primates. This belief forces me to find an explanation that can convince myself, at least.

To my consolation, a review paper published after I wrote the above paragraph supports my speculation. When discussing the failure of many imaging studies to find activity in the preoptic area during sexual arousal, it is stated that the very low spatial extension of this region and the relatively low spatial resolution of neuroimaging techniques may be the cause of this failure (Mouras, 2006). A few studies purportedly finding activation of the hypothalamus are cited in the Mouras

(2006) paper in order to make the human imaging data consistent with the large amount of non-human animal data that have firmly established a role for the preoptic area. However, one of these studies found low, unilateral hypothalamic activation in response to a pornographic movie, while far larger activations were observed in several areas of the cerebral cortex (Arnow *et al.*, 2002). Moreover, what is not mentioned in the review is that other recent studies fail to find hypothalamic response to sexual stimuli causing erection (Georgiadis and Holstege, 2005) or find only a transient response (Ferretti *et al.*, 2005) as discussed above.

I have not reviewed all the published imaging studies. Rather than doing that, I have essentially limited the discussion to those including an objective measure of sexual arousal, erection. This seems essential if we are to talk about changes in brain activity caused by sexual arousal and not by stimuli that some researcher imagines as sexually arousing. In conclusion, it seems that imaging studies of sexual arousal only coincide with regard to enhanced activity in the insular cortex. In all studies employing visual stimuli, the cingular cortex is also activated. The meaning or interest of these observations remains obscure for the moment.

2 The female

2.1 Introduction

The neural control of lordosis in the female rat has been elucidated by Don Pfaff in a series of brilliant studies summarized in two books (Pfaff, 1980, 1999). Most of the results obtained in the rat have been confirmed in studies performed in cats (van der Horst and Holstege, 1998). In this section, I will only give a brief outline of what is known about the lordosis. The aim of the section on females is not to penetrate all that is known about the neural control of lordosis but rather to provide a basis for the issues that will be discussed in subsequent chapters. As we all know, lordosis is not a prominent constituent of sexual behaviors in women. Rather than discussing lordosis, I have focused on other elements of female sexual behavior. However, all the information required for a complete appreciation of the arguments made in this book is included. For a more profound knowledge of brain mechanisms involved in the control of the non-human sexual behaviors, particularly lordosis, the publications mentioned above are strongly recommended.

The groundbreaking studies of Don Pfaff (1968) and Walter Stumpf (Stumpf *et al.*, 1975) showed that estrogen receptors are widely distributed in the mammalian brain. The medial amygdala, the lateral septum, the medial preoptic area, the bed nucleus of the stria terminalis, the ventromedial nucleus of the hypothalamus, the arcuate nucleus and the ventral premamillary nucleus are among the regions where estrogen receptors are localized. The other ovarian hormone important for female sexual behavior, progesterone, has receptors located in the medial preoptic area, the ventromedial nucleus of the hypothalamus, the paraventricular nucleus, the medial amygdala and some other structures (e.g. Sar and Stumpf, 1973;

Warembourg, 1978). Although the distribution of progesterone receptors may be somewhat more limited than that of estrogen receptors, it is far too extensive to give any specific clues as to exactly in which structures the ovarian hormones act to control female sexual behavior. It is interesting to note that estrogen and progesterone receptors frequently are located in the same brain areas and even in the same cells (e.g. Blaustein *et al.*, 1988). This is not surprising since, as was mentioned a few pages above, sexual behavior is produced, at least in rodents, through the combined action of estrogens and progestins. Nevertheless, the sole knowledge of receptor distribution is not enough for any founded speculations as to brain sites necessary and/or important for female sexual behavior.

2.2 The ventromedial nucleus of the hypothalamus

I have already mentioned that rather old studies had shown that estrogen implanted at some hypothalamic sites in ovariectomized animals could reactivate female sexual behavior in cats, rabbits and ewes. Many more recent studies have confirmed these observations and offered additional data with regard to the exact site of action (e.g. Pleim *et al.*, 1989). There is no doubt that activation of estrogen and progestin receptors within the ventromedial nucleus of the hypothalamus is sufficient for the induction of complete sexual receptivity in female rats. This has not only been found by implanting minute amounts of estradiol followed by progesterone into that area, but also through local administration of estrogen or progesterone receptor antagonists or antisense oligonucleotides directed against the estrogen or progesterone receptors. After implantation of estradiol, alone or combined with systemic or local progesterone, sexual behavior is enhanced in ovariectomized subjects as we already know quite well. On the contrary, if the content of estrogen or progesterone receptors within the ventromedial nucleus is reduced by administration of an antisense oligonucleotide directed against the estrogen receptor and/or progesterone receptor genes, treatment with estrogen alone or estrogen + progesterone has a reduced capacity to stimulate sexual behavior (Etgen and Barfield, 1986; Ogawa *et al.*, 1994). All these elegant experiments have clearly shown that hormone actions within the ventromedial nucleus of the hypothalamus are both sufficient and necessary for the induction of sexual behavior in ovariectomized subjects.

It is important to note that proceptive behaviors are suppressed as much as lordosis following interference with hormone actions in this nucleus. For example, Ogawa *et al.* (1994) reported an 80% reduction of proceptive behaviors in rats after local infusion of an antisense oligonucleotide directed against the progesterone receptor. The effectiveness of the antisense was evidenced by a substantial reduction in the number of progesterone-receptor immunoreactive cells within the ventromedial nucleus. This number was unaffected in another structure of potential importance for female sexual behaviors, the medial preoptic area. More recent data from mice have confirmed the importance of the ventromedial nucleus for both proceptive and receptive behaviors. A short hairpin iRNA targeting the estrogen receptor α coupled to an adeno-associated virus vector was infused into this nucleus. The expression levels of the estrogen receptor as well as of the estrogen-inducible progesterone receptor were much reduced by the treatment. Likewise, lordosis and proceptivity were almost abolished (Musatov *et al.*, 2006).

The necessity of the ventromedial nucleus of the hypothalamus for the activation of female sexual behavior by ovarian hormones has been confirmed in several lesion studies in rats, guinea-pigs and hamsters (Law and Meagher, 1958; Goy and Phoenix, 1963; la Vaque and Rodgers, 1975; Malsbury *et al.*, 1977; Mathews and Edwards, 1977; Pfaff and Sakuma, 1979a; Okada *et al.*, 1980; Richmond and Clemens, 1988). The typical procedure employed in all these studies was to perform the lesion in already ovariectomized females and, after a varying recovery period, ovarian hormones were administered in doses that normally induce sexual receptivity. The stimulating effects of the hormone treatment were consistently reduced in subjects with extensive damage to the ventromedial nucleus.

Additional evidence for a role of this nucleus was obtained in a study where it was electrically stimulated. A reasonable prediction is that electrical stimulation of a brain site should have effects on a particular behavior opposite to those of a lesion, if the site indeed were of importance for that particular behavior. This prediction was confirmed in rats (Pfaff and Sakuma, 1979b). Furthermore, a study in sexually receptive female monkeys found that electrical stimulation of the ventromedial nucleus produced a presentation posture. This behavior consists of lifting the hip and turning the hindquarters toward the male. The males found the presenting posture most inciting, because they usually responded with touching, mounting and thrusting (Koyama *et al.*, 1988). Interestingly, the stimulation did not activate any presentations in the absence of a male. This fact indicates that the ventromedial nucleus does not control the motor output underlying the posture. If that had been the case, then stimulation would have had the same effects in the presence and in the absence of a male. Rather it seems that the responsiveness of the ventromedial nucleus to specific sensory cues was enhanced by stimulation. The reader with good memory may remember that the sensitivity to incentive stimuli is determined by the abstract concept of 'central motive state'. Supposing that the observations by Koyama *et al.* (1988) are correct, we can propose that some part of the central motive state controlling sexual motivation is materialized in the ventromedial nucleus of the primate hypothalamus.

It is not only in monkeys, rats and mice that we have evidence suggesting that the ventromedial nucleus is of importance for phenomena related to sexual incentive motivation in addition to controlling the display of lordosis, under the condition that we indeed accept the intensity of proceptive behaviors as indicative of degree of responsiveness to sexual incentives, obviously. A study in female hamsters reported that ventromedial nucleus lesions not only reduced lordosis but also proceptivity (Floody, 2002). This observation makes it possible to include the hamster among the species in which the ventromedial nucleus may control incentive motivational phenomena besides lordosis.

To conclude our analysis of the ventromedial nucleus of the hypothalamic site, we can state that the data obtained in lesion and stimulation studies coincide nicely with those from the studies of the effects of manipulations of ovarian hormone actions. Lordosis (receptivity) as well as proceptivity seem to be controlled by neurons in the ventromedial nucleus.

2.3 The preoptic area

Another brain site containing estrogen and progesterone receptors is the medial preoptic area. Here, lesions affect sexual receptivity, lordosis, in a way exactly opposite to that of ventromedial nucleus lesion. Thus, ineffective doses of estradiol were found to become effective after preoptic lesion (Powers and Valenstein, 1972). These data suggest that the preoptic area exerts a tonic, inhibitory influence on female rat sexual receptivity. This proposal has been confirmed in a large number of experiments (see Xiao *et al.*, 2005 for a recent example of a nicely conducted study). Neurons with cell bodies within the preoptic area are probably crucial, because the axon-sparing neurotoxin, ibotenic acid, has effects very similar to those of electrolytic lesion (Hoshina *et al.*, 1994). The facilitatory effect is not specific to rats. Lesion of the preoptic area of female guinea-pigs also enhances responsiveness to ovarian hormones (Olster, 1995, 1998; Olster and Paulson, 1997). In contrast to the facilitatory effects on lordosis, lesions of the preoptic area with ibotenic acid have been reported to reduce proceptivity in female rats (Hoshina *et al.*, 1994). This observation was not confirmed in another study, also employing ibotenic acid for lesioning this area. There was no deleterious effect of the lesion on proceptive behaviors (Guarraci and Clark, 2006). Thus, it appears that preoptic lesion consistently has facilitatory effects on lordosis while its effects on proceptive behaviors are unclear in rats.

It may be possible to explain why the effects of preoptic lesions on proceptive behaviors are inconsistent. To do that, we must assume that stimuli emitted by a male may sometimes activate proceptive behaviors at a distance. The stimulus control of proceptive behaviors is unclear, but it is known that a distant male can activate such behaviors. Furthermore, we must assume that the preoptic area is sensitive to distant sexual incentives in the female as well as it is in males. In the next paragraph I will, in fact, present experimental evidence for that statement. In any test situation, a variable proportion of the total number of proceptive behaviors displayed is activated by distant stimuli. How large or small this proportion is depends on the specific situation. The fact that some studies have found effects of preoptic lesion on proceptive behaviors while others have not can probably be attributed to differences in testing situation and consequent differences in the proportion of proceptive behaviors activated by distant stimuli. If that proportion were large, then we would expect a substantial reduction after preoptic lesion since the preoptic area reacts to distant stimuli. On the other hand, if the proportion of proceptive behaviors activated by distant stimuli were small, then we would not expect an effect of preoptic lesion. Most unfortunately, none of the studies of the effects of preoptic lesion on proceptive behaviors has taken the trouble to distinguish proceptive behaviors in response to tactile stimulation from the male from those in response to distant stimuli. There is, then, no way to verify my proposed explanation of contradictory results.

The preoptic area is important for the capacity of distant sexual incentive stimuli to activate approach behaviors. Axon-sparing ibotenic acid lesions of this area have been found to reduce the intensity of approach to an inaccessible male as well as to a male in a situation where copulatory interactions were possible (Guarraci and Clark, 2006). Moreover, the sexual incentive properties of the odor of a male are abolished by radiofrequency lesion of the medial preoptic area (Xiao *et al.*, 2005). These interesting data show that, in the female rat, this area is crucial for the distant sexual incentive properties of a male. It was, in fact, the Guarraci and Clark (2006) and the Xiao *et al.* (2005) data that made me believe that the preoptic area is also important for the display of proceptive behaviors in response to distant stimuli.

Some data from female primates show that the preoptic area or its monkey equivalent may be important for proceptivity in species other than the rat. In marmosets, lesion of this area almost eliminates proceptive behaviors without having any effect on receptivity (Kendrick and Dixson, 1986). Tongue-flicking, a rhythmic tongue protrusion that is supposed to invite the male to copulate, was almost abolished after the lesion. Similarly, proceptive stares, described as 'flattened ear tufts and fixed stare at the male', were reduced. A complication with these data is that tongue-flicking is also performed while the female is actually copulating with the male. These tongue-flicks are called receptive and they were also displayed at a lower frequency after the lesion. Nevertheless, the authors concluded that receptivity was not affected. These observations were confirmed in a later study where ibotenic acid was used for lesioning preoptic area neurons, again in female marmosets. It may be remembered that ibotenic acid selectively destroys neurons without causing much damage to passing axons. In view of this, the data were interpreted as showing that neurons within the preoptic area are important for female proceptivity but not for receptivity (Dixson and Hastings, 1992). Another study, conducted in one female *Macaca mulatta* and one *Macaca fuscata*, revealed that electrical stimulation of the preoptic area led the females to display a presenting posture if a male was present (Koyama *et al.*, 1988). This same study also reported that stimulation of the ventromedial nucleus stimulated the display of the presentation posture, as mentioned in the preceding section. Nevertheless, the effects of stimulation are exactly those that would be predicted from the results of the lesion studies described above. Whenever a lesion at a brain site reduces a behavior, stimulation of the same site should enhance the same behavior. Thus, it would appear that the preoptic area indeed is of importance for proceptive behaviors in female primates.

2.4 On the importance of other brain sites inferred from studies of copulation-induced brain activation

There are several brain structures in addition to the ventromedial hypothalamus and the preoptic area that might be of importance for female sexual behaviors.

In the beginning of this section I mentioned that the overlapping distribution of estrogen and progesterone receptors was too vast to offer specific cues as to crucial structures. Rather than using this distribution as a basis for speculations, some scientists have reasoned that brain regions activated by the execution of sexual behaviors or by sexually relevant stimuli somehow must be involved in the control of these behaviors or responses to these stimuli. As we know, there are reasonably accurate, non-invasive procedures for determining brain activity. We may recall that the favorite one among those working with rats is the *post-mortem* determination of the number of neurons expressing the immediate early gene *c-fos*. This has obviously been done in a large number of experiments. Typically, sexually receptive females are allowed different amounts of sexual interaction with males, or they are allowed to smell male odors. Sometimes, the male has been replaced by a syringe plunger. This object can easily be introduced into the vagina and the cervix can be stimulated in a controlled way. I will not review the large number of studies on the subject. A more interesting point to discuss is whether the activation of a particular brain structure during the execution of a particular behavior really means that the structure is important for that behavior. *A priori* there is no reason to expect that. Many brain structures may be active without being specifically involved in the behavior being executed. Thus, *fos* activation by itself does not necessarily mean anything. Additional data need to be presented before a convincing case can be made for the importance of any particular brain structure in any particular behavior.

A beautiful illustration to the point of view presented in the preceding paragraph can be found in a splendid little paper by Fay Guarraci and colleagues (Guarraci *et al.*, 2004). To appreciate fully their ingenuity some background information is required. Vaginocervical stimulation occurs every time the male performs an intromission. In the female rat, this sensory stimulation has many consequences, as was mentioned in Chapter 3. It is, for example, important for the pacing of sexual interactions. It is also important for the physiological consequences of copulatory behavior, like the induction of pseudopregnancy. This is a state where the corpora lutea do not regress as fast as usual, but remain active for about 14 days. If the copulation had been fertile, the placenta would have started to produce large amounts of estrogens and progesterone long before that time, making the corpora lutea unnecessary for the maintenance of pregnancy. Anyway, vaginocervical stimulation is an important event associated with female copulatory behavior and the brain structures showing enhanced *fos* expression as a consequence of it have been well described. They include, in addition to the obvious sites of the ventromedial nucleus and the medial preoptic area, the amygdala and the bed nucleus of the stria terminalis (Pfaus *et al.*, 1993).

Ibotenic acid lesion of the medial amygdala failed to modify female rat sexual behavior in paced mating tests. Receptivity and proceptivity as well as all aspects of pacing were unaltered by the lesion. Similar results were obtained following lesion of the bed nucleus of the stria terminalis. Lesion of the medial preoptic area, in contrast, enhanced the duration and probability of escapes from the male following intromission and ejaculation (Guarraci *et al.*, 2004). This was interpreted as suggestive of an increased impact of vaginocervical stimulation in lesioned animals, a conclusion already reached in a similar study from another laboratory a few years earlier (Yang and Clemens, 2000). The important point here is that the destruction of structures showing intense *fos* expression following copulation does not alter copulatory behavior. It seems quite reasonable to suggest, then, that these structures are not of any particular importance for that behavior. Furthermore, it should also be evident that, exactly as I stated many lines ago, activation of a brain structure during a behavior does not by itself imply that the structure is crucial or even important at all for the behavior. This annoying conclusion should be remembered when interpreting the very popular imaging studies in human subjects.

2.5 Conclusion concerning the neural control of female sexual behavior in non-human mammals

The ventromedial nucleus is crucial for female sexual behaviors in all species studied. Enhanced activity in this nucleus facilitates the occurrence and/or enhances the intensity of sexual behaviors. The medial preoptic area may have the opposite effects, but data on this are either not entirely consistent (rodents) or too scarce (primates) for any final conclusion. The many other brain structures now and then suggested to be of importance may perhaps be so in some particular contexts, but their role is probably marginal.

It might be interesting to speculate about the exact function of the crucial brain sites mentioned above. It is evident that the ventromedial nucleus is important for lordosis. Without it, this response to tactile stimulation is severely reduced. Proceptive behaviors, frequently also a response to tactile stimulation from the male, are also reduced when ventromedial nucleus function is impaired. Thus, it can be proposed that the ventromedial nucleus is important for responses to proximal stimuli, mainly tactile. Such stimuli activate the motor pattern of lordosis as well as those patterns forming the proceptive behaviors. On the other hand, the preoptic area is important for the activation of approach behaviors to distant, sexually relevant stimuli, like odor, sound or sight of a male.

We must recall that the abstract mechanism determining an individual's responsivity to an incentive stimulus is called the central motive state. Since the preoptic area is necessary for the approach response to a sexual incentive stimulus, it can be suggested that the central motive state regulating the probability of response to a remote sexual incentive is materialized in this structure. Some lines ago I suggested that the central motive state regulating the display of lordosis and perhaps also of proceptive behaviors in response to proximate sexual incentive stimuli, notably tactile, may be located in the ventromedial nucleus. The combined action of the preoptic area and the ventromedial nucleus, then, controls all aspects of female sexual behavior, from the approach to a mate to the execution of copulatory reflexes. It is obvious that other brain regions are involved in the sensory input to these structures and the motor output from them. What I propose is not that the rest of the brain could be eliminated as far as sex behavior is concerned. What I suggest is that the transmission of sensory input to motor output takes place in these structures. If active, then appropriate sensory input excites motor output. If not, nothing happens.

The sensory input generated by distant stimuli needs to be elaborated somewhere else and given sexual significance before it is directed to the preoptic area. I have suggested that the medial prefrontal cortex may be the site where this stimulus elaboration occurs (Ågmo *et al.*, 1995). Proximal sensory input, like tactile stimulation of the flanks or the vaginocervical region, may arrive directly at the ventromedial nucleus and at the preoptic area without any elaboration. These speculative proposals coincide with all available data. It is not seriously contradicted by any known observations. However, it is undoubtedly in need of some sophistication and some unequivocal experimental data before it can be transformed from speculation into fact. Only further, systematic research can provide the basis for such an endeavor.

2.6 A reciprocal inhibitory relationship between the ventromedial nucleus of the hypothalamus and the preoptic area: neural circuits forming the basis for the bisexual potential

We have now arrived at a point where it is most useful to refresh our recollections of male sexual behavior. It is probably not an abuse to suggest that any reader who has forgotten that the medial preoptic area is crucial for male sexual behavior should return to the section on males in this chapter and read it with some attention. The rather exciting point here is that the structure that stimulates male sexual behavior at the same time inhibits female sexual behavior, at least in rodents. In other words, the brain site in which nervous activity enhances the likelihood that male sexual behavior will be displayed reduces the likelihood that female sexual behavior will be displayed. We have already learned that lesion of the ventromedial nucleus of the hypothalamus reduces female sexual behavior and we learned in the previous section that lesion of this nucleus facilitates male sexual behavior. Again, the brain site making female sexual behavior more likely at the same time makes male behavior less likely. There seems to be a kind of reciprocal relationship between male and female sexual behavior with regard to its neural control. Low activity in the preoptic area combined with high activity in the ventromedial nucleus is associated with high likelihood for the display of female sexual behavior. This fact applies to males as well as to females. The attentive reader may remember that preoptic lesion also facilitates lordosis in males and that ventromedial nucleus lesion reduces it in males as much as it does in females.

If we now turn to the opposite situation, low activity in the ventromedial nucleus and high activity in the preoptic area, we find that it is associated with a high likelihood for the display of male sexual behavior. The interesting questions that appear immediately are whether preoptic lesion abolishes mounting in females in the same way as it does in males and whether ventromedial lesion facilitates mounting in females just as it does in males. The answer to these questions is unavailable at present. An abstract presented at the Society for Neuroscience meeting in Orlando, Florida, in 2002 contains the only existing data concerning the effects of lesions of the preoptic area and ventromedial nucleus of the hypothalamus on mounting behavior in female rats. Most unfortunately, the abstract itself only states that, 'The effects of these treatments are currently being evaluated and will be presented' (Afonso *et al.*, 2002). My memory of the poster is somewhat vague after 4 long years, but I recollect that the data shown were most preliminary. However, it appears that a complete report will be submitted for publication before the writing of this book is finished. I can only advise any interested reader to look for it. Nevertheless, despite the uncertainty about the central nervous control of mounting in females, I dare to suggest that its control is similar to that in males. The difference between males and females is, then, extremely simple: in the male, high activity in the preoptic area and low activity in the ventromedial nucleus make the display of male copulatory behaviors more likely than that of female copulatory behaviors. In the female, high activity in the ventromedial nucleus and low activity in the preoptic area make the display of female copulatory behaviors more likely than that of male copulatory behaviors. However, both males and females have the potential for expressing behaviors typical of the opposite sex in addition to the potential of expressing behaviors typical of their own sex. The only difference between males and females is the likelihood of expression of a behavior, a quantitative difference. There is no qualitative sex difference in behavior, as far as I can see. This important issue will be addressed in more depth in Chapter 8.

A caveat needs to be added here. The reasoning in the preceding paragraph is based on rat data and it is limited to the probability of expressing the copulatory reflexes, basically mount and lordosis. Since we do not have the necessary data from other species, we cannot determine whether the principle of reciprocal inhibitory control of male and female copulatory motor patterns is valid in any species other than the rat. Furthermore, the activation of sexual incentive motivation by distant stimuli seems to depend on an intact preoptic area in both male and female rats. If the observations justifying this statement were confirmed in more experimental studies, then we could conclude that the preoptic area has a double function. It would then control the propensity to express male copulatory behaviors *and* the reactivity to distant sexual incentives in both sexes. On the contrary, the only function of the ventromedial nucleus would be the control of female copulatory behaviors in both sexes.

2.7 Central nervous control of sexual behavior in the human female

We now turn our attention to the human female. The distribution of androgen and estrogen receptors has been studied in a few subjects, both males and females. Receptors are found in the areas reported to be important for female sexual behavior in rodents and non-human primates, viz. the medial preoptic area and the ventromedial nucleus (Donahue *et al.*, 2000; Fernández-Guasti *et al.*, 2000). In addition, gonadal hormone receptors are found in structures unrelated to sexual behaviors, like the temporal lobe (Sarrieau *et al.*, 1990) or the hippocampus (Beyenburg *et al.*, 2000). Thus, the receptor distribution data offers no reason for believing that the brain structures important for sexual behaviors in females of other species should not be so also in women. We do not have, for many reasons, any data on the consequences of lesions in the ventromedial nucleus or medial preoptic area. Similarly, we do not have any data from studies where these structures have been electrically stimulated. What we have, though, are some imaging data. I have already expressed some reservations against this kind of data, but since they have become so popular, and since some people have a strange tendency to believe in them, I cannot avoid mentioning these data. Furthermore, it would be unfair not to do so, in view of the fact that I mentioned several imaging studies in the section on males.

In one study, women were exposed to a fragment of a pornographic movie and a documentary as control. The pornographic movie caused activation of the thalamus, caudate nucleus, globus pallidus, the inferior frontal lobe, insula, corpus callosum, the anterior cingulated gyrus and the inferior temporal lobe. In fact, the highest activation was observed in the temporal lobe (Park *et al.*, 2001a). It is not entirely clear what kind of meaningful information these data might contain. None of the structures showing enhanced blood flow during exposure to the pornographic movie has been attributed any particular role in female sexual behavior in non-human females. On the contrary, those structures known to be important for non-human female sexual behaviors failed to show increased blood flow. Another, very complex experiment involving disgust-evoking pictures in addition to sexually arousing ones, and employing both male and female subjects, reported that sexually arousing pictures caused enhanced blood flow in the right temporal/occipital lobe, the right ventral striatum, the right amygdala, the right hypothalamus and the right thalamus as well as bilateral activation of the brain stem. A questionnaire evaluation of sexual arousal found that pornographic pictures were equally effective in men and women. Furthermore, there was no sex difference in hemodynamic response (Stark *et al.*, 2005). This is one of the few studies reporting effects in the hypothalamus. Unfortunately, anatomical resolution was too poor for determining where in the right hypothalamus the enhanced blood flow occurred. Again, it is rather difficult to see how this elaborate study contributes to our understanding of sexual functions in women.

The third study (Archer *et al.*, 2006) I would like to mention has a more interesting design. Women in surgical menopause (mean age when the study was performed was 50 years) were exposed to pornographic video fragments and to neutral videos as controls. The women were first untreated, then they were give estrogen replacement therapy and finally estrogen + testosterone. They were exposed to the videos under each of these treatment conditions. A group of premenopausal women was also used (mean age 43 years). When the menopausal women were untreated, the pornographic video enhanced blood flow more than the neutral video only in the left middle frontal gyrus. Under estrogen replacement therapy, these women displayed more enhancement of blood flow in the left thalamus, medial frontal and right cingulate and parahippocampal gyri in response to the pornographic video than in response to the neutral one. When given testosterone in addition to estrogen, blood flow enhancement was larger in these same areas. There was no other difference between estrogen alone and estrogen + testosterone. In the premenopausal women, the pornographic movie enhanced blood flow in the thalamus, amygdala and medial frontal gyrus more than the neutral movie did. Furthermore, it was found that the response to both the neutral and pornographic videos was larger in premenopausal than in untreated menopausal women.

It is not immediately evident how the observed changes in cerebral blood flow reported in the studies mentioned above can contribute to our understanding of women's response to visual sexual incentives. Those knowledgeable of imaging studies will also notice that there is a host of non-sexual situations that enhance blood flow in the brain regions responding to sexual incentives. What these hemodynamic responses really represent can only be a matter of speculation. The exciting question of whether such speculations ever can be considered as informed or if they are intrinsically fabled cannot be answered at present.

Meaningful data concerning the brain sites involved in sexual behaviors in women are, as we have seen, extremely scarce. In fact, I feel forced to conclude that we do not have enough human data for any conclusion at all. However, this state of affairs can probably be remedied if we make use of data from the rather large number of studies in non-human animals. Among mammals, including non-human primates, it appears that the ventromedial nucleus of the hypothalamus and the medial preoptic area are of importance in every species studied. I can see no reason for doubting that this is also the case in women. In addition, structures involved in sensory input and motor output are, as in other animals, necessary. Furthermore, the exceptional flexibility of human sexual behaviors repeatedly mentioned before makes it likely that additional brain regions are involved in some aspect or another of human female sexual behaviors. Despite these undisputable facts, I do believe that the neural model based upon non-human data presented in the preceding section applies equally well to women. The main differences between the human, man or woman, and other animals is probably to be found in the stimulus control of sexual motivation and behavior and in the motor patterns associated with the execution of behaviors leading to sexual reward, not in the brain structures where hormones determine the responsivity of the central motive state.

3 General conclusion

We have seen that two hypothalamic structures are essential for the activation of sexual motivation and for the execution of sexual behaviors in males as well as females in all mammalian species studied. We have not, in this chapter, seen how other brain areas might modify the response to sexually relevant stimuli. The role of the olfactory bulbs was more appropriately discussed in the section corresponding to the stimulus control of sexual behavior since these structures are more involved in sensory mechanisms than in the integration of stimulus input and motor output. Furthermore, while data concerning the two hypothalamic structures are quite consistent, the role of the olfactory bulbs is still unclear. Data regarding the many other brain sites now and then suggested to be involved in the control of sexual behavior are still more confusing and insufficient for proposing any reasonably well founded conclusion. That has been my reason for avoiding any discussion of them.

There are undoubtedly still many mysteries associated with the exact functions of the medial preoptic area and the ventromedial nucleus in the control of sexual behaviors. Both these sites are involved in many behaviors in addition to sexual and also in some important autonomic functions. It is probable that rather small cell populations within each of these structures occupy themselves with sexual behaviors while other cell populations are involved in other behaviors or in autonomic function. We do not know if the population in charge of sex behaviors is formed by all cells expressing gonadal hormone receptors, or if cells not expressing these receptors also are involved, or if some gonadal hormone-expressing cells are involved in non-sexual behaviors. The first case would be the easiest for experimental manipulations of these cells, but I fear that the latter cases are not unlikely at all. There is still a way to go before we understand how neurons within the preoptic area and ventromedial nucleus connect sensory inputs to motor outputs and what the gonadal hormones do when determining the likelihood of such connections. Nevertheless, there is no doubt that we sooner or later will be in a position to have a complete understanding.

Summary

1 In males of all species studied, a functional medial preoptic area is necessary for the display of sexual behavior. There are good reasons for believing that this area is crucial for sexual incentive motivation activated by distant stimuli. It is also possible that the copulatory motor patterns are under preoptic control.
2 The ventromedial nucleus of the hypothalamus may exert some inhibitory action on male sexual behaviors.
3 Male rats display lordosis in addition to mount. Lesion of the preoptic area facilitates the expression of lordosis, while lesion of the ventromedial nucleus inhibits it.
4 Imaging studies in the human male exposed to sexually relevant stimuli have yielded trivial results.
5 Female rats show reduced lordosis behavior after lesion of the ventromedial nucleus and enhanced lordosis after lesion of the preoptic area.

6 The ventromedial nucleus is also important for the display of proceptive behaviors. The role of the preoptic area in the control of these behaviors is unclear in the rat. On the contrary, in primates it seems to determine the display of proceptive behaviors.
7 There is a reciprocal inhibitory relationship between the preoptic area and the ventromedial nucleus with regard to the display of male and female copulatory responses in male as well as in female rats.
8 The central nervous structures involved in the control of sexual behavior in women are poorly known.

References

Ågmo, A. and Villalpando, A. (1995). Central nervous stimulants facilitate sexual behavior in male rats with medial prefrontal cortex lesions. *Brain Res* **696**, 187–193

Ågmo, A., Villalpando, A., Picker, Z. and Fernández, H. (1995). Lesions of the medial prefrontal cortex and sexual behavior in the male rat. *Brain Res* **696**, 177–186

Afonso, V.M., Lehmann, H. and Pfaus, J.G. (2002). Brain regions associated with female mounting behavior in the rat. *Soc Neurosci Abstr* 482.1

Archer, J.S., Love-Geffen, T.E., Herbst-Damm, K.L., Swinney, D.A. and Chang, J.R. (2006). Effect of estradiol versus estradiol and testosterone on brain-activation patterns in postmenopausal women. *Menopause* **13**, 528–537

Arnow, B.A., Desmond, J.E., Banner, L.L. *et al.* (2002). Brain activation and sexual arousal in healthy, heterosexual males. *Brain* **125**, 1014–1023

Balthazart, J. and Surlemont, C. (1990). Androgen and estrogen action in the preoptic area and activation of copulatory behavior in quail. *Physiol Behav* **48**, 599–609

Beach, F.A. (1940). Effects of cortical lesion upon the copulatory behavior of male rats. *J Comp Psychol* **29**, 193–244

Beyenburg, S., Watzka, M., Clusmann, H. *et al.* (2000). Androgen receptor mRNA expression in the human hippocampus. *Neurosci Lett* **294**, 25–28

Blaustein, J.D., King, J.C., Toft, D.O. and Turcotte, J. (1988). Immunocytochemical localization of estrogen-induced progestin receptors in guinea pig brain. *Brain Res* **474**, 1–15

Bressler, S.C. and Baum, M.J. (1996). Sex comparison of neuronal fos immunoreactivity in the rat vomeronasal projection circuit after chemosensory stimulation. *Neuroscience* **71**, 1063–1072

Broadbent, N.J., Squire, L.R. and Clark, R.E. (2004). Spatial memory, recognition memory, and the hippocampus. *Proc Nat Acad Sci USA* **101**, 14515–14520

Bunnell, B.N., Friel, J. and Flesher, C.K. (1966). Effects of median cortical lesions on the sexual behavior of the male hamster. *J Comp Physiol Psychol* **61**, 492–495

Chateau, D., Chabli, A. and Aron, C. (1987). Effects of ventromedial nucleus lesions on the display of lordosis behavior in the male rat. Interactions with facilitatory effects of male urine. *Physiol Behav* **39**, 341–345

Christensen, L.W., Nance, D.M. and Gorski, R.A. (1977). Effects of hypothalamic and preoptic lesions on reproductive behavior in male rats. *Brain Res Bull* **2**, 137–171

Davidson, J.M. (1966). Activation of the male rat's sexual behavior by intracerebral implantation of androgen. *Endocrinology* **79**, 783–794

Dixson, A.F. and Hastings, M.H. (1992). Effects of ibotenic acid-induced neuronal degeneration in the hypothalamus upon proceptivity and sexual receptivity in the female marmoset (*Callithrix jacchus*). *J Neuroendocr* **4**, 719–726

Donahue, J.E., Stopa, E.G., Chorsky, R.L. *et al.* (2000). Cells containing immunoreactive estrogen receptor-α in the human basal forebrain. *Brain Res* **856**, 142–151

Dörner, G., Döcke, F. and Hinz, G. (1969). Homo- and hypersexuality in rats with hypothalamic lesions. *Neuroendocrinology* **4**, 20–24

Dudley, C.A., Rajendren, G. and Moss, R.L. (1992). Induction of fos immunoreactivity in central accessory olfactory structures of the female rat following exposure to conspecific males. *Mol Cell Neurosci* **3**, 360–369

Edwards, D.A. and Einhorn, L.C. (1986). Preoptic and midbrain control of sexual motivation. *Physiol Behav* **37**, 329–335

Erskine, M.S. and Hanrahan, S.B. (1997). Effects of paced mating on c-fos gene expression in the female rat brain. *J Neuroendocr* **9**, 903–912

Etgen, A.M. and Barfield, R.J. (1986). Antagonism of female sexual behavior with intracerebral implants of antiprogestin RU38486: correlation with binding to neural progestin receptors. *Endocrinology* **119**, 1610–1617

Everitt, B.J. (1990). Sexual motivation: a neural and behavioral analysis of the mechanisms underlying appetitive and copulatory responses of male rats. *Neurosci Biobehav Rev* **14**, 217–232

Everitt, B.J. and Stacey, P. (1987). Studies of instrumental behavior with sexual reinforcement in male rats (*Rattus norvegicus*): II. Effects of preoptic area lesions, castration, and testosterone. *J Comp Psychol* **101**, 407–419

Fernández-Guasti, A., Kruijver, F.P.M., Fodor, M. and Swaab, D.F. (2000). Sex differences in the distribution of androgen receptors in the human hypothalamus. *J Comp Neurol* **425**, 422–435

Fernández-Guasti, A., Larsson, K. and Beyer, C. (1986). GABAergic control of masculine sexual behavior. *Pharmacol Biochem Behav* **24**, 1065–1070

Ferretti, A., Caulo, M., del Gratta, C. *et al.* (2005). Dynamics of male sexual arousal: distinct components of brain activation revealed by fMRI. *Neuroimage* **26**, 1086–1096

Ferris, C.F., Snowdon, C.T., King, J.A. *et al.* (2001). Functional imaging of brain activity in conscious monkeys responding to sexually arousing cues. *Neuroreport* **12**, 2231–2236

Floody, O.R. (2002). Time course of VMN lesion effects on lordosis and proceptive behavior in female hamsters. *Horm Behav* **41**, 366–376

Georgiadis, J.R. and Holstege, G. (2005). Human brain activation during sexual stimulation of the penis. *J Comp Neurol* **493**, 33–38

Goy, R.W. and Phoenix, C.H. (1963). Hypothalamic regulation of female sexual behaviour: establishment of behavioural oestrous in spayed guinea-pigs following hypothalamic lesions. *J Reprod Fert* **5**, 23–40

Gray, P.A., Janczewski, W.A., Mellen, N., McCrimmon, D.R. and Feldman, J.L. (2001). Normal breathing requires preBötzinger complex neurokinin-1 receptor-expressing neurons. *Nature Neurosci* **4**, 927–930

Greco, B., Edwards, D.A., Michael, R.P. and Clancy, A.N. (1998). Androgen receptors and estrogen receptors are colocalized in male rat hypothalamic and limbic neurons that express fos immunoreactivity induced by mating. *Neuroendocrinology* **67**, 18–28

Guarraci, F.A. and Clark, A.S. (2006). Ibotenic acid lesions of the medial preoptic area disrupt the expression of partner preference in sexually receptive female rats. *Brain Res* **1076**, 163–170

Guarraci, F.A., Megroz, A.B. and Clark, A.S. (2004). Paced mating behavior in the female rat following lesions of three regions responsive to vaginocervical stimulation. *Brain Res* **999**, 40–52

Hansen, S., Köhler, C., Goldstein, M. and Steinbusch, H.V.M. (1982). Effects of ibotenic acid-induced neuronal degeneration in the medial preoptic area and the lateral hypothalamic area on sexual behavior in the male rat. *Brain Res* **239**, 213–232

Harding, S.M. and McGinnis, M.Y. (2005). Microlesions of the ventromedial nucleus of the hypothalamus: effects on sociosexual behaviors in male rats. *Behav Neurosci* **119**, 1227–1234

Heimer, L. and Larsson, K. (1966). Impairment of mating behavior in male rats following lesions in the preoptic-anterior hypothalamic continuum. *Brain Res* **3**, 248–263

Hennessey, A.C., Wallen, K. and Edwards, D.A. (1986). Preoptic lesions increase the display of lordosis by male rats. *Brain Res* **370**, 21–28

Hillarp, N.Å., Olivecrona, H. and Silfverskiöld, W. (1954). Evidence for the participation of the preoptic area in male mating behavior. *Experientia* **10**, 224–225

Hoshina, Y., Takeo, T., Nakano, K., Sato, T. and Sakuma, Y. (1994). Axon-sparing lesion of the preoptic area enhances receptivity and diminishes proceptivity among components of female rat sexual behavior. *Behav Brain Res* **61**, 197–204

Hull, E.M., Lorrain, D.S., Du, J. *et al.* (1999). Hormone–neurotransmitter interactions in the control of sexual behavior. *Behav Brain Res* **105**, 105–116

Hull, E.M., Meisel, R.L. and Sachs, B.D. (2002). Male sexual behavior. In *Hormones, brain and behavior*, vol. 1 (D.W. Pfaff, A.P. Arnold, A.M. Etgen, S.E. Fahrbach and R.T. Rubin, eds), pp. 3–137. Academic Press: New York

Hurtazo, H.A., Paredes, R.G. and Ågmo, A. (2003). Inactivation of the medial preoptic area/anterior hypothalamus (MPOA/AH) reduces sexual incentive motivation in male rats. *Soc Neurosci Abstr* 404.6

Keast, J.R. (1999). The autonomous nerve supply of male sex organs – an important target of circulating androgens. *Behav Brain Res* **105**, 81–92

Keast, J.R. and Gleeson, R.J. (1998). Androgen receptor immunoreactivity is present in primary sensory neurons of male rats. *Neuroreport* **9**, 4137–4140

Keast, J.R. and Saunders, R.J. (1998). Testosterone has potent, selective effects on the morphology of pelvic autonomic neurons which control the bladder, lower bowel and internal reproductive organs of the male rat. *Neuroscience* **85**, 543–556

Kendrick, K.M. and Dixson, A.F. (1986). Anteromedial hypothalamic lesions block proceptivity but not receptivity in the female common marmoset (*Callithrix jacchus*). *Brain Res* **375**, 221–229

Koyama, Y., Fujita, I., Aou, S. and Oomura, Y. (1988). Proceptive presenting elicited by electrical stimulation of the female monkey hypothalamus. *Brain Res* **446**, 199–203

Koyama, Y., Satou, M., Oka, Y. and Ueda, K. (1984). Involvement of the telencephalic hemispheres and the preoptic area in sexual behavior of the male goldfish, *Carassius auratus*: A brain lesion study. *Behav Neural Biol* **40**, 70–86

Larsson, K. (1962). Mating behavior in male rats after cerebral cortex ablation. I. Effects of lesions in the dorsolateral and the median cortex. *J Exp Zool* **151**, 167–176

Larsson, K. (1964). Mating behavior in male rats after cerebral cortex ablation. II. Effects of lesions of the frontal lobes compared to lesions in the posterior half of the hemispheres. *J Exp Zool* **155**, 203–214

La Vaque, T.J. and Rodgers, C.H. (1975). Recovery of mating behavior in female rat following VMH lesion. *Physiol Behav* **14**, 59–63

Law, T. and Meagher, W. (1958). Hypothalamic lesions and sexual behavior in the female rat. *Science* **128**, 1626–1627

Lloyd, S.A.C. and Dixson, A.F. (1988). Effects of hypothalamic lesions upon the sexual and social behaviour of the male common marmoset (*Callithrix jacchus*). *Brain Res* **463**, 317–329

Malsbury, C.W., Kow, L.M. and Pfaff, D.W. (1977). Effects of medial hypothalamic lesions on the lordosis response and other behaviors in female golden hamsters. *Physiol Behav* **19**, 223–237

Mathews, D. and Edwards, D.A. (1977). Involvement of the ventromedial and anterior hypothalamic nuclei in the hormonal induction of receptivity in the female rat. *Physiol Behav* **19**, 319–326

Merari, A. and Ginton, A. (1975). Characteristics of exaggerated sexual behavior induced by electrical stimulation of the medial preoptic area in male rats. *Brain Res* **86**, 97–108

Morgantaler, A. and Crews, D. (1978). Role of the anterior hypothalamus-preoptic area in the regulation of reproductive behavior in the lizard, *Anolis carolinensis*: Implantation studies. *Horm Behav* **11**, 61–73

Moser, M.B., Moser, E.I., Forrest, E., Andersen, P. and Morris, R.G.M. (1995). Spatial learning with a minislab in the dorsal hippocampus. *Proc Nat Acad Sci USA* **92**, 9697–9701

Mouras, H. (2006). Neuroimaging techniques as a new tool to study the neural correlates involved in human male sexual arousal. *Curr Med Imaging Rev* **2**, 71–77

Musatov, S., Chen, W., Pfaff, D.W., Kaplitt, M.G. and Ogawa, S. (2006). RNAi-mediated silencing of estrogen receptor α in the ventromedial nucleus of the hypothalamus abolishes female sexual behaviors. *Proc Nat Acad Sci USA* **103**, 10456–10460

Ogawa, S., Olazábal, U.E., Parhar, I.S. and Pfaff, D.W. (1994). Effects of intrahypothalamic administration of antisense DNA for progesterone receptor mRNA on reproductive behavior and progesterone receptor immunoreactivity in female rat. *J Neurosci* **14**, 1766–1774

Okada, R., Watanabe, H., Yamanouchi, K. and Arai, Y. (1980). Recovery of sexual receptivity in female rats with lesions of the ventromedial hypothalamus. *Exp Neurol* **68**, 595–600

Olster, D.H. (1993). Ibotenic acid-induced lesions of the medial preoptic area/anterior hypothalamus enhance the display of progesterone-facilitated lordosis in male rats. *Brain Res* **626**, 99–105

Olster, D.H. (1995). Progesterone-facilitated lordosis in medial preoptic area-lesioned juvenile guinea pigs. *Horm Behav* **29**, 519–530

Olster, D.H. (1998). Lordosis-enhancing medial preoptic area lesions do not alter hypothalamic estrogen receptor- or progestin receptor-immunoreactivity in prepubertal female guinea pigs. *Brain Res* **790**, 254–263

Olster, D.H. and Paulson, K.C. (1997). Effects of medial preoptic area and septal lesions on puberty in female guinea pigs. *Biol Reprod* **56**, 731–738

Paredes, R.G. (2003). Medial preoptic area/anterior hypothalamus and sexual motivation. *Scand J Psychol* **44**, 203–212

Paredes, R.G. and Ågmo, A. (1992). Facilitation of sexual behavior shortly after electrolytic lesion of the medial preoptic area. What does it mean? *Brain Res Bull* **29**, 125–128

Paredes, R.G. and Baum, M.J. (1995). Altered sexual partner preference in male ferrets given excitotoxic lesions of the preoptic area anterior hypothalamus. *J Neurosci* **15**, 6630

Paredes, R.G. and Baum, M.J. (1997). Role of the medial preoptic area/anterior hypothalamus in the control of masculine sexual behavior. In *Annual Review of Sex Research* (R.C. Rosen, C.R. Davis and H.J. Ruppel Jr, eds), pp. 68–101. Society for the Scientific Study of Sexuality: Allentown

Paredes, R.G., Haller, A.E., Manero, M.C. and Ågmo, A. (1990). Medial preoptic area kindling induces sexual behavior in sexually inactive male rats. *Brain Res* **515**, 20–26

Paredes, R.G., Highland, L. and Karam, P. (1993). Socio-sexual behavior in male rats after lesions of the medial preoptic area: evidence for reduced sexual motivation. *Brain Res* **618**, 271–276

Paredes, R.G., Tzschentke, T. and Nakach, N. (1998). Lesions of the medial preoptic area/anterior hypothalamus (MPOA/AH) modify partner preference in male rats. *Brain Res* **813**, 1–8

Park, K., Kang, H.K., Seo, J.J., Kim, H.J., Ryu, S.B. and Jeong, G.W. (2001a). Blood-oxygenation-level-dependent functional magnetic resonance imaging for evaluating cerebral regions of female sexual arousal response. *Urology* **57**, 1189–1194

Park, K., Soe, J.J., Kang, H.K., Ryu, S.B., Kim, H.J. and Jeong, G.W. (2001b). A new potential of blood oxygenation level dependent (BOLD) functional MRI for evaluating cerebral centers of penile erection. *Int J Impot Res* **13**, 73–81

Pfaff, D.W. (1968). Uptake of estradiol-17β -^{3}H in the female rat brain. An autoradiographic study. *Endocrinology* **82**, 1149–1155

Pfaff, D.W. (1980). *Estrogens and brain function: neural analysis of a hormone-controlled mammalian reproductive behavior*. Springer: New York

Pfaff, D.W. (1999). *Drive. Neurobiological and molecular mechanisms of sexual motivation*. MIT Press: Cambridge

Pfaff, D.W. and Sakuma, Y. (1979a). Deficit in the lordosis reflex of female rats caused by lesions in the ventromedial nucleus of the hypothalamus. *J Physiol* **288**, 203–210

Pfaff, D.W. and Sakuma, Y. (1979b). Facilitation of the lordosis reflex of female rats from the ventromedial nucleus of the hypothalamus. *J Physiol* **288**, 189–202

Pfaus, J.G., Kleopoulos, S.P., Mobbs, C.V., Gibbs, R.B. and Pfaff, D.W. (1993). Sexual stimulation activates c-fos within estrogen-concentrating neurons of the female rat forebrain. *Brain Res* **624**, 253–267

Pleim, E.T., Brown, T.J., MacLusky, N.J., Etgen, A.M. and Barfield, R.J. (1989). Dilute estradiol implants and progestin receptor induction in the ventromedial nucleus of the hypothalamus: correlation with receptive behavior in female rats. *Endocrinology* **124**, 1807–1812

Powers, B. and Valenstein, E.S. (1972). Sexual receptivity: Facilitation by medial preoptic lesions in female rats. *Science* **175**, 1003–1005

Richmond, G. and Clemens, L.G. (1988). Ventromedial hypothalamic lesions and cholinergic control of female sexual behavior. *Physiol Behav* **42**, 179–182

Sar, M. and Stumpf, W.E. (1973). Neurons of the hypothalamus concentrate [^{3}H]-progesterone or its metabolites. *Science* **183**, 1266–1268

Sarrieau, A., Mitchell, J.B., Lal, S., Olivier, A., Quirion, R. and Meaney, M.J. (1990). Androgen binding sites in the human temporal cortex. *Neuroendocrinology* **51**, 713–716

Satou, M. (1984). Telencephalic and preoptic areas integrate sexual behavior in hime salmon (landlocked red salmon, *Oncorynchys nerka*): Results of electrical brain stimulation experiments. *Physiol Behav* **33**, 441–447

Schmidt, R.S. (1968). Preoptic activation of frog mating behavior. *Behaviour* **30**, 239–257

Shimura, T., Yamamoto, T. and Shimokochi, M. (1994). The medial preoptic area is involved in both sexual arousal and performance in male rats: re-evaluation of neuron activity in freely moving animals. *Brain Res* **640**, 215–222

Slimp, J.C., Hart, B.L. and Goy, R.W. (1978). Heterosexual, autosexual and social behavior of adult male rhesus monkeys with medial preoptic-anterior hypothalamic lesions. *Brain Res* **142**, 105–122

Södersten, P., de Jong, F.H., Vreeburg, J.T. and Baum, M.J. (1974). Lordosis behavior in intact male rats: Absence of correlation with mounting behavior or testicular secretion of estradiol-17β and testosterone. *Physiol Behav* **13**, 803–808

Södersten, P. and Larsson, K. (1974). Lordosis behavior in castrated male rats treated with estradiol benzoate or testosterone propionate in combination with an estrogen antagonist, MER-25, and in intact male rats. *Horm Behav* **5**, 13–18

Södersten, P. and Larsson, K. (1975). Lordosis behavior and mounting in male rats: Effects of castration and treatment with estradiol benzoate or testosterone propionate. *Physiol Behav* **14**, 159–164

Soulairac, A. and Soulairac, M.L. (1972). Action des substances neurostimulantes sur le comportement sexual du rat mâle après lésions du cortex cérébral. *J Physiol (Paris)* **65**, Suppl. 3, 504

Stark, R., Schienle, A., Girod, C. *et al.* (2005). Erotic and disgust-inducing pictures – differences in the hemodynamic responses of the brain. *Biol Psychol* **70**, 19–29

Stumpf, W.E., Sar, M. and Keefer, D.A. (1975). Anatomical distribution of estrogen in the central nervous system of mouse, rat, tree shrew, and squirrel monkey. *Adv Biosci* **15**, 77–88

Truitt, W.A. and Coolen, L.M. (2002). Identification of a potential ejaculation generator in the spinal cord. *Science* **297**, 1566–1569

Truitt, W.A., Shipley, M.T., Veening, J.G. and Coolen, L.M. (2003). Activation of a subset of lumbar spinothalamic neurons after copulatory behavior in male but not female rats. *J Neurosci* **23**, 325–331

van der Horst, V.G.J.M. and Holstege, G. (1998). Sensory and motor components of reproductive behavior: pathways and plasticity. *Behav Brain Res* **92**, 157–167

Warembourg, M. (1978). Radioautographic study of the brain and pituitary after [^{3}H]-progesterone injection into estrogen-primed ovariectomized guinea pig. *Neurosci Lett* **7**, 1–5

Wersinger, S.R., Baum, M.J. and Erskine, M.S. (1993). Mating-induced fos-like immunoreactivity in the rat forebrain: a sex comparison and a dimorphic effect of pelvic nerve transection. *J Neuroendocr* **5**, 557–568

Xiao, K., Kondo, Y. and Sakuma, Y. (2005). Differential regulation of female rat olfactory preference and copulatory pacing by the lateral septum and medial preoptic area. *Neuroendocrinology* **81**, 56–62

Yang, L.Y. and Clemens, L.G. (2000). MPOA lesions affect female pacing of copulation in rats. *Behav Neurosci* **114**, 1191–1202

CHAPTER

7

Learning and sex: Sexual activity as reinforcement and reward

1 An ultrashort introduction to the principles of learning

Basic for any understanding of the apparent complexities of sexual behavior is the issue of how learning can modify the stimulus control of that behavior. Of equal importance is the issue of how learning may modify the motor patterns leading to sexual reward. The role of learning for sexual behaviors becomes paramount as soon as we want to understand human sexual behaviors, but it may also have considerable importance for non-human mammals. In this chapter, I try to show that the execution of sexual acts, more precisely acts involving genital stimulation, is involved in both classical conditioning and operant learning. Moreover, sexual responses and the stimuli controlling them may be the subject of non-associate forms of learning, notably habituation. The role of non-associative learning will also be mentioned. In the human, social learning is the main determinant of sexual behaviors, both with regard to the stimulus control of these behaviors and with regard to the motor patterns involved. I will introduce the fundamental principles of social learning and add a few words about its crucial role.

Classical conditioning, usually understood as a learning process by which one stimulus acquires the capacity to replace another, is the main process involved in modifications of the stimulus control of sexual activities in non-human mammals. In primary classical conditioning, the unconditioned stimulus, a stimulus that spontaneously activates a behavior, becomes associated with a previously neutral stimulus because of repeated presentations of the two stimuli in conjunction. A simple example may be a light that is presented to a hungry rat for a few seconds before a food pellet appears in the food cup. Initially, the rat will show no reaction to the light, more than perhaps a simple orienting reaction, i.e. turning the head toward the stimulus and looking at it. The food pellet, however, causes the rat to approach the food cup and subsequently ingest the pellet. When the light has preceded the food pellet a sufficient number of times, the rat will approach the food cup as soon as the light appears. The light has become a conditioned stimulus, evoking an important part of the response, approach, previously evoked only by the unconditioned stimulus, the food pellet. The light has replaced the food pellet as the stimulus causing the rat to display the behavior pattern of approach. We already know that any stimulus activating approach behaviors is a positive incentive. The neutral light stimulus has become a positive incentive because of classical conditioning. Instead of a light, any neutral stimulus could have been presented for a few seconds before delivery of the food pellet, like a tone or an odor. Likewise, instead of using food to a hungry rat as the unconditioned stimulus, we could have used water to a thirsty rat or any other stimulus causing approach without previous learning. A sexual example would be to make a light precede the appearance of a sexually receptive female rat to a male. After a sufficient number of pairings of the light and the unconditioned sexual incentive of a female, the light should acquire sexual incentive properties.

Rather than replacing one stimulus with another through the process of classical conditioning, we may arrange a situation such that an act produces a desirable stimulus. A hungry rat may be put into a cage where a nose poke in a small opening in the upper right corner produces delivery of a food pellet at the bottom of the opposite corner. If this occurs for a sufficient number of times, that rat will immediately put her nose in the opening when introduced into the cage after a couple of hours of food deprivation. In other words, an act, in this case poking the nose in a hole at the upper right corner of the cage, which produces a desirable consequence, in this case a food pellet, tends to be repeated with increasing frequency. We say that the rat has learned to nose poke. This kind of conditioning, in which the execution of a motor pattern precedes a desirable stimulus or event, is called operant conditioning.

Any event or stimulus able to enhance the probability of repetition of the act immediately preceding its appearance is called a reinforcer. Whether a stimulus or event has the quality of being a reinforcer or not can only be determined by observing its behavioral consequences. The circularity of the definition of a reinforcer has been the subject of much debate over the years, but learning theorists have long since solved the problem. We do not care about that here. What we have to care about, though, is the distinction between a reinforcer and a reward. That distinction is not always made, and the reason for this confusion can be attributed to Thorndike. The great learning theorist considered the terms reinforcement and reward as synonyms (Thorndike, 1932). However, reinforcement refers to effects on learning. The stimulus (e.g. food) or event (e.g. access to a new space) enhances the probability that the act producing it will be repeated. Each time a reinforcer appears, learning is strengthened, reinforced. Reward refers to some kind of affective response to the stimulus or event and has no obligatory relationship to learning. The distinction between reward and reinforcement has been brilliantly analyzed in a somewhat old paper (White, 1989). In most cases, a stimulus or event having the property of reinforcer also has the property of reward, but there is no necessary or logical association. This makes it important to evaluate both the affective reactions produced by a stimulus or event and the effects it may have on learning. We will see that there are experimental procedures for doing that.

Classical and operant conditioning are two forms of associative learning. In classical conditioning, one stimulus is associated with another and in operant conditioning, an act is associated with an outcome. Besides associative learning, there are non-associate forms of learning. One of these is called habituation, which is usually defined as a reduction of the intensity of a response or of the probability of appearance of the response as a consequence of repeated presentations of a constant stimulus. The reduction must be a result of processes within the central nervous system, independent of adaptation in the sensory organs or fatigue in the effector systems. A typical example of habituation is our response to a loud noise. The first time we hear an unexpected, loud noise from a construction across the street we normally jump but, if the same noise is repeated a hundred times within one hour, we will no longer react to it. This lack of reaction is neither because I cannot hear it, nor because I am tired of jumping.

Instead of habituating to a constant stimulus, some people react in the opposite way. When my neighbor turns on his unpleasant music, a collection of Norwegian folk music tunes where the violins whine in a manner extremely aggressive to the human ear, I may not react for the first 15 minutes. Then I start to feel anxious and an intense activity in my sympathetic nervous system is evident. If it does not stop within another 15 minutes, I knock at the neighbor's door and ask him to turn the noise off. My reaction to a constant stimulus increased with repeated exposure. Please observe that all the folk music tunes sound identical to any non-expert, thereby satisfying the criterion of constant stimulus. This kind of learning is called sensitization. Another form of sensitization is observed when a completely irrelevant but strong or noxious stimulus is presented. The most famous case is certainly the enhanced gill and siphon withdrawal observed in response to a water jet in the sea hare, *Aplysia californica*, when a weak electric shock is applied to the head or tail.

The last form of learning I want to mention is the least understood. It is called social learning and is characterized by modifications of behavior as a result of social interactions, including observations of the behavior of other individuals, and in the absence of reinforcement. One form of social learning is called observational learning. It consists of observing someone and copying her behavior. A couple of conditions need to be satisfied if observational learning is to occur. These include the rather obvious necessity of paying attention to the person being observed and a detailed retention of the observed behavior in memory. The observer must also have the motor capacities for a replication of the behavior that was observed. This is not enough, though, for the socially learned behavior to manifest itself. Appropriate motivation is necessary and appropriate circumstances are required. A simplistic example may illustrate some of these principles. Imagine that we come to an airport for the first time in life. We are obviously overloaded with heavy luggage and we have no idea of what to do. In order to find out, we start to observe what other people do. We notice that most people entering the departure hall walk straight up a flight of stairs carrying their suitcases. A common reaction would be to suppose that the check-in counters are located on a higher level of the departure hall. We start to walk toward the stairs, only to find out that our suitcases were heavier than our muscle strength could cope with. We had paid attention, we had memorized, we had the motivation and the appropriate context, but not the motor capacities to execute the behavior pattern we had acquired through social learning.

It is not necessary directly to observe other, living people for acquiring knowledge by social learning. Instead of watching a living person we can watch, for example, a pornographic movie and acquire the behavior patterns shown there through social learning, or we can read a pornographic novel and do likewise.

Learning specialists know a lot with regard to both classical and operant conditioning. They know, for example, that both these forms of learning are as important in the human as in other mammals. They also know a lot about reinforcement and how to arrange reinforcement in relation to the preceding act in order to maximize the rapidity of learning, or maximize the amount of effort expended for getting the reinforcement, or how to make the learned act particularly resistant to extinction, and many other things. Habituation and sensitization are also quite well known. Social learning is the least understood of all forms of learning and it is probably of limited importance for non-human sexual behaviors. On the contrary, human sexuality is heavily determined by social learning, as pointed out already. Anyone interested in learning more about learning should consult any current textbook, of which there are plenty. The subject is fascinating indeed but, despite this fact, I will not enter into any elaborate analysis of learning theory. This brief and probably wanting introduction is enough for our purposes.

2 Habituation of sexual responses

2.1 Studies in non-human mammals

I defined habituation as a decrease in the intensity of a response because of repeated exposure to a constant stimulus caused by processes in the central nervous system and unrelated to sensory adaptation or effector fatigue. In male rats and other male mammals, prolonged exposure to a sexually receptive female and the associated execution of copulatory behavior leads eventually to an end of sexual activity. It is said that the animal is sexually exhausted. The cause of the disappearance of sexual behavior can obviously be that the male is physically exhausted, in other words fatigue in the effector systems. Another cause may be that his sensory organs have adapted and no longer register the stimuli emitted by the female. If we replace the original female partner with a new one, the male will reinitiate copulation. This phenomenon has been termed the Coolidge effect in honor of the 30th president of the USA, Calvin Coolidge. It is not entirely clear how the term was established, but there is a little anecdote that seems to be in the public domain and which I can tell. It is said that the president and his wife visited a farmer who had a substantial number of hens but only one rooster. The following conversation is said to have occurred.

Mrs Coolidge: Does the rooster copulate more than once each day?
Farmer: Dozens of times.

Mrs Coolidge: Please tell that to the president.
President Coolidge: Same hen every time?
Farmer: Oh no, Mr President, a different one each time.
President Coolidge: Tell that to Mrs Coolidge.

Whether the anecdote is true or not is unimportant. It is good enough either way. To return to more serious matters: by replacing the female with a new one we have removed a constant stimulus (the original female) and introduced a new one, at least somewhat different from the first (provided rats have some capacity for recognition of individuals, which in fact they seem to have). Remember that habituation is specific to a particular stimulus. Now, since the male starts to copulate again, he cannot have suffered from fatigue in effector systems. If he had been physically exhausted, a change of female should have no effect. Adaptation and disadaptation of sensory receptors is certainly a possibility, but it seems quite remote. There is, actually, no reason to believe that a new female would have an immediate effect on receptor adaptation. Dishabituation because of the introduction of a new stimulus is probably the most convincing explanation for the reactivation of behavior. The Coolidge effect can be considered as evidence for the existence of habituation of sexual responses in male rats. A similar effect has been described in some other rodents (reviewed in Dewsbury, 1981) and in rams (Beamer *et al.*, 1969), bulls (Bailey *et al.*, 2005) and in the male rhesus monkey (Michael and Zumpe, 1978), suggesting that habituation/dishabituation of sexual responses is widespread among male mammals.

As is so frequently the case with phenomena of sexual behavior, there are very few data on a Coolidge effect in females. One study (Krames, 1971) exposed females to a new male when the original male had ceased to copulate, with the purpose of determining if the female accepted as many ejaculations from the new partner as she had with the original one. She did. The authors did not care much about the females' behavior, since they were interested in finding out whether the Coolidge effect in males could be attributed to changes in female behavior rather than to exhaustion/habituation of the male. They reasoned that the male might stop copulating because of reduced sexual motivation on the part of the female. This reasoning could be tested by replacing the original male with a new one. If the new male performed as many ejaculations as the original male had done, then it could be concluded that changes in female behavior was not the cause of cessation of male behavior. This observation is certainly interesting, but it does not tell us anything about a possible existence of a Coolidge effect in females, although some scientists cite this paper as evidence for that.

Something most similar to a Coolidge effect has been reported in female hamsters, though. In an interesting experiment, hamsters copulated until they no longer displayed lordosis in response to the male's mounts, or until they actively fought him away. At that moment, some of the females were given a new male. They started to display lordosis again (Lisk and Baron, 1982). This appears to be equivalent to what has been reported in males. However, the display of lordosis in response to the male's mount is not at all the same as actively to search sexual contact. The male Coolidge effect is exactly that and not the display of a somatic reflex in response to stimulation provided by the partner. Although suggestive, it is not self-evident that the hamster data show that the female actively seeks sexual contact with the new male in the way a male pursues and mounts a new female.

There is only one study in female rodents that unequivocally shows a Coolidge effect. That study was performed by two undergraduate students in my laboratory a few years ago (Ågmo *et al.*, 2002). They tested female rats in the pacing procedure. If a female maintained herself inaccessible to the male for a continuous period of more than 20 minutes following an ejaculation, we considered her to be sexually exhausted or to have habituated to the male. This occurred after having received between three and five ejaculations. The original male was either removed and then put back in the mating test cage (control) or he was removed and replaced by a new male (experimental). The females did not respond to the reintroduction of the original male in any particular way, but the new male rapidly provoked the display of proceptive behaviors and the female's return to the male compartment. The 'new' male was sometimes a rested male (no immediately preceding sexual activity) or a male that was in his fifth post-ejaculatory refractory interval when being put in the mating test cage. The availability of such males was assured by having several males copulating in another part of the observation room during the experiments with the females. Thus, it seems that the Coolidge effect exists also in female rats. The results of this experiment are shown in Figure 7.1. There are, to my knowledge, no other data in female non-human animals.

Although the Coolidge effect frequently, if not usually, is interpreted as a consequence of habituation to the original female, there is one alternative explanation that must be considered. The prolonged consumption of or commerce with a reinforcer, like an extremely tasty sweet solution or a very funny computer game, leads to a reduction of its value. This effect has been well documented and is called

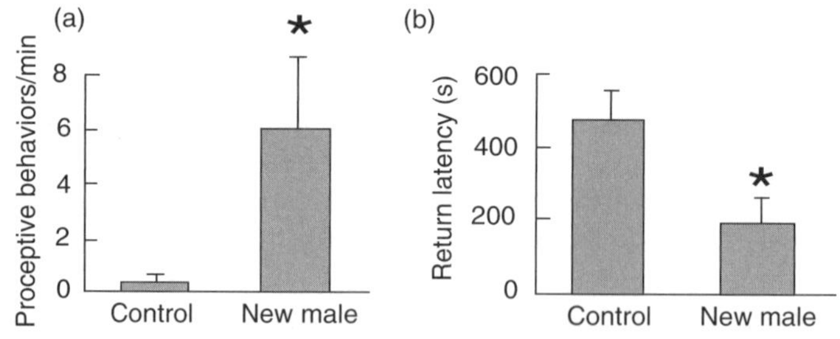

FIGURE 7.1 (a) Number of proceptive behaviors per minute displayed after the introduction of a new male to sexually exhausted females compared to that observed when the original male was removed and then replaced. The recording of proceptive behaviors stopped when copulation was reinitiated. (b) Length of the latency to enter the male's area in sexually exhausted female rats when a new male had been introduced compared to that recorded when the original male was removed and then replaced. *, different from control, $P < 0.05$. (Data from an unpublished study by Turi, Ellingsen and Ågmo.)

negative alliesthesia (Cabanac and Lafrance, 1990; Cabanac, 1992; Zhao and Cabanac, 1994). For example, if a human or a rat was asked to taste a small amount of a rather concentrated sucrose solution and then indicate the degree of pleasantness of the taste, they would rate it highly, at least the rats. Now, if we perform the same test immediately after having consumed a large amount of sucrose solution, both the human and the rat would rate the pleasantness of the taste far lower than prior to consumption. It is this reduction in pleasantness of a stimulus as a result of consumption or other commerce with it that is called negative alliesthesia. After prolonged copulation, the pleasantness of sex would be reduced and the incentive value of the stimulus indicating the availability of more sex would also be reduced. The decline in sexual activity after extensive copulation could, then, easily be explained by negative alliesthesia rather than habituation. The reactivation of copulatory activity by a new female, however, cannot be accounted for by alliesthesia. In rats, copulatory behavior is stereotyped, meaning that the copulatory acts performed with one female are extremely similar to those performed with any other female. Thus, the introduction of a new female should have no effect, since that could not signal the availability of activities different from those already performed. Negative alliesthesia would not be affected by a change of female. Habituation remains the most likely explanation.

If a Coolidge effect existed in the human, then we would have serious problems distinguishing alliesthesia from habituation. Human copulatory behaviors are not stereotyped. An example will illustrate the consequences of this. A man stops performing penile-vaginal intercourse after two ejaculations. The cessation of sexual activity may be attributed to either habituation or negative alliesthesia. Now, imagine that another woman appears and the man reinitiates sexual behavior with her. The only difference is that he performs cunnilingus on her rather than having penile-vaginal intercourse with her. The observation of the man's behavior seems unambiguous enough, but the explanation becomes quite intricate. First, we can maintain that the new woman was a new stimulus and, therefore, a cause of dishabituation. Sexual activity was consequently reinitiated. This coincides with our behavioral observations. Second, we can suggest that the new woman somehow signaled that she would not like penile-vaginal intercourse but rather cunnilingus. The negative alliesthesia affecting penile-vaginal intercourse would not affect the propensity to perform cunnilingus and the man would reinitiate sexual activity. In other words, the man can have reactivated his copulatory behavior either because of a new stimulus (dishabituation) or because he anticipated a new kind of sexual activity (disappearance of negative alliesthesia). The observed behavior is compatible with both explanations. One possible solution to this dilemma would be to ask the man.

The discussion of alliesthesia is slightly out of context here. Alliesthesia refers to the affective consequences of a stimulus or a behavior and this subject will be discussed in another section of this chapter. However, since alliesthesia is a potential explanation for the Coolidge effect, I considered it acceptable to mention it here. Furthermore, by applying the concept to human sexual behavior, I got an opportunity to remind the reader of the fundamental differences between copulatory behavior in the human and that of non-human animals. After this spontaneous digression from our main theme, I will immediately return to the issue of habituation.

2.2 Studies in humans

As was mentioned above, something similar to a Coolidge effect has been reported in the male rhesus monkey. This is not the only primate in which habituation/dishabituation of sexual responses has been shown. We have a respectable amount of data from men and, surprisingly, also some data from women. However, these data do not refer to reduced intensity of copulatory behavior as a consequence of extended copulation, but exclusively to reduced arousal in response to sexual incentives. Let us look at some of these data.

In one of the first, if not *the* first, study specifically designed for evaluating habituation of sexual arousal in the human, 40 young male students were exposed to slides depicting heterosexual couples engaged in oral or genital sexual activity. Arousal, as always defined as erection, was quantified by penile plethysmography. One group of subjects was shown the same slide several times, either until there was no penile response or up to 27 stimulus presentations. Another group was shown five different slides an equal number of times. Results revealed a rapid decrease in response intensity in the group shown the same slide, while the response decreased much more slowly in the group exposed to five different slides (O'Donahue and Geer, 1985). The fact that the decline was slower in the group shown five different slides than in the group shown the same slide excludes the possibility that the reduced response was due to fatigue in effector systems. Both groups made the same number of responses, so if fatigue would have been the only cause of response intensity reduction, the decrease should have been similar in both groups. Since this was not the case, it can be concluded that the reduced response was due to habituation. A posterior study, employing audiotapes with sexually arousing content as stimuli, extended these data from habituation within one single experimental session to lasting habituation over several sessions. This design allowed also for the detection of typical habituation phenomena like partial spontaneous recovery between sessions. As predicted from data on habituation of other responses, the spontaneous between-session recovery gradually diminished. When the penile response to the audiotape showed complete habituation, different audiotapes were used and the response recovered (O'Donohue and Plaud, 1991; Plaud *et al.*, 1997). We will soon see that long-term habituation of sexual arousal responses to a constant stimulus has applications far beyond the laboratory.

Habituation of sexual responses in men seems to be a reliable finding. In addition to the studies mentioned above employing slides and audiotapes as stimuli, several other researchers have reported reduced penile response after repeated exposure to pornographic movies. A hypothesis concerning the mechanism causing the response reduction has also been proposed. According to this hypothesis, repeated viewing of the same pornographic video leads to a weakened attention to the movie, and this weakening of attention appears to be related to the diminished erection. Indeed, when a measure of attention is included in analyses of covariance of the penile response, the habituation effect is much attenuated (Koukounas and Over, 1993, 1999, 2001). This sounds quite realistic. However exciting a pornographic movie or an audiotape or a slide may be, and however intensely we focus our attention on it at the first exposure, we would probably be less attentive after viewing the same one minute fragment 10 times or more. A question not addressed by the hypothesis is whether our attention is weakened because we find the stimulus less arousing, or whether it arouses us less because we attend it less. This question is not easily answered because I see no way to obtain unequivocal experimental data permitting the elimination of one of these alternatives. This, in turn, means that the hypothesis does not further our understanding of the habituation of sexual responses in any significant way.

In women, habituation of sexual responses has been less studied than in men and results are unclear. The first study performed in women exposed eight undergraduates to a one minute fragment of a pornographic video showing penile-vaginal intercourse with the man on top. Vaginal photoplethysmography was used in order to estimate vaginal blood flow, the typical measure of sexual arousal in women. The video fragment was repeated 18 times in a single session. After that, two new video fragments were shown, one illustrating penile-vaginal intercourse with the woman on top and the other showing intercourse with rear entry. There was a gradual decrease in vaginal blood flow with repeated presentations. After the 18th presentation of the constant stimulus, the new videos were shown. The vaginal response was much enhanced by the new stimulus, a typical example of dishabituation. This fact nicely demonstrates that the response reduction was not due to fatigue in the vaginal tissues or in the nerves controlling the vaginal blood flow (Meuwissen and Over, 1990). This study suggests that women just as men easily habituate to a sexually relevant stimulus. Unfortunately, another study failed to find any evidence for a reduction of vaginal blood after repeated presentation of a slide with a sexually arousing picture or pornographic video fragments (Laan and Everaerd, 1995). This negative study gives the impression of being well controlled and carefully executed, making the absence of any sign of habituation still more frustrating. The conclusion is also rather unattractive: men habituate faster and to a greater degree than women. This sounds very much like social prejudice turned into a scientifically respectable argument. In my own opinion, available data are not sufficient for proposing any founded hypothesis with regard to habituation of sexual responses in women. However, I would be reluctant to admit that women are any different from men in this respect.

2.3 A note on habituation in long-term relationships

In the preceding paragraphs, we have seen that men habituate to a constant sexual stimulus in a laboratory setting. A very important question is whether habituation also occurs outside the laboratory. One way to approach an answer to that question would be to determine the sexually arousing properties of a stable female partner. Humans and some other mammals have the habit of forming relationships of such a kind that all or at least most sexual activity is performed with one single partner for considerable periods of time. One consequence of this is that the members of the pair are repeatedly exposed to each other, perhaps hundreds or thousands of times if the relationship lasts very long. Repeated exposure to a constant stimulus should lead to habituation, as we know by now. Although women have the habit of changing hair color, perfumes and soaps with an amazing frequency, thereby changing some of their potential sexual incentive characteristics, I have the impression that most men do not notice any of these changes. They rather regard their partner as a reasonably constant stimulus. Most men change hairstyle, after shave and hair color less frequently and can be expected to constitute a constant stimulus for their partner. This constancy should, logically, lead to habituation. Indeed, some scientists have suggested that the reduction in the intensity of sexual behaviors occurring in most couples as a function of the duration of the relationship is caused by habituation (Klusmann, 2002).

Explanations of diminished sexual activity as the duration of relationship increases in terms of habituation need some qualifications. The most important is that there is no habituation to a stimulus containing useful information. No animal will habituate to a light signaling the proximate availability of food, for example. No car driver will habituate to a red light at a busy intersection, independently of how frequently he passes that crossing. In the experiments on habituation of sexual responses, the arousing stimulus was a pornographic movie, or an audiotape describing sexual activities, or a slide with some exciting content. None of these stimuli had any consequence, except that of producing erection. As it turned out, the erection was useless since no sexual activity followed the presentation of the stimuli. In other words, sexual incentives were repeatedly presented in the absence of sexual reward. Within the stable couple, sexual reward will be obtained, frequently or occasionally, depending on the couple. The incentive stimuli emitted by the partner will, consequently, be associated with sexual reward with a frequency determined by the intensity of sexual activity in the couple. When that reward is obtained sufficiently often, the incentive stimuli emitted by the partner will signal a rewarding event and no habituation will occur because these stimuli carry meaningful information. On the contrary, if the association between stimuli emitted by the partner and sexual reward is infrequent or entirely absent, then these stimuli cease to carry meaningful information and habituation can occur.

There is not necessarily any reciprocity between the members of the couple with regard to the obtention of sexual reward and the ensuing habituation or absence of it. A man may obtain sexual reward (orgasm) at every single one of the frequent sexual encounters with his partner, while the partner may never do so. This means that the man will never habituate, but the partner may show a fast progression in habituation until the man has lost his sexual incentive properties.

It is possible to maintain that the notion of habituation is irrelevant for sexual incentives. Accepting that all sexual incentives are learned, any reduction in the propensity to respond to a sexual incentive as a consequence of non-reward is *not* habituation but extinction of a conditioned response. In the human, the concept of habituation of sexual responses is perhaps unnecessary. Extinction and negative alliesthesia can explain behavioral changes in all the situations where habituation has been employed as explanatory element. In non-human animals, the mechanisms of habituation may be of some importance.

3 Sensitization of sexual responses

Sensitization, a facilitation of behavior produced by strong or noxious irrelevant stimuli, or by repeated presentations of certain kinds of stimuli, has been described with regard to sexual behavior in rats. Electric shock to the skin or tail pinch have repeatedly been found to facilitate some aspects of male copulatory behavior or to induce copulatory behavior in sexually inactive males (Barfield and Sachs, 1968; Wang and Hull, 1980; Leyton and Stewart, 1996). Again, I know of no data from female rats. Nevertheless, there is no doubt that irrelevant stimuli may enhance the intensity of sexual behavior in male rats. There may well be similar studies in other non-human species, but I have not invested the effort needed for finding them in any search. With regard to the mechanisms behind sensitization, it is generally assumed that the irrelevant stimulus increases general arousal, making the animal more reactive to any stimulus in the environment. Since the salient stimulus in all experimental procedures for studying male sexual behavior is the receptive female, and since a likely response to this stimulus normally is copulation, the enhanced arousal takes the expression of facilitated copulatory behavior.

Regarding sensitization in the human, we are in a most unusual situation. The only experimental studies I know of were performed in women! The purpose was not at all to study sensitization of sexual responses, but instead to determine the effects of activation of the sympathetic nervous system or the influence of anxiety on sexual arousal. The stimuli or actions employed for generating sympathetic activation and/or anxiety were completely irrelevant as far as sex is concerned. Yet they were found to enhance responses to sexually relevant stimuli. This coincides perfectly with the definition of sensitization given earlier in this chapter. Thus, although the scientists performing the studies I soon will mention did not interpret their results in terms of learning, I will use their data as evidence in favor of sensitization of sexual responses.

The oldest paper containing observations relevant to the issue of sensitization that I have been able to localize reports a study in which women viewed either an anxiety-producing film or a neutral travel film shortly before viewing a pornographic movie. The anxiety sequence 'depicted in vivid detail the aftermath of several tragic automobile accidents, including occupants' death cries' and the neutral sequence depicted a travelogue of Nova Scotia. Changes in vaginal blood volume in response to the pornographic movie were, not unexpectedly, taken as indicators of sexual arousal. The women having viewed the anxiety-producing movie showed a larger vaginal response to the pornographic movie than women having watched the neutral movie (Hoon *et al.*, 1977). Similar results were obtained in another study (Palace and Gorzalka, 1990). Again, an anxiety-evoking film fragment preceding exposure to a pornographic film segment enhanced the vaginal response. Some kind of semimystical account in terms of emotion effects could certainly be used for explaining these data, but a more parsimonious explanation would be to suggest that the anxiety-inducing movie simply had sensitized the women to the sexual stimuli in the ensuing pornographic movie. By saying that, we limit ourselves to a pure description of behavior (enhanced vaginal response to sexually arousing stimuli following exposure to an irrelevant stimulus) without any speculation as to cause. We just employ the label corresponding to the behavioral effects observed. If we would like to go a step further and speculate about the physiological mechanisms behind the behavioral effect, we could, for example, propose that activation of the sympathetic nervous system during exposure to the anxiety-inducing movie was the cause for the heightened response. This possibility was actually mentioned by Palace and Gorzalka (1990).

Another study offers some information relevant to the hypothesis that sympathetic activation sensitizes the organism to sexual incentives. Thirty-eight women answered a questionnaire evaluating 'trait and state anxiety' after having seen a pornographic movie. 'Trait anxiety' is something like a stable individual characteristic reflecting a dispositional tendency to experience anxiety, while 'state anxiety' is an acute emotional response consisting of feelings of apprehension and enhanced sympathetic nervous system activity. The vaginal response to the movie was assessed by photoplethysmography. It was found that women scoring moderately high on the 'state anxiety' scale showed a larger increase in vaginal blood flow than women having either low or high scores. These data were interpreted in terms of sympathetic influences on sexual arousal in response to a sexual stimulus (Bradford and Meston, 2006).

As it turns out, the role of sympathetic activation in sexual arousal in women has been evaluated more directly in a series of well-designed studies. In order to achieve a rather intense sympathetic activation, women were asked to exhaust themselves doing physical exercise, always in the form of cycling. The demanded speed and adjusted workload was such that the poor women exercised at a constant 70% of their maximum oxygen uptake. Fifteen minutes after a 20-minute exercise a pornographic movie was shown and the vaginal response was recorded with photoplethysmography. On another occasion, the women saw a pornographic movie without preceding exercise. It was found that the vaginal response to the pornographic movie was significantly larger after exercise than after rest. Curiously, exercise had no effect on the women's subjective rating of the intensity of sexual arousal. The authors concluded that the results provide indirect support for the notion of a facilitatory influence of sympathetic activation on physiological sexual arousal (Meston and Gorzalka, 1995, 1996; Meston, 2000). I would add that they also show that a sexual response may be sensitized.

Further evidence for sympathetic activation as the cause underlying at least some forms of sensitization of sexual arousal by irrelevant stimuli comes from pharmacological studies. The adrenergic agonist ephedrine or its vehicle was administered to women in a double-blind repeated measures design. About 40 minutes after taking the drug or vehicle, the women watched first a short neutral film and then a pornographic movie while the vaginal response was recorded with photoplethysmography. Ephedrine had no effect on vaginal blood flow during the neutral film. However, the response to the pornographic movie was larger after ephedrine than after the vehicle (Meston and Heiman, 1998). For those not familiar with basic physiology, I just mention that the sympathetic nervous system releases noradrenaline, which acts at adrenergic receptors throughout the body. Ephedrine should, then, simulate some of the effects of sympathetic activation. The results obtained could, therefore, be considered as supporting the idea that enhanced sympathetic activity stimulates the response to sexual stimuli. The latter is what we have called sensitization and we can therefore conclude that sensitization of sexual responses is mediated by the sympathetic nervous system.

4 Classical conditioning of sexual responses

4.1 Non-human studies

Classical conditioning with sexual unconditioned stimuli has not been much studied in rodents. Almost all experimental data we have today stem from birds rather than from rodents. To be more exact, from studies performed in the Japanese quail. Michael Domjan and his group at the University of Texas in Austin, have performed a brilliant series of studies on sexual conditioning in the quail. Since I have decided to limit this book to an analysis of mammalian sexual behaviors, I will not review, nor even summarize, this brilliant work. Suffice to say that an arbitrary stimulus like a light or a sound can acquire sexual incentive properties through classical conditioning in male quails (see Domjan *et al.*, 1986, 1992; Domjan, 1994). Furthermore, copulatory behavior can be conditioned to an inanimate object under the condition that this object has some resemblance to a quail. A most interesting feature of this conditioning is that there is no extinction. Repeated exposure to the inanimate object without any concurrent exposure to a living quail does not lead to a disappearance of copulatory behavior, nor even to a reduction of its intensity (Köksal *et al.*, 2004). We will return to this beautiful experiment in Chapter 12.

The earliest rat studies (Zamble *et al.*, 1985, 1986; Cutmore and Zamble, 1988) of classical conditioning with sexual unconditioned stimuli are not easy to interpret. The design was too complicated and the only conclusion that can be drawn from these studies is that some kind of sexual incentive property might become conditioned to contextual cues. This is not unknown to people working with sex behavior in rats. A typical observation is that experienced male rats often start to mount each other as soon as they arrive at the room where sex behavior tests are performed. It is most rare that rats without sexual experience do so. The only reasonable explanation for this change in behavior related to sexual experience is that some cues in the observation room somehow have become conditioned incentives acting on the rats' central motive state to enhance its excitability to a level such that the normally inefficient stimuli emitted by its cage partner(s) suddenly become able to release mounting behavior.

Complicated designs continue to be used in experiments on classical conditioning of sexual responses in rats. A series of studies began with an experiment where an odor was added either to a sexually receptive female or to a non-receptive female before sexual interaction with a male. Other females were unscented. After several trials of this kind, the males were given access to one scented and one unscented female. The males trained with a scented, receptive female preferred to ejaculate with the scented female. Males trained with a scented non-receptive female as well as males trained with an unscented receptive female preferred to ejaculate with the unscented female. There was no difference at all with regard to latencies to mount or intromit the scented and unscented females and the number of pre-ejaculatory mounts and intromissions received was similar for both females (Kippin *et al.*, 1998). The interpretation of these data is not easy. The unconditioned stimulus activating sexual responses was undoubtedly the female. I would even dare to suggest that it was the receptive female. The odor added to the female was the conditioned stimulus. The unconditioned response was approach to and execution of copulatory acts with the receptive female. After conditioning, at the test, we had the rather atypical situation of simultaneous exposure to both the unconditioned and conditioned stimulus (receptive female + odor). Normally, conditioning is evaluated as the intensity of response to the conditioned stimulus alone. Furthermore, the conditioned response is normally similar to the unconditioned response (approach and copulation). In the Kippin *et al.* (1998) study, the conditioned response was entirely new (preference for a particular female). This means that it is extremely difficult to figure out what was learned during the training trials and what kind of learning might have been involved. Still more disturbing is the fact that the males trained with the unscented, sexually receptive female displayed an ejaculatory preference for the unscented female. Here, we have no conditioned stimulus at all, yet the subjects learned exactly the same response as the subjects in the 'conditioning group', preferring to ejaculate with females similar to those that were used during training rather than with an unknown female. This means that the added odor was of no particular importance for the behavior displayed at the test. This, in turn, means that there is no reason for suggesting that the odor had become a conditioned stimulus for anything at all. There is no need for speculating as to what might have been learned during the training sessions, but it is evident that the added odor had nothing to do with it.

Notwithstanding these problems of interpretation, a few subsequent papers employing a similar conditioning procedure were dedicated to a search for the nature of the olfactory conditioning underlying 'conditioned ejaculatory preference' (Kippin and Pfaus, 2001a, 2001b; Kippin *et al.*, 2001). Control procedures, like repeated copulation with a consistently unscented female, were omitted from these later papers. Despite this minor problem, the data from all these studies show that male rats prefer to ejaculate with females smelling similarly to females with whom they have ejaculated earlier. Moreover, the establishment of this preference requires that the female is present during the post-ejaculatory interval. However, these data do not show that a previously neutral stimulus acquires control over an unconditioned response. They have no relevance, in fact, for the problem we are trying to analyze, namely the possibility of classical conditioning involving a sexual unconditioned response. I mentioned these studies only because some colleagues might have been led to believe that they in fact represented an example of sexual conditioning similar to those performed in quails.

Whereas the few existing studies of classical conditioning employing a sexual unconditioned stimulus (like a conspecific of the opposite sex) and the execution of copulatory behaviors (like intromission or ejaculation) as unconditioned response mentioned above have suffered from flaws in the methods, another approach to conditioning has apparently been more successful. An interesting study in the ewe (Gelez *et al.*, 2004) provides some data with regard to an olfactory conditioned stimulus and an endocrine unconditioned response. As we learned in Chapter 3, release of gonadotrophins from the pituitary and ensuing release of gonadal hormones are parts of the visceral reactions to a sexual incentive. A ram elicits LH release of similar magnitude in experienced and inexperienced anestrus ewes, both with regard to the maximum blood concentration and to the number of pulses during 4 hours of exposure. This shows that the ram is an unconditioned stimulus for the unconditioned response of LH release. In a second experiment, ewes were repeatedly exposed to males during the breeding season. Some of these males were scented with lavender while others were unscented. Two months later, when the breeding season had ended and the ewes had entered into anestrus, they were exposed to fleece scented with lavender and the LH response was recorded. It turned out that the females that had been exposed to a lavender smelling male during the breeding season showed a larger response than females exposed to an unscented male or to a lavender smelling female. Thus, the lavender had acquired the capacity to elicit a response similar to the response elicited by the unconditioned stimulus 'ram' in the conditioned group. In other words, conditioning had been successful.

An old study in rats (Graham and Desjardins, 1980) analyzed classical conditioning of LH release in a way similar to the ewe experiments mentioned in the preceding paragraph. Here, the unconditioned stimulus was simply 'placing the male in a cage with a sexually receptive female for 15 min'. This very complex stimulus consisted undoubtedly of all cues emitted by the female as well as the execution of copulatory acts. We are told nothing about the possible sexual interactions occurring during the 15 minutes, so some males may have displayed no sexual activity and others may have made a couple of mounts or intromissions, or they may all have achieved at least one ejaculation. Deficiencies in the report of behavioral data are not uncommon, but that does not make this bad habit less annoying. In this particular case, the consequence is that we do not know the exact nature of the unconditioned stimulus. The conditioned stimulus was the odor of methylsalicylate. Some subjects were exposed to this odor immediately before they were transferred to the cage containing a receptive female. Others, the controls, were exposed to the same odor without any ensuing exposure to a female, or they were exposed to the odor several hours before being placed together with the receptive female. At test, the subjects were exposed only to the odor of methylsalicylate. The conditioned group, methylsalicylate exposure + immediate access to the female, showed a robust LH release in response to the odor while the other groups did not. The Graham and Desjardins (1980) data clearly show that conditioning had occurred. As mentioned, the only problem is that we do not know the exact nature of the unconditioned stimulus. Nevertheless, it is not unreasonable to conclude that the endocrine response to a sexual stimulus can be conditioned to a neutral stimulus through the process of classical conditioning.

4.2 Human studies

There are no rat data concerning classical conditioning of erection or other sexual arousal responses and, likewise, there are none concerning conditioning of copulatory reflexes like mount with pelvic thrusting. In the absence of useful rodent data on these viscerosomatic responses we have to turn to studies in humans. Classical conditioning of sexual responses is actually one of the few problems in the field of experimental sex research that has been more thoroughly studied in the human than in the rat. In fact, the first studies of conditioning of human sexual arousal (erection) were performed long before similar studies were undertaken in rats or other non-human animals.

In one of the early studies, three adult psychologists were employed as subjects rather than the habitual psychology students. They were exposed to pairings of pictures of nude women (unconditioned stimulus) and a pair of black boots (conditioned stimulus). The unconditioned, as well as conditioned response, was erection as evaluated by penile plethysmography. (Observe that when I write erection I do not necessarily mean full erection, a maximally enlarged and rigid penis. In the term 'erection' I include also small increases in penile circumference without any noticeable rigidity. The correct term should probably be tumescence, but I find that term cumbersome and sonorically unattractive. Therefore I prefer erection.) After several pairings, the boots alone produced erection. Moreover, the response to the boots was generalized to similar boots. Finally, repeated presentations of the boots without the ensuing picture of the nude females led to a reduced response, an effect called extinction in learning theory. Generalization as well as extinction are typical features of conditioned responses and the ensemble of the data from this study appears to show convincingly that sexual arousal can be conditioned to a neutral stimulus through the process of classical conditioning (Rachman, 1966). It can be argued that data from three subjects is not enough for any firm conclusion and it can be argued that there was no test of a lack of response to the boot picture before conditioning, and it can also be argued that there was no adequate control for pseudoconditioning. All this is certainly true but, at the same time, it must be accepted that most pioneering studies turn out to be lacking in some respect or another. The passage of time and the associated changes in scientific fashions, as well as the accumulation of additional knowledge, often make old experiments look primitive and faults not known at the time they were done become apparent. Notwithstanding the potential criticisms of the Rachman (1966) report, an additional experiment (Rachman and Hodgson, 1968), again using boots as the conditioned stimulus and employing five subjects, essentially replicated the results of the previous one.

A few more recent studies have reported similar results, although conditioning effects have been rather small. For example, one study employing pornographic movie segments as unconditioned stimuli and pictures of half naked women as neutral stimuli succeeded in demonstrating a modest conditioning effect (Lalumière and Quinsey, 1998). A substantial drawback with this study is that the conditioned stimulus already produced some response before conditioning. The reasoning behind this unfortunate stimulus choice was that the stimulus should be biologically relevant. It would certainly have been possible to find a neutral biologically relevant stimulus, but that was not considered necessary. Another experiment employing nine male psychology students used pictures of half naked or nude women as unconditioned stimuli and erection, as always determined by penile plethysmography, as the unconditioned response. A picture of a penny jar was the neutral, conditioned stimulus. In one group, the conditioned stimulus was presented for 15 seconds followed by the unconditioned stimulus for 30 seconds. There was a one second overlap between the stimuli. In another group, the unconditioned stimulus was presented before the conditioned. This is called backward conditioning, a procedure that does not lead to any learning according to a huge literature. A third group was randomly exposed to both stimuli. This does not produce any learning, but it is a control for the effects of repeated exposure. After 15 pairings it turned out that the first group responded with erection to the conditioned stimulus while there was no response in the other groups (Plaud and Martini, 1999). This nice paper confirms once more that sexual arousal indeed can be classically conditioned in the human male. A later, rather complex study (Hoffman *et al.*, 2004) does not add any further knowledge regarding conditioned sexual arousal in males but contributes to some confusion because of its overly ambitious design.

Women have been less attractive than men for scientists interested in conditioning. I do not really know why and I do not want to speculate around such a sensitive issue. The first report on sexual conditioning in women I have been able to find dates from 1997, which is more than 30 years after the first studies on men were published. Groups of six or seven young women, all students, were exposed to an amber light as conditioned stimulus and 2 min excerpts from pornographic movies depicting either penile–vaginal intercourse or cunnilingus were used as unconditioned stimuli. I use the plural form here because a total of 22 movie excerpts were used. The idea was that repeated exposure of the same sexual act for many times could lead to habituation and a lessened unconditioned response. The conditioned response was enhanced vaginal blood flow as assessed by photoplethysmography. Fifty conditioning trials were performed. It turned out that there was no significant conditioning (Letorneau and O'Donohue, 1997). This is not the place to make conjectures about possible causes for this negative result, but it would be premature to conclude that women are different from men.

Some years later, another report on classical conditioning of sexual responses in women was published. This time, the researchers had conditioned a neutral stimulus, a gun, to 30 seconds segments of a pornographic movie. Curiously enough, another neutral stimulus, a picture of a male abdomen, turned out to be ineffective (Hoffman *et al.*, 2004). These rather unclear results do not allow for any kind of conclusion. My guess is that more well designed studies sooner or later will reveal that sexual arousal can be conditioned just as well in women as it can in men. History shows that many of the prejudice-based hypotheses concerning sexual functions in women and females of other species initially are confirmed in some studies, only to be rejected in the near future. In the present case, the prejudice, conscious or unconscious, should be that women are less excited by sexual stimuli than men, and therefore more unlikely to show conditioned arousal.

In all the conditioning studies mentioned above, the unconditioned stimulus was a sexual incentive producing the unconditioned response of arousal, understood as erection or enhanced vaginal blood flow. The conditioned stimulus was usually a neutral stimulus with no effect at all by itself, as should be the case in conditioning studies. Entirely different procedures have also been employed in many studies. As always, a sexual incentive producing erection was presented, but instead of presenting the stimulus in conjunction with an initially neutral conditioned stimulus, an aversive unconditioned stimulus was presented in conjunction with the sexually arousing stimulus. The idea was that this aversive stimulus should give rise to an unconditioned avoidance response and the avoidance response should become associated with the sexual stimulus. Examples of the unconditioned aversive stimuli that were used are electric shock, usually to a finger, a foul and irritating odor, or drugs causing nausea, like apomorphine. The persons having the privilege to participate in this kind of studies were rarely psychology students. Rather they were individuals responding with sexual arousal to stimuli judged as inadequate by contemporary society. The main group of subjects in all these studies was men responding with sexual arousal to stimuli emitted by other men, in other words male homosexuals. Some pedophiles and fetishists also participated, like the homosexuals with unknown and unreported enthusiasm. The purpose of these conditioning experiments was not to enhance our understanding of how sexual learning functions, but to make the subjects respond less to, and eventually be less attracted to, what were considered unacceptable stimuli. In the case of homosexuals, the situation has changed and it is no longer considered as urgent to make them abandon sexual reactions in response to other men as it was some years ago, but for pedophiles and fetishists there is still little or no social acceptance. However, this latter seems also to be the case for conditioning procedures of the kind just described. This means that available data are quite old and that there is little reason to believe that more data will be collected in the foreseeable future.

A review of most of the old studies on aversive conditioning concluded that it may be effective in reducing erection in response to the stimulus used. Long-term effects are unclear (Quinsey and Marshall, 1983). One possible cause of the modest effects usually found is the employment of inadequate learning procedures. These were sometimes mixtures of classical and operant conditioning, and emetic drugs were administered before the presentation of the sexual stimulus, a kind of extremely inefficient backward conditioning. Sometimes the subjects were asked to read sentences with homosexual content that were shown on slides and after reading was finished a shock was given to a finger. This could have produced an aversion to the act of reading rather than to the content of the sentences. For examples of some of these rather exotic procedures, the reader is kindly referred to an arbitrary, short selection of the original publications (Bancroft, 1969; McConaghy, 1969, 1975). The deficiencies characteristic of many of the aversion conditioning studies can perhaps be attributed to the likely fact that moral zeal constituted a more powerful foundation for these studies than scientific curiosity and search for factual knowledge ever did. Some familiarity with and understanding of learning theory would, perhaps, have been helpful in the design of the conditioning procedures.

Even if we may disagree with the purpose of the aversion conditioning studies, I cannot avoid regretting that they were not performed in a proper way. Some data from rats convincingly show that an aversive stimulus can become associated with an unconditioned sexual incentive, making that incentive lose its positive incentive properties and even turn it into a negative incentive. Logically, both approach to and copulation with the devalued incentive will be absent (Ågmo, 2002). The appearance of an aversive event contingent on copulation had previously been found to eliminate copulatory behaviors in male rats (Koch and Peters, 1987; Peters *et al.*, 1989). In view of these unequivocal data in rats, there is no reason to believe that humans would be immune of the same type of learning. My prediction is that sexual incentives *can* lose their positive incentive properties through conditioning of one kind or another in both men and women. As we will see, this hypothesis has some important consequences when we try to understand some kinds of human sexual dysfunctions.

5 Sexual activity as reinforcement for operant learning

5.1 Studies in non-human mammals

The possibility that the execution of sexual acts could be used for the reinforcement of learning started to be evaluated in the early 1950s. Research on this question was not at all prompted by an interest in sex. Rather it was a result of theoretical discussions among the dominating learning theorists. Some theorists argued that a reduction of biological need was required for reinforcement, while others maintained that the reinforcement of learning was possible in the absence of need reduction. What was then called 'incomplete sexual behavior' was not believed to reduce any biological need. For example, it was generally agreed that a mount without intromission could not afford need reduction. Similarly, non-ejaculatory intromissions were not considered to reduce any biological need. Contemporary prejudice posed that only ejaculation reduced a kind of organismic need for sex. We now know that this kind of reasoning is completely unfounded, but that does not lessen the interest of the data obtained in those early studies. The rationale for a study and the interpretation of the data may both become hopelessly obsolete with the passage of time and changes in the dominating theoretical frameworks, but *data* obtained in a carefully designed and executed experiment never lose their value.

The earliest study that tried to determine whether sex could reinforce operant learning used an extremely simple operant task, running in a straight runway with a sexually receptive female at the end. Other rats found a male at the end. In both cases, the goal animal was introduced at the moment the experimental subject entered the goal box. Those given a female in the goal box were allowed to make two copulations (old fashioned language among people unaccustomed to talk about sexual behavior). With copulation was meant either a mount or an intromission. Those given a male in the goal box were left together with the companion for 2 minutes. A third group of rats was used. This group consisted of males given access to a female in the goal box, but never displaying any copulatory behavior. These males were adequately called non-copulators. After 28 trials performed in blocks of two trials every second day, the rats given a female in the goal box ran faster than those given a male. The slowest animals were the non-copulators (Sheffield *et al.*, 1951).

In the Sheffield *et al.* (1951) experiment, the subjects had neither acquired a new response nor displayed an increased frequency of an existing response, the normal criteria for learning. They simply executed an already existing response faster. However, for some reason, it was at the time rather common to use running speed as an indicator of response strength and increases in response strength were thought to be caused by learning. Independently of the veracity of this notion, it is interesting to note that the execution of copulatory acts had behavioral consequences that social interaction by itself did not have. It is tempting to pose, as the authors indeed did, that copulatory activity functions as a reinforcement.

Enhanced likelihood of performance of a response caused by sexual reinforcement was shown in a study in a T-maze. On one side there was a sexually receptive female and on the other side a male was found. Some experimental males ejaculated with the female at all trials. Some other males were removed from the female after two intromissions. It turned out that the group allowed to ejaculate with the female rapidly learned to choose the female arm of the T-maze. Those given only two intromissions also learned fast, but they soon lost their interest in that arm and choice became random. Furthermore, the ejaculating group showed a substantial reduction of running time, while the group given only two intromissions showed only a modest reduction (Kagan, 1955). This study confirms that sexual activity functions as reinforcement and it also shows that ejaculation is a more efficient reinforcement than two intromissions. These observations were extended some years later in another experiment, again employing a T-maze. The amount of sexual interaction was none in one group, one mount in another, one intromission in a third and four intromissions in a fourth. The last group showed learning superior to all other groups. The one intromission group also learned, albeit less then the four intromission group, while animals allowed to display only one mount failed to learn anything at all (Whalen, 1961). It appears that mounting lacks reinforcing capacity. Ejaculation is superior to intromission, and four intromissions are superior to one with regard to capacity to reinforce learning. Perhaps I might propose that the reinforcing effect of copulatory behavior is directly proportional to the amount and to the quality of behavior.

The fact that the amount of copulatory activity displayed is proportional to the amount of reinforcement obtained certainly suggests that it is the execution of copulatory acts that is the reinforcing event. Another question that might be of some importance for assuring that the reinforcement indeed is sexual is whether it is dependent on gonadal hormones or not. The hormone-dependency of the reinforcing value of copulatory acts was evaluated in an experiment employing a straight runway similar to the one used by Sheffield *et al.* (1951). Exactly as occurred in that study, running time to a receptive female decreased over successive trials. The males were then castrated and tested in the runway. Running time showed a dramatic increase after castration and testosterone replacement reduced it to precastrational levels. The intensity of copulatory acts showed exactly the same pattern (Beach and Jordan, 1956). These data strongly support the notion that the sexual acts actually are the reinforcing event.

An ingenious operant learning procedure was used in one old experiment (Denniston, 1954). Male rats were trained to turn a wheel one revolution in order to make a sexually receptive female appear. He was then allowed to perform five 'copulations' before the female was removed. Another revolution of the wheel made the female reappear. The latency to wheel turning was gradually reduced from almost 100 seconds on the first trial to about 44 seconds on the last trial. This reduction is a clear indication of learning and, as everyone knows, learning is not possible without reinforcement. Despite all these data, it may be so that some reader is still not convinced that the execution of copulatory acts can reinforce learning. This doubt will certainly disappear if I add some studies that employed procedures more well known to a contemporary reader than straight runways, T-mazes or turning wheels.

Responding in the most popular of all operant learning procedures, the operant chamber or Skinner box, can also be controlled by sexual reinforcement. If the availability of a receptive female is made contingent on any kind of response, a male subject will learn it. Several nice experiments have established that (Schwartz, 1956; Jowaisas *et al.*, 1971; Everitt *et al.*, 1987; Woodson and Balleine, 2002). There is no doubt, then, that the execution of copulatory acts can reinforce operant learning in exactly the same way as food and drink, the favorite reinforcements among learning researchers. Most unfortunately, the Skinner box studies have systematically employed copulation until one ejaculation as reinforcement in one way or another. Some experiments were so designed that the male gained access to a female and kept that access until ejaculation, as soon as a pre-established number of responses, or a pre-established time with responding, had occurred. Others allowed the rat to perform one intromission after a fixed number of responses. Immediately after intromission, the female was removed and the male could make more responses and gain access to the female for another intromission, and so on until ejaculation occurred. Then the session was ended. This means that at least one series of response per session became reinforced with ejaculation. Because of that, we do not know if the execution of one or several mounts without intromissions has any reinforcing effect. Neither do we know whether intromission in the absence of ejaculation is reinforcing or not. A result of this is that we cannot determine if there is a direct relationship between amount of copulatory acts performed and the amount of reinforcement obtained in the Skinner box. In the absence of evidence to the contrary, I suggest that the relationship established in the runway and maze studies mentioned above also holds for other operant procedures.

So far we have only discussed operant learning in males. There are few runway or maze studies employing females, at least so far as I know. A reason for the scarcity of female data is probably that copulatory behavior was thought to have no or even aversive consequences in female rats. We owe that notion to the old study of Gordon Bermant that I mentioned in the context of female sexual behavior. To the benefit of those with memory deficiencies, I just repeat that female rats escape from males after a sexual interaction if they have the opportunity to do so. Furthermore, the more intense the interaction the longer is the time the female stays away (Bermant, 1961; Peirce and Nuttall, 1961). The fact that the female withdraws from the male means that the male has been transformed into something aversive, a negative incentive according to incentive motivation theory. Now, if execution of copulatory acts has some aversive consequence transforming the male into a negative incentive, then there is absolutely no reason to imagine that such acts should be able to function as positive reinforcers.

Despite the notion that sexual acts are aversive to females, the possibility that sexual interaction with a male is reinforcing was evaluated in a quite elegant experiment. A straight runway was used. In the goal box, a sexually active or an inactive male appeared on the female's arrival. Some ovariectomized females were tested when sexually receptive because of hormone treatment and others were not given hormones and were consequently non-receptive. In the goal box, the females were removed either after the first sexual response on part of the male, be it mount, intromission or ejaculation, or after 10 seconds if no sexual interaction occurred. The elegance of this experiment resides in the fact that there was an intertrial interval of 10 minutes. As might be recalled, the aversive consequences of sexual interaction, even of receiving an ejaculation, dissipates spontaneously in a time far less than 10 minutes. Thus, in this experiment the females were never tested during the period in which the male was a negative incentive. The results showed that the running time was reduced as the number of trials increased. The reduction was largest in the sexually receptive females, but it was equally large independently of whether they met a sexually active or inactive male in the goal box (Bolles *et al.*, 1968). The fact that all groups ran faster suggests that they had learned something. The event that reinforced learning must be the social contact with the male, since a sexually active and a sexually inactive male were equally reinforcing. If copulatory acts had been reinforcing, then the females would have run faster to the sexually active male since he would have provided sexual in addition to social reinforcement. Not surprisingly, then, the authors concluded that sexual acts are not reinforcing for the female. This study has become a kind of classic and is frequently mentioned in textbooks of all kinds. Its popularity may be related to the fact that the notion that females (including women) should not like sex was dominant in traditionalist social groups at the time the study was published and it may still be surviving among some fundamentalist groups and at some places.

Another study (Drewett and Spiteri, 1979), again using a straight runway, reports results entirely different from the Bolles *et al.* (1968) experiment. Castrated male Wistar rats, either untreated or given dihydrotestosterone alone or dihydrotestosterone + estradiol provided different kinds of reinforcement. As we learned in Chapter 4, Wistar males do not copulate after treatment with dihydrotestosterone, but the hormone has most of the peripheral actions that testosterone has. Thus, these males most probably smelled and looked like intact rats. When estradiol is added to dihydrotestosterone, the males not only look and smell like intact rats, but they also copulate like intact rats. The females running to a castrated male ran faster after training. This replicates the finding that a sexually inactive male is a reinforcer. The females running to the dihydrotestosterone treated males ran faster than those running to the castrated male. This suggests that a male looking and smelling like an intact male is more reinforcing than a castrated male, although no sexual acts were performed. Finally, those finding a sexually active male in the goal box, one of those treated with dihydrotestosterone + estradiol, ran faster still. The conclusion must be that this male was the most reinforcing. Since he looked and smelled like the non-copulating, dihydrotestosterone treated male, his enhanced reinforcing capacity must be due to the fact that he performed sexual acts such as mounts and intromissions. Here we find ourselves in the most unpleasant situation where two well-designed experiments give contradictory results. I will not try to invent an explanation. It is beyond my intellectual capacity.

Sexual reinforcement also seems to work in other procedures. In a T-maze experiment where female rats found a sexually active male in one arm and a sexually inactive male in the other, they rapidly learned to run to the arm with an active male in the goal box. They were allowed to stay for 1 minute with the animal in the goal box, regardless of the amount of sexual interaction taking place. The active males employed in this experiment were quite proficient, performing a mean of 3.5 intromissions per minute. The inactive males never displayed any copulatory behavior at all. After the females had reached the criterion of learning, eight consecutive choices of the active male, the position of the males was changed. The reversal task was learned by all females (Drewett, 1973). This experiment clearly shows that copulatory acts also reinforce learning in a T-maze.

Further support for the notion that copulatory acts are reinforcing in female rats were obtained in a most elegant study (Matthews *et al.*, 1997). First, the consequence of introducing the nose in a hole was exposure to the odor of male bedding. This stimulus functions as a positive incentive, as we learned in the chapter on non-human, olfactory incentives. However, the odor did not enhance the frequency of nose-poking. Then the odor of an extract of the preputial gland from a male was tested. Those with good memory may remember that it has been proposed that the site of production of the odor with sexual incentive properties is the preputial gland. It failed to enhance responding. Then, nose-poking caused the appearance of a sexually active male, rapidly performing an intromission. The response requirements were then gradually increased until the females had to perform 25 nose-pokes before the male appeared. This worked excellently and the females continued to respond under these conditions. This beautiful experiment establishes, beyond doubt, that the receipt of one intromission has reinforcing consequences for the female. The possible effect of a mount without intromission was tested neither in this experiment nor in any other where copulatory acts were used as reinforcement. This is a pity, since the female's own copulatory behavior is extremely similar before and during a mount and an intromission. In both cases, she displays lordosis and the male's approach activates proceptive behaviors with the same frequency regardless of whether he eventually will perform a mount without intromission or a mount with intromission. The point here is that we cannot determine if it is the female's own execution of copulatory acts that is reinforcing, or of it is the vaginocervical stimulation received from the male at intromission that is reinforcing, or if a combination of both is necessary. It is only to be hoped that someone takes the trouble to perform the adequate experiment. To my judgment, the question of whether the female's own behavior is reinforcing or not is quite interesting.

While the experiments with female rats, with one exception, have shown that copulatory acts do reinforce learning while non-sexual social contact does not or does it far less than copulatory acts, data from female mice show that social contact is as powerful a reinforcer as copulatory acts (Matthews *et al.*, 2005). An operant procedure was used and the female subjects were either tested when fully receptive or when non-receptive. The reinforcing event was the appearance of either a sexually active male or a female contingent on a bar press. Initially, the response requirement was a fixed ratio of 1. When the response rate exceeded 10 per session, this requirement was changed to a gradually increasing variable interval. The final response requirement was VI 30. The rate of acquisition of the bar-pressing response and the final rate of responding did not differ between receptive and non-receptive females. Likewise, there was no difference in acquisition or final response rate between female subjects reinforced with access to a male or with access to a female. These data show that social contact + copulatory acts are no more reinforcing than social contact alone. Thus, copulatory acts have no reinforcing value by themselves. This was exactly the same conclusion arrived at in the old Bolles *et al.* (1968) experiment with female rats. In view of the many positive results obtained in later studies in female rats, Matthews *et al.* (2005) apparently felt obliged to offer some kind of explanation for their negative data. They consider several possibilities and reject them all. Two quite likely possibilities are not considered, though. One is that their procedure contained too many extraneous elements. For example, the event following the correct response was not only appearance of a male or a female, but also a tone. Moreover, release of the reinforcing animal consisted in removal of a floor-less cage inside the operant chamber. The end of access to the reinforcer animal was signaled by a buzzer and then the floor-less, reinforcer holding cage was introduced. So many things were going on that the formation of an association between lever press and copulatory acts became too difficult. Perhaps the tone combined with release of the reinforcer was the critical event, or perhaps the disturbance associated with the introduction of the holding cage in the midst of copulation had some aversive property neutralizing the reinforcing effects of copulatory acts.

The other possible explanation not considered is that mice are not small rats. There may indeed be a species difference with regard to reinforcing value of copulatory acts. Or there may be a species difference with regard to appropriate behavioral procedures. Personally, I do not believe in the former explanation. I favor the latter. While the complex procedure employed may have worked in rats, there is no *a priori* reason to believe that it should work in the same way in mice. Thus, I do not take the negative results of the Matthews *et al.* (2005) study as clear-cut evidence for a lack of reinforcing capacity of copulatory acts in mice. On the contrary, I believe that future studies will show that such acts can reinforce learning in female mice in the same way as it does in female rats.

To conclude this section on reinforcement of operant learning by the execution of copulatory acts, I state that sex is an efficient reinforcer of learning in males and females. Rat data are clear, while the very limited mouse data are less so. It must immediately be mentioned that there are a lot of studies pretending to have studied sexual reinforcement without actually doing so or that are cited in support for the notion that sexual activity is reinforcing. I have not written a word about these studies. The reason is that they do not report data concerning a change in behavior because of experience. Change in behavior is the basic element in all learning and the reinforcer is the stimulus/event causing this change. In other words, a behavior that initially is unlikely becomes likely when it is repeatedly followed by a reinforcer. The initially unlikely behavior can be running down a straight runway or pressing a lever in a Skinner box. If the behavior is performed more frequently or more rapidly in a subject when it has acquired experience (obtained the reinforcer contingent on the behavior performed), then we can reasonably say that the behavior has been reinforced. A study not reporting *change* in behavior caused by repeated exposure to a potential reinforcer cannot pretend to have studied reinforcement. Some examples of this misuse of the notion of reinforcement can be found in Bermant (1961), Peirce & Nuttall (1961), Law and Gerbrandt (1967a, 1967b), French *et al.* (1972), Meyerson and Lindström (1973). The fact that these papers do not provide data relevant to the issue of sexual reinforcement does not in any way reduce their scientific merit. Most of them report nicely done experiments and some of them are even excellent. Generally, they show that potential sexual partners are positive incentives and some also show that such partners are incentives only for animals in the appropriate hormonal state. This is important and useful information in itself. Many other interesting phenomena are also addressed in them, but they are of no concern to us at this moment.

5.2 Studies in humans

It is quite easy to note that I have mentioned no data from the human or any other primate. There are very few studies performed in non-human primates. Some data from the female rhesus monkey certainly suggest that operant learning with sexual reinforcement works in monkeys just as it does in other mammals, but the experiments in which the data were collected were not designed for analyzing learning (Bonsall *et al.*, 1978; Michael *et al.*, 1978). Therefore, I do not find it meaningful to discuss them. Turning to the human, we can proceed with a very rapid presentation of experimental data. The reason why it is rapid is that there are no data to mention, at least not as far as I know. For some reason, experiments on operant learning with sexual reward are not done in the human. This is certainly not because of a lack of volunteer subjects. I am convinced that many of our male students would participate with pleasure in experiments where the reward was having sex for free with a young and beautiful girl. I am equally sure that many of our female students would happily accept to be subjects in such studies. The lack of volunteers is certainly not the cause for the lack of empirical studies on the subject. There must be some other reason, which I cannot figure out.

6 Social learning of sexual responses

With regard to classical conditioning, operant conditioning, habituation and sensitization of sexual responses, we have a respectable amount of experimental data. Certainly more than enough for assuring that there is nothing special with sexual responses, at least not so far as their modifiability by learning. They behave exactly like other responses such as eating, drinking, exploration and many others. On the contrary, when we come to the last form of learning on our list, social or observational learning, we do not have any experimental data at all. Yet it is most likely that human sexuality is, essentially, a social construction achieved by a combination of social learning and formal instruction. More basic learning mechanisms can and do modulate human sexual behaviors, but their role is probably much less prominent. The situation is the contrary in non-human animals. There, modifications of sexual behaviors are brought about by basic learning mechanisms, in the extent that experience modifies these behaviors at all, and very little if anything can be attributed to social learning. This fact quite obviously impedes most basic and all comparative research on social learning of sexual behaviors. If the human is the only animal in which this kind of learning is of any real importance, then how could we study it in a comparative perspective? Furthermore, the immense power obtained through rigorous experimental control is lost when our subjects of study are limited to the human species. Additionally, most of the methods of modern neuroscience employed in non-human animal research are of little use in the human. Finally, we cannot forget that many societies have no interest in the scientific study of human sexuality and particularly not in studies that may challenge the notion of penile–vaginal intercourse as its basic and only permissible activity. The consequence of all this and of some other factors not mentioned is that our ignorance about social learning of human sexual behavior is almost total.

In case someone does not believe in the fundamental social nature of human sexual behaviors, I will briefly mention some examples of socially learned aspects or rules about sexuality acquired through formal instruction. First, with whom we can have sex. There is an age limit for the display of sexual behaviors in most societies, either imposed by law or by custom. In many human groups there are prohibitions against having sex with our parents and with our children, and frequently also with our siblings, and sometimes even more distant relatives are excluded from the potential choice of partner. Persons of the same sex have only recently been accepted as sexual partners and that only in the most advanced societies. All these arbitrary rules are imposed upon us through formal instruction. Second, where we can have sex. Some people would like to have sex under the table between the entrée and the dessert when dining at a fancy restaurant. If they materialized their wish, they would be thrown out and prosecuted by criminal justice. Some like sex outside and some like it in the dark, some in the light. Whatever our choice may be, it is a result of social learning, perhaps combined with operant learning. At a student party, some newly formed couple might discover an unused bedroom. The guy observing them when they sneak away will soon take his boyfriend to the same room with the same purpose. We all learn where to have sex, essentially through formal instruction or social learning. Third, the kind of motor patterns displayed when having sex are also determined by a combination of social and operant learning. Some men are extremely fond of fellatio, while some others prefer penile–anal intercourse. Some want to use an ultrathin strawberry-tasting green condom, while others prefer a black variant with rough surface and a taste of banana. These different copulatory behaviors have certainly been acquired by social learning. It is unlikely that the pertinent information was provided as part of formal instruction and it is equally unlikely that the individuals in the example spontaneously had discovered the pleasures of fellatio and anal intercourse. It is far more likely that they had seen these behaviors on a late TV-show or an early porn session with rented movies at a friend's house. A woman's taste for partners equipped with extravagant condoms was perhaps acquired at a girls-only party, where condom taste happened to be one of the subjects of conversation. There is no need to continue with this rather infamous list of examples. Anyone can make it grow with innumerable own examples of socially acquired elements of sexual behaviors.

It is also important to note that the sexual incentive properties of external stimuli are learned in the human. One of the very few scientists having dedicated some effort to this issue suggests that *all* sexual incentives are learned (Hardy, 1964). Initially, no external stimulus has any sexual meaning, according to his reasoning. However, mechanical stimulation of the genitalia is proposed to have reinforcing consequences. Thus, any act leading to such stimulation is reinforced and acquired through operant learning, and any external stimulus can become associated with genital stimulation and acquire sexual incentive properties through classical conditioning. However, albeit classical conditioning of incentive properties is possible, as we have seen earlier in this chapter, the extent to which it explains human sexual incentives is unclear. It is most likely that an important proportion of these incentives is learned through social learning. The cultural diversity as well as carefully documented changes over time in the stimuli with sexual incentive properties is suggestive of a crucial role for social learning.

The lack of experimental data on social learning of sexual responses obliges me to end this section without any strong conclusion. I will, however, make some publicity for a brilliant, old book, originally published in 1973 and out of print for many years. A few years ago, it was reprinted and some new material was added in a foreword and some more in an afterword (Gagnon and Simon, 2002). It is the only insightful account of the role of social learning in human sexuality that I know of. I find it extremely interesting and thought provoking, although it is basically speculative. But what else could it be in the absence of data?

7 Conclusions regarding sexual responses and learning

I have presented data showing that neutral stimuli may acquire sexual incentive properties through classical conditioning. We have also seen that the execution of copulatory acts can reinforce operant learning. Furthermore, mammals, including the human male and probably also the female, habituate to sexual stimuli. Such stimuli can also be sensitized. Although all these kinds of learning unquestionably assure important modifications of sexual behavior through experience, and although they are the only learning mechanisms of any real importance in non-human animals, their role in human sexual behavior is less certain. Many aspects of this behavior are probably also determined by these learning mechanisms in the human, but their importance compared to that of social learning is difficult to determine. I will make no guess.

What is quite evident to anyone with some elementary knowledge of human sexual behavior and human sexual incentives is that the importance of inborn determinants is negligible. The secrets of human sexual functions can never be unraveled through a search for purely biological determinants. An understanding of the behavioral mechanisms, including learning, by which external stimuli exert their control of the activation of sexual motivation and the display of sexual behaviors, is absolutely essential.

As a final comment on learning of sexual responses I will cite a few lines from a classic 1950 book. The purpose with that is to show that realization of the supreme importance of learning is not new.

> General observations on man and also on animals show that sexual excitement can readily be learned as a conditioned response to previously neutral cues and result in a considerable increase in drive. Similarly, such observations indicate that previously neutral cues can acquire learned-reinforcement value by association with sexual reinforcements. . . . That the components can be learned is also illustrated by the extreme variability of cultural standards of sexual desirableness from society to society and even during different historical periods in our own society (Dollard and Miller, 1950, p. 85).

8 The sexual reward

8.1 The positive affect: behavioral data

Sex causes positive affect. Some people would even say that few things cause a positive affect as intense as that experienced at orgasm. This positive affect produced by the execution of copulatory behaviors is what we understand with sexual reward. The role of affect shall not be underestimated. It is generally believed that positive incentive stimuli activate approach behavior because they induce some fraction of the affect produced by the event they predict. The odor of a receptive female rat, for example, activates approach behaviors in most male rats. Approach to the incentive is normally followed by copulatory acts culminating in ejaculation. Ejaculation is associated with a rather long-lasting positive affect in the male and in the female receiving it, as we soon will see. The female's odor predicts a proximate positive affect. The quality of predictor of positive affect is fundamental for all positive incentives.

It seems that the simple exposure to an incentive, without any further interaction with it, can produce positive affect. A male or a female rat will approach an inaccessible conspecific of the opposite sex day after day and week after week, even when sexual interaction never occurs. It seems as if the stimuli emitted by the potential sexual partner are intrinsically rewarding. The same can happen with food. A hungry rat may approach an inaccessible food pellet, day after day and week after week, even if she is never allowed to taste it. These approach behaviors would be completely unexplainable if we did not assume that the simple exposure to the incentives have some rewarding consequence. What is lacking at this moment is experimental support for the notion that the simple exposure to an incentive is rewarding. Please observe that there is no reason to believe that such an exposure could be used for reinforcing learning. Here we have a good example of the importance of distinguishing reward from reinforcement.

Even if the rewarding properties of the exposure to a sexual incentive have not been experimentally established, we know that the execution of copulatory acts until ejaculation is rewarding, meaning that it induces positive affect. We do not, regrettably, have any objective data from the human showing that the execution of sexual acts until orgasm produces positive affect. Perhaps we can trust introspective data on this matter. In other mammalian species, however, we have experimental data confirming the belief that sex is rewarding. I will first explain how we can look into the mind of a rat and figure out if he or she feels good or not, and after that I will turn to the data.

Many years ago scientists became interested in searching for the causes of drug addiction. I ignore why such a search seemed interesting, but one of the hypotheses that appeared stated that some people get addicted to drugs because they experience positive affect, or pleasure to use more vulgar language, after taking the drugs. In order to test the hypothesis in animal models, a procedure able to detect affective consequences of drugs was needed. A procedure able to do that was soon described (Beach, 1957a, 1957b). However, its routine use turned out to be rather impractical and it did not have much success. A more user friendly variant of the original procedure was described some years later (Rossi and Reid, 1976) and now it became a success. It has been used in hundreds of studies since then and it continues to be popular among those interested in positive affect and drug addiction. It has also been adapted for use with natural stimuli/events supposed to induce positive affect. Several reviews listing all the drugs and events employed as source of positive affect are available (Schechter and Calcagnetti, 1993, 1998; Bardo *et al.*, 1995; Tzschentke, 1998).

The basic setup consists of two compartments differing in several stimulus modalities, usually odor, color and texture. For example, one compartment may have a floor covered with sawdust smelling of pine and black walls, while the other may have a steel grid floor, white walls and an odor of diluted vinegar. When affective consequences of a drug are the matter of interest, the drug is usually administered just before the subject is confined to one of these compartments. A following day, the drug vehicle is administered and the subject is then confined to the other compartment. The alternating drug and vehicle administrations are usually repeated a couple of times. At the test, the subjects are allowed to move freely between the compartments and the time spent in each of them is recorded. No drug is given before the test session.

The idea behind the procedure is that potential affective consequences of the drug become associated with the environmental cues present in the compartment through classical conditioning. Consequences of the vehicle injection should be associated with the corresponding compartment. If successful, the conditioning procedure should have transformed the initially neutral environmental cues in the compartment associated with drug administration into conditioned positive incentives. At test, these cues should activate approach behavior like any incentive. The cues in the compartment associated with vehicle administration should, at least in the ideal situation, remain neutral even after conditioning. Thus, at test, the experimental subjects should spend more time in the drug-associated compartment than in the vehicle-associated compartment. If that should happen to occur, then we would say that they show a conditioned place preference. An identical procedure may be employed for studying drugs or stimuli/events having aversive effects. In that case, a conditioned place aversion is produced.

It could be argued that we do not know if conditioned place preference really is a behavioral expression of drug-induced positive affect. Initially, the arguments in favor of an explanation in terms of affect were based on anthropomorphism. All drugs known to produce positive affect in the human can be used for conditioning place preference, while drugs perceived as neutral in humans are ineffective. Likewise, all compounds having aversive effects in the human produce a conditioned place aversion. I have criticized the use of anthropomorphisms, and I must accordingly consider the preceding reasoning as faulty. Rather I maintain that learning theory excludes any explanation other than conditioning of positive affect. What else producing approach behavior could possibly be conditioned in this particular situation? Many experimental controls have been employed over the years and all alternative explanations have been ruled out.

A natural stimulus/event may replace a drug for producing positive affect. We tested the affective consequences of the execution of copulatory behavior in male rats many years ago. The basic idea was that copulation until ejaculation should induce a positive affect outlasting the sexual interaction itself. Conversations with colleagues and friends had convinced me that the pleasure of having an orgasm did not dissipate instantaneously. Supposing this to be true, we could replace the injection of a drug causing positive affect with an opportunity to copulate until ejaculation. This we did. Male rats were allowed to ejaculate once with a sexually receptive female. Immediately after ejaculation, we introduced the male into one of the compartments of a place preference box. On alternate days, the male was taken directly from its home cage and placed in another compartment. We repeated this three times and then we performed the test. It turned out that the males preferred the compartment associated with the after-effects of ejaculation. Another group of males were exposed to a non-receptive female for a time equal to that needed for achieving one ejaculation in the first group and the animals were then placed in the place preference compartment. At test, this group displayed no preference. This simple experiment suggests that the execution of copulatory behavior indeed induces a state of positive affect in male rats (Ågmo and Berenfeld, 1990). Similar results have later been reported by others (Martínez and Paredes, 2001; Camacho *et al.*, 2004).

While copulation has been believed to produce positive affect in males long before any experimental demonstration existed, it was for a long time discussed whether this was the case also in females. In my discussion of paced mating in the female rat, I mentioned that scientists have been fascinated by possible aversive consequences of sexual interaction. The proposal that sex was aversive to the female rat was made seriously and experimental data were interpreted as supporting it. I already insinuated that this idea coincided nicely with typical social prejudice dominant at the time. Many groups were still maintaining that sex was a conjugal obligation, which women had to endure for the sake of preservation of the species. The idea that women could find any pleasure in sexual acts was reserved for prostitutes, nymphomaniacs and other perverts. In that context, the rat data came in very handy. However, we now know that copulatory acts produce as much positive affect in females as they do in males. This has convincingly been shown in the conditioned place preference procedure. Female rats were allowed to copulate until the male ejaculated, some in a procedure where the male controlled sexual interactions, some in the pacing procedure. It turned out that only paced mating produced positive affect (Paredes and Alonso, 1997; Martínez and Paredes, 2001). Male-controlled mating was affectively neutral, producing neither preference nor aversion in the procedure. These data could easily be interpreted in a feminist way, saying that they demonstrate that females enjoy sex only when they, and not the male, are in control. Some more recent data do not coincide with the feminist view. In two nicely controlled studies, it was reported that the execution of copulatory acts was rewarding for female rats regardless of the mating procedure employed (Meerts *et al.*, 2005; Meerts and Clark, 2006; Yamada *et al.*, 2006). Thus, with or without control, females show positive affect after copulating.

We are now in a position to conclude that experimental data show that sexual activity can induce a state of positive affect in male and female rats. It can be assumed that this is also the case in other mammals, including the human.

8.2 The positive affect: neurotransmitters

Neuroscientists and many others are convinced that psychic events, like mood states, are nothing more than processes within the brain. Such a conviction makes it legitimate to ask for the neurochemical substrates of, for example, mood states. Positive affect, or pleasure or happiness or whichever vulgar expression we prefer to use, is a theme attracting attention from people even outside the circle of neuroscientists. Because of the double audience, it might be worth the trouble to make some comments on this topic here and now.

The only central nervous mechanisms we know something about in relation to sex-induced positive affect is transmitter release. Assuming that the execution of copulatory behaviors culminating in orgasm is an efficient affect-producing procedure, we can start by looking at the neurotransmitters whose release is enhanced around orgasm. This has been done in a couple of studies, but most of them have looked at one single transmitter. Knowing that we already have dozens of transmitters in the brain, and that new ones appear every year, it is unfortunate that attention has been so extremely focused on only one. Still worse, many studies have reported exactly the same data, which seems like a kind of waste. The reason for this uniformity in thought and work is that dominant prejudice at the time these studies were popular had it that the neurotransmitter dopamine was crucial for the experience of positive affect. It was even believed that the positive affect was localized to the nucleus accumbens, one of the projection areas of the mesolimbic dopamine system (Wise, 1982; Wise and Bozarth, 1987).

Dopamine is released from the moment a male or a female is exposed to an individual of the opposite sex until ejaculation. Thereafter release declines rapidly (Pfaus *et al.*, 1990, 1995; Pleim *et al.*, 1990; Phillips *et al.*, 1991; Mermelstein and Becker, 1995). Dopamine is also released in rats subjected to tail pinch, foot-shock, aggressive encounters and similar unpleasant events (Doherty and Grattton, 1992; Tidey and Miczek, 1996; Takahashi *et al.*, 1998; Ferrari *et al.*, 2003). The variety of events able to enhance dopamine release in the accumbens was shown in an elegant experiment in 1986. Aggressive encounters, tail pinch and encounters with a female shared the capacity to release accumbens dopamine (Louilot *et al.*, 1986). The fact that dopamine is released in all situations enhancing arousal and general activity, regardless of the kind of affective reaction associated with the enhanced arousal, suggests that dopamine in the nucleus accumbens subserves motor mechanisms. This notion stems from a brilliant paper concerning the function of the nucleus accumbens published in 1980 (Mogenson *et al.*, 1980). All observations pointed in the direction of a function related to motricity. Thus, already in the mid-1980s, it was evident that dopamine release in the accumbens could not have much to do with positive affect. This has been pointed out many times since then (Robinson and Berridge, 1993; Salamone *et al.*, 1997; Berridge and Robinson, 1998; Ikemoto and Panksepp, 1999). Despite the clear evidence to the contrary, the hypothesis of accumbens dopamine release as an event underlying positive affect continues to circulate in some badly informed textbooks and among many amateur scientists.

After excluding dopamine as a transmitter associated with positive affect, we need to find an alternative. This is easily done, since the endorphins had been implicated in reward states long before the contradictory dopamine data had started to accumulate (Belluzzi and Stein, 1977). In view of this, we evaluated the capacity of an opiate receptor antagonist to block the positive affect produced by ejaculation in male rats. To that end, the place preference procedure was used. Before each session, some rats were injected with naloxone, then allowed to ejaculate. Immediately thereafter they were placed in the place preference chamber. After a few sessions, the test was made. Control rats showed a robust place preference, but there was no trace of a preference in rats treated with naloxone. On the contrary, a group of rats injected with a dopamine antagonist displayed a clear place preference after conditioning (Ågmo and Berenfeld, 1990). The conclusion was quite clear: the positive affect caused by sexual interaction is an opioid-dependent event. Similar data have been obtained in female rats (Paredes and Martínez, 2001). As was the case with the male, a dopamine antagonist does not block the positive affect induced by sexual interaction in female rats (Garcia-Horsman and Paredes, 2004). Exactly as in males, positive affect caused by sexual interaction is an opioid-dependent event.

After having shown that sexually induced positive affect is opioid dependent, it was quite natural to ask where in the brain the sex-induced opioid release could be localized. At the time, there was no procedure available to measure rapid changes in opioid release, so we had to employ indirect approaches. Some data already existing suggested that the medial preoptic area could be an important site. We proceeded to analyze a possible copulation-induced release in the preoptic area with help of the kindling procedure. Electrical kindling consists of the repeated application of short pulses of a current of low intensity to a brain site. The first couple of stimulations have no visible behavioral effect, but they excite neurons in the vicinity of the stimulation electrode. After several stimulations, the neuronal excitation is increased both in amplitude and duration and, eventually, behavioral manifestations are seen. When kindling is complete, the previously ineffective electrical stimulation produces full-blown convulsions. After a kindling-induced convulsive crisis, the rat remains immobile for a while. This state is called post-ictal behavioral depression. Administration of an opiate antagonist strongly reduces the duration of the post-ictal behavioral depression and it has become habitual to consider that state as opioid-dependent. We reasoned that kindling with the electrode placed in the preoptic area and determinations of the duration of the opioid-dependent post-ictal behavioral depression observed in completely kindled animals could be used as indirect evidence for opioid release during copulation. If opioids indeed were released during the execution of copulatory reflexes and especially at ejaculation, this release should enhance the duration of the post-ictal behavioral depression, and that enhancement should be blocked by an opiate antagonist. In line with this reasoning, we applied electrical stimulation to some rats 1 minute after ejaculation. Another day we applied an identical electrical stimulus in the absence of sexual activity. The post-ictal behavioral depression was dramatically prolonged when the stimulus was applied 1 min after ejaculation in rats with the electrode within the preoptic area. This increase was blocked by naloxone. In rats with the stimulating electrode in the basolateral amygdala, sexual activity did not modify the duration of the post-ictal behavioral depression, suggesting that the opioid release was specific to the preoptic area (Paredes *et al.*, 1992). Recent studies employing far more sophisticated techniques have confirmed that sexual activity indeed releases opioids within the preoptic area (Coolen *et al.*, 2004).

There is still one element missing from a complete picture of sexually induced positive affect. We know that it is opioid-dependent and we know that opioids are released in the preoptic area, but we do not know if opioid receptors within that area are involved in the generation of positive affect. This question was answered in a simple experiment many years ago. We administered an enkephalin to the preoptic area before place preference conditioning. The results were quite clear. Even a very small dose of enkephalin produced place preference when infused into that area (Ågmo and Gómez, 1991). Finally, a group of rats were allowed to copulate until they had achieved one ejaculation after infusion of a minute amount of the opiate antagonist naloxone into the preoptic area. No place preference could be established, while controls as always showed a strong preference. Naloxone infused into the nucleus accumbens was, obviously, ineffective (Ågmo and Gómez, 1993). Taken together, all these data make it likely that the positive affect caused by sexual activities is dependent on opioid release in the preoptic area.

9 General conclusion

There is no doubt that the execution of sexual acts can be reinforcing and rewarding. Likewise, neutral stimuli may become sexual incentives through classical conditioning. We have seen considerable evidence in favor of this notion. Habituation to sexual incentives may be important in non-human mammals, but in humans it seems more reasonable to talk about extinction or negative alliesthesia when response intensity is reduced because of repeated exposure to a constant stimulus. Responses to sexual incentives may also be enhanced by sensitization. Most of human, in contrast to non-human, sexual behaviors are acquired by social learning or formal instruction. The main consequence of all this is that human sexual behaviors can be molded by society and individual experience to an extent far superior to that seen in any other mammal. This fact has paramount importance for any analysis of dysfunctional sexual behaviors. I am completely convinced that the enormous moldability of human sexual behaviors by learning is the reason why sexual dysfunctions are surprisingly common in the human and utterly rare among non-human animals.

We have also seen that sexual activity can produce positive affect. Experimental data on this issue stem from non-human studies, but there is no reason to believe that humans do not experience positive affect when engaging in sexual behaviors. It is not extravagant to suggest that sexual incentives are stimuli predicting the potential obtention of sexual reward. Accordingly, sexual reward is the fundamental process directly or indirectly determining all sexual behaviors.

Summary

1 There are several kinds of learning. Habituation and sensitization are forms of non-associative learning. Classical and operant conditioning as well as social learning are forms of associative learning. Learning theory can teach us much about these different forms of learning.
2 Sexual responses can habituate with repeated presentations of a constant stimulus and they can be sensitized by irrelevant stimuli. Negative alliesthesia is another process that may temporarily affect a stimulus' capacity to activate sexual approach and copulatory behaviors.
3 There is much evidence showing that a neutral stimulus can become a sexual incentive through classical conditioning. Data from humans and non-human mammals coincide on this.
4 Operant conditioning can be established with sexual reinforcement. All kinds of responses can be acquired. There are few data on operant learning with sexual reinforcement in primates, including the human. However, there is no reason to believe that primates are different from other mammals in this respect.

5 Social learning is the main determinant of human sexual behaviors, both with regard to the incentives controlling sexual motivation and with regard to the motor patterns executed during copulation. Unconditioned (unlearned) sexual responses are rare in the human.
6 In non-human animals, social learning is of slight importance for sexual behaviors. This difference has many important implications, one of which is that we should not consider the human as just another mammal.
7 The positive affect caused by sexual behaviors, notably orgasm, is in all likelihood a phenomenon dependent on opioid release in the central nervous system. Some data suggest that this release may be located to the preoptic area.

References

Ågmo, A. (2002). Copulation-contingent aversive conditioning and sexual incentive motivation in male rats: evidence for a two-stage process of sexual behavior. *Physiol Behav* **77**, 425–435

Ågmo, A. and Berenfeld, R. (1990). Reinforcing properties of ejaculation in the male rat: role of opioids and dopamine. *Behav Neurosci* **104**, 177–182

Ågmo, A. and Gómez, M. (1991). Conditioned place preference produced by infusion of met-enkephalin into the medial preoptic area. *Brain Res* **550**, 343–346

Ågmo, A. and Gómez, M. (1993). Sexual reinforcement is blocked by infusion of naloxone into the medial preoptic area. *Behav Neurosci* **107**, 812–818

Ågmo, A., Ellingsen, E., Turi, A.L. and Larsson, K. (2002). A female Coolidge effect. Annual Meeting of the International Behavioral Neuroscience Society, Capri, Italy

Bailey, J.D., Anderson, L.H. and Schillo, K.K. (2005). Effects of novel females and stage of estrous cycle on sexual behavior in mature beef bulls. *J Anim Sci* **83**, 613–624

Bancroft, J.A. (1969). Aversion therapy of homosexuality. *Br J Psychiat* **115**, 1417–1431

Bardo, M.T., Rowlett, J.K. and Harris, M.J. (1995). Conditioned place preference using opiate and stimulant drugs: a meta-analysis. *Neurosci Biobehav Rev* **19**, 39–51

Barfield, R.J. and Sachs, B.D. (1968). Sexual behavior: stimulation by painful electric shock to skin in male rats. *Science* **161**, 392–395

Beach, F.A. and Jordan, L. (1956). Effects of sexual reinforcement upon the performance of male rats in a straight runway. *J Comp Physiol Psychol* **49**, 105–110

Beach, H.D. (1957a). Morphine addiction in rats. *Can J Psychol* **11**, 104–112

Beach, H.D. (1957b). Some effects of morphine on habit function. *Can J Psychol* **11**, 193–198

Beamer, W., Bermant, G. and Clegg, M.T. (1969). Copulatory behaviour of the ram, *Ovis aries*. II: Factors affecting copulatory satiation. *Anim Behav* **17**, 706–711

Belluzzi, J.D. and Stein, L. (1977). Enkephalin may mediate euphoria and drive-reduction reward. *Nature* **266**, 556–558

Bermant, G. (1961). Response latencies of female rats during sexual intercourse. *Science* **133**, 1771–1773

Berridge, K.C. and Robinson, T.E. (1998). What is the role of dopamine in reward: hedonic impact, reward learning, or incentive salience? *Brain Res Rev* **28**, 309–369

Bolles, R.C., Rapp, H.M. and White, G.C. (1968). Failure of sexual activity to reinforce female rats. *J Comp Physiol Psychol* **65**, 311–313

Bonsall, R.W., Zumpe, D. and Michael, R.P. (1978). Menstrual cycle influences on operant behavior of female rhesus-monkeys. *J Comp Physiol Psychol* **92**, 846–855

Bradford, A. and Meston, C.M. (2006). The impact of anxiety on sexual arousal in women. *Behav Res Ther* **44**, 1067–1077

Cabanac, M. (1992). Pleasure: the common currency. *J Theor Biol* **155**, 173–200

Cabanac, M. and Lafrance, L. (1990). Postingestive alliesthesia – the rat tells the same story. *Physiol Behav* **47**, 539–543

Camacho, F., Sandoval, C. and Paredes, R.G. (2004). Sexual experience and conditioned place preference in male rats. *Pharmacol Biochem Behav* **78**, 419–425

Coolen, L M., Fitzgerald, M.E., Yu, L. and Lehman, M.N. (2004). Activation of μ opioid receptors in the medial preoptic area following copulation in male rats. *Neuroscience* **124**, 11–21

Cutmore, T.R.H. and Zamble, E. (1988). A pavlovian procedure for improving sexual performance of noncopulating male rats. *Arch Sex Behav* **17**, 371–380

Denniston, R.H. (1954). Quantification and comparison of sex drives under various conditions in terms of learned responses. *J Comp Physiol Psychol* **47**, 437–440

Dewsbury, D.A. (1981). Effects of novelty on copulatory behavior: The Coolidge effect and related phenomena. *Psychol Bull* **89**, 464–482

Doherty, M.D. and Grattton, A. (1992). High-speed chronoamperometric measurements of mesolimbic and nigrostriatal dopamine release associated with repeated daily stress. *Brain Res* **586**, 295–302

Dollard, J. and Miller, N.E. (1950). *Personality and psychotherapy. An analysis in terms of learning, thinking, and culture*. McGraw-Hill: New York

Domjan, M. (1994). Formulation of a behavior system for sexual conditioning. *Psychonom Bull Rev* **1**, 421–428

Domjan, M., Huber-McDonald, M. and Holloway, K.S. (1992). Conditioned copulatory behavior to an artificial object: efficacy of stimulus fading. *Anim Learn Behav* **20**, 350–362

Domjan, M., Lyons, R., North, N.C. and Bruell, J. (1986). Sexual Pavlovian conditioned approach behavior in male japanese quail (*Coturnix coturnix japonica*). *J Comp Psychol* **100**, 413–421

Drewett, R.F. (1973). Sexual behaviour and sexual motivation in the female rat. *Nature* **242**, 476–477

Drewett, R.F. and Spiteri, N.J. (1979). The sexual attractiveness of male rats: olfactory and behavioral components. *Physiol Behav* **23**, 207–209

Everitt, B.J., Fray, P., Kostarczyk, E., Taylor, S. and Stacey, P. (1987). Studies of instrumental behavior with sexual reinforcement in male rats (*Rattus norvegicus*): I. Control by brief visual stimuli paired with a receptive female. *J Comp Psychol* **101**, 395–406

Ferrari, P.F., van Erp, A.M.M., Tornatzky, W. and Miczek, K.A. (2003). Accumbal dopamine and serotonin in anticipation of the next aggressive episode in rats. *Eur J Neurosci* **17**, 371–378

French, D., Fitzpatrick, D. and Law, O.T. (1972). Operant investigation of mating preferences in female rats. *J Comp Physiol Psychol* **81**, 226–232

Gagnon, J.H. and Simon, W. (2002). *Sexual conduct: the social sources of human sexuality*, 2nd edn. AldineTransaction: New Brunswick

Garcia-Horsman, P. and Paredes, R.G. (2004). Dopamine antagonists do not block conditioned place preference induced by paced mating behavior in female rats. *Behav Neurosci* **118**, 356–364

Gelez, H., Archer, E., Chesneau, D., Campan, R. and Fabre-Nys, C. (2004). Importance of learning in the response of ewes to male odor. *Chem Senses* **29**, 555–563

Graham, J.M. and Desjardins, C. (1980). Classical conditioning: Induction of luteinizing hormone and testosterone secretion in anticipation of sexual activity. *Science* **210**, 1039–1041
Hardy, K.R. (1964). An appetitional theory of sexual motivation. *Psychol Rev* **71**, 1–18
Hoffman, H., Janssen, E. and Turner, S.L. (2004). Classical conditioning of sexual arousal in women and men: effects of varying awareness and biological relevance of the conditioned stimulus. *Arch Sex Behav* **33**, 43–53
Hoon, P.W., Wincze, J.P. and Hoon, E.F. (1977). A test of reciprocal inhibition: are anxiety and sexual arousal in women mutually inhibitory? *J Abnormal Psychol* **86**, 65–74
Ikemoto, S. and Panksepp, J. (1999). The role of nucleus accumbens dopamine in motivated behavior: a unifying interpretation with special reference to reward-seeking. *Brain Res Rev* **31**, 6–41
Jowaisas, D., Taylor, J., Dewsbury, D.A. and Malagodi, E.F. (1971). Copulatory behavior of male rats under an imposed operant requirement. *Psychonom Sci* **25**, 287–290
Kagan, J. (1955). Differential reward value of incomplete and complete sexual behavior. *J Comp Physiol Psychol* **48**, 59–64
Kippin, T.E. and Pfaus, J.G. (2001a). The development of olfactory conditioned ejaculatory preferences in the male rat. I. Nature of the unconditioned stimulus. *Physiol Behav* **73**, 457–469
Kippin, T.E. and Pfaus, J.G. (2001b). The nature of the conditioned response mediating olfactory conditioned ejaculatory preference in the male rat. *Behav Brain Res* **122**, 11–24
Kippin, T.E., Samaha, A.N., Sotiropoulos, V. and Pfaus, J.G. (2001). The development of olfactory conditioned ejaculatory preferences in the male rat. II. Parametric manipulation of conditioning session number and duration. *Physiol Behav* **73**, 471–485
Kippin, T.E., Talianakis, S., Schattman, L., Bartholomew, S. and Pfaus, J.G. (1998). Olfactory conditioning of sexual behavior in the male rat (*Rattus norvegicus*). *J Comp Psychol* **112**, 489–499
Klusmann, D. (2002). Sexual motivation and the duration of partnership. *Arch Sex Behav* **31**, 275–287
Koch, P.C. and Peters, R.H. (1987). Suppression of adult copulatory behaviors following LiCl-induced aversive contingencies in juvenile male rats. *Dev Psychobiol* **20**, 603–611
Köksal, F., Domjan, M., Kurt, A. *et al.* (2004). An animal model of fetishism. *Behav Res Ther* **42**, 1421–1434
Koukounas, E. and Over, R. (1993). Habituation and dishabituation of male sexual arousal. *Behav Res Ther* **31**, 575–585
Koukounas, E. and Over, R. (1999). Allocation of attentional resources during habituation and dishabituation of male sexual arousal. *Arch Sex Behav* **28**, 539–552
Koukounas, E. and Over, R. (2001). Habituation of male sexual arousal: effects of attentional focus. *Biol Psychol* **58**, 49–64
Krames, L. (1971). Sexual responses of polygamous female and monogamous male rats to novel partners after sexual cessation. *J Comp Physiol Psychol* **77**, 294–301
Laan, E. and Everaerd, W. (1995). Habituation of female sexual arousal to slides and film. *Arch Sex Behav* **24**, 517–541
Lalumière, M.L. and Quinsey, V.L. (1998). Pavlovian conditioning of sexual interests in human males. *Arch Sex Behav* **27**, 241–252
Law, O.T. and Gerbrandt, L.K. (1967a). Sexual preference in female rats: I. Choices in tests with copulation. *Psychonom Sci* **8**, 493–494
Law, O.T. and Gerbrandt, L.K. (1967b). Sexual preferences in female rats: II. Choices in tests without copulation. *Psychonom Sci* **8**, 495–496

Letorneau, E.J. and O'Donohue, W.T. (1997). Classical conditioning of female sexual arousal. *Arch Sex Behav* **26**, 63–78

Leyton, M. and Stewart, J. (1996). Acute and repeated activation of male sexual behavior by tail-pinch: opioid and dopaminergic mechanisms. *Physiol Behav* **60**, 77–85

Lisk, R.D. and Baron, G. (1982). Female regulation of mating location and acceptance of new mating partners following mating to sexual satiety: The Coolidge effect demonstrated in the female golden hamster. *Behav Neural Biol* **36**, 416–421

Louilot, A., LeMoal, M. and Simon, H. (1986). Differential reactivity of dopaminergic-neurons in the nucleus-accumbens in response to different behavioral situations – an in vivo voltammetric study in free moving rats. *Brain Res* **397**, 395–400

Martínez, I. and Paredes, R. G. (2001). Only self-paced mating is rewarding in rats of both sexes. *Horm Behav* **40**, 510–517

Matthews, T.J., Abdelbaky, P. and Pfaff, D.W. (2005). Social and sexual motivation in the mouse. *Behav Neurosci* **119**, 1628–1639

Matthews, T.J., Grigore, M., Tang, L., Doat, M., Kow, L.M. and Pfaff, D.W. (1997). Sexual reinforcement in the female rat. *J Exp Anal Behav* **68**, 399–410

McConaghy, N. (1969). Subjective and penile plethysmograph responses following aversion-relief and apomorphine aversion therapy for homosexual impulses. *Br J Psychiat* **115**, 723–730

McConaghy, N. (1975). Aversive and positive conditioning treatments of homosexuality. *Behav Res Ther* **13**, 309–319

Meerts, S.H., Christensen, A.E. and Clark, A.S. (2005). Parameters of mating-induced conditioned place preference in female rats. *Soc Neurosci Abstr* 418.7

Meerts, S.H. and Clark, A.S. (2006). Induction of a conditioned place preference by non-paced mating behavior in female rats. Society for Behavioral Neuroendocrinology 10th Annual Meeting, Pittsburgh, P4

Mermelstein, P.G. and Becker, J.B. (1995). Increased extracellular dopamine in the nucleus accumbens and striatum of the female rat during paced copulatory behavior. *Behav Neurosci* **109**, 354–365

Meston, C.M. (2000). Sympathetic nervous system activity and female sexual arousal. *Am J Cardiol* **86**, 30F–34F

Meston, C.M. and Gorzalka, B.B. (1995). The effects of sympathetic activation on physiological and subjective sexual arousal in women. *Behav Res Ther* **33**, 651–664

Meston, C.M. and Gorzalka, B.B. (1996). The effects of immediate, delayed, and residual sympathetic activation on sexual arousal in women. *Behav Res Ther* **34**, 143–148

Meston, C.M. and Heiman, J.R. (1998). Ephedrine-activated physiological sexual arousal in women. *Arch Gen Psychiat* **55**, 652–656

Meuwissen, I. and Over, R. (1990). Habituation and dishabituation of female sexual arousal. *Behav Res Ther* **28**, 217–226

Meyerson, B.J. and Lindström, L.H. (1973). Sexual motivation in the female rat. A methodological study applied to the investigation of the effect of estradiol benzoate. *Acta Physiol Scand* **Suppl 389**, 1–80

Michael, R.P. and Zumpe, D. (1978). Potency in male rhesus monkeys: effects of continuously receptive females. *Science* **200**, 451–453

Michael, R.P., Bonsall, R.W. and Zumpe, D. (1978). Consort bonding and operant-behavior by female rhesus-monkeys. *J Comp Physiol Psychol* **92**, 837–845

Mogenson, G.J., Jones, D.L. and Yim, C.Y. (1980). From motivation to action – functional interface between the limbic system and the motor system. *Progr Neurobiol* **14**, 69–97

O'Donohue, W.T. and Geer, J.H. (1985). The habituation of sexual arousal. *Arch Sex Behav* **14**, 233–246

O'Donohue, W. and Plaud, J.J. (1991). The long-term habituation of sexual arousal in the human male. *J Behav Ther Exp Psychiat* **22**, 87–96

Palace, E.M. and Gorzalka, B.B. (1990). The enhancing effects of anxiety on arousal in sexually dysfunctional and functional women. *J Abnormal Psychol* **99**, 403–411

Paredes, R.G. and Alonso, A. (1997). Sexual behavior regulated (paced) by the female induces conditioned place preference. *Behav Neurosci* **111**, 123–128

Paredes, R.G., Manero, M.C., Haller, A.E., Alvarado, R. and Ågmo, A. (1992). Sexual behavior enhances postictal behavioral depression in kindled rats – opioid involvement. *Behav Brain Res* **52**, 175–182

Paredes, R.G. and Martínez, I. (2001). Naloxone blocks place preference conditioning after paced mating in female rats. *Behav Neurosci* **115**, 1363–1367

Peirce, J.T. and Nuttall, R.L. (1961). Self-paced sexual behavior in the female rat. *J Comp Physiol Psychol* **54**, 310–313

Peters, R.H., Blythe, B.L., Koch, P.C. and Kueker, C.A. (1989). Copulation-illness association in male rats: Lithium chloride dose and delay manipulations. *Behav Neurosci* **103**, 117–123

Pfaus, J.G., Damsma, G., Nomikos, G.G. *et al.* (1990). Sexual behavior enhances central dopamine transmission in the male rat. *Brain Res* **530**, 345–348

Pfaus, J.G., Damsma, G., Wenkstern, D. and Fibiger, H. C. (1995). Sexual activity increases dopamine transmission in the nucleus accumbens and striatum of female rats. *Brain Res* **693**, 21–30

Phillips, A.G., Pfaus, J.G. and Blaha, C.D. (1991). Dopamine and motivated behavior: Insights provided by *in vivo* analysis. In *The mesocorticolimbic dopamine system: from motivation to action* (P. Willner and J. Scheel-Krüger, eds), pp. 199–224. Wiley: Chichester

Plaud, J.J., Gaither, G.A., Henderson, S.A. and Devitt, M.K. (1997). The long-term habituation of sexual arousal in human males: a crossover design. *Psychol Rec* **47**, 385–398

Plaud, J.J. and Martini, J.R. (1999). The respondent conditioning of male sexual arousal. *Behav Modificat* **23**, 254–268

Pleim, E.T., Matochik, J.A., Barfield, R.J. and Auerbach, S.B. (1990). Correlation of dopamine release in the nucleus accumbens with masculine sexual behavior in rats. *Brain Res* **524**, 160–163

Quinsey, V.L. and Marshall, W.L. (1983). Procedures for reducing inappropriate sexual arousal: an evaluation review. In *The sexual aggressor. Current perspectives on treatment* (J.G. Greer and I.R. Stuart, eds), pp. 267–289. Van Nostrand Reinhold: New York

Rachman, S. (1966). Sexual fetishism: an experimental analogue. *Psychol Rec* **16**, 293–296

Rachman, S. and Hodgson, R.J. (1968). Experimentally induced 'sexual fetishism': Replication and development. *Psychol Rec* **18**, 25–27

Robinson, T.E. and Berridge, K.C. (1993). The neural basis of drug craving: an incentive sensitization theory of addiction. *Brain Res Rev* **18**, 247–291

Rossi, N.A. and Reid, L.D. (1976). Affective states associated with morphine injections. *Physiol Psychol* **4**, 269–274

Salamone, J.D., Cousins, M.S. and Snyder, B.J. (1997). Behavioral functions of nucleus accumbens dopamine: empirical and conceptual problems with the anhedonia hypothesis. *Neurosci Biobehav Rev* **21**, 341–359

Schechter, M.D. and Calcagnetti, D.J. (1993). Trends in place preference conditioning with a cross-indexed bibliography; 1957–1991. *Neurosci Biobehav Rev* **17**, 21–41

Schechter, M.D. and Calcagnetti, D.J. (1998). Continued trends in the conditioned place preference literature from 1992 to 1996, inclusive, with a cross-indexed bibliography. *Neurosci Biobehav Rev* **22**, 827–846

Schwartz, M. (1956). Instrumental and consummatory measures of sexual capacity in the male rat. *J Comp Physiol Psychol* **49**, 328–333

Sheffield, F.D., Wulff, J.J. and Backer, R. (1951). Reward value of copulation without sex drive reduction. *J Comp Physiol Psychol* **44**, 3–8

Takahashi, H., Takada, Y., Nagai, N., Urano, T. and Takada, A. (1998). Effects of nicotine and footshock stress on dopamine release in the striatum and nucleus accumbens. *Brain Res Bull* **45**, 157–162

Thorndike, E.L. (1932). *The fundamentals of learning*. Columbia University Press: New York

Tidey, J.W. and Miczek, K.A. (1996). Social defeat stress selectively alters mesocorticolimbic dopamine release: an in vivo microdialysis study. *Brain Res* **721**, 140–149

Tzschentke, T.M. (1998). Measuring reward with the conditioned place preference paradigm: A comprehensive review of drug effects, recent progress and new issues. *Progr Neurobiol* **56**, 613–672

Wang, L. and Hull, E.M. (1980). Tail pinch induces sexual behavior in olfactory bulbectomized rats. *Physiol Behav* **24**, 211–215

Whalen, R.E. (1961). Effects of mounting without intromission and intromission without ejaculation on sexual behavior and maze learning. *J Comp Physiol Psychol* **54**, 409–415

White, N.M. (1989). Reward or reinforcement: what's the difference? *Neurosci Biobehav Rev* **13**, 181–186

Wise, R.A. (1982). Neuroleptics and operant behavior: the anhedonia hypothesis. *Behav Brain Sci* **5**, 39–87

Wise, R.A. and Bozarth, M.A. (1987). A psychomotor stimulant theory of addiction. *Psychol Rev* **94**, 469–492

Woodson, J.C. and Balleine, B.W. (2002). An assessment of factors contributing to instrumental performance for sexual reward in the rat. *Quart J Exp Psychol* **55B**, 75–88

Yamada, K., Horibe, S. and Kondo, Y. (2006). Effects of opiate antagonist on mating-induced conditioned place preference in female rats. Society for Behavioral Neuroendocrinology 10th Annual Meeting, Pittsburgh, P28

Zamble, E., Hadad, G.M., Mitchell, J.B. and Cutmore, T.R.H. (1985). Pavlovian conditioning of sexual arousal: first- and second-order effects. *J Exp Psychol Anim Behav Proc* **11**, 598–610

Zamble, E., Mitchell, J.B. and Findlay, H. (1986). Pavlovian conditioning of sexual arousal: parametric and background manipulations. *J Exp Psychol Anim Behav Proc* **12**, 403–411

Zhao, C. and Cabanac, M. (1994). Experimental study of the internal signal of alliesthesia induced by sweet molecules in rats. *Physiol Behav* **55**, 169–173

CHAPTER

8

Preference for same sex partners: Basic concepts and its occurrence in non-human mammals

1 Concepts

1.1 The concept of preference

Preference without a choice is not possible. I cannot prefer anything if I have no alternative. So with sexual preference. Whenever we talk of a preference for sexual partners of one sex or another, we make the underlying assumption that individuals of the opposite sex are also available as partners. This is an important point, because in some situations there is no choice. For example, women locked up in a women's prison usually have no access to male partners. If some of these women were to engage in sex with each other, there would be no reason to maintain that they showed a preference for individuals of their own sex. Likewise, if the soldiers in the Great War had sex with each other in the trenches of Flanders, we cannot say that they manifested any preference for other men. It was simply that no women were available in the trenches.

We also need to introduce the concept of preference into an appropriate theoretical context, making it subjectable to meaningful experimental analysis. I can see no better way to give 'preference' a meaning than to analyze the concept in terms of incentive motivation theory. Indeed, basic to the notion of sexual preference is the idea that one sex has superior sexual incentive properties relative to the other. Translated into observable behavior, this means that an experimental subject approaches one incentive more than another. When we talk about preference, then, we talk of a difference in the intensity of approach behaviors between one incentive and another. We assume that this difference is caused by a difference in incentive value. It is important to keep in mind that the term 'preference' has no sense if not anchored in a behavior with known causation. By employing incentive motivational concepts, we give it a sense.

1.2 The concept of homosexual behavior: a source of much confusion

Preference for having sex with partners of one's own sex is called homosexuality in contemporary society. It becomes immediately urgent to differentiate this preference from a strange concept frequently used in both the scientific and the lay literature, namely the concept of homosexual behavior. If we look at the behaviors in humans preferring to have sex with their own sex, so to say, we find that many of these behaviors are exactly the same as those displayed in sexual activities with the opposite sex. For example, many men have anal intercourse with their female partners (20–50% of 'heterosexuals' according to surveys such as those conducted by Kinsey *et al.* (1948, 1953) and by Laumann *et al.* (1994)), while others have it with their male partners, and some with both. Independently of the partner's sex, the behavior consisting of insertion of the erect penis into the rectum through the anus is always the same. We cannot call this behavior homosexual. It is simply sexual.

Some behaviors displayed in same-sex encounters cannot be displayed in opposite-sex encounters. A man can never fellate a woman, while he easily can fellate a man. If we change our verbal habits and employ the term oral sex rather than fellatio, then a man can perform oral sex with a female as well as with a male partner. The motor patterns would probably be somewhat different, but the act of stimulating the partner's genitals with the mouth and tongue presumably produces the same consequence, a pleasurable sensation, in both cases. We can, according to the reasoning of one of the most influential behavior theorists of all times, Skinner, consider the differences in specific motor patterns as entirely irrelevant. In the Skinner box, where the task is to press a lever, it is of no importance whatsoever if the subject presses the lever with his right front paw, left back paw, nose or tail. Each of these motor patterns produces the same result, and which of them the subject chooses is of no concern (Skinner, 1935). We can, then, consider fellatio and cunnilingus as similar responses. The consequence is that oral sex is oral sex whatever the sex of the partner. It would be possible to extend the list of examples, but instead of wasting my and your time with that, I will rapidly conclude that most sexual behaviors performed by a human are exactly the same independently of the sex of the partner. This means that the concept of homosexual behavior is of limited utility. It is, however, a source of much intellectual and moral confusion and it should preferably be left to enter the oblivion of history.

Skinner's (1935) notions about the nature of a response could, in fact, be stretched somewhat further than I did in the preceding paragraph, particularly with regard to human sexual behaviors. Human sexual motor patterns are, so we know, far more varied than those of any other mammal. Knowing that the execution of sexual acts and particularly the experience of orgasm are both reinforcing and rewarding (see Chapter 7 if you have forgotten this), we can consider them as equivalent to the food or water reinforcements frequently used in operant tasks, including lever pressing in the Skinner box. Now, any behavior pattern that by some means involves a lever press will produce reinforcement, and will in effect be considered 'a response'. No further classification is needed. Exactly the same reasoning can be applied to all operant tasks and to species other than rodents, for example *Homo sapiens*. A 'response' is any motor pattern producing reinforcement. With regard to sexual activities, we could reasonably argue that any behavior leading to orgasm (reinforcement and reward) is a 'sexual response'. We could also maintain that no further classification is needed. Moreover, the behavior of the rat in the Skinner box is determined by the contingencies of reinforcement. In fact, all relevant aspects of the rat's behavior are determined by these contingencies. This holds also for other species, including the human (Skinner, 1969). It occurs to me that one possible way to describe and understand sexual behaviors could be to describe the contingencies of reinforcement (orgasm) and forget about the specific behavior patterns leading to it. In such a context, terms like homosexual or heterosexual would have no meaning. However, many social conventions and religious rules make strong distinctions between permissible and forbidden sexual motor patterns and to lump all these motor patterns together in the category 'sexual response' would undoubtedly provoke a moral outcry. After this digression into the Skinnerian paradise, we return to the specific subject at hand.

Some sexual behavior patterns in non-human animals are also similar regardless of the sex of the partner. A sexually receptive female rat will respond with lordosis to a male's mount, to another female's mount or the experimenter's fingers touching the flanks and the perineum. There is no known difference between a lordosis activated by a male, by another female or by an experimenter. Thus, a lordosis is a lordosis regardless of the source of the stimulation activating it. A male rat mounts females with much enthusiasm, but he may also mount other males. In both cases, his behavior, including the thrusting pattern, is the same. Again, mounting is mounting independently of who the mountee is. In fact, it appears that sexual behaviors with individuals of the same sex are identical to those displayed when having sex with individuals of the opposite sex.

1.3 Another confusing concept: sexually dimorphic behaviors

A point that must be taken into consideration, even if the only purpose is to reject it, is the fact that popular, and frequently not so popular, beliefs maintain that some sexual behaviors are sexually dimorphic in both the human and other animals. There is, indeed, one sexual behavior that we must accept not only as *typical* of one sex but also as *possible* only for one sex. I am referring to a class of behaviors usually called penetrative sex and within that class to the acts of the penetrator. The act of sexual penetration requires an erect penis on the part of the penetrator and only men and other males are properly equipped for displaying that kind of behavior pattern. If we exclude women employing plastic surrogates for penetrating other women or men, we need to conclude that the penetrator, or inserter as some like it, is always a male. However, the human male can easily penetrate other males, so being the penetrated, or the insertee as some like it, is not a privilege of women. While inserter behavior is sexually dimorphic, behaviors displayed by the insertee are not necessarily so. Again, just as was the case with oral sex, where it had no importance whether it was directed toward male or female genitalia, penetrating one opening is, in principle, no different from penetrating another, at least not as far as the penetrator (inserter) is concerned. It appears, then, that the only dimorphic sexual behavior is penile penetration. All other sexual behaviors may be displayed by both men and women. It is very tempting to speculate that it is exactly this sexual dimorphism that has made penetrative sex the only legitimate for millennia. Historians may have thought of that possibility.

In non-human animals we have, not surprisingly, another sexual behavior category that might be sexually dimorphic, that of accepting penetration or being the insertee, if we should happen to prefer that expression. There are few reports of male non-human mammals penetrating other males, but it may occur in rams as we will see in a few pages. Nevertheless, it is very unusual. As I pointed out long ago, this is most likely a consequence of the stereotyped nature of copulatory behavior in non-human mammals. While the human male can make use of basic anatomical knowledge to localize usable body openings in other men or in women, non-human males seem to have a weaker knowledge of anatomy. Furthermore, it is not sufficient to localize the opening. The penetrating organ needs to be put in contact with it, and entry must frequently be ascertained with the help of hands and sometimes also of artificial lubrication. All these activities are possible for the human male because of his extremely flexible copulatory behavior, combined with his motor abilities and intellectual creativity. On the contrary, none of the necessary supplementary behaviors are easily available to rats, and not even to monkeys and apes, because of a far more stereotyped copulatory behavior and less creativity.

In addition to penetrative sex, some other sex-related behaviors may be sexually dimorphic in several species. For example, hop-darting and ear wiggling are not common in male rats. There are no data as to their frequency in males, but I have to admit that it is rather unusual if not outright exceptional to see a male performing hop-darting or ear wiggling. It is possible that these behaviors should also be considered sexually dimorphic. Independently of this, it is most likely that there are more behavioral sexual dimorphisms in non-human animals than in the human. This can probably be attributed to the stereotyped, not to say reflexive, nature of copulatory behavior in the former.

Although only a few sexual behaviors are dimorphic in the sense that they are displayed by one sex but not by the other, we must admit that some behavior patterns are far more frequent in one sex than in the other. This could be considered as a statistical dimorphism. For example, the likelihood that a sexually receptive female rat will display lordosis when mounted is far above the likelihood that a male rat will display lordosis when mounted (Figure 8.1). Similarly, the likelihood of displaying a mount when exposed to a receptive female is higher in a male than in a female exposed to another female. It is certainly this difference in probability of occurrence that makes some scientists consider mounts as male-typical behaviors and lordoses as typical of females. For convenience I will frequently do the same.

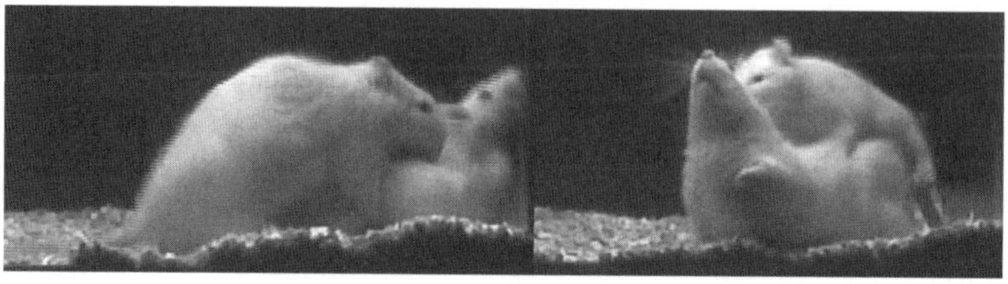

FIGURE 8.1 Left: A male rat mounting a sexually receptive female. Note the female's lordosis posture. Right: The male rat seen in the left panel is now mounted by another male. Please note the male mountee's lordosis posture. Both pictures are still images from a video, making the quality suboptimal.

The probability of performing a response is not only determined by the sex of the individual, but also by the environmental stimuli present at any moment. No sexual behavior is displayed in the absence of a sexual incentive. It is probable that the sex of the incentive is important for the likelihood of performing a male-typical or a female-typical response. The frequency of mounting is probably lower in male rats exposed to other males than in males exposed to receptive females. The display of lordosis in females is probably less likely when mounted by a female than when mounted by a male. Likewise, female proceptive behaviors are probably more likely when exposed to males than when exposed to other females. Experimental data unequivocally supporting these suppositions are not available, but this is perhaps not crucial for the moment. The important point here is that the sex of the incentive, in addition to the sex of the individual reacting to the stimuli emitted by the incentive, may determine the likelihood of occurrence of a particular behavior.

The rather complicated interaction between incentive and the behavior displayed as well as a classification of these behaviors was the subject of a brilliant chapter by Frank Beach (Beach, 1979). He suggested that the behavior displayed by a male mounting a female should be called normal, regardless of whether the female was receptive or not. On the contrary, the behavior displayed by a male mounting another male should be called homosexual. The mounted male's behavior was called normal if he resisted or remained passive and homosexual if he responded with lordosis. A female mounting another female shows homosexual behavior independently of whether the mounted female responds with lordosis or not. Finally, the behavior shown by a female mounting a male is given no name at all by Beach. This is also the case for behaviors shown by a male in response to a female's mount even if he should happen to respond with lordosis. Beach goes on suggesting that the label 'homosexual' is not very illuminating. I do agree. Instead, the 'sex' of the motor pattern should be considered. With that is understood that a mount is a male motor pattern while lordosis is a female motor pattern. I do not agree. It is probably true that mounts are more frequent in males than in females, and it is certainly true that lordoses are more frequent in females, something I already insisted upon. Despite the differences in frequency of appearance, I find it more convincing to call a mount a mount regardless of the sex displaying it. It is particularly so since a mount performed by a male and a female are identical motor patterns, as was made evident in the section describing sexual behaviors in Chapter 2. This same argument applies to lordosis. The behavior pattern should be called lordosis independently of the sex of the individual performing it. There is no reason to believe that a lordosis performed by a female is any different from that performed by a male. I admit that Beach's (1979) proposal to move away from the term homosexual was a good one. However, the replacement terms he suggested, 'homotypical' or 'heterotypical', do not contain any useful information, so far as I can judge. The time has arrived to move away from these expressions and stick to terms describing the behaviors themselves without adding any value-laden labels.

We may intuitively have noted that the assumption on which terms like 'homosexual behavior' or 'homotypical behavior' rest is that sexual behaviors indeed are sexually dimorphic. We have noted that, in the human, this is correct only for the behavior displayed by the inserter while inserting. In other mammals, that behavior plus the one displayed by the insertee while being inserted are the only indisputable cases of sexually dimorphic behaviors in the strict sense of dimorphism. We forget about proceptive behaviors for the moment. If we accept the notion of a statistical dimorphism, we may indeed accept that the main elements of non-human, particularly rodent, copulatory behaviors, mount and lordosis, are dimorphic. At the time of Frank Beach's (1979) analysis, it was entirely respectable to regard mount and lordosis as dimorphic behaviors because of sex differences in the frequency of occurrence. However, since then we have learned much more regarding lordosis in male rats and mounting in female rats. Freud's visionary notion of a fundamental, universal bisexuality has received so much empirical support in empirical studies of non-human animals that the concept of dimorphisms in sexual behaviors is, as I have argued for a couple of paragraphs by now, not really meaningful anymore. With the demise of that concept, terms like homotypical and heterotypical must also disappear. We will soon turn to one example of the kind of experimental observations that has contributed to this. Actually we will do that now.

Most valuable data concerning the issue of classification of sexual behaviors in old fashioned categories like homosexual and heterosexual or homotypical and heterotypical come from a series of exciting studies by Jim Pfaus and colleagues at Concordia University in Montréal. They decided to analyze female mounting behavior in more detail than has ever been done before. The requisite for being able to do so was, obviously, the use of a situation where female mounting is readily shown. Not all experimental situations permit the female to express mounting behavior. For example, during sexual interactions with an active male, the female is normally kept busy displaying proceptive behaviors and lordosis in response to the male's approaches and mounts. Moreover, sexually active males do not easily accept being mounted. This means that there is little opportunity for the female to present mounting behaviors. An alternative would be to evaluate female mounting behavior in same sex pairs. This has been done in many studies and, although female mounting is not unusual in this situation, its frequency is rather low. In contrast, females show high levels of mounting when exposed to non-copulating males. Such males are easily obtained by castrating a group of animals and then selecting those that do not show any copulatory behavior at all after a few weeks. When ovariectomized females are paired with non-copulating males, almost no female mounting is observed. However, after treatment with estradiol with or without progesterone, the number of mounts increased substantially. Progesterone without estradiol was ineffective. These results show that mounting is dependent on estrogens. The role of progesterone is uncertain, but an estrogen + progesterone synergy cannot be excluded until low doses of estradiol have been tested. Data from a group of cycling females confirmed the importance of ovarian hormones. There was a sharp increase in the number of mounts at proestrus. What is particularly important in the present context is the conclusion, which I will cite literally: 'Thus, mounting behavior may not be sexually dimorphic but rather guided by certain external and internal stimuli that both sexes are sensitive to' (Afonso and Pfaus, 2006, pp. 36–37). In subsequent papers, it was shown that the external stimuli determining the occurrence of female mounting were similar to those controlling mounting in males (Afonso *et al.*, 2006a, 2006b). This series of studies suggests that Beach's (1979) refusal to put a label on the behavior shown by a female mounting a male was unwarranted. There is no reason why it should not be called mounting. The addition of the adjective heterosexual to the noun mounting does not seem to add any further information. Moreover, the results of the studies from the Pfaus group certainly reinforce the notion I have already put forth several times, namely that neither mounting nor lordosis should be considered as sexually dimorphic behaviors.

1.4 Preferences for sexual motor patterns

Having got rid of the useless term 'homosexual behavior' and limited the term 'sexual dimorphism' to the rather insignificant cases where a real dimorphism exists, we can now start to concentrate on the subject of this chapter, sexual preferences. The expression could, in principle, refer both to the sex of the partner we choose for our sexual activities and to the specific choice of motor pattern. We will begin this section with a brief discussion of the latter choice.

Someone may choose penile–vaginal intercourse with the woman in supine position whenever a choice is available, while someone else may choose fellatio as soon as the partner accepts it. The choice of a specific motor pattern for the execution of sexual acts should, in principle depend on the individual's history of reinforcement and reward. The pattern most likely leading to the obtention of sexual reinforcement and/or to the most intense sexual reward should be selected, according to behavior theory. However, other factors may also influence the choice.

For example, some prostitutes charge more for fellatio than for penile–vaginal intercourse. Thus, although a client might have multiple experiences telling him that fellatio provides a larger reward than penile–vaginal intercourse, he may choose the latter for economical reasons. Likewise, a woman may have experiences telling her that cunnilingus is the activity that most likely leads to orgasm and, although she would happily choose that behavior pattern, she engaged in anal intercourse because her partner insisted on that. Another woman, attending a tea party with her female friends, watched a pornographic movie depicting a woman being penetrated simultaneously by two men in different orifices. At the earliest subsequent occasion, she tried to organize a similar situation for herself because of social learning. In addition to personal experience, social and moral norms regarding sexual behaviors may also exert profound influences on the choice of sexual motor pattern. Thus, although the basic mechanism determining choice of sexual activity should be the history of reinforcement, the immediate choice is constantly modified by other factors. Some, but far from all, of these factors were illustrated in the preceding vulgar examples.

I did not use the word preference in the preceding paragraph. However, whenever we choose one sexual motor pattern over another one, we show a preference. It could be posed that the motor pattern we most frequently choose among all available patterns is the preferred one. Absent from the preceding paragraph was also any mention of non-human mammals. This is a consequence of the fact that the number of possible sexual motor patterns is most limited among these mammals. In a male rat, for example, the choices are limited to perform a mount or a lordosis. It can even be questioned whether there is any choice at all, since both mount and lordosis are motor patterns reflexively activated by external stimuli, essentially tactile as I have insisted upon already. The lack of choice of which sexual motor patterns to display is not limited to rats. It is only in some primate species that we can find any real variability in sexual motor patterns. This means that the choice of or preference for a particular sexual motor pattern is a question of little concern for those studying sexual behavior in non-human mammals. More surprising is that we have very little reliable information on preferences in the human. Curiously enough, a question of no scientific importance seems to occupy most of those studying the choice of sexual motor patterns in the human: whether the chosen activity includes the use of a condom or not. This scientifically uninteresting matter may be fundamental to condom producers and perhaps also to those trying to prevent the spread of sexually transmitted diseases. In the present context, though, this kind of research is not very illuminating. We must conclude that empirical studies of preferences for the many possible and available sexual motor patterns are almost non-existent in the human. The many surveys giving us data on the proportion of people engaging in anal sex, oral sex, animal sex, etc, do not offer any data making it possible to estimate preferences for any of these motor patterns. I will not enter into any discussion of the methods that might allow us to determine which is an individual's preferred sexual motor pattern. I simply end this discussion by writing that we do not know.

1.5 Beyond dimorphisms and homosexual behaviors

The preceding few paragraphs had the purpose of convincing us that the term 'sexually dimorphic behavior' has very little utility. I had already concluded that the expression 'homosexual behavior' is also useless as a description of sex behaviors. Neither that term nor the notion of sexual dimorphism contributes to our understanding of these behaviors in any way.

As so many times before in this book, we have arrived at a surprisingly simple conclusion: behaviors should be described as such and not by the invention of fancy terms with vague or no empirical content. This applies to human as well as to non-human sexual behaviors. Similarly, there is no need for classifying or grouping behaviors in essentially arbitrary categories such as 'homotypical' or 'heterotypical'. I will not argue against all classifications of behaviors into categories, but with regard to sexual behaviors we have much to gain and little to lose by abstaining from the creation of categories that are based more on social conventions than on careful description and analysis of behavior patterns.

2 Preference for a particular sex: non-human mammals

2.1 Introduction

When we talk about sexual preferences, we usually think of a preference for a partner of a particular sex. As we all know, we can, at any moment, prefer having sex with someone of our own sex or with someone of the opposite sex. I already explained that preference inevitably means that we have a choice, so we can reformulate our subject of inquiry into a question of choice of one alternative over others available at the same time. In most of relevant non-human research on choice of sexual partner, the alternatives offered the experimental subject is between a male and a female conspecific. The subject can, then, choose one individual of its own sex or one of the opposite sex. The choice is, as we can easily imagine, determined by the incentive values of the individuals, so that the one with the larger incentive value will activate more intense approach behaviors on the part of the experimental subject than the one with the lower incentive value.

In addition to studies comparing the relative incentive value of individuals of the two sexes, there is a huge literature comparing the incentive value of individuals of the same sex. Notably females are given the possibility to approach two or more males. Occasionally, a male is offered the possibility to approach several females. These kinds of experimental procedures, where one individual can choose between several individuals of the opposite sex, are grouped under the name of mate choice tests. Results of such tests are rarely interpreted in terms of incentive motivation. Rather they employ far-fetched speculations as to the adaptive value of approaching one individual rather than another, normally wrapped in the esoteric fabric of sociobiology. They offer no useful information regarding the subject on hand, and probably not on any other subject either. I mention their existence only because of my open spirit.

2.2 Studies in intact mammals

General

A substantial literature on sexual preference in rats shows that males prefer females over other males, and sexually receptive females prefer males over other females. This was discussed at some length in Chapter 3, and I will neither repeat the experimental data supporting this conclusion nor provide any more references. Rather I will briefly mention some of our own unpublished data that may be of relevance here. We have, during the last decade, observed many hundreds of male rats in a test for sexual incentive motivation, where the male subject can choose between approaching a sexually receptive female and another male. On one particular test, a small proportion of males (<15%) may approach the male more than the female. However, we have never found a male that consistently approaches another male more than a female. We have also tested males that display lordosis and even ear wiggling when mounted by other males. The number of such males that we have tested is low (five displaying lordosis and one showing ear wiggling in addition to lordosis) and the results must be regarded as suggestive at most. Like other males, these males approach the female more than the male in the sexual incentive motivation test, showing that the female has a larger incentive value than the male also in this group. An inquiry among colleagues running similar tests has confirmed that male rats systematically preferring another male over a female probably are non-existent or extremely rare.

We have not observed many females given the choice between a male and another female, so it is impossible to determine whether females preferring other females are as rare as males preferring other males are. However, inquiring among colleagues, I have learned that nobody has data showing the existence of females systematically preferring other females. I venture suggesting that preference for individuals of the same sex is extremely unusual among rats, if it exists at all. Not even non-copulating males, that is males never displaying any sexual behavior at all during a large number of tests with receptive females, show preference for another male. They simply have no preference (Portillo and Paredes, 2004).

It does not seem that rats are exceptional in any way. In fact, there are no data from other rodents suggesting a systematic preference for individuals of the same sex. In fact, a reasonably recent review of same-sex partner preference among unmanipulated mammals found only a handful of species in which such a preference has been reported (Vasey, 2002a). Furthermore, as we will soon see, data are convincing for only two mammalian species and in one of them the same-sex preference applies only to a specific subpopulation. I do not exaggerate when maintaining that same-sex preferences are extremely unusual among mammals.

I am aware of the fact that there are innumerable anecdotes on dogs and cats and other domestic animals preferring their own sex. Stories about dogs being homosexual or lesbian cats are legion wherever pet enthusiasts gather and I fear similar stories about horses, cows and pigs circulate whenever male farmers meet out of reach of the censorship of their wives. I would not for a moment doubt as to the veracity of all these anecdotes, but I am not entirely convinced about their significance. Pets and domestic animals are raised and live in environments that are extremely remote from their natural habitats. The animals themselves are extremely remote from their wild ancestors. And above all, they have social, interspecies relationships of a most unusual kind. Their behavior is a result of all these peculiar circumstances. Our laboratory rats and mice are definitely more representative of a 'natural' animal than a Belgian Blue cow or a Chinese crested powder puff dog.

Do cows sporadically prefer other cows?

One of the mammalian species in which something similar to a same-sex preference has been reported is the domestic cow. There is one report, from New Zealand of all places, of a few cows that were engaged in female-female mounting in the presence of a potent bull. However intensely the bull tried to attract the attention of these females, he did not succeed and the females continued to mount each other for hours (Kilgour *et al.*, 1977). The fact that cows mount each other is well known, and is another example of the inherent bisexuality of which we have talked several times. The interesting point here is not that at all, it is rather the observation that even when these cows had the choice to copulate with a willing and waiting bull, they did not. They preferred to carry on the sexual interaction with another female. Having the possibility of choice, and still choosing the female, is an eloquent example of same-sex preference. Whether similar behavior exists in cows at other places than New Zealand is not known. Furthermore, it is not known whether these cows systematically preferred to copulate with females or if the observation was a kind of one night stand. The Kilgour *et al.* (1977) report is certainly interesting, but I await additional data before concluding or even proposing that same-sex preference is an established phenomenon among cows.

Rams sometimes prefer to have sex with other rams

There are more data on preference for individuals of the same sex among rams. The first description of a same-sex preference in rams was based on a total of eight subjects, four of which copulated with ewes with the intensity that rams usually show, while the other four showed little interest. All animals were then tested in an enclosure where a sexually receptive ewe and a ram were tethered to opposite sides. It turned out that the four copulating rams approached the ewes much more than the rams, while the reverse was the case for the four rams that did not copulate with the ewes in the preceding test. It was concluded that rams displaying low or no sexual activity with ewes do so because they prefer rams (Zenchak *et al.*, 1981).

All the subjects in this study had been raised in all male groups. An interesting observation is that the ram-preferring rams had engaged in significantly more sexual activities with other rams than those preferring ewes (Zenchak and Anderson, 1980). This observation may be more important than it might appear to be at this moment.

Several later studies have confirmed that some rams exclusively mount other rams even when receptive ewes are available. Data from a total of 584 rams show that the proportion exclusively mounting other males was 9.5%. About 12% did not show any sexual behavior at all, while 55.6% exclusively mounted ewes. The remaining 22% copulated both with rams and ewes (Roselli *et al.*, 2004b). It is not surprising that several attempts to find the cause of the preference for mounting other males rather than females have been made. One quite evident possible cause is hormonal alterations. In fact, it has been reported that rams exclusively mounting other rams have lower plasma concentrations of estradiol, estrone and testosterone (Resko *et al.*, 1996). The differences do not seem to be particularly reliable, since a subsequent study failed to replicate these observations. Males exclusively mounting other males were no different from males exclusively mounting females (Pinckard *et al.*, 2000). In view of these results, it seems very unlikely that some alteration in gonadal hormones is the cause for the habit of mounting other rams rather than ewes. Another finding is a lower number of estrogen receptors in the amygdala of rams mounting other rams. The amount of estrogen receptors in areas that might be of some importance for sexual behavior, like the medial preoptic area and hypothalamic regions was similar in male-mounting and in female-mounting rams. Nevertheless, it was suggested that the lower number of estrogen receptors in the amygdala might cause these rams to mount males and not females (Perkins *et al.*, 1995). This suggestion is bizarre in many ways. The amygdala has only a marginal influence on copulatory activity in rodent males, as far as can be judged from lesion studies (see Hull *et al.*, 2002). The effects of lesions are usually limited to a temporary enhancement of mount latencies and some prolongation of ejaculation latency. Supposing that estrogen receptors in the amygdala somehow modify nervous activity, these modifications could hardly be more dramatic than those produced by destruction of the tissue.

There are more reasons than those exposed above for not believing that the amygdala controls copulatory behavior and that estrogen receptors within it could influence preference for one sex over another. When observing the behavior of rams mounting other rams, it is usually found that they perform a larger number of mounts per test than those mounting females (e.g. Pinckard *et al.*, 2000). The reason is probably that the rams mounting females perform a larger number of ejaculations than those mounting males. Please remember that each ejaculation is also followed by a period of inactivity in rams. Anyway, there is nothing to suggest that the sexual behavior is less intense in male-mounting than in female-mounting rams. Assuming that the amygdala indeed is of some importance for copulatory behavior, then any functionally relevant alteration of the activity of amygdala neurons should manifest itself in a change in the intensity of this behavior. This is not the case. Furthermore, the many manipulations of the amygdala that have been performed in male rodents have never given the slightest indication of changes in sexual preferences. Finally, there are data showing that estrogens act in the preoptic area when stimulating sexual behaviors in rams (Parrott and Baldwin, 1984). There is a long way from the preoptic area to the amygdala. The possibility that alterations in the amount of estrogen receptors in the amygdala should cause a ram to mount other rams instead of ewes is remote indeed.

Another explanation that has been put forward for the habit of mounting rams instead of ewes is a reduced activity of aromatase in the preoptic area (Resko *et al.*, 1996). This observation was made in the rams that were also found to have reduced serum concentrations of testosterone, estrone and estradiol, a finding that was not replicated in later studies, as mentioned. Perhaps the reduced activity of aromatase had some relationship to the reduced availability of substrate, or perhaps the observation will turn out as ephemeral as the other data reported in the same paper. However, for the moment, we can imagine that the observation is replicable. It makes some sense, because there is good evidence telling us that sheep belong to those mammals in which aromatization, as well as 5α-reduction of testosterone, is necessary for the activation of sexual behaviors (e.g. Parrott, 1978; Crichton *et al.*, 1991). Thus, a reduced activity of a crucial enzyme at a brain site crucial for male sexual behaviors may reasonably affect those behaviors. Another question is whether it also may affect the sex towards which those behaviors are directed.

There is not any foundation for believing that altered hormonal activity in the preoptic area of an adult animal would change its preference for a particular sex. All kinds of manipulations, like lesion, electrical stimulation, hormone implantations, or implantations of hormone antagonists, have been performed in the preoptic area of male rats (see Chapter 6 for a summary of some of the relevant studies). Sexual behaviors may be reduced or disappear altogether, or they may be enhanced above the level shown by unmanipulated rats. However, in no case have these manipulations led to a preference for sexual interaction with other males or to a more intense approach of males than females. What can be imagined, though, is that the lower aromatase activity found in adulthood was present also during fetal development and brain differentiation. In rams, the sexual differentiation of the brain occurs well before birth. A deficient aromatase activity during that period could, eventually, lead to a lack of brain differentiation in the male direction. Below I will mention rat data showing that this may be the case. However, when pregnant ewes were treated with an aromatase inhibitor during most of the critical period of pregnancy, and their male offspring were tested for sexual preferences, it turned out that the incidence of same-sex preference was not different from that of controls (Roselli *et al.*, 2002). The dose of the aromatase inhibitor employed in this study reduced the activity of the enzyme by 85% (Roselli *et al.*, 2003). This reduction was much larger than that observed in the adult rams studied earlier. Thus, there is no support for the notion that reduced estrogen formation during fetal development is the cause of same-sex preferences in adult rams.

In a last effort to find some convincing cause for same-sex preferences in rams, brains were cut and a small region called the ovine sexually dimorphic nucleus of the medial preoptic area–anterior hypothalamus was described. The size of this nucleus, expressed either as length or volume, was larger in male rams exclusively mounting females than in rams exclusively mounting males. The latter had a size of this nucleus similar to females. Similar differences were found with regard to the number of cells (Roselli *et al.*, 2004a). These data contrast to another study, where cell sizes in several brain regions were determined in rams displaying a 'normal' level of sexual activity with ewes, in low-performing rams, and in rams exclusively mounting other rams. A comparison of cell sizes with habitual statistical procedures failed to detect any group difference. However, discriminant analysis revealed that low-performing subjects differed from 'normal' subjects with regard to cell sizes in the medial amygdala, preoptic area, bed nucleus of the stria terminalis and the ventromedial nucleus of the hypothalamus. There was no difference between males mounting other males and 'normal' males (Alexander *et al.*, 2001b). Furthermore, the expression of *fos* in the preoptic area and bed nucleus of the stria terminalis in response to exposure to a receptive female or another male does not differ between rams mounting females and rams mounting males (Alexander *et al.*, 2001a). In view of the conflicting data reported, no conclusion can be drawn. In any case, and as mentioned in the preceding paragraph, there is no reason to believe that small differences in the cell numbers of the preoptic area can account for the differences in choice of sex of the partner. If this were the case, then small lesions of the preoptic area should modify partner preference. Such an effect of incomplete preoptic lesion has never been reported and, in view of the many studies in which such lesions have been made, it appears extremely unlikely that it exists.

This review of the search for a cause for the habit of some rams exclusively to mount other rams should have convinced us that there is no convincing explanation for that habit as yet. Hormone concentrations in blood are not different from rams exclusively mounting females, the report of reduced number of estrogen receptors in the amygdala needs to be replicated and a credible rationale for attributing the difference in sexual partner preference to this potential fact needs to be proposed, the aromatase deficiency hypothesis has been disproved by experimental data, and the preoptic cell size/count differences are unreliable and, if confirmed, unable to explain the choice of sexual partner. I would like to suggest that the search for a cause undertaken so far has followed a most unfortunate path. Rather than searching for observable differences in the endocrine, enzymatic and nervous systems, a simple behavioral explanation can be proposed. It can be subjected to and founded upon several kinds of experimental tests in living animals, rather than being based on *post hoc* descriptions like cell counts in dead brains or estimations of enzyme activities in brain homogenates.

Rams mounting other rams now and then achieve ejaculation. So far as I can understand, ejaculation is associated with anal penetration. I deduce this from a paper where rams mounting other rams achieved a mean of 1.9 ejaculations during three tests of 30 minutes each. For comparison, I mention that rams copulating with ewes achieved a mean of 14.5 ejaculations during similar tests. In that paper we are told that the reason for the rather low number of ejaculations in the male-preferring animals was that '. . . the angle and anatomy of male-oriented rams' preferred sexual partner, the teaser ram, was not conducive to intromission, thus precluding successful ejaculatory attempts . . .' (Pinckard *et al.*, 2000, p. 1951). This is extremely interesting and important. Interesting, because it is one of the very rare examples of sodomy in non-human mammals. Important, because it shows that the ram mounting other rams now and then receives the same reward and reinforcement as he could have achieved by mounting and intromitting a female. Those having read the discussion of sexual reward and reinforcement in Chapter 7 will immediately remember that there is abundant data demonstrating that intromission/ejaculation can reinforce learning and produce a state of positive affect, reward. There is no reason to imagine that a ram is so perceptive and has such an exquisite taste that he distinguishes ejaculation in the rectum of a male from ejaculation in a female's vagina. These two events can undoubtedly be considered as equally reinforcing and rewarding. From that point of view, the sex of the partner is uninteresting.

The achievement of ejaculation is less likely when copulating with a ram than when copulating with an ewe. If I abuse the data from the Pinckard *et al.* (2000) paper to calculate the probability of ejaculation upon mounting, I find that it is about 0.34 for the rams copulating with females. The corresponding value for the rams copulating with other rams is 0.04. From the comment about the difficulty of intromitting the teaser ram in this particular study, it can perhaps be concluded that this is an unusually low figure but, even if we make room for this potential error, we need to accept that the probability of achieving intromission/ejaculation is far lower for rams mounting other rams than for those mounting females. The differential likelihood of reinforcement should make the female preferred over the male. This is exactly what happens in about 55% of rams. Only about 10% persists in the less favorable behavior of copulating exclusively with the male. As mentioned long ago, 22% copulates with both, but presumably they copulate much more frequently with females than with males. We also need to pay some attention to the fact that even the rams considered as exclusively mounting males in fact sire litters almost as frequently as female-preferring rams do (Stellflug *et al.*, 2006). Their reproductive capacity does not seem to suffer much from their engagement in same-sex activities during the test designed for determining sexual preference. This shows that the exclusive mounting of other males is situation-specific, and not an unmodifiable behavioral trait. It could be argued that the difference between male-mounting and female-mounting rams is simply a difference in probability of partner choice. The likelihood of choosing to copulate with a female is lower in what is called male-mounting rams than in female-mounting rams. In some situations, like the sexual preference test, this reduced probability takes the appearance of an exclusive choice of same-sex partner, while in other situations a female may be chosen, as evidenced by the frequent fathering of offspring.

The proposal that the ram-mounting males may have *learned* to prefer other males may be perceived as far-fetched by those who like to explain preferences for individuals of the same sex as an inborn characteristic. I will dedicate a few paragraphs to argue in favor of it. First of all, we need to know that all ram-mounting males known have been raised similarly. At puberty, they were separated from females. From that moment on they spent their time together with other rams. There is nothing special in that. Most sheep breeders use this procedure. The young rams like to mount each other. In fact, mounting is a rather frequent activity. It is not impossible that some of the many mounts performed are associated with anal penetration/ejaculation. Whenever this happens, the mounter is reinforced and rewarded. If this occurs a sufficient number of times, the stimuli present at the moment will become associated to the positive affect produced by ejaculation through classical conditioning. They will become conditioned sexual incentives in the way explained in Chapter 7. The most salient stimuli in the moment are certainly those emitted by the ram which is mounted. Thus, these stimuli will acquire the capacity to activate approach behaviors on future occasions. Moreover, the motor pattern leading to intromission/ejaculation will be reinforced as in any operant learning procedure and their likelihood of repetition will gradually increase. In the preference test, usually performed after months or years of cohabitation with and mounting and intromitting/ejaculating in other rams, the stimuli emitted by the teaser ram will have sexual incentive properties and the subject will approach them, and then mount. At the same time, the ewe will emit unconditioned sexual incentives. The rams exclusively mounting the teaser ram will be those where conditioning to male stimuli has been particularly strong. So strong, in fact, that stimuli emitted by the female have lost their capacity to evoke mounting in the test for sexual preference. Evidently, they have not lost that capacity elsewhere, since the male-mounting males sire offspring. Here it must be observed that even the male-preferring rams do respond to the female with some approach behaviors during the sexual preference tests, although they do not mount her. In other situations, mounting, as mentioned, follows this approach.

Some empirical support for the hypothesis outlined in the preceding paragraph comes from a study where it was shown that rams mounting other rams had engaged in significantly more sex-like behaviors with other rams in the all-male rearing environment than rams mounting females had (Zenchak and Anderson,1980). When I first mentioned that study a couple of pages ago, I promised that we later should discover its importance. We will see it now. It is not too far-fetched to assume that the rams mounting most also are the rams obtaining the largest number of intromissions/ejaculations. As we know, intromission/ejaculation is a reinforcing event, and the larger the number of reinforcing events, the more intense the association between male stimuli and sexual approach and activity becomes. In consequence, these rams should be more likely to mount other rams. Thus, the Zenchak and Anderson (1980) study supports an explanation of the preference for mounting other rams in terms of learning. The support would have been considerably more convincing if data on the frequency of intromission/ejaculation in the rams had been reported rather than the frequency of sex-like behaviors. This term is rather vague. Here we have another example of the unfortunate habit of not reporting complete behavioral data.

A somewhat more recent study failed to find any relationship between home-pen mounting of other males and subsequent preference for males (Price *et al.*, 1988). However, the results are reported as correlations between mounting in the home-pen and the proportion of mounts directed towards the male in the sexual preference test. This is quite different from a comparison of males *exclusively* mounting other males during the test with males *exclusively* mounting the female as was done in the Zenchak and Anderson (1980) study, and I do not consider these data as contradicting the arguments exposed in the preceding paragraph. Regrettably, this study also failed to report complete behavioral data. It is almost certain that the crucial variable is the number of ejaculations achieved with other males in the home pen, but no data on this are reported.

It is easy to summarize this rather long discussion of rams. Beyond doubt, a small proportion of rams mount exclusively other rams at a test designed for determining sexual preferences. They even do so at several tests. These male-mounting rams appear to have blood concentrations of gonadal hormones that are not different from those of rams exclusively mounting females. They may have less estrogen receptors in the amygdala, but that proposal is based on a single study. Likewise, they may have less active aromatase in the preoptic area, but again this affirmation is based on a single study. There is one study reporting differences in cell number and size in a small subregion of the preoptic area, while another study failed to find any difference between male-mounting and female-mounting rams. There does not seem to exist any convincing explanation for the preference of other rams in the test for sexual preferences. Furthermore, the male-mounting rams impregnate females at a rate not much different from the female-mounting rams, showing that the preference for other rams is specific to a particular test situation. An explanation for the preference of other rams in some situation can be put forth in terms of learning. Basic to this is that rams mounting other rams occasionally achieve anal penetration (intromission) and ejaculation, events known to be both reinforcing and rewarding. This fact not only provides a basis for learning, it also explains why rams belong to the extremely rare groups of non-human animals where individuals having a sexual preference for their own sex have been described. Because of some anatomical/motor coincidence, rams can achieve intromission while copulating with other rams, exactly as the human male copulating with other males can. The ram can learn to associate stimuli emitted by other rams with sexual reward, and to execute copulatory responses (mounting) in response to these stimuli by classical and operant conditioning, respectively. Some limited, indirect support for this notion exists.

Female snow monkeys like to have sex with other females

There are numerous accounts of sexual behaviors between females of a primate species called *Macaca fuscata*, Japanese macaque, or snow monkey (for a review see Vasey, 2002a). There is nothing peculiar with this. Sexual activities between individuals of the same sex are very common among mammals, as mentioned so many times that I start to feel ashamed of repeating it once more. However, one peculiarity with female snow monkeys is that they form temporary female-female associations lasting from less than one hour up to over a week. These associations have been termed 'consortships' (Wolfe, 1983). Typical of these consortships is a rather intense sexual activity where the females mount each other. One study reports a mean of 31 mounts per hour of observation (Vasey, 2002b). Furthermore, while 55% of the females formed this kind of sexual association with other females, only 45% formed sexual associations with males. These data demonstrate that there is a substantial extent of sexual activities between female snow monkeys. Without having quantitative data from other species, I nevertheless assume that such intense same-sex activities are fairly unusual. What is still more peculiar with these monkeys, though, is that there is evidence showing that the females prefer to have sex with each other when an attractive male is also available and even insisting on sex.

The experimental procedures for determining sexual preferences have been varied, but one of the more direct approaches consisted of observing responses to invitations from a male in pairs of females engaged in homosexual activity. These invitations were directed towards one female in the pair, and she had, then, the opportunity to choose between accepting the invitations from the male or continue mounting the other female. In 92.5% of the situations of this kind, the female chose to continue mounting the other female. The male was chosen in only 2.5% of the situations (Vasey, 1998). These data quite convincingly show that female snow monkeys, at least in the population studied, prefer sexual interactions with other females when given a choice. However, this does not mean that females preferring to mount other females will never accept copulating with males. In fact, all females will have one or more sexual interaction with males during the breeding season (Vasey, 2002a). Exactly as was the case with the male-mounting rams, a clear preference for the own sex is evident in certain situations, but that preference does not impede sexual relations with members of the opposite sex in other situations.

It is not at all surprising that the intense same-sex activity described in some populations of female snow monkeys has prompted speculations as to the cause. The most obvious according to popular prejudice, some unusual production of gonadal hormones during early and/or adult life, has not attracted any attention, at least not to my knowledge. On the contrary, cell numbers have been counted in a small region in the vicinity of the preoptic area, the dorsocentral portion of the anterior hypothalamic nucleus. This region, of unknown function, was selected because it is sexually dimorphic in the rhesus macaque (Byne, 1998). With dimorphic is here understood that it is larger in males than in females, not that it is exclusive to one sex or the other. It is a good example of a statistical dimorphism. The reasoning was that the difference between males and females might be smaller in the snow monkey population where females show high levels of same-sex behavior than in the rhesus monkey, a species where female same-sex activity is low or entirely absent. Results showed that the size as well as the neuron counts differed as much between males and females in the snow monkey as it does in the rhesus (Vasey and Pfaus, 2005). It appears, then, that there is no structural peculiarity in a region known for its importance in male sexual behaviors and perhaps also for proceptive behaviors in the female (see Chapter 6).

Rather than attributing the female snow monkeys' habit of same-sex mounting to endocrine alterations or structural quirkiness in some region of the central nervous system, an explanation in terms of learning has been proposed. It is based on a careful analysis of the females' behavior while mounting other females. During mounting, prolonged clitoral contact with some part of the mountee's body is common. This occurs, for example, when performing a sitting mount. This is a mount where the mounter sits on the mountee's back in a jockey-like position, while grasping the mountee's upper back with her hands and the mountee's lower back with her feet. The mounter repeatedly and vigorously rubs her clitoris against the back of the mountee. At another kind of mount, the double-foot clasp, the mounter grasps with her feet between the mountee's ankles and hips, and with her hands on the mountee's back. In that way, the mounter's groin is in contact with the mountee's perineum, thereby stimulating the clitoris of the latter. In addition to procuring these kinds of clitoral stimulation to mounter and mountee, respectively, both females frequently masturbate by rubbing their clitorises with their tails. This short description of same-sex mounts in female snow monkeys is based on accounts by Vasey (1998, 2002) and Wolfe (1983). The important point here is that female-female mounting and associated masturbation provide efficient clitoral stimulation. We may assume that this stimulation produces a state of positive affect and that it can reinforce learning. There are no experimental data showing this to be true, but I would suggest that most women would accept that clitoral stimulation indeed is pleasurable. The fact that the female snow monkey engages in clitoral stimulation should by itself suggest that the effect in macaques is similar to that in women. Otherwise it would be difficult to explain why the females display such an intense mounting activity.

Provided that the clitoral stimulation obtained while mounting other females *and* while being mounted by another female is rewarding and reinforcing, motor patterns leading to this stimulation should acquire enhanced probability of repetition through operant learning. Likewise, stimuli present during the pleasurable experience should acquire sexual incentive properties through classical conditioning. It is not impossible that stimuli emitted by the female partner are salient in the situation. Thereby they are most likely to get sexual incentive properties. If monkeys emit stimuli such that an individual can be distinguished from other individuals, then a particular female would become the sexual incentive. This is exactly what seems to occur in the consortships mentioned earlier. With the passage of time and the execution of a considerable amount of sexual responses, the incentive properties of this particular female may get reduced because of negative alliesthesia or some other behavioral process. The consortship would then end.

The reasoning presented in the preceding paragraphs is not much different from what Vasey (2002a) proposes. 'Once females evolved the capacity to mount, and the capacity to derive sexual gratification during mounts through clitoral stimulation, they could do so just as easily by mounting females, as males' (Vasey, 2002a, p. 173).

In this context, it may be useful to draw attention to the most interesting results of a comparison of the motor patterns employed by female snow monkeys when mounting other females with those employed by males when mounting females. It was found that the males' motor pattern was much more stereotyped than the females' pattern. This was explained by suggesting that the males' mounts were constrained by the requirements imposed by penile–vaginal penetration. There are not so many mounting positions and movements available that assure successful intromission. All motor patterns not leading to intromission can be expected to become extinguished because of lack of reward and reinforcement, leaving only those most likely to produce the rewarding and reinforcing event of intromission. The females, at difference, can attain clitoral stimulation in a variety of ways. Their copulatory motor patterns were frequently far more complex than the males', employing combinations of forward, sideward and upward movements. Presumably, these motor patterns assure abundant clitoral stimulation. They were never observed in males (Vasey *et al.*, 2006).

One conclusion that immediately can be drawn from the exciting study of copulatory motor patterns in the snow monkey just described is that primates have a potentiality for modifying these motor patterns to an extent not seen in other mammals. We can also note that learning may shape these motor patterns in a very important way. It is evident that learning would not be able to have this effect in the absence of some degree of motor flexibility, so the latter rather than disparities in learning ability is probably the cause of this basic difference between primates and other mammals. Furthermore, the snow monkey is the only non-human species where all evidence points to an explanation of same-sex preferences in terms of learning. I suggested that the ram-mounting rams also have acquired their preference through learning, but admittedly there is no experimental evidence supporting that proposal.

The snow monkey is the last example of same sex preferences in unmanipulated animals I will mention. It is so because I know of no other mammalian species where this kind of preferences has been described in a scientifically acceptable way. We might find it amusing to consider the fact that in the two species where we have solid data showing that a significant proportion of a population shows a sexual preference for the same sex, the execution of copulatory acts (mounting) with individuals of the same sex leads to a sexually rewarding and reinforcing event (intromission/ejaculation in the ram, clitoral stimulation in the snow monkey).

2.3 Search for experimentally induced same-sex preferences

Many experimental manipulations have been employed in efforts to alter the sex-typical likelihood for displaying mount or lordosis in some way or another. The early studies concentrated on the display of sexual behaviors considered to be typical of the opposite sex. The question of preferring one sex over the other was not addressed at all. This situation has changed little, and our knowledge regarding same-sex preferences is still rather limited. Some rather recent data which might appear to be of relevance to the issue at hand here were presented in Chapter 5, during the discussion of the effects of natural or human-made deletions of genes coding for the gonadal hormone receptors. As pointed out in Chapter 5, gene deletions affect the organism throughout the life-span, including the period of sexual differentiation of the brain, making the results of this research of potential relevance for the topic of sexual preferences. However, I decided to include the gonadal hormone receptor gene deletion studies in the context of endocrine regulation of sexual behaviors rather than in the context of sexual preferences for a simple reason. Mammals, including the human, showing same-sex preferences do not lack any gene coding for gonadal hormone receptors. Therefore, organisms lacking one or more of these receptor genes may be useful experimental tools contributing to an understanding of the endocrine regulation of sexual behaviors, but they are of no real relevance for explaining the origin of same-sex preferences.

Studies in non-primate female mammals

All manipulations effective in changing the likelihood for the display of sex-typical and/or sex-atypical behaviors and preferences involve the endocrine system and they are efficient only when performed at an early age, normally at the age of maximum brain growth and differentiation. The typical procedure has been to administer androgens to females both perinatally and in adulthood or estrogens to males at these same times. Sometimes the sex-typical hormone was administered in adulthood instead of or in addition to hormones of the opposite sex. A typical example may be a classical guinea-pig study published in 1959 (Phoenix *et al.*, 1959). Pregnant females were given testosterone during the gestation period. The offspring were gonadectomized in adulthood and then treated with estrogen + progesterone. The females showed a reduced capacity to display lordosis. Then the subjects were treated with testosterone and it was found that the prenatally treated females mounted more than controls. There was no effect of prenatal hormone treatments on the behavior of males.

Another example is a study in which male rats were castrated shortly after birth. Some of the males were given estradiol on postnatal day 5 and some others were given oil. When these animals were treated with estradiol + progesterone in adulthood, it turned out that those given oil neonatally responded with lordosis to a male's mount more frequently than those given estradiol (Feder and Whalen, 1965). As we already know, this kind of study gave rise to the notion of an organizing effect of steroids early in life and an activational effect in adult individuals. There is a huge literature on this subject, which I will not discuss. Any textbook of behavioral endocrinology will analyze the issue at length. I simply conclude that the rodent data show that neonatal endocrine manipulations can alter the likelihood of presentation of specific sexual behaviors, so that neonatal gonadal hormones increase the probability of mounting in both sexes, while the lack of such hormones increases the probability of presenting lordosis in both sexes. The fact that the probabilities of the display of certain motor patterns can be altered by gonadal hormones early in life has led to the rather widespread but unfounded notion that early endocrine alterations can be a cause of the condition of homosexuality in the human.

We have already concluded that the motor patterns underlying human sexual behaviors are neither male nor female. Thus, the analogy suggesting that enhanced mounting in female rats, or enhanced lordosis in male rats, has something to do with human homosexuality is flawed. However, if we could show that perinatal endocrine manipulations affect the preference for one sex over the other, then we could perhaps assume that they somehow may be related to homosexuality. As so elegantly put elsewhere: 'The essence of partner preference lies not in what the animal does, but in which individuals it does it with' (Adkins-Regan *et al.*, 1997, p. 497). Experimental data evaluating the effects of early endocrine manipulations on preference for a particular sex are not abundant, and results are, as we will see, not always crystal clear. We begin this short review of relevant data with a discussion of studies in female rats. In the oldest of these, females were either treated with androgen or with vehicle perinatally. Some subjects were ovariectomized before puberty while some were left intact. When adult, they were given testosterone and then tested for the intensity of approach behaviors towards a male or a sexually receptive female. The perinatally androgen-treated subjects that had been ovariectomized before puberty displayed high levels of typically male behaviors like mount and intromission and approached a female more than a male. The effect of perinatal androgen treatment was weaker in the females that were ovariectomized as adults rather than prepubertally. In this very complex experiment, some other females, perinatally androgenized or vehicle-treated, were ovariectomized prepubertally. They were implanted with an estradiol-filled Silastic® capsule which they carried between 30 and 90 days of age. After removal of the estradiol implant, treatment with testosterone was initiated. The females given vehicle perinatally + estrogen from day 30 to 90 displayed more intromission-like behaviors than females treated with testosterone perinatally + estrogen from day 30 to 90. Both groups approached a receptive female more than a male (de Jonge *et al.*, 1988).

The de Jonge *et al.* (1988) experiment reveals several interesting aspects of hormone effects on sexual behaviors. First, perinatal treatment with androgen makes a female show a behavior similar to an unmanipulated male when treated with androgens in adulthood. Second, estrogen treatment during puberty has an effect very similar to perinatal treatment with androgens. Third, the combined treatment with androgen perinatally + estrogen during puberty has no larger effect than treatment with vehicle perinatally + estradiol during puberty. These results show that the notion of sexual differentiation as a phenomenon limited to the period of maximal brain growth early in life is overly simplified. Hormone actions at puberty are also able to influence adult brain functions in a significant way. We will return to this later in this section.

Before leaving the de Jonge *et al.* (1988) experiment, I feel obliged to express some regrets regarding the choice of hormone treatments. Giving androgens to newborn female pups is certainly a very unphysiological manipulation. It is equally unphysiological to treat adult females with large doses of androgens. Rats do not produce androgens in the adrenals (Bélanger *et al.*, 1989; van Weerden *et al.*, 1992), and the only androgen source in females is the ovary. Physiological blood concentrations are, therefore, very low in females throughout the life span. The exposure of females to large doses of androgens is an event that is very unlikely to occur outside the laboratory and it is not easy to see how the effects of this kind of hormone manipulation can tell us much about the normal endocrine determination of sexual behaviors. A still more serious problem is that perinatal or pubertal hormone manipulations are not at all necessary for producing a male-like sexual preference in female rats as shown in a nice study (Merkx, 1984b). Females were ovariectomized when adult, and then subcutaneously implanted with a testosterone-containing Silastic® capsule. They were then housed in a residential maze, where they could approach an ovariectomized female. Occasionally, this female was brought into estrus by sequential administration of estradiol and progesterone. The experimental subjects approached this female more when she had received the hormone treatment than at other times. In fact, the experimental females approached the hormone treated female incentive as much as male rats did (Merkx, 1984b). This very important, but largely ignored, study shows that the preferred sex is dependent on the gonadal hormone circulating in adulthood and not predetermined early in life. Please remember that the females in the de Jonge *et al.* (1988) study were treated with testosterone when adult. I must admit, though, that treating female rats with a rather large dose of testosterone is very unphysiological. Whether the Merkx (1984b) experiment has any relevance outside the laboratory environment is not clear at present. Its results appear to be reliable, though. In another study, also employing the residential maze, females ovariectomized as adults, were treated with dihydrotestosterone + estradiol and allowed to choose to spend their time close to either an inaccessible non-receptive female or an inaccessible receptive female. The hormone treatment produced a preference for the receptive female, exactly as it does in castrated males (Merkx *et al.*, 1989). Neither dihydrotestosterone nor estradiol alone had this effect. Again, it seems that the adult endocrine environment determines the preferred sex.

While the data reported from studies in the residential maze are rather consistent, other procedures may give completely different results. In Meyerson's open field test for sexual motivation, females treated neonatally with oil and ovariectomized as adults approached a male more than a female regardless of whether they were treated with estradiol or testosterone. However, females given a neonatal injection of testosterone responded in the opposite way to hormone treatments. Both estradiol and testosterone produced more intense approach towards the female in these animals (Meyerson *et al.*, 1979). This seems to be a rather clear indication that hormones early in life indeed differentiate sexual approach behaviors. Unfortunately, these data contradict the results obtained in the earlier mentioned de Jonge *et al.* (1988), Merkx (1984b) and Merkx *et al.* (1989) experiments. If we try to arrive at some kind of conclusion amidst these contradictory observations, it must be rather vague. A tentative proposal maintaining that the role of hormones for the sexual differentiation of the stimuli provoking approach behaviors in the adult female rat is unclear would be the most appropriate. However, some more recent data may be of help.

Matuszczyk (2003) determined the effects of a rather reasonable endocrine manipulation in prenatal female rats. The dams were injected with an estrogen receptor antagonist on days 13 to 19 of pregnancy. At 75 days of age, the female offspring were ovariectomized and then treated with estradiol benzoate and progesterone. The hormone-treated females were allowed to approach either a male or a sexually receptive female. They approached the male more than the female independently of prenatal treatment. It seems, then, that blockade of estrogen receptors during the last days of pregnancy is without effect on female rats' choice of sexual partner. The rationale for this experiment was that estrogens have been shown to be produced in rat fetuses towards the end of pregnancy (Habert and Picon, 1984). Thus, blocking their receptors should conceivably affect sexual differentiation, provided that estrogens are involved in that process. Results did not support this notion. I consider these data as supporting the notion that prenatal gonadal hormones are of slight importance for determining which sex will be the most attractive when being an adult female rat. Nevertheless, more data are needed before any final conclusion can be accepted or perhaps even proposed.

Ferrets are not like rats. In one experiment, female ferrets were treated with testosterone between days 27 and 39 of gestation (delivery was on day 40) and the offspring were injected with testosterone within 5 minutes of delivery. At postnatal day 5, the females were implanted with a testosterone-filled Silastic® capsule, which was removed at postnatal day 20. Finally, when adult they were treated with estradiol. These females displayed a preference for another female rather than for a male (Baum *et al.*, 1990). It may be observed that the rather violent treatment employed assured that the female ferrets were exposed to as much androgens as males normally are, at least. Less extensive perinatal treatments with testosterone are ineffective. Furthermore, adult ovariectomized female ferrets treated with testosterone showed no preference for a male. In fact, they showed no preference at all (Stockman *et al.*, 1985). We may remember that adult female rats treated with testosterone prefer a sexually receptive female over a non-receptive. Whether these results are due to a real species difference between rats and ferrets or to differences in perinatal hormone treatment is not possible to determine at present. Regardless of that, it is probably safest to conclude that the early differentiation of sexual partner preference in females is not well understood. The role of hormones in adulthood is not better understood. Part of our ignorance is probably due to the unorthodox endocrine manipulations employed. Exposing females, either during early development or in adult life, to androgen concentrations within the range typical of males does not seem to be a particularly exciting approach. Such conditions are extremely rare outside the research laboratory and may have little relevance for the physiological processes of sexual differentiation. Reduction or elimination of hormone availability has been a useful technique during much of endocrinology's history, but the contrary procedure may have limited value.

Studies in male non-primate mammals

Turning to males, the situation is somewhat clearer. This is probably because we have more data on males than on females, as is so often the case. If a male rat is castrated shortly after birth, on postnatal day 0 to be exact, and treated with estradiol + progesterone when adult, he will not only display lordosis, ear wiggling and hop-darting with the same frequency as a female, bur he will also approach a sexually active male more than a receptive female. When these males received a testosterone implant in adulthood, they changed their behavior and approached a sexually receptive female more than a male (Matuszczyk *et al.*, 1988). This fascinating result suggests that male rats are similar to females in the way that adult hormonal environment determines the preferred sex, while early hormone exposure or lack of such is of minor importance. Similar results were obtained in a study by Meyerson *et al.* (1979). Their males, castrated shortly after birth, also approached a male more than a female when given estradiol and they approached a female more than a male when given testosterone. Another experiment, this time performed in the residential maze, which was popular in the Netherlands some years ago as we might have noticed by now, obtained slightly different results. Males were castrated within 24 hours of birth and implanted with Silastic® capsules containing either testosterone, dihydrotestosterone or estradiol when adults. In the maze, they could approach an inaccessible female made receptive with sequential treatment of estradiol and progesterone, or an ovariectomized, non-treated female. Animals implanted with capsules containing testosterone or estradiol approached the receptive female more than the non-receptive. Dihydrotestosterone was ineffective (Merkx, 1984b). This result is somewhat different from those reported by Matuszczyk *et al.* (1988) and Meyerson *et al.* (1979), but the choices were also different. In the latter studies, the males chose between a receptive female and a male, as might be remembered, which is not at all the same as choosing between females differing in hormonal status. Had the subjects in the Merkx (1984b) study had the same choice as those in the Matuszczyk *et al.* (1988) and the Meyerson *et al.* (1979) experiments, results might have been similar. What is perhaps more important is that the neonatally castrated animals in Merkx (1984b) study responded to hormone treatments in a way identical to that of males castrated in adulthood (Merkx, 1984a). The obvious conclusion was that neonatal testicular hormones have little effect on adult sociosexual preference behavior. This coincides with the conclusion of the Matusczcyk (1988) and Meyerson *et al.* (1979) studies.

Instead of subjecting the experimental subjects to neonatal castration, rats can be treated with an aromatase inhibitor from birth until puberty. Since there is abundant data suggesting that estrogens are crucial for sexual differentiation in male rats (see Chapter 5), this treatment should be equivalent to castration with regard to central nervous effects. Peripheral effects of testosterone should not be blocked, though, because these effects are mediated by dihydrotestosterone. When such an experiment was done, it turned out that the male rats treated with the aromatase inhibitor showed the same preference for odors of receptive females compared to odors of males as control rats did. The test was performed in adulthood and all subjects had been left intact. Furthermore, no group difference was found when these animals could choose between an inaccessible male and an inaccessible receptive female. On the contrary, copulatory behaviors were affected by the aromatase inhibitor. All treated males displayed lordosis every time they were mounted and they had a reduced intensity of male sexual behaviors (Bakker *et al.*, 1996).

Similar results were obtained in another experiment (Domínguez-Salazar *et al.*, 2002). Here, the experimental males could interact sexually either with a male or with a receptive female during the test for sexual preference. Moreover, a group treated with an androgen receptor antagonist was included in addition to a group treated with an aromatase inhibitor. None of the treatments affected the males' intensity of approach to a receptive female, although the androgen antagonist impaired male-like copulatory behaviors. On the contrary, the aromatase inhibitor failed to modify male-like copulatory behaviors, while lordosis was much enhanced. These studies nicely coincide in suggesting that neonatal hormone manipulations have little or no influence on sexual preferences in adult animals.

Further support for this notion comes from an experiment in which dams were treated with an estrogen receptor antagonist, an androgen receptor antagonist, or vehicle. The male offspring were tested for sexual behavior and sexual preferences when adult both when unmanipulated and after treatment with testosterone. Both antagonists had a minor effect on the intensity of approach to a receptive female, but in no case did the treated males show any preference for the male. Male copulatory behavior was affected by both antagonists. A small reduction of the proportion of animals displaying intromission and ejaculation was observed. With regard to female sexual behavior, the estrogen antagonist enhanced the propensity to respond with lordosis to a mounting male. In fact, almost every mount induced the lordosis posture (Matuszczyk and Larsson, 1995). Again, we find marked effects of early hormone manipulations on copulatory behaviors but only minor effects on the intensity of approach to a female.

I have been unable to find any study evaluating the effects of early hormonal manipulations on sexual approach behaviors in the male ferret. There are a number of studies analyzing the effects of such manipulations on copulatory behaviors, but they are not of any relevance for the present discussion. Nevertheless, I do not believe that there is any need to insist on a discussion of more data. We should already be convinced that, whereas adult copulatory behaviors may be quite heavily influenced by neonatal hormone treatment, the choice of sexual partner is much less so. This probably means that the propensity to respond to certain stimuli with the display of a stereotyped motor pattern is determined during sexual differentiation of the brain early in life. These responses are, or are similar to, reflexes and conceivably dependent on pre-established brain circuitry. The sexual incentive properties of distant stimuli do not seem to be differentiated in the same way. The responses to distant sexual incentives are, as we know quite well by now, arbitrary. There is no reason to assume that the extremely varied approach responses activated by sexual incentives should be entirely controlled by preestablished brain circuitry, not even in rats.

Studies in primates

We can, perhaps, propose that the degree to which the early hormonal environment determines adult copulatory behaviors is directly proportional to the degree of stereotypy of these behaviors. The more stereotyped or reflex-like the behavior pattern is, the more likely is its dependence on neuronal circuitry differentiated early in life. We can easily turn this hypothesis the other way round, posing that the less stereotyped the copulatory motor patterns are, the less they will depend on preestablished circuitry. Considering that primates have much less stereotyped copulatory motor patterns than rodents or carnivores, we should expect them to be less determined by early hormone actions. With regard to distant sexual incentives activating arbitrary approach behaviors, we have already seen that they are rather insensitive to early hormone actions in rodents. They may be completely independent of such actions in primates. These proposals could easily be submitted to a confrontation with reality but, as we soon will see, it is not that simple. While there are a number of studies that have evaluated the effects of early endocrine manipulations on copulatory behaviors in the rhesus monkey, there is no study analyzing preferences for a particular sex after perinatal hormone manipulations in primates. Nevertheless, we will now turn to a short examination of existing data on copulatory behaviors.

In the rhesus monkey, the androgen-producing activity of the fetal testes is initiated around day 40 of the about 170 days of gestation. It continues at peak level until around day 75, when it starts to decline. There is a second peak around day 140. Male fetuses have higher concentrations of testosterone than female fetuses, but there is no difference in dihydrotestosterone or androstenedione (Resko and Ellinwood, 1981; Resko, 1985). Androgen receptors are mainly localized to the fetal preoptic area/anterior hypothalamus, the amygdala and the cerebellum. There is no difference in receptor concentration between males and females, except in the amygdala where males have more receptors than females (Pomerantz and Sholl, 1987). These results were confirmed and extended in a study of androgen uptake in the fetal rhesus brain. Androgen uptake was largest in the preoptic area, followed by the amygdala. Interestingly, 75% of the androgen taken up into the preoptic area was aromatized to estradiol. In the amygdala, about 40% was aromatized (Michael *et al.*, 1989). It is believed that the fetal ovaries do not synthesize gonadal hormones. This means that the hormonal environment during the fetal development differs between males and females only with regard to the availability of testosterone and its metabolites. Thus, it can be supposed that any hormone dependent sexual differentiation of the brain is a consequence of direct or indirect testosterone actions. Postnatally, the male testis reduces androgen production on the day of birth, but thereafter it secretes large amounts of androgens for about 2–3 months.

Any study of the role of hormones in sexual differentiation in the rhesus monkey would either need to increase the concentration of testosterone in the circulation of female fetuses during the period of activity in the male fetal testis and during the postnatal period of androgen secretion or block the actions of androgens in the male fetus as well as postnatally during a corresponding period. This has, as far as I have been able to find out, not been done. There are, however, several reasonably good approximations.

In a couple of early studies, large doses of testosterone were administered to rhesus females between days 35 to 75 of pregnancy. It is likely that female fetal blood concentration of testosterone was in the normal range for males. Consequently, these females developed male external genitalia, with a penis of normal size for males and an empty scrotum. They have ovaries, but the cyclic function is severely disrupted. Moreover, they have no vaginal orifice. This kind of rhesus female is frequently called a pseudohermaphrodite. Some studies report that prepubertal pseudohermaphrodites show enhanced mounting of other females, an observation that has been interpreted as supporting the idea that testosterone has differentiated the brain in a male direction. Taking the unusual external genitalia of pseudohermaphrodites into account, it would be strange if they did not mount more than other females. The potentially rewarding and reinforcing sensory feedback obtained from the penis should assure that. In prepubertal, unmanipulated females the clitoris does probably not receive as much tactile stimulation during mounting as the pseudohermaphrodites' penis. Likewise, some changes in play behavior observed in pseudohermaphrodites can easily be attributed to their unusual physical appearance.

As an example of one of the early studies, I choose to summarize a report by Phoenix and Chambers (1982). Adult ovariectomized pseudohermaphrodites, castrated males and 'normal' ovariectomized females were treated either with testosterone or estradiol. They were then tested with estrogen-treated females. Whereas males increased mounting, intromission and ejaculation in response to testosterone, the females did not, independently of whether they had received prenatal testosterone or not. Treatment with estradiol had no significant effect on copulatory behavior in any group. These results suggest that prenatal androgen treatment of female rhesus monkeys does not modify their copulatory behavior in response to other females. Their brains do not seem to have been differentiated in a male direction, despite the fact that their external genitalia were manifestly masculinized. An interesting fact is that these same females, when tested as infants and juveniles, had shown male-like behavior patterns. The fact that they did not copulate with females when adults demonstrates that the early modifications of behavior were not a result of a definitely masculinized brain. The suggestion I made above, that these early changes were due to the altered external genitalia and perhaps other peripheral effects of fetal androgen exposure, seems to be strengthened by the Phoenix and Chamber (1982) study.

A rather similar experiment, but including a group of intact males for comparison, gave essentially the same results (Pomerantz *et al.*, 1986). Testosterone treatment of adult pseudohermaphrodites did not activate copulatory behavior in response to an estrogen-treated female. The behavior of the female partners was also analyzed. They displayed a high level of proceptive behaviors both in response to intact and castrated males and very few to the 'normal' females. Likewise, the pseudohermaphrodites induced few proceptive responses on the part of their female partners. In fact, they were not different from 'normal' females in that respect. Finally, while the partner females displayed high levels of receptivity to the intact males, they did so less frequently to the castrated males and less frequently still to the pseudohermaphrodites. These data suggest that prenatal androgen treatment has negligible effects on adult copulatory behavior. Furthermore, other monkeys perceive the pseudohermaphrodites as similar to 'normal' females and not to males. Again, we need to conclude that prenatal treatment with testosterone does not masculinize the brain. Moreover, it does not affect the sexual incentive properties of the treated individual.

An annoying problem with both studies summarized in the preceding paragraph was that they compared pseudohermaphrodites, 'normal' females and males during sexual interaction with females. The reason for doing that was obviously that the hypothesis being tested posed that prenatal androgen masculinizes the brain. After having discovered that it does not, another question that might be asked is whether prenatal androgen defeminizes the brain. With that is meant that the females might have lost their capacity to display typical female copulatory behaviors. The problem with this proposal is, according to the experts, that there is no typical female rhesus copulatory behavior. Primates do not show a lordosis posture like rodents or stand still like pigs or ewes when being mounted. Indeed, there does not seem to exist any sexually dimorphic copulatory behavior in primates. In contrast, some proceptive behaviors activated by the presence of a male are considered as sensitive indicators of the female's willingness to engage in copulatory interaction (Wallen, 1990). According to this proposal, a comparison of the frequency of proceptive behaviors in response to a male and in response to a female could provide some indication with regard to sexual preferences. If more proceptive behaviors were displayed in response to a male than to a female, it could be concluded that the male was the preferred partner. In the opposite case, it could be concluded that the female was the preferred partner. If this latter occurred in a pseudohermaphrodite, it could be inferred that the brain had been defeminized by the androgen treatment.

A very neat experiment analyzed both female proceptive behaviors in response to a tethered male and the male's responses to the females (Pomerantz *et al.*, 1985). This allowed for a simultaneous evaluation of the male's sexual incentive properties for the females and the females' sexual incentive properties for the males. The mothers of one group of females had been treated with testosterone from day 42 until day 96 or 97 of gestation, while mothers of another group were given dihydrotestosterone from day 42 until sometime between day 97 and 102 of gestation. A group of unmanipulated females as well as a group of castrated males were also included. All females were implanted with an estradiol-containing Silastic® capsule assuring blood concentrations close to those found in cycling monkeys around the time of ovulation. Behaviors like female approach and solicitations per unit time were much reduced by prenatal androgens. The number of presentations as well as receptivity (acceptance of the male's invitations to copulate) did not differ between control and neonatally androgenized females. Thus, it appears that the male evoked less intense approach behaviors in the androgenized females than in controls. When looking at the male's behavior, we find a similar pattern. The male's frequency of contacts as well as responses to presentations were lower when exposed to an androgenized female than when exposed to a control female. His mount frequency was also much lower with androgenized females. These data clearly establish that neonatally androgenized females evoke less intense approach behavior in males than control females. The authors conclude two things:

1 since testosterone and dihydrotestosterone had similar effects, aromatization can be of no importance
2 prenatal androgenization leads to a defeminized behavior in adulthood.

I have no problems with the first conclusion. However, the second seems more problematic. I will immediately explain why. The androgenized females had male external genitalia with a penis similar to that found in neonatally castrated males and with scrota of variable size. Vaginal orifices were obliterated (Pomerantz *et al.*, 1985). In the section on the role of olfactory stimuli as sexual incentives in Chapter 3, I quite extensively treated the role of vaginal secretions for the attractiveness of female rhesus monkeys. The conclusion was that such secretions strongly influenced a male's behavior toward the female. Without vaginal openings and a functional vaginal epithelium, it seems difficult for the androgenized females to produce these secretions and to get them out to the exterior of their body. It is, then, not in the slightest surprising that they are less attractive than unmanipulated females. Furthermore, female proceptive behaviors are never displayed in the absence of a male. This means that stimuli emitted by the male are the determinant of the occurrence or non-occurrence of these behaviors. If the male seeks less contact and responds less to the female's presentations because of her lack of attractivity, it can immediately be predicted that the female's responses in the form of proceptive behaviors will also be less frequent. Sexual behaviors are always a result of an interaction and, if one individual for some reason modifies her behavior, then the other will also modify his. My proposal is that all the effects seen in the Pomerantz *et al.* (1985) experiment can be attributed to peripheral actions of the prenatal androgens and that the results do not in any way provide evidence for an androgen-dependent sexual differentiation of the brain. Nevertheless, a subsequent study employing a slightly different procedure replicated the observation that androgenized females displayed less proceptive behaviors (Thornton and Goy, 1986). Again, I propose that the cause is to be found in the peripheral actions of androgen.

The argument for a peripheral explanation of the effects of neonatal androgenization that I proposed in the preceding paragraph is not new. It appeared almost immediately upon the publication of the Pomerantz *et al.* (1985) and Thornton and Goy (1986) papers. In an attempt to address this criticism, a study was performed where some female rhesus monkeys received prenatal androgens starting much later than in the earlier studies. The duration of treatment was also reduced, so that one group of pregnant mothers was treated between days 115 and 139 and another between days 40 and 64. The late treatment does not masculinize the external genitalia. Juvenile behaviors and the behavior of the mothers towards their offspring were evaluated and few differences were found between females exposed to androgens early and females exposed late in prenatal life (Goy *et al.*, 1988). These data are completely irrelevant, since the crucial variables of attractiveness, proceptive behaviors and copulatory behaviors displayed in adulthood, are missing. In addition, the fact that external genitalia were not *masculinized* does not necessarily mean that other peripheral consequences of androgen treatment, like an enlarged clitoris, could have had behavioral consequences. Play behavior, for example, is probably not entirely determined by the genitals. Muscular strength, running speed and many other potentially androgen-dependent peripheral phenomena may also have a say.

As I might have mentioned already, monkeys are peculiar in many ways. One of the most salient is that females have attracted more attention than males, at least with regard to studies looking upon the effects of early endocrine manipulations of sexual differentiation. Perhaps the dominantly male scientific community is more attracted to female primates than to females of any other mammalian order. Independently of this speculative proposal, we are in a situation where data on male rhesus monkeys are extremely scarce. I know of only a few studies where the actions of prenatal androgens have been blocked. In one, the androgen receptor antagonist flutamide was administered to pregnant females either between days 35 and 70 or between days 110 and 145. The group receiving flutamide early developed poorly masculinized external genitalia, while those receiving the antagonist late appeared 'normal'. Nevertheless, the deandrogenized males mounted receptive females as much as control males did. Because of the poor shape of the penis, the early deandrogenized males made fewer intromissions and achieved less ejaculations than the other groups. Nevertheless, the conclusion was that blockade of androgen receptors during prenatal life does not modify the sexual differentiation of the brain, at least not with regard to copulatory behaviors (Wallen, 2005). It would seem, then, that males are similar to females in the way the sexual differentiation of their brain is largely independent of prenatal gonadal hormones. There is also an experiment in which male rhesus monkeys were treated with a GnRH agonist from birth until they were 4 months old (Eisler *et al.*, 1993). This treatment efficiently suppresses pituitary LH release and testicular androgen release. When adult, these males were tested for copulatory behavior with receptive females. There was no difference between GnRH treated males and controls when tested in the breeding season. For some reason, there was a difference in the non-breeding season. This latter observation seems quite absurd, since there is no reason to make the poor males copulate in a season where they normally do not. Although the authors concluded that they had found evidence for a postnatal, organizing function of androgens, I must say that this evidence is slight indeed as far as copulatory behavior is concerned.

There is something absent from the monkey literature and that is the aspect of choice. Despite a substantial investment of time and effort, I have not been able to find a singly study where androgenized females, or deandrogenized males for that, have been presented with a choice between a male and a female partner. We saw that procedures of this kind have not been legion in rodent studies, but at least we have a few. In primates, none. An inevitable consequence of this lack of data is that we do not know if prenatal hormones, in some way or another, affect the choice of sexual partner. It is extremely risky to make predictions of any kind in the absence of data, so I should refrain from doing so. I cannot avoid, though, expressing my dissatisfaction with the progress within this field of research. Study after study has reported effects or lack of such on copulatory behaviors in situations without choice. It seems as if scientists in some way became obsessed with the execution of copulatory acts with an imposed, rather than chosen, partner.

I have reviewed only a fraction of the rather extensive literature on pseudohermaphrodites. The comparative perspective has almost got lost, since I have only mentioned data from the rhesus monkey. This was not a choice of mine but a consequence of the fact that the rhesus has been the most popular species. There is very little information from other species and nothing that would add something substantial to the picture I have drawn here. I will now summarize the discussion of primates in a rather radical suggestion.

My personal conviction is that there is no hormone-dependent sexual differentiation of the monkey brain with regard to copulatory behavior. These behaviors are not sexually dimorphic in any strict sense of the word, so there is no reason why they should be differentiated. Whatever differences there may be between female and male behavior patterns, they can be attributed to differences in external genitalia. The modes by which sexual reward, caused by tactile stimulation of the genitals, is obtained depend on the form and localization of these genitals. The obtainment of tactile stimulation of an erect penis requires motor patterns that are quite different from those required to obtain tactile stimulation of a clitoris and adjacent tissue. Since sufficiently intense genital stimulation is both rewarding and reinforcing, monkeys will learn which motor patterns lead to such stimulation through a process of operant conditioning. Since many motor patterns can cause genital stimulation, many motor patterns will be learned. This extraordinary simple explanation for the acquisition of copulatory motor patterns explains, without any problem whatsoever, both the variability seen in primate copulatory motor patterns and the differences we might observe between sexes. The function of the hormones during fetal and early postnatal life with regard to future sex behavior is, according to this reasoning, merely to ensure proper differentiation of the external genitals. No action within the central nervous system is required.

We know that androgen receptors are located in many regions of the fetal brain and we know that they are already functional in the fetus. Those believing that nature is perfect and, at the same time fond of teleological explanations, would immediately insist upon an obligatory function. I do not agree. The receptors will be useful later in life, anyway. They are needed for the activation of sexual behaviors in adulthood. If no hormones act on the central motive state, then there will be no reaction to sexual incentives. Furthermore, we know that aromatase is active during the prenatal period and we know that a large proportion of the testosterone entering the fetal preoptic area is aromatized. Yet we know that aromatization is not necessary for androgen actions on sexual behavior in the rhesus monkey. Both the androgen receptors and aromatase may have important functions in fetal life unrelated to sexual behaviors. Perhaps the brain has something to say about functions other than sex.

2.4 Endocrine events during puberty and their possible importance for sexual preferences

Sexual differentiation has for a long time been considered as an event taking place before, around or shortly after birth. During the last few years, it has become apparent that this notion is overly simplified. Something must happen in the brain sometime between the late postnatal period and the end of puberty. For example, sexual behavior cannot be activated to adult levels in prebubertal male hamsters, rats or ferrets even when treated with large doses of androgens. Similarly, absence of gonadal hormones during puberty leads to a reduced responsiveness to these hormones when adult. Male hamsters deprived of androgens during puberty mount, intromit and ejaculate less in a test for copulatory behavior than hamsters exposed to physiological concentrations of these hormones. Not only do pubertal androgens enhance male-typical sexual behaviors in response to androgen treatment in adulthood, but they also reduce the expression of female-typical behaviors. Again, male hamsters not exposed to androgens during puberty display the lordosis response as easily as females. On the contrary, hamsters exposed to androgens during puberty are less likely to respond with lordosis (reviewed in Sisk and Zehr, 2005; Schulz and Sisk, 2006). All these observations strongly suggest that, in male hamsters, pubertal androgens facilitate the occurrence of male-typical responses while making the display of female-typical responses less likely. About the same conclusion was reached with regard to the consequences of perinatal hormone treatments. There are no data concerning pubertal hormone effects on the choice of sexual partner. As always, scientists are far more interested in the probability of display of copulatory reflexes than in factors determining the choice of partner. Nevertheless, we should not forget that gonadal hormone actions on the brain during puberty may have consequences for adult sexual behaviors including the sex preferred for displaying these behaviors with. The field of pubertal hormone actions and subsequent adult sexual behaviors is very young. I predict that many interesting findings will emerge in this field over the years to come.

Summary

1 Preference without a choice is impossible. Whenever two or more incentives are present simultaneously, the one with the largest incentive value will cause more approach behaviors than others. We say that it is the preferred incentive.

2 Copulatory motor patterns are difficult to classify in terms of homosexual or heterosexual. A mount is a mount, regardless of the sex of the mountee. It is suggested that terms like 'homosexual behavior' should be abandoned.

3 Few sexual behaviors are sexually dimorphic. A mount displayed by a male is very similar to a mount displayed by a female and a lordosis presented by a female is indistinguishable from that presented by a male. The only sexually dimorphic behavior is that of inserter. The inserter, or penetrator, is always a male.

4. In contrast to the unique *real sexual dimorphism*, meaning that a behavior pattern exists in one sex but not in the other, there are many examples of *statistical dimorphisms*, meaning that the frequency of a particular behavior is higher in one sex than in another. The interest of this kind of dimorphisms is variable.

5. Sexual preference is an expression that normally refers to the preferred sex of the partner. However, it could just as well refer to the motor pattern preferred for expressing copulatory responses.

6 There are not many examples of same-sex preferences among non-human mammals. Indeed, such preferences have been firmly established only in one subpopulation of rams and in one primate species.

7 Explanations of the same-sex preference in terms of alterations in the endocrine system have failed and there is no convincing evidence for a peculiar structure of the brain in rams or monkeys preferring to have sex with members of their own sex.

8 It appears that learning can explain same-sex preferences in non-human mammals.

9 Endocrine manipulations during the period of sexual differentiation of the brain in rodents have reliable consequences for the probability of display of copulatory reflexes. Perinatal gonadal hormones reduce the likelihood of display of lordosis and enhance the likelihood of mount. The effects on sexual preferences are unclear, but some data suggest that adult gonadal hormones determine this preference. Estrogens given in adulthood cause preference for males, while androgens given to adult animals cause preference for females.

10 Pseudohermaphroditic female rhesus monkeys do not show clear-cut alterations in their copulatory behaviors and sexual preferences that cannot be attributed to peripheral effects of early hormone treatments.

11 Gonadal hormones may be important for the differentiation of neural circuitry involved in copulatory reflexes like mount and lordosis. More complex sexual behaviors, like the arbitrary motor patterns used for approaching a partner, do not seem to be much affected by hormones secreted early in life.

12 In primates, the main role of gonadal hormones is to assure appropriate reactivity of the central motive state in response to sexual incentives.

References

Adkins-Regan, E., Mansukhani, V., Thompson, R. and Yang, S. (1997). Organizational actions of sex hormones on sexual partner preference. *Brain Res Bull* **44**, 497–502

Afonso, V.M. and Pfaus, J.G. (2006). Hormonal and experiential control of female – male mounting in the female rat. *Horm Behav* **49**, 30–37

Afonso, V.M., Bablekis, V. and Pfaus, J.G. (2006a). Sensory mediation of female – male mounting in the rat: II. Role of tactile and conspecific cues. *Physiol Behav* **87**, 863–869

Afonso, V.M., Woehrling, A. and Pfaus, J.G. (2006b). Sensory mediation of female – male mounting in the rat: I. Role of olfactory cues. *Physiol Behav* **87**, 857–862

Alexander, B.M., Rose, J.D., Stellflug, J.N., Fitzgerald, J.A. and Moss, G.E. (2001a). Fos-like immunoreactivity in brain regions of domestic rams following exposure to rams or ewes. *Physiol Behav* **73**, 75–80

Alexander, B.M., Rose, J.D., Stellflug, J.N., Fitzgerald, J.A. and Moss, G.E. (2001b). Low-sexually performing rams but not male-oriented rams can be discriminated by cell size in the amygdala and preoptic area: a morphometric study. *Behav Brain Res* **119**, 15–21

Bakker, J., van Ophemert, J. and Slob, A.K. (1996). Sexual differentiation of odor and partner preference in the rat. *Physiol Behav* **60**, 489–494

Baum, M.J., Erskine, M.S., Kornberg, E. and Weaver, C.E. (1990). Prenatal and neonatal testosterone exposure interact to affect differentiation of sexual behavior and partner preference in female ferrets. *Behav Neurosci* **104**, 183–198

Beach, F.A. (1979). Animal models for human sexuality. In *Sex hormones and behaviour*, (CIBA Foundation Symposium, new series, 62), pp. 113–143. Excerpta Medica: Amsterdam

Bélanger, B., Bélanger, A., Labrie, F., Dupont, A., Cusan, L. and Monfette, G. (1989). Comparison of residual C-19 steroids in plasma and prostatic tissue of human, rat and guinea-pig after castration: unique importance of extratesticular androgens in men. *J Steroid Biochem* **32**, 695–698

Byne, W. (1998). The medial preoptic and anterior hypothalamic regions of the rhesus monkey: cytoarchitectonic comparison with the human and evidence for sexual dimorphism. *Brain Res* **793**, 346–350

Crichton, J.S., Lishman, A.W., Hundley, M. and Amies, C. (1991). Role of dihydrotestosterone in the control of sexual behavior of castrated male sheep. *J Reprod Fert* **93**, 9–17

de Jonge, F.H., Muntjewerff, J.W., Louwerse, A.L., and van de Poll, N. (1988). Sexual behavior and sexual orientation of the female rat after hormonal treatment during various stages of development. *Horm Behav* **22**, 100–115

DomInguez-Salazar, E., Portillo, W., Baum, M.J., Bakker, J. and Paredes, R.G. (2002). Effect of prenatal androgen receptor antagonist or aromatase inhibitor on sexual behavior, partner preference and neuronal Fos responses to estrous female odors in the rat accessory olfactory system. *Physiol Behav* **75**, 346

Eisler, J.A., Tannenbaum, P.L., Mann, D.R. and Wallen, K. (1993). Neonatal testicular suppression with a GnRH agonist in rhesus monkeys: effects on adult endocrine function and behavior. *Horm Behav* **27**, 551–567

Feder, H.H. and Whalen, R.E. (1965). Feminine behavior in neonatally castrated and estrogen-treated male rats. *Science* **147**, 306–307

Goy, R.W., Bercovitch, F.B. and McBrair, M.C. (1988). Behavioral masculinization is independent of genital masculinization in prenatally androgenized female rhesus macaques. *Horm Behav* **22**, 552–571

Habert, R. and Picon, R. (1984). Testosterone, dihydrotestosterone, and estradiol-17β levels in maternal and fetal plasma and in fetal testes in the rat. *J Steroid Biochem* **21**, 193–198

Hull, E.M., Meisel, R.L. and Sachs, B.D. (2002). Male sexual behavior. In *Hormones, brain and behavior*, vol. 1 (D.W. Pfaff, A.P. Arnold, A.M. Etgen, S.E. Fahrbach and R.T. Rubin, eds), pp. 3–137. Academic Press: New York

Kilgour, R., Skarsholt, B.H., Smith, J.F., Bremner, K.J. and Morrison, M.C.L. (1977). Observations on the behavior and factors influencing the sexually active group in cattle. *Proc New Zealand Soc Anim Prod* **37**, 128–135

Kinsey, A.C., Pomeroy, W.B. and Martin, C.E. (1948). *Sexual behavior in the human male*. Saunders: Philadelphia

Kinsey, A.C., Pomeroy, W.B., Martin, C.E. and Gebhard, P.H. (1953). *Sexual behavior in the human female*. Saunders: Philadelphia

Laumann, E.O., Gagnon, J.H., Michael, R.T. and Michaels, S. (1994). *The social organization of sexuality. Sexual practices in the United States*. University of Chicago Press: Chicago

Matuszczyk, J. (2003). Enhanced feminine sexual behavior and infertility in female rats prenatally treated with an antiestrogen. *Scand J Psychol* **44**, 251–256

Matuszczyk, J.V., Fernández-Guasti, A. and Larsson, K. (1988). Sexual orientation, proceptivity, and receptivity in the male rat as a function of neonatal hormonal manipulation. *Horm Behav* **22**, 362–378

Matuszczyk, J.V. and Larsson, K. (1995). Sexual preference and feminine and masculine sexual behavior of male rats prenatally exposed to antiandrogen or antiestrogen. *Horm Behav* **29**, 191–206

Merkx, J. (1984a). Effect of castration and subsequent substitution with testosterone, dihydrotestosterone and oestradiol on sexual preference behaviour in the male rat. *Behav Brain Res* **11**, 59–65

Merkx, J. (1984b). Effects of neonatal testicular hormones on preference behaviour in the rat. *Behav Brain Res* **12**, 1–7

Merkx, J., Slob, A.K., and van der Werff ten Bosch, J.J. (1989). Preference for an estrous female over a non-estrous female evinced by female rats requires dihydrotestosterone plus estradiol. *Horm Behav* **23**, 466–472

Meyerson, B.J., Eliasson, M. and Hetta, J. (1979). Sex-specific orientation in female and male rats: development and effects of early endocrine manipulation. *Adv Biosci* **25**, 451–460

Michael, R.P., Bonsall, R.W. and Rees, H.D. (1989). The uptake of [^{3}H]testosterone and its metabolites by the brain and pituitary gland of the fetal macaque. *Endocrinology* **124**, 1319–1326

Parrott, R.F. (1978). Courtship and copulation in prepubertally castrated male sheep (wethers) treated with 17β-estradiol, aromatizable androgens, or dihydrotestosterone. *Horm Behav* **11**, 20–27

Parrott, R.F. and Baldwin, B.A. (1984). Sexual and aggressive behaviour of castrated male sheep after injection of gonadal steroids and implantation of androgen in the hypothalamus: a preliminary study. *Theriogenology* **21**, 533–542

Perkins, A., Fitzgerald, J.A. and Moss, G.E. (1995). A comparison of LH secretion and brain estradiol receptors in heterosexual and homosexual rams and female sheep. *Horm Behav* **29**, 31–41

Phoenix, C.H. and Chambers, K.C. (1982). Sexual behavior in adult gonadectomized female pseudohermaphrodite, female, and male rhesus macaques (*Macaca mulatta*) treated with estradiol benzoate and testosterone propionate. *J Comp Physiol Psychol* **96**, 823–833

Phoenix, C.H., Goy, R.W., Gerall, A.A. and Young, W.C. (1959). Organizing action of prenatally administered testosterone propionate on the tissues mediating mating behavior in the female guinea pig. *Endocrinology* **65**, 369–382

Pinckard, K.L., Stellflug, J. and Stormshak, F. (2000). Influence of castration and estrogen replacement on sexual behavior of female-oriented, male-oriented, and asexual rams. *J Anim Sci* **78**, 1947–1953

Pomerantz, S.M., Goy, R.W. and Roy, M.M. (1986). Expression of male-typical behavior in adult female pseudohermaphroditic rhesus: comparisons with normal males and neonatally gonadectomized males and females. *Horm Behav* **20**, 483–500

Pomerantz, S.M., Roy, M.M., Thornton, J.E. and Goy, R.W. (1985). Expression of adult female patterns of sexual behavior by male, female, and pseudohermaphroditic female rhesus monkeys. *Biol Reprod* **33**, 878–889

Pomerantz, S.M. and Sholl, S.A. (1987). Analysis of sex and regional differences in androgen receptors in fetal rhesus monkey brain. *Dev Brain Res* **36**, 151–154

Portillo, W. and Paredes, R.G. (2004). Sexual incentive motivation, olfactory preference, and activation of the vomeronasal projection pathway by sexually relevant cues in non-copulating and naive male rats. *Horm Behav* **46**, 330–340

Price, E.O., Katz, L.S., Wallach, S.J.R. and Zenchack, J.J. (1988). The relationship of male-male mounting to the sexual preferences of young rams. *Appl Anim Behav Sci* **21**, 347–355

Resko, J.A. (1985). Gonadal hormones during sexual differentiation in vertebrates. In *Handbook of Behavioral Neurobiology, vol. 7, Reproduction* (N. Adler, D.W. Pfaff and R.W. Goy, eds), pp. 21–42. Plenum: New York

Resko, J.A. and Ellinwood, W. (1981). Testicular hormone production in fetal rhesus macaques. In *Fetal Endocrinology* (M.J. Novy and J.A. Resko, eds), pp. 253–267. Academic Press: New York

Resko, J.A., Perkins, A., Roselli, C.E., Fitzgerald, J.A., Choate, J.V.A. and Stormshak, F. (1996). Endocrine correlates of partner preference behavior in rams. *Biol Reprod* **55**, 120–126

Roselli, C.E., Larkin, K., Resko, J.A., Stellflug, J.N. and Stormshak, F. (2004a). The volume of a sexually dimorphic nucleus in the ovine medial preoptic area/anterior hypothalamus varies with sexual partner preference. *Endocrinology* **145**, 478–483

Roselli, C.E., Larkin, K., Schrunk, J.M. and Stormshak, F. (2004b). Sexual preference, hypothalamic morphology and aromatase in rams. *Physiol Behav* **83**, 233–245

Roselli, C.E., Resko, J.A. and Stormshak, F. (2003). Estrogen synthesis in fetal sheep brain: effect of maternal treatment with an aromatase inhibitor. *Biol Reprod* **68**, 370–374

Roselli, C.E., Schrunk, J.M., Stormshak, F., Resko, J.A. and Stellflug, J.N. (2002). Reproductive development and behavior of lambs exposed in utero to an aromatase inhibitor during a critical period for sexual differentiation. *Biol Reprod* **66**, Suppl. 1, 302

Schulz, K.M. and Sisk, C.L. (2006). Pubertal hormones, the adolescent brain, and the maturation of social behaviors: lessons from the syrian hamster. *Mol Cell Endocrinol* **254/255**, 120–126

Sisk, C.L. and Zehr, J.L. (2005). Pubertal hormones organize the adolescent brain and behavior. *Front Neuroendocrinol* **26**, 163–174

Skinner, B.F. (1935). The generic nature of the concepts of stimulus and response. *J Gen Psychol* **12**, 40–65

Skinner, B.F. (1969). *Contingencies of reinforcement: a theoretical analysis.* Appleton Century Crofts: New York

Stellflug, J., Cockett, N.E. and Lewis, G.S. (2006). Relationship between sexual behavior classifications of rams and lambs sired in a competitive breeding environment. *J Anim Sci* **84**, 463–468

Stockman, E.R., Callaghan, R.S. and Baum, M.J. (1985). Effects of neonatal castration and testosterone treatment on sexual partner preference in the ferret. *Physiol Behav* **34**, 409–414

Thornton, J. and Goy, R.W. (1986). Female-typical sexual behavior of rhesus and defeminization by androgens given prenatally. *Horm Behav* **20**, 129–147

van Weerden, W.M., Bierings, H.G., van Steenbrugge, G.J., de Jong, F.H. and Schröder, F.H. (1992). Adrenal glands of mouse and rat do not synthesize androgens. *Life Sci* **50**, 857–861

Vasey, P.L. (1998). Female choice and inter-sexual competition for female sexual partners in Japanese macaques. *Behaviour* **135**, 579–597

Vasey, P.L. (2002a). Same-sex sexual partner preference in hormonally and neurologically unmanipulated animals. *Ann Rev Sex Res* **13**, 141–179

Vasey, P.L. (2002b). Sexual partner preference in female Japanese macaques. *Arch Sex Behav* **31**, 51–62

Vasey, P.L., Foroud, A., Duckworth, N. and Kovacovsky, S.D. (2006). Male-female and female-female mounting in Japanese macaques: a comparative study of posture and movement. *Arch Sex Behav* **35**, 117–129

Vasey, P.L. and Pfaus, J.G. (2005). A sexually dimorphic hypothalamic nucleus in a macaque species with frequent female-female mounting and same-sex sexual partner preference. *Behav Brain Res* **157**, 265–272

Wallen, K. (1990). Desire and ability: hormones and the regulation of female sexual behavior. *Neurosci Biobehav Rev* **14**, 233–241

Wallen, K. (2005). Hormonal influences on sexually differentiated behavior in nonhuman primates. *Front Neuroendocrinol* **26**, 7–26

Wolfe, L.D. (1983). Japanese macaque female sexual behavior: a comparison of Arashiyama East and West. In *Female primates: Studies by women primatologists* (M.F. Small, ed.), pp. 141–157. Alan R. Liss: New York

Zenchak, J.J. and Anderson, G.C. (1980). Sexual performance levels of rams (*Ovis aries*) as affected by social experience during rearing. *J Anim Sci* **50**, 167–174

Zenchak, J.J., Anderson, G.C. and Schein, M.W. (1981). Sexual partner preference of adult rams (*Ovis aries*) as affected by social experiences during rearing. *App Anim Ethol* **7**, 157–167

CHAPTER

9

Preference for same sex partners in the human

1 The search for a biological foundation for preferences for the same sex

When we talk of sexual preferences or sexual orientation in the human, we employ a kind of euphemism for homosexuality. I will discuss both these expressions quite extensively in this chapter. For the sake of simplicity, I will also employ the old-fashioned terms 'homosexuality' and 'homosexual'. These notions are by themselves problematic in many ways. They have been discussed and analyzed by all kinds of people, including cardinals, Baptist ministers, politicians of all inclinations, judges of the Supreme Court of the United States of America and similar institutions in many other countries, gay and lesbian activists, historians, sociologists, philosophers and even by a few neuroscientists. At the root of many of these analyses are moral judgments, outright prejudice, political opportunism, academic career building, good faith and much more. Instead of entering the conceptual discussion right away, I will remain on a very low level of abstraction and summarize the research that has been done with the purpose of finding a neurobiological explanation for homosexuality.

All behavior is caused by activities in the nervous system. When a male rat approaches a sexually receptive female, for example, we know that he does so partly because of olfactory input. The firing of neurons in the main and/or olfactory accessory bulbs somehow contributes to activate motor patterns that reduce the distance between the male and the female. Likewise, when a human male chooses to approach a woman for executing sexual responses, he does so because of activities in the central nervous system. If he chose to approach another man instead, that would also be caused by activities within the nervous system. In some way or another, stimuli emitted by the sexual incentive, regardless of whether it is a man or a woman, modify the firing rate of neurons somewhere in the brain, and the result will be approach. We could call the nervous processes underlying the approach behavior its biological cause or biological basis. In the search for a biological basis for approach to individuals of one's own sex, the term 'biological basis' does not have that meaning. Instead of referring to a set of nervous responses to particular stimuli, it refers to constant predispositions of the nervous system causing it to respond more to stimuli emitted by one's own sex than to stimuli emitted by the opposite sex. These constant predispositions may be caused by genes exclusive to people responding to the own sex more than to the opposite, or by endocrine events during brain differentiation, again exclusive to those responding to individuals of the own sex, or by particular blood concentrations of gonadal hormones in adulthood. There may certainly be other factors causing a permanent predisposition for responding to sexual incentive stimuli emitted by members of the own sex, but the three mentioned here are those that have attracted most attention.

1.1 The homosexual gene

Proposals concerning a biological basis for homosexuality were put forth by some physicians during the latter half of the 19th century. It was even thought that this particular condition was congenital, caused by some inherited element much in the same way as eye color was inherited. Among those favoring this view was the famous von Krafft-Ebing (1886). The difference between inherited eye color, for example blue, and inherited homosexuality was that the former was normal, while the latter was pathological. For many years, scientists searched for endocrine imbalances or other signs of abnormalities that might explain the disease of homosexuality, but success was limited. It is now firmly established that there are no endocrine differences between adult homosexuals and heterosexuals. I will not review the abundant literature on the subject, but if someone is curious I can recommend any biologically oriented treatise on homosexuality. In view of all the negative data, an explanation of homosexuality in terms of adult endocrine alterations is now untenable. However, speculations concerning hormonal alterations during the period of sexual differentiation of the brain still continue to be advanced, and such alterations are occasionally suggested as cause for the disease. In addition, some scientists have tried and are still trying to find proof of the hereditary component proposed during the 19th century, with variable success (see e.g. Kallman, 1952). The search for a hereditary component has been a tradition with other diseases, like schizophrenia and depression, so it is not surprising that similar studies also have been and are performed on homosexuals.

Homosexuality is no longer considered a disease. The young generation may even ignore that homosexuality was a disease until 1973 in the USA and until 1979 in Sweden. I do not know when it was removed from the disease list in other countries, but it was probably around the same time, that is during the 1970s. Despite the removal of homosexuality from the disease list, search for a biological explanation has continued to be pursued in much the same way as it was during the disease period. There have only been some discursive modifications, like replacing the term 'homosexuality' with the expression 'sexual orientation', and the term 'disease' with the term 'condition'. The fact that research on the causes of homosexuality is still pursued is most curious. Search for the causes of a disease has always the explicit or implicit rationale that if a cause is found, then a cure can be proposed. This was certainly also the rationale for most of the homosexuality research performed at the time when it was considered a disease. However, when the disease label was removed, meaning that homosexuality came to be considered as nothing more than a normal variant of human sexual behaviors, then there is no need for a cure. In this condition, finding out the biological basis of homosexuality would be no more interesting than finding out the biological cause for some people's preference for having sex with a fat partner as opposed to those preferring a thin partner, or for preferring a blond partner rather than a dark. Personally, I cannot feel much enthusiasm for the search for a biological cause for homosexuality. Nevertheless, any discussion of this issue would be incomplete without mentioning at least some of the research that has been done. Therefore, we will now look at some of the research performed after that the disease label was removed. I will begin with the search for a hereditary component.

One of the first studies addressing the issue of heritability of homosexuality in the non-disease era comes from a group at the Psychology Department of the Northwestern University in Illinois. They employed a rather simple procedure. Groups of 'predominantly homosexual' and 'predominantly heterosexual' men were recruited and the incidence of homosexuality among the siblings of the homosexual men was compared to that among the siblings of the heterosexual men. It was found that homosexual men had about four times as many homosexual brothers as would be predicted from the national average of prevalence of homosexuality. In the heterosexual group, the number of homosexual brothers was close to the expected number, based on national prevalence estimates. It was concluded that there is a significant familial component to male homosexuality (Pillard and Weinrich, 1986). This conclusion was wise, because the data do not demonstrate that there is a hereditary component. The siblings shared at least part of the childhood environment, in addition to having some common genes. In a more sophisticated study from the same group, the favorite subjects of behavioral genetics were employed: twins and adoptive siblings. It turned out that if one monozygotic twin was homosexual, the likelihood for the other to be the same was 52%. The corresponding figure for dizygotic twins was 22% and for adoptive siblings it was only 11%. Curiously enough, for real siblings the figure (9.2%) was lower than for the adoptive siblings. Now it was concluded that there is some component of heritability in homosexuality (Bailey and Pillard, 1991). The indefatiguable Northwestern University group has also directed its attention toward female homosexuals. If a woman is homosexual, then the likelihood for a monozygotic twin to also be homosexual is 48%, for a dizygotic twin it is 16%, and for adoptive sisters it is 6%. The non-twin, real sisters of the homosexual women had a likelihood of 14% for being homosexual. Again it was concluded that heritability is a significant contributing factor in homosexuality (Bailey *et al.*, 1993). A more recent study from another research group, reported quite similar results (Kendler *et al.*, 2000).

The data from the twin studies are in line with those reported for many other conditions and are neither particularly surprising nor particularly exciting. A study employing much more sophisticated genetic techniques, however, found its way to the headlines in newspapers throughout the Western world. The pedigree of male homosexuals was analyzed and it was found that the incidence of homosexuality among brothers was 13.5%, far above the estimated population incidence, which in this study was found to be 2%. Among other relatives, only maternal uncles and sons of maternal aunts (I think they would be called cousins in a more habitual language) displayed an incidence of homosexuality above the expected value, but below that found in brothers. After convoluted genetic reasoning, it was concluded that a tendency to homosexuality was inherited and that the heritable element was located on the X chromosome. That would explain why maternal uncles and sons of maternal aunts had an unexpectedly high incidence of homosexuality, while no such incidence was observed in fathers of homosexuals. Having concluded that the inheritable element was located on the X chromosome, it proceeded with an elegant analysis of markers spanning this chromosome. A linkage between homosexuality and markers in the distal portion of Xq28, the subtelomeric region of the long arm of the X chromosome, was found (Hamer *et al.*, 1993). The statistical procedures employed in the establishment of the linkage are quite complex and it is far beyond my capacity to judge whether they are adequate, reliable and trustworthy. Others, however, have that capacity. It first turns out that it is far from certain that the putative inheritable element is located on the X chromosome. The data supporting it are compatible with a range of alternative explanations. Furthermore, the statistical significance of the linkage between X chromosome markers and homosexuality is based on a series of unproven assumptions. The most doubtful of them is considered to be the supposed existence of a major gene accounting for homosexuality. This supposition is basic for the statistical model used. It is argued that a major gene underlying such a common trait as homosexuality would be subjected to strong selective pressure against it, making its existence unlikely (Risch *et al.*, 1993).

In a posterior study, the supposition of an X chromosome-linked transmission was dropped and markers were distributed all over the genome. The linkage between the Xq28 region and homosexuality was not replicated in the new sample. In fact, no significant linkage between any marker and homosexuality was found (Mustanski *et al.*, 2005). A failure to find any association between the Xq28 region and homosexuality had already been reported in a Canadian sample (Rice *et al.*, 1999). There is no doubt that the sum of existing data suggest that there is no genetic difference between homosexuals and heterosexuals. This conclusion clearly contradicts the data from the twin studies, which quite consistently suggested an inheritable element in homosexuality. However, speculations on heritability based on data from twin studies are far less convincing than a direct analysis of the human genome. Calculations of heritability of a trait is a pure statistical exercise in the spirit of Sir Francis Galton (1869), without any necessary correspondence to the transmission of genes. In fact, we do not inherit behavioral traits but the structure of proteins. I wonder which protein structure corresponds to heterosexuality.

1.2 The homosexual brain

Having concluded that there is no convincing reason for imagining that homosexuality is determined by our genes, we need to look for other explanations for some people's habit of having sex with individuals of their own sex. One possible cause for that habit is differences in brain structure. At least some neuroscientists believe so, although I have never been able to understand why a difference in size of some brain structure should cause people to prefer their own sex for sex. Anyway, a difference in size of the suprachiasmatic nucleus has been reported between homosexual and heterosexual men. The rationale for choosing to compare the size of the suprachiasmatic nucleus in the brains of these groups of men was curious indeed. It is generally accepted that this nucleus is important for biological rhythms of many kinds. Rhythms are often disturbed in people with Alzheimer's disease. Thus, it was not too far-fetched to imagine that dead Alzheimer patients may have alterations in the suprachiasmatic nucleus. This turned out to be the case. These patients had a smaller nucleus than people who died without having shown symptoms of Alzheimer's disease. However, those dead Alzheimer patients who had suffered from AIDS while alive did not have a smaller suprachiasmatic nucleus. Supposing that people dying from AIDS generally are homosexuals, it was reasoned that homosexuality may be associated with an enlarged suprachiasmatic nucleus. The initially big suprachiasmatic nucleus of a homosexual would attain normal size when struck by Alzheimer's disease. This hypothesis was tested by comparing the volume, the number of cells and the number of vasopressin containing cells in the suprachiasmatic nucleus in heterosexual and homosexual AIDS victims. In addition, the volume and number of cells in the sexually dimorphic nucleus of the preoptic area were compared between homo- and heterosexual men. This latter is quite understandable, because we all know that the preoptic area is crucial for sexual behaviors. Results showed that the suprachiasmatic nucleus had a larger volume, contained more cells and had more vasopressin containing cells in homosexuals than in heterosexuals. No difference was found in the sexually dimorphic nucleus in the preoptic area. It was concluded that the functional implications of the difference in size of the suprachiasmatic nucleus were momentarily not clear (Swaab and Hofman, 1990).

A lesion study performed in male rats showed that the absence of a suprachiasmatic nucleus did not affect their preference for a sexually receptive female rat in a test where they could choose between approaching her and approaching another male (Kruijver *et al.*, 1993). Based on the data from the dead AIDS patients, it had been predicted that a lesion would enhance the preference for the female. If an enlarged suprachiasmatic nucleus makes humans like their own sex, a diminished or absent nucleus should have the opposite effect. The fact that the prediction was not confirmed is not surprising in any way. The suprachiasmatic nucleus has never been implicated in the regulation of sexual behaviors.

While Swaab and Hofman (1990) failed to find any difference in size of the sexually dimorphic nucleus in the preoptic area between hetero- and homosexuals, a part of a structure in the anterior hypothalamus, known as the interstitial nuclei, was found to be much smaller in homosexuals than in heterosexuals. Again, the brains were taken from homosexual men who had died from AIDS and from some purportedly heterosexual AIDS victims. In addition, the brains from a couple of supposedly heterosexual men who died from other causes were analyzed. Some female brains were also included. The particular interstitial nucleus, which was found to be small in homosexual men, the INAH 3, was also small in women. These data were interpreted as showing that a large INAH 3 produces a sexual preference for women, and a small nucleus, found in women and homosexual men, causes a sexual preference for men (LeVay, 1991). This study was, like the one reporting the initial Xq28 results, widely commented in newspapers and magazines. However, it is not evident that the difference between homosexuals and heterosexuals reported by LeVay (1991) is reliable. Although data from larger groups of homo- and heterosexual men as well as from women confirm that women have a smaller INAH 3 than men, no difference between homo- and heterosexual men was found. Based on these results, the conclusion was that sexual orientation cannot be predicted on the basis of INAH 3 volume or cell number (Byne *et al.*, 2001). Here we have another example of how data supporting a hypothesis concerning the causes for homosexuality first provokes a considerable enthusiasm, but later turn out to be impossible to replicate.

In addition to the lack of replicability, studies discovering differences in sizes of brain structures between homo- and heterosexual men suffer from the problem of translating these differences into causes for sexual preferences. Data from rats showed that the suprachiasmatic nucleus is not involved in sexual preferences, and a huge amount of data show the same with regard to preoptic–anterior hypothalamic structures. As was mentioned in Chapter 6, there are literally hundreds of papers reporting the effects of lesions of these areas in many mammalian species including primates and, with extremely rare exceptions, no alteration of sexual preferences has been reported. If a reduced size of these structures makes individuals prefer males over females, then it should have been observed on a regular basis. The attentive reader will undoubtedly recall, though, that preoptic lesions do facilitate the expression of lordosis in male as well as in female rats. A reasonable interpretation of this observation is that female sexual behavior is facilitated by preoptic lesion. That was actually the conclusion I proposed in Chapter 6. However, we also know that preoptic lesions abolish proceptive behaviors in females. These behaviors are often interpreted as expressing the motivation to engage in copulatory behaviors. Thus, while the copulatory reflex of lordosis is facilitated both in males and females, the motivational aspects of sexual behavior seem to be disrupted. Preference for a sexual partner is determined by incentive motivational mechanisms and unrelated to the execution of copulatory reflexes. The rat data, then, suggest that the preoptic area is crucial for sexual motivation in both sexes, but there are no data showing that it determines preference for a particular sex.

There are three studies, all from the same research group (Paredes and Baum, 1995; Kindon *et al.*, 1996; Paredes *et al.*, 1998) reporting that lesion of the preoptic area in male rats and ferrets indeed changes preference from a female to a male. However, these data can be criticized on several grounds. For example, using procedures similar to those employed in these studies, even castration of a male rat produces a preference for the male (Xiao *et al.*, 2004). What these data actually show is that a male, under certain circumstances, is a superior *social* incentive for males with no or much reduced sexual motivation. On the contrary, in a well-established test for sexual incentive motivation, we have shown that inactivation of the medial preoptic area strongly reduces the sexual incentive value of a sexually receptive female without altering that of another male (Hurtazo *et al.*, 2003).

Proposing that the suprachiasmatic nucleus could be a site where preference for one's own sex is caused was quite creative in view of the lack of importance of this nucleus for sexual behaviors. However, the rationale, which was described above, was in a way logical, although not convincing. Proposing that the anterior commissure is the site of homosexuality is astonishing. I exaggerate a little bit here, but it was actually suggested that:

> factors operating early in development differentiate sexually dimorphic structures and functions of the brain, including the anterior commissure and sexual orientation, in a global fashion (Allen and Gorski, 1992, p. 7199).

The background for that statement was the observation that the area of the anterior commissure was found to be 34% larger in homosexual than in heterosexual men. The reason for determining that area in brains of dead homo- and heterosexuals was some reports suggesting that women have a larger commissure than men. Now, since popular prejudice has it that homosexuals are generally effeminate, it was reasonable to suppose that they also have a larger commissure, although there is no reason at all to assume that it is of any importance for sexual behaviors. As seems to be a consistent phenomenon in studies of differences between homo- and heterosexual men, the finding of a larger anterior commissure in homosexuals could not be replicated (Lasco *et al.*, 2002). Furthermore, the supposed difference in commissure size between men and women did not appear. This will be the last example of the efforts made for finding a structural difference between hetero- and homosexual men.

The preceding brief review of the several studies pursuing potential causes for homosexuality has revealed that this pursuit has been futile. The general pattern of events is that a report of some difference between homo- and heterosexual men appears, attracts attention and then circulates for years as a kind of established fact. Sooner or later a replication will be made, and that replication will fail to reproduce the original results. However, failure attracts far less attention than success, so the initial report will continue to circulate as unchallenged fact for many years after it has been shown to be ephemeral. Nevertheless, I do not think it is unfair to propose that to date there is no convincing evidence for a genetic or structural difference between homosexual and heterosexual men. This conclusion is not a result of my personal critical spirit. Others who have taken the trouble seriously to evaluate the published research on biological determinants of homosexuality arrive at the same negative conclusion (e.g. Byne and Parsons, 1993; Mbugua, 2003). A somewhat acid evaluation of the entire field of biological determinism of human sexual behavior was recently supplied by a social scientist. He considered the belief in biological determinants as an instance of fetishistic or magical thinking (Lancaster, 2006). It is not self-evident that he is entirely wrong.

It is noteworthy that all the studies of brain structures have been performed on male brains. Women's brains have only been used for comparison with the homosexual men and then an effort was made to include only heterosexual women. This means that we have no information regarding possible structural differences between homo- and heterosexual women. Despite this, I would predict that a search for structural differences in women's brains would not be more successful than it has been in men's brains. It is not only in studies of brain structure that homosexual women have been ignored, but they have also been so in the search for the crucial gene causing homosexuality. Again, if homosexual women would some day attract the scientists' attention, I suspect that results would be exactly as meager as the results from homosexual men have been.

1.3 The homosexual scalp and hand

Besides the research on central nervous differences between homosexuals and heterosexuals summarized above, some scientists have ventured into far more extravagant endeavors. For example, rather than exploring differences inside the skull, structural differences on the skull's outside have been proposed as signs of homosexuality. When the direction of the hair-whorl rotation on the top of the head was determined in homosexuals it was found that 29.8% of them had a counterclockwise rotation. This occurred only in 8.2% of heterosexual men. A noteworthy feature of this study was that the gay subjects were recruited while relaxing on Rehobot Beach, Delaware, and the direction of rotation was determined by discretely observing the heads of the subjects. Bald, long-haired and individuals wearing a cap were excluded for obvious reasons. The randomly chosen heterosexual men were observed in shopping malls, grocery stores, barber shops, gas stations and fast-food restaurants and at three airports in Maryland, New Hampshire and Massachusetts. This creative study concluded that there must be some biologic/genetic factor contributing to homosexuality since such factors determine the direction of rotation of the hair-whorl (Klar, 2004). A replication of these spectacular findings is needed before their real importance can be determined.

Handwriting may be sexually dimorphic and it has been suggested that prenatal hormonal influences can affect it, at least in women (Beech and Mackintosh, 2005). No surprise, then, that the handwriting of homosexual men has been compared to that of heterosexual men. Curiously enough, homosexual and heterosexual men were found to have identical handwriting (Lester *et al.*, 1977). The stereotype of homosexual men as effeminate in all aspects was, exceptionally, found to be false.

Homosexual men are more non-right handed than heterosexual men (Lalumière *et al.*, 2000). In my ignorance, I thought that non-right handed was a convoluted way of saying left-handed, but it is not so easy. There are also the categories of mixed-handers and non-consistent right handers. Anyway, the excess of non-right handed homosexuals is attributed to early neurodevelopmental disruptions that affects handedness as well as sexual differentiation. There is even a report on associations between handedness and masculinity and femininity (Lippa, 2003b), which fits nicely with the idea of the effeminate homosexual. There is much to say about these studies, but I cannot see how they can offer any information of relevance to the subject being discussed in this chapter, and probably not to any other subject either.

As a last piece of entertainment, I will mention that there is a sexual dimorphism with regard to the ratio between the ring finger and the index. Men are more likely to have the ring fingers longer than index fingers, while the opposite is true for women. Here, we have another example of a quantitative dimorphism of small magnitude. As always, this vital dimorphism is considered to be caused by differences in prenatal hormone concentrations between men and women. Homosexual men are similar to women in finger length (Lippa, 2003a). Another nice example confirming the stereotype that gay men are generally effeminate. However, homosexual women were no different from heterosexual women. These observations could not be replicated in a later study. In fact, the opposite was found. Finger length ratios in women but not in men were related to homosexuality (Kraemer *et al.*, 2006). These results of tremendous scientific and practical implications must await confirmation before they can revolutionize human society.

Some other still more unbelievable studies on differences between homosexuals and heterosexuals have been published, but I will give them the benefit of ignorance. Instead, I will return to the first couple of studies I mentioned and to their failure to find a reliable difference between homosexual and heterosexual men. There are many potential causes for this failure. These same causes would most likely apply to studies on homosexual women, if they would be undertaken some day or another. During the next few sections of this chapter, I will discuss some of the potential causes. In addition to that, I will approach the mystery of why there is a search for a 'biological' cause at all. I already mentioned the lack of a convincing rationale a couple of pages ago. This lack becomes still more evident if the most limited relevance of homosexuality as a problem of public health is considered. Besides preferences for having sex with a certain kind of people, humans have preferences for things that may be far more important. Some people have a preference for fat or otherwise unhealthy foods, while others show a marked preference for healthy food. The consequences of food preferences for public health is certainly far more important than preferences for sexual partners, yet there is no search for differences in brain structure between high-fat and low-fat preferring people. A story could certainly be built in support for ideas about perinatal determinants of later food choice in the same way as it has been done with regard to sexual partner choice. Why has it not been done? In order to answer questions of this kind, the importance of which should not be underestimated, we need to put homosexuality in a social context. That is one aim of the following sections.

2 Preference versus orientation: the role of will and other issues

The concept of homosexuality is highly value-loaded. An immediate consequence of this is that almost any opinion expressed on the subject is prone to be criticized, perhaps even violently, by some group or another. This is perhaps not of major importance for a courageous scientist, so long as funding is not jeopardized. From a scientific point of view, a more serious problem is that the concept is vague and understood in different ways by different individuals. I have tried to avoid using the words homosexuality or homosexual and replaced them with sexual preference when discussing studies of non-human animals. However, as soon as we turn to the human, even the concept of preference may be victim of violent attack. Some want the expression 'sexual preference' to be replaced by the expression 'sexual orientation'. This seemingly trivial proposal is related to how we conceive the causes for having sex with same-sex individuals and this is a highly contentious issue. For reasons which I ignore, the notion of preference is considered as founded on the belief that the sex of the preferred partner is decided by volitional mechanisms, while orientation is regarded as being determined by mechanistic processes over which the individual's will has no influence. Furthermore, the notion of 'preference' does not necessarily refer to an immutable personal characteristic, while the notion of 'orientation' seems to do exactly that. The opposition between these two terms is reflected in opposing views on the very concept of homosexuality. A view favored by many holds that homosexuality is a stable individual characteristic, just like sexual orientation is, while others consider homosexuality as a temporary, unstable state. I will begin this section with a humble attempt to analyze these issues, with the hope that such an analysis will contribute to reduce confusion and make the subsequent parts of this chapter understandable without excessive intellectual effort. I will first present some thoughts concerning the role of volition as a determinant of sexual preference. This is crucial in order to contrast preference and orientation, which will be the issue that follows.

In Chapter 8, I defined preference. For the sake of clarity, I will repeat that a preference for something always requires a possibility of choice. Without a choice, there is nothing to prefer. Furthermore, I restated preference in terms of relative incentive value of available alternatives. An individual will approach the incentive with the largest incentive value among those available at the moment. For convenience, I employed the verb 'choose' for the action and the noun 'choice' for the process. These words are simple labels put on an observed behavior and do not imply anything at all with regard to the dynamics of the internal processes determining the behavior. When a male rat chooses to approach a sexually receptive female rather than another male, we say that the male made a choice of the female or that he chose the female. What actually happened was that the incentive stimuli emitted by the female had a more powerful action on the central motive state than the incentive stimuli emitted by the other male. Thus, the central motive state activated motor patterns leading to approach to the female. In this process, there is no need to assume that the rat made a conscious choice between the available alternatives, and much less does it imply that the approach to the female was determined by the rat's free will. The behavior displayed by the rat was entirely determined by the interaction between incentive stimuli and the central motive state. However, when we want to describe the rat's behavior, we may say that he chose the female rather than the male. The use of the verb 'choose' might seem contradictory, but if we keep in mind that it is only a convenient label put on an observed behavioral sequence, there is no contradiction whatsoever. It simply means that the rat approached one incentive among those available. This same reasoning can obviously be applied to the human. Therefore, when I talk of a preference for one sex over another, I mean that an individual approached one sex and not the other in a situation where both were available. This occurred because the incentive stimuli emitted by a particular individual had a larger impact on the central motive state than the incentive stimuli emitted by other persons present in the situation. We can, if we like, regard this as a perfectly deterministic process in which the individual's free will, if such a thing exists, had no influence at all. The notion that the relative incentive value of available incentives determines which of them is approached has important corollaries. One was already mentioned, namely that the behavior displayed is not a result of volition. Another is that the sex of the individual with the largest incentive value in the situation is quite uninteresting. Whether it is a man or a woman emitting the superior incentive stimuli is of no concern in this deterministic analysis of preference. For the moralists and those unfamiliar with behavior theory, though, this irrelevant aspect may be a question of utmost importance.

Any discussion of preference as a result of differences in relative incentive value between available incentives is incomplete without some mention of the many factors determining incentive value. In Chapter 4, I discussed the issue of human sexual incentives at length. The main conclusion was that we know little. Concerning the factors determining the momentaneous incentive value of a sexual incentive, we know still less. Nevertheless, there is no doubt that both our remote and immediate past experiences with the incentive are decisive. Immediate past experience may have produced habituation or sensitization to an incentive, or negative alliesthesia may be at work. More remote experiences may have effects through classical or operant conditioning, or through social learning. Internalized rules evoking mental representations of aversive consequences of commerce with an incentive may devaluate it, while mental representations of future rewarding events attainable only through commerce with the incentive may heighten its value. These are probably the most important factors, but there may be others. We know very little about their operation in relation to sexual incentives, but there is a considerable amount of data showing how some of these factors influence the value of other kinds of incentives. To review these data here is out of the question. They are scattered over several psychological fields, some of which I am almost completely ignorant.

Despite limited present knowledge, hypotheses regarding the exact influence of any of these factors on sexual incentive value can be subjected to experimental test. This fact is rather important, and should by itself show how fruitful the incentive motivational approach to sexuality is.

Most behavioral scientists are not opposed to the notion that preference of one thing over others, or choice of a thing or a particular behavior, is a deterministic process. In fact, any scientist studying behavior is convinced that it obeys a set of rather simple laws and principles. If not, it could not be the object of science. If yes, there is no room for a free will. A free will could never be predicted and never subjected to any law or principle. Far more eloquent arguments showing that the notion of a free will is unnecessary and even incompatible with science are found in a brilliant analysis of the determination of behavior which was published many years ago (Skinner, 1971). It is still worth reading by anyone with the slightest interest in behavior theory. A more recent analysis of the issue of will can be found in a magnificent paper (Ainslie, 2005). In this paper, we are also shown how the experience of a will corresponds to certain motivational patterns.

The demise of the will as a scientific concept was much accelerated by Freud's conception of the human as a victim of forces from the id fighting against the censorship of the superego. In that fight, there was no room for a will. A consequence of this demise is that the subject of volition has become marginal, or survives as a kind of exoticism, in behavioral science. However, outside the circle of behavioral scientists, the concept of will and all of its philosophical, moral and political implications is still in good health. Because of that, it has been argued that the term 'sexual preference', suggesting a willful choice, should be abandoned and replaced by the term sexual orientation. According to those favoring this suggestion, orientation is necessarily deterministic. However, as we have seen, this distinction is untenable. Furthermore, the concept of sexual orientation gives no clue at all to the processes involved in orienting, while preference has both a theoretical and a behavioral content. When asked to give a cause in terms of orientation for a male rat's approach to a female rather than to a male, we would have to say that he does that because he is so oriented. To my simple mind, such a statement has no sense. It would be necessary to add another question, namely what it means to be so oriented. The answer to that question must sooner or later refer to behavior, which was that the male rat approached the female rather than the male and therefore his orientation is toward the female. This circular argument is meaningless and it continues to be so until we add something. That thing must be the question 'why?'. Why does the male approach the female instead of the male? The answer, because he is so oriented, may seem to explain the rat's behavior. However, as we know since Chapter 1, the question of 'why?' has no place in science. Likewise, the answer I just provided does not explain anything at all.

In order to account for the rat's behavior in the situation described above, we would have to find the cause which makes him approach the female rather than the male and that cause cannot be 'he is so oriented', as we learned a few lines ago. If we use different relative incentive value as the cause of the rat's behavior, we have given a general behavioral principle as the cause. It is valid for all kinds of alternative incentives and in all kinds of situations. On the contrary, the explanation 'because he is so oriented' refers in fact to a particular set of alternatives in a particular situation. A test with the same female and male incentives the following day, for example, could reveal that the male was approached more intensely than the female. We would then be forced to conclude that our subject had changed his orientation overnight. If we use the incentive value explanation instead, we would simply state that on this occasion, the male had larger incentive value than the female. The general principle of relative incentive value used for explaining behavior remains valid, while the circumstantial explanation of orientation turned out to be false. Incidentally, the female's reduced incentive value on the second test may be a consequence of the fact that she was no longer sexually receptive and therefore smelled and behaved differently. Since 'orientation' is believed to refer to a stable, inner characteristic, changes in the incentive stimulus can have no effect on this stable characteristic.

I could invent many other examples of the uselessness of the orientation concept for explaining behavior. Rather than doing that, I will simply point out that a behavioral concept becomes useful only when it is possible to integrate into a theoretical body of some generality. Sexual orientation refers specifically to the choice of sexual partner and has no generality. Of course, we could imagine that this choice obeys behavioral mechanisms that are not at work in any other choice. However, then we would need to fill our brain with thousands of choice mechanisms, or orientations. I, for example, am very fond of a Swedish nut chocolate with the name Schweizernöt. Whenever available, I choose it among all the other chocolates. Am I equipped with a chocolate orientation making me choose Schweizernöt? Or is it simply so that I choose the chocolate with the largest incentive value? And how about the choice of Bourbon? I always choose Knob Creek when it is available. Have I a Bourbon orientation? Or do I simply choose the one with the largest incentive value? The concept of orientation may be of help in the exercise of the sport of orienteering but not in behavioral science.

Some critical reader may consider that my treatment of the concept of orientation has been overly negative. There is no reason for being unfair, so I will try to approach that concept from an angle different from the one of behavior theory. One possible alternative is to take a look at the non-scientific or commonsensical meanings of the term. According to the online version of the Oxford English Dictionary, one meaning that appears to be appropriate in the present context is:

> 'a person's basic attitude, beliefs, or feelings; a person's emotional or intellectual position in respect of a particular topic, circumstance, etc.; (now) *spec.* sexual preference (downloaded from http://dictionary.oed.com/cgi/entry/00333795?single=1&query_type=word&queryword=orientation&first=1&max_to_show=10 on September, 25, 2006).

These approximations to a definition of orientation refer to inner, mental states and not to observable behavior. Even if attitudes, beliefs or feelings could be measured, and an individual's emotional or intellectual position in respect to a particular topic could be determined, it is most uncertain how such measures or determinations relate to actual behavior. A long series of unproved suppositions regarding the relationship between attitudes, beliefs, feelings and behavior would have to be made. Despite all the good will I can muster, I still find the concept of 'orientation' extremely cumbersome compared to the concept of 'preference' whenever we seek to explain the mechanisms making an individual perform sexual activities with a person of one sex rather than the other.

In view of the clear inferiority in explanatory power of the concept of 'orientation' compared to the concept of 'preference', there must be some extrascientific reasons justifying its use. The notion that 'orientation' refers to a life-long characteristic makes it possible to argue that it is inborn and that the individual has no influence over his own orientation. It could, perhaps, be maintained that those preferring the term orientation are not necessarily interested in understanding the neural or psychological mechanisms by which a human approaches one individual rather than another. Their main interest is to make evident that this approach is not determined by free will. Consequently, the individual making a particular approach cannot be held responsible for it. She is driven by forces outside of voluntary control. This notion has a considerable importance in the popular debate concerning homosexuality, as we will see in the next section.

3 Acceptability of same-sex behaviors and the role of willfulness

I have repeatedly pointed out that all sexual activities not involving penile–vaginal intercourse inevitably are non-reproductive. From the Christian perspective, they are sins. From the biological perspective, they are inadaptive. I have also repeatedly mentioned that many sexual activities other than penile–vaginal intercourse are not only sins for the churches and inadaptive for the biologists, but they have also been, and some are still, included in the penal codes of many countries. All these facts need to be added to the social opprobrium still provoked by some sexual activities. While cunnilingus and fellatio appear to be generally accepted as legitimate sexual behaviors in most civilized countries, this is not always the case for sex with individuals of the same sex, and particularly not for sex with prepubertal individuals. Still worse evidently is sex with prepubertal individuals of one's own sex. Sex between adult individuals of the same sex appears to be well accepted in Europe, at least Western Europe, while homosexuality seems to continue to evoke passion on the west side of the Atlantic Ocean. The exception might be Canada, while Latin America and the USA appear to persist in a late 19th century approach to same-sex sexual behaviors. Since the USA has a considerable output of research reports, most of the research on same-sex behaviors comes from this country. The USA is also a dominant player in the media business, and news reports and comments on research findings are distributed over a large portion of the world by American companies. This fact has certainly contributed to a discussion of homosexuality in terms of a willful or unavoidable inclination even in countries where neither protestant fundamentalists nor biologists are particularly influential. As we will see, the notion of willfulness is of much concern to those approaching homosexuality from a moral point of view, and it has also contaminated some research efforts.

An illuminating study of US media coverage of research reports on homosexuality reveals that the dominant theme in newspaper, magazine, television and radio comments on such reports was a dichotomy between being born gay and choosing to be gay. Furthermore, the notion of being born gay includes the supposition of biological determinism, a conceptualization of homosexuality and heterosexuality as mutually exclusive categories, and the conviction that political and social change is dependent on proof that homosexuality is an innate and unchanging characteristic (Wilcox, 2003). By making homosexuality inborn, homosexuals can argue that it is the way God made them, and being a homosexual cannot, in consequence, be a sin. The following quote illustrates this:

> Is it a sin if you can't help yourself, if you are just playing the genetic hand God dealt you? That is the question at the heart of moral debates about whether homosexuality is a lifestyle choice or a matter of genetic predisposition. (Reeves, 1993, p. A9).

This very important comment allows us to understand much of the current discussion around sexual preferences. If it could be shown that homosexuality, understood as a permanent and exclusive characteristic, is inborn and consequently part of the genes which God has been kind enough to design for us, then we could not consider it a sin. On the contrary, if it is a result of an act of free will, it is not only a sin, but it can also be changed. Thus, it is evident that American homosexuals who want to be completely integrated in Midwestern society are forced to adhere to the cannon that they were born gay. The conclusion to be drawn here is that many of the points of view upon homosexuality that have gained some prominence have their origin in the efforts of American homosexuals to get rid of the label of sinners. A biological determinant of homosexual behavior would, according to this reasoning, contribute to give same-sex relationships a status equal to that of heterosexual relationships.

It must be admitted that this reasoning is not false. Data show that people having the belief that homosexuality is biologically determined are more tolerant to same-sex relationships than those believing that such relationships are the result of an act of will. This is true in the USA (Ernulf *et al.*, 1989; Whitley, 1990), and also in Canada (Altemeyer, 2001) and Sweden (Landén and Innala, 2002). In the latter countries, religious fundamentalism is probably less common and the concept of sin less popular than in the USA, and homosexuals and heterosexuals share essentially the same privileges. It is difficult to see, then, why there should be a relationship between belief in biological determinants and tolerance in Canada and Sweden. However, it is also known that obese individuals are more negatively evaluated by those attributing obesity to willful overeating as opposed to physiological disorder, presumably outside of the control of will (de Jong, 1980). Perhaps the human mind in general is more tolerant to uncontrollable behaviors than to acts of will. Another explanation is that the relationship between tolerance and the acceptance of a biological explanation in fact is the inverse of what is believed, in the way that those who are tolerant to homosexuality at the outset are more favorable to biological determinism than those less tolerant are (Hegarty, 2002). According to this latter point of view, the attribution of cause is by itself of slight importance.

It appears that the importance given to the distinction between homosexuality as a condition caused by a willful decision on one hand and as an inevitable consequence of uncontrollable events on the other is great in conservative, religious groups. This is amazing, since people belonging to such groups should be well aware of the fact that Saint Augustine considered sexuality as outside the control of the will. Indeed, that conviction justified the establishment of innumerable rules for sexual behaviors, in the dark ages summarized in publications called the Penitentials (see Hawkes, 2004). Since sexuality was outside the control of the will, the Holy Roman Catholic and Apostolic Church felt obliged to help people control their sexuality through a set of rules that, if respected, would prevent people from entering into situations where uncontrollable sexual urges might appear. The Penitentials are now forgotten, but their teachings still have profound influences on the way many parents raise their children and on interpersonal relationships among people in general. Instructing children to avoid sexually exciting situations is not unusual and instructing spouses to do the same when going unaccompanied to a party is not exceptional. Considering that some rules and the spirit of the Penitentials are still influential, it is curious that the rationale for those rules and that spirit has been forgotten. Despite contemporary society's amnesia, the notion that sexuality is *not* controlled by the will has dominated European thinking for more than a thousand years. Consequently, a proposal saying that the choice of sexual partner is outside the control of the will regardless of the causes for that choice is not anything new. Already the Fathers of The Church knew that.

The reasoning making Saint Augustine conclude that sexuality is uncontrollable by human will is only one line of support for this notion. I will, in fact, conclude this section by remarking that the contradiction between being born gay and having chosen to be gay, so important in the public discourse on homosexuality in the USA, is only apparent. When we choose something, meaning that we express a preference for one thing over others depending on their relative incentive value, that choice is always determined by something. It may be by our genes, by the concentration of hormones in the blood, by our experience, by the position of the moon or whatever, but it is *always* determined by something. Otherwise choice would be random. The statement 'being born gay' is one possible cause for choosing to perform sexual activities with the same-sex, but it is not, in principle, different from any other cause. As soon as we assume a scientific attitude and toss the notion of free will into the place where it should have been for centuries, the cemetery for sterile ideas, any cause for the choice of a particular sexual partner becomes just as respectable as any other cause. With that in mind, we can turn to an examination of possible causes for choice of same-sex incentives without the unpleasant burden of prejudice. However, first we need to address the ambiguities inherent in the concept of homosexuality. These ambiguities may explain the modest success obtained in the search for biological causes in terms of stable, inborn predispositions.

4 Problems with the use of the concept of homosexuality and the category of homosexual in scientific research

4.1 Uncertainties associated with classification and the notion of stability

In popular and quasiscientific analyses of homosexuality, it is frequently argued that the category of homosexual acquires meaning only when used in opposition to the category of heterosexual. I am not familiar with all the arguments for and against this opinion, but I am not convinced that it enlightens us much. Likewise, many social scientists and very few neuroscientists are of the opinion that the concept of homosexuality is a social construction. It is, according to this point of view, a socially created category with obscure correspondence in nature. I am not sure that this belief is helpful either. In fact, all concepts are products of reason and created by humans living in society. Nevertheless, it may be of some help briefly to examine the basic concepts employed both in the research on same-sex behaviors and in the non-scientific discussions of them.

At the beginning of this chapter, I have already mentioned that the concept of homosexuality as well as the category of homosexual is problematic. With concept I mean here *an idea of a class of objects*, not the objects themselves. The concept of homosexuality, then, refers to all mental representations evoked by the word. It is, in a way, an intrapsychic abstract event. On the contrary, a category is an entity consisting of a class of objects sharing some property. The way to define a category is to describe the property shared by all objects belonging to it. Thus, individuals belonging to the category of homosexual have at least one property in common. In order to make the category meaningful, all members must have that property, and no individual having it can be outside of the category. In the case of homosexuality, the defining property is controversial. On one extreme, we can consider that any individual ever having had sex with a person of the own sex is a homosexual. On the other extreme, we can consider that only individuals who exclusively have had sex with their own sex are homosexual. There is no agreement on this point. An additional uncertainty inherent in categories formed by living organisms is that some properties may change over time. Whenever this occurs, we are faced with an unstable category. Individuals can move out from it and into it at any moment. Some individual properties, like eye color or sex are stable. Others, like hair color or body weight may change over time. The category formed by individuals with green eyes can be considered as stable, while the category of slender is unstable. In the case of the category of homosexual, there is a considerable debate with regard to its stability. I will pursue the problem of categorization with a couple of additional examples and I will also introduce the contentious issue of search for causes of the defining property.

Instead of classifying people in the categories of homosexual and heterosexual we may, for the sake of an innocent example, classify them either as Steelers fans or Dodgers fans. Whether these categories are meaningful or not depends on the use we want to make of them. If I just want to figure out how to mention the Steelers' shameful loss against the Dodgers last week in a conversation with the barber, it is helpful to know if he is a Steelers fan or not. If I should happen to offend him, he might use his knife to cut my throat. Although vitally important in the moment, this use of a category is quite trivial. On the contrary, if I want to start a research project to find out the differences in brain structure between a Steelers fan and a Dodgers fan, the categorization of people in these groups should have some relevance, practical or scientific. Furthermore, we have to ensure that the classification of individuals as either Steelers or Dodgers fan is correct. Having done that, we can look at the brains, find a difference and propose an explanation for differential preference in terms of that difference in brain structure. This is straightforward, even if the coincidence between a difference in brain structure and a difference in football preference not necessarily establishes a causal relationship of any kind. That would require a series of additional studies, but we leave that issue in peace for the moment. However, it might so happen that my research hypothesis states that the future preference for a football team was established perinatally, during the period of brain differentiation and axonal growth. If I assume that, then I must also assume that football preferences are stable throughout lifetime. Although not an expert on sports, I have no reason to doubt that this might be the case. If it indeed were, then my research project could proceed and perhaps discover the crucial difference in brain structure between a Steelers fan and a Dodgers fan. Obviously, even if such a difference were to be found, it needs to be shown that it is significant for preference for a particular football team rather than an accidental observation. Now let us imagine that I am wrong with regard to stability of preference, and that a Steelers fan suddenly may become a Saints or a Giants fan. It might even happen that a Dodgers fan becomes a Steelers fan. The category of fan would, then, be unstable. In this case, a search for a perinatally established preference for a football team would be absurd. A search for the momentaneous brain differences would still be valid, provided my classification into categories is correct at the moment the brains are analyzed. However, an attribution of these differences to some perinatal factor would clearly be contrary to reason.

The preceding example should have illustrated the quandaries of research on homosexuals. First, we need to ask whether the category of homosexual is of any particular interest. The answer is probably more dependent on moral and/or sociopolitical convictions than on facts of science. Second, the classification as homosexual, or heterosexual for that, may be based on different criteria, making it unclear what the category actually means. This is a small problem, though, compared to the third issue that needs to be solved: the problem of stability. Is a homosexual always a homosexual, or may she/he change into heterosexual at any moment? Likewise, is heterosexual a stable category? These questions become of fundamental importance for many of the studies searching for a neurobiological cause for homosexuality. Rather than trying to give an immediate answer, we will first explore some sources of information regarding homosexuality and homosexual.

4.2 Homosexual behaviors: a lesson from history and some observational data pertinent to the issues of classification and stability

Support for the arbitrariness inherent in the notion of homosexuality comes from history (see e.g. Bullough, 1979; Spencer, 1995). It seems that the concept of homosexuality was created as late as the 19th century, while descriptions of copulatory behaviors with individuals of the same sex were already common in the distant past. We have no written records of homosexual activities from ages earlier than Greek and Roman antiquity, but considering that some monkeys show what we now call homosexual behaviors, we can assume that humans, too, have engaged in such behaviors for as long as there has been a human species. Despite the existence of the behavior, there was no need to create a particular label for the individuals engaging in it until quite recently. There are, just to take one example, many descriptions of same-sex activities in ancient Greece and Rome, but the concept of homosexuality does not appear in any of them. Literary sources tell us about the close association between men favored in the armies of Elis (the city-state where the Olympic games were founded in 776 BC), Thebes and Sparta. It was thought that close friendship, including having sex with each other, fostered solidarity among the men (Burg, 2002). It was probably not a stupid idea, although most European and American generals would disagree. As subversive tongues have it, the army is the only place where you are decorated for killing your fellow men but expelled from the force if you dare to love one. Anyway, in those ancient times sexual activities between individuals of the same sex were socially accepted and there was no need to create a particular category for those engaging in these behaviors. Furthermore, no clear distinction was made between having sex with other men and having sex with women. This is probably a consequence of the fact that most antique discourse on sexuality concentrates on penetrative sex, either penile–vaginal intercourse or sodomy. The difference between one orifice and another was regarded as quite insignificant. In contrast to the slight importance given to the sex of the partner, the motor pattern displayed during copulation was significant. Being the inserter was regarded as an expression of manliness, while the role of the insertee was less glamorous. Whether this distinction is real or created in the minds of contemporary historians interpreting classical texts is impossible for me to judge, but the important point is that if there existed any social judgment with regard to copulatory acts, it referred to the motor pattern expressed and *not with whom* this pattern was expressed. Because people were not categorized according to the sex of their preferred partner but according to their preferred motor pattern, a category of homosexuality could not have any meaning. Furthermore, the men having sex with other men also had sex with their wives, as part of their marital duties. The contention that having sex with one's own sex somehow excludes the possibility of engaging in similar activities with the opposite sex never occurred to the Greeks. This is another nice illustration of the fact that the partner's sex was of very little concern. Contemporary notions of homosexuality are, to a large degree, based on the idea of exclusiveness.

A short comment on the choice of motor pattern for defining the sexual categories of inserter and insertee is warranted. Those reading this book carefully might remember, from Chapter 8, that the motor pattern associated with penetration is the only sexually dimorphic copulatory motor pattern. Penetration is the only behavior that can be performed by one sex and not by the other. Curiously enough, it is the privilege of men. Thus, the Greeks chose the only copulatory motor pattern exclusive for men as criterion for establishing sexual categories. Not surprisingly, Greek society considered the category of inserter as superior to the category of insertee. That the act of inserting was considered as an expression of manliness is entirely logical, since only men can perform it. On the contrary, being the insertee is equally possible for both men *and* women, so it cannot be regarded as an expression of manliness. We may agree or disagree with the notion of superiority of being the inserter, but we cannot deny the rationality of the contention that it is a male privilege.

From a scientific point of view, the Greek categories inserter–insertee are as problematic as homosexual–heterosexual. In the case of women, the category insertee is unequivocal. A woman can never be the inserter, so for women there is no doubt with regard to their belonging to the category of insertee. Likewise, there is no doubt as to its stability. A woman can never become an inserter. For men, however, the situation is unclear. Any man can be either inserter or insertee, so categorization is not automatic. Furthermore, any man can change from inserter to insertee at any moment, and he can change back to inserter whenever convenient. The category is unstable. Furthermore, the category insertee may contain both men and women. In the case of men, they are unstable members, while women are stable. Although the Greek method of categorizing copulatory activities may seem perfectly logical, its scientific use is no easier than the contemporary categories of homosexual and heterosexual. Nevertheless, it is important to remember that sexual activities may be categorized according to essentially arbitrary criteria and that the dichotomy homosexuality–heterosexuality is only one of many possible classification schemes.

The preceding account suffices as an example of the many criteria we can use for establishing sexual categories. It should also illustrate that independently of the criteria we choose, there will always appear problems in determining to which category an individual belongs. Likewise, independently of the criterion we choose, there will be the problem of stability. Having established this, we can now try to approach the first part of the drama, that of assigning individuals to a category. As is evident to everyone, this needs to be done on the basis of observations of behavior of one kind or another.

The accounts of sexuality in ancient times are inadequate for scientific analysis. All our knowledge of ancient sexual behaviors is based on hearsay, which should be considered as entirely anecdotal. We have no quantitative data on the prevalence of different sexual activities and no direct observations performed by observers with verifiable credibility. Personally, I am always extremely skeptical to anecdotal material. I fear that what is judged to merit recordings in the form of literary descriptions, paintings or sculptures are events outside the ordinary, or perhaps manifestations of the artist's creativity without any immediate connection to reality. One could fear that this fascination for the extraordinary becomes a key element in descriptions of sexuality. Thus, *a priori*, there is no reason to believe that the accounts of Greek and Roman authors represent everyday behaviors. In fact, there is no way to determine the prevalence of same-sex behaviors in ancient Greece and Rome. Likewise, the behavioral criteria for categorizing someone as inserter or insertee are unknown and the stability of these categories is impossible to estimate. However, these societies' views upon the behaviors described are probably reproduced correctly, since they are most consistent between one source and another. Nevertheless, it would be far more reassuring to have some direct observational data on same-sex behaviors and on the individuals performing them. Although such data from antiquity are non-existent, we do have some from contemporary society. They come from an exceptional study performed in the late 1960s and reported in a charming book (Humphreys, 1970).

In an undisclosed American city, there was a public convenience visited by men interested in having sex with other men, in addition to those who just needed to satisfy a physiological demand for voiding the bladder or the bowels. It was localized in a park, so most of the men arrived by car. In order to observe the behavior of the men looking for sexual interaction, Humphreys pretended to be one of them. He assumed the task of lookout, making it possible for him to stay in the toilet without attracting attention and without interrupting the normal activities. Since the police now and then raided the toilet, the function of lookout was a necessity and raised no suspicion. Besides observing the men's behavior, Humphreys noted the license plates of the cars used by some of the most assiduous visitors. He then went to the car register to get the addresses of these men. For some reason, Humphreys soon secured an employment as life insurance salesman. With the pretext of proposing a life insurance, he proceeded to visit the men at their home. When selling life insurances, it is customary to ask questions about marital status, number of children, general health, annual income, and so on. In that way, Humphreys obtained a considerable amount of information about the men coming to the toilet for sex. Finally, he revealed the purpose of his presence in the public toilet to some men, who accepted to be interviewed. This extraordinary research is one of the very few cases where we actually have observational data concerning same-sex behavior and reliable personal information about the individuals engaging in it.

Most of the toilet visitors stopped by on the way to or from work. According to a strict and silent ritual they demonstrated their willingness to have sex by displaying a full erection while standing at the pissoir. When they had received signals of acceptance from another man, both disappeared into one of the stalls. There fellatio was performed and, immediately afterwards, both men left without a word. Sodomy was exceptional. Many intricate details of the interactions in the toilet are described in the book, but they are not necessary for our present purpose. What is more interesting, though, are the data obtained during the home visits. It turned out that a substantial proportion of the men were married and had children. Some had mistresses. The reason they frequently gave for visiting the toilet and having oral sex there was that they needed an ephemeral distraction or wanted sexual activity in addition to that obtained in their current relationship, without much expense of money and/or emotional involvement. They could just as well have visited a cheap brothel. However, that kind of service had been closed because of violation of morality laws. The most appropriate alternative was a public convenience.

The data from this study make evident the difficulty inherent in questions of classification. It is far from clear whether we should consider these men as homosexual or not. Most of them also had sex with women, so they do not fit the most popular stereotype of a homosexual. Indeed, some of them may have had sex with a man on the way home from work and sex with their wife later in the evening. Thus, the sex of their partner may be highly variable. Another question is whether these men have a preference for other men or for women. The answer to that question is impossible to give, since they did not have access to both sexes in the toilet. In the absence of a possibility of choice, we can have no idea of their preference. However, if the comment stating that they could just as well have visited a cheap brothel is true, then we must conclude that they had no particular preference. They wanted to execute copulatory responses, regardless of the sex with whom they executed them. This attitude seems quite similar to the one I described for ancient Greek society.

Many years have passed since Humphreys (1970) collected his data. In the period from the late 1960s up to present, the social acceptance of homosexuality has apparently increased. Public conveniences may no longer be a meeting place for homosexuals. In fact, in Oslo and probably at many other places, it appears that sauna clubs have taken over the public toilets' function as places for having fast and anonymous sex between men. A recent newspaper article revealed that the local health authorities were concerned about the risk of HIV infection in these clubs. They even suggested that the 'dark rooms', small rooms where men meet in absolute darkness for having sex without knowing with whom, should be closed. One argument for that was that many of these men also had unprotected sex with women, thereby spreading infections not only between men but also to the opposite sex (Lillebø, 2006). The force of that argument does not seem to be overwhelming, but the description of the typical behavior in the sauna clubs is surprisingly similar to the behavior displayed in the public conveniences, according to Humphrey's (1970) account. I will not return to the sauna stories and I will not use them in support for any further assertions. After all, they come from a newspaper report and not a scientific study.

The Humphreys (1970) data illustrate very nicely the problem of assigning individuals to a category. It also illustrates that theoretical discussion of categories is one thing and applying them to real data is another. I would like to substantiate this assertion by mentioning additional data from carefully done observational studies of sexual behavior outside the laboratory. However, I know of no other study where information has been gathered by direct observation combined with interviews and other means of obtaining personal information. It is, in fact, most unfortunate that we have so little observational data on homosexual women and men. In contrast, we dispose of a huge amount of questionnaire data. Tragically enough, most of these data have been gathered because of the preoccupation associated with the risk of HIV infections and they are of no relevance either for questions regarding the behavioral basis for categorizing individuals as homosexuals or for questions regarding the stability of this behavioral basis. Instead of being able to present real data concerning the question of stability, I am forced to trust an opinion expressed by Kinsey many years ago:

> . . . both homosexual and heterosexual activities may occur coincidentally in a single period of life of a single individual; and . . . the exclusive activities of one type may be exchanged, in the brief span of a few days or a few weeks, for an exclusive pattern of the other type, or into a combination pattern that embraces the two types (Kinsey, 1941, p. 428).

In addition to Kinsey's own data supporting his contention of homosexuality as an unstable condition, there are more recent reports evaluating the prevalence and incidence of change from homosexuality to heterosexuality (e.g. Cameron and Cameron, 2002). They confirm that such change is not unusual. Thus, anecdotal evidence from ancient Greece, interview data collected by Kinsey, observational data obtained by Humphreys (1970) and a questionnaire study asking the appropriate questions (Cameron and Cameron, 2002) all suggest that homosexuality is not a stable category.

The preceding arguments can explain why studies of a genetic basis as well studies of differences in brain structure between homosexuals and heterosexuals have given inconsistent results. There is no need to point out that if classification in either the category of homosexual or that of heterosexual is incorrect, then no reliable data can be obtained. In the case of the twin studies and the more sophisticated genetic studies, classification was based on questionnaires or direct questioning of the participants. This information can perhaps be considered as trustworthy with regard to the participants' sexual preferences at the moment they were questioned, but neither we nor the researchers have any knowledge concerning their sexual history. It is not impossible that some of them had belonged to the other category at some earlier point in life. In the brain structure studies, the participants were dead at the moment they were included. Classification as homosexual or heterosexual was based on their medical record. The reliability of these records is impossible to estimate. Thus, it is not surprising that the results of these studies are difficult to replicate. Finally, considering the overwhelming evidence that homosexuality is not a stable condition, the entire logical underpinning of the genetic and brain structure studies disappears. Please remember that data from the non-human mammals in which homosexuality has been described, rams and the snow monkey, coincide with the human data in showing that the sexual character is unstable.

The arguments expressed above are not new in any way. The dilemmas which have to be faced by those trying to understand homosexuality as a biologically caused, stable entity have eloquently been identified by many others (e.g. De Cecco and Parker, 1995) and exposed over a period of many years.

5 Sexual preferences can be explained by incentive motivation theory without any reference to the concept of homosexuality

In Chapter 8, I dismissed the concept of homosexual behaviors as basically useless. Now I am going a step further and dismissing the category of homosexual as largely useless for scientific purposes. It may well be of importance in doctrines of faith or in discussions of morality, but I cannot really see how it can help us to understand human sexual behavior. Likewise, the category of heterosexual can be dismissed from scientific use on the same grounds. I am well aware of the fact that it is always easy to criticize an idea and dismiss it as being of little use. It is frequently more difficult to propose a better alternative. However, in this particular case, I will propose an alternative that may prove to be far more fruitful than the categories of homosexual and heterosexual have been. I suggest replacing both these categories not with other categories but with an entirely different concept taken directly from behavior theory. It is probably not difficult to guess that I will propose to conceive same-sex behavior as a result of the action of sexual incentives, in exactly the same way as heterosexual behaviors are a result of the action of such incentives. As already suggested, the sex of the incentive is not of any particular importance. It can be argued that this shift of terminology has no direct consequence, because we would anyway have to explain why some individuals give a larger incentive value to individuals of the same sex than to those of the opposite sex. This argument has not much force, though. It could just as well be asked why some people give a larger incentive value to thin people than to fat people, or to fair-haired rather than dark-haired people. No need to mention that preference is situational and not a stable characteristic of the individual.

In order to illustrate how general incentive motivational concepts can be employed for explaining same-sex preferences, I suggest that we return to the public convenience where Humphreys (1970) obtained his data. Our first task will be to reveal the factors that made the men come to the toilet in the first place. The most common reason impelling humans to visit a toilet, urgencies of the bladder or bowels, can be excluded. Instead of that, we must figure out which sexual incentive brought them to the toilet. Since they did not come there because of physiological necessity, and thereby discovered the possibility of sexual interaction by accident, they must have had a mental representation of some sexual activity, which functioned as incentive. The origin of this mental representation must be their experience acquired at earlier visits or social learning. Someone might have told them about this toilet and mentioned the activities taking place there. Thus, on the way to or from work, the mental representation of fellating or being fellated by another man is an efficient sexual incentive producing a detour to the toilet. Once inside, prior experience or social learning determined the ritualized behaviors displayed in order to establish sexual contact. If successful, sexual reward in one form or another was obtained and the purpose of the toilet visit was attained. The sequence of behaviors leading up to the sexual reward was reinforced and the habit strength increased through the mechanisms of operant learning. If the establishment of sexual interaction and the ensuing reinforcement occurs only on some visits, the partial reinforcement effect may enter into action. In behavior theory, partial reinforcement is the condition where only a fraction of the responses is rewarded, for example one visit of three to the toilet leads to sexual reinforcement, or one lever press of 10 leads to a food pellet in the Skinner box. Responses acquired during the condition of partial reinforcement are particularly resistant to extinction. If sexual reinforcement were obtained irregularly, the habit of making a detour to the toilet would become extremely fixed.

It is likely that most of the married toilet men returned home following an after-work visit. There, the wife may emit stimuli turning her into an efficient sexual incentive before or at bedtime. The man may approach and initiate sexual interaction. Furthermore, in the Humphreys (1970) study, some men, either married or single, had a mistress. On some occasions, the mental representation of having sex with the mistress may appear at work, for example shortly before lunch. In this moment, the mistress turns into a sexual incentive, activating the behavior of inviting her for lunch and a subsequent visit to a cheap hotel.

The preceding account of the incentive motivational and learning mechanisms that might have determined the behavior of the toilet men is extremely simple. Every assertion made in the account can, in principle and probably also in practice, be submitted to experimental test. That is an important advantage compared to other kinds of hypotheses that must await the death of the individual before being tested. Furthermore, there is no need for employing vague categories as causes for behavior and there is no need to suppose stability of preferences. The dominant incentive at each moment is determined by processes such as habituation, negative alliesthesia etc, as mentioned before. Thus, the extraordinary flexibility of human behavior can easily be accommodated in the explanation.

The notion of situational preference is, in fact, the most reasonable explanation for the behavior displayed by the toilet men. It is probably the most reasonable explanation for the behavior of any woman or man.

It does not demand any extraordinary intellectual talent to discover that the incentive motivational explanation proposed above poses several questions that are impossible to answer at the moment. Among these is the issue of the nature of the stimuli functioning as sexual incentives in the human. Already, in Chapter 4, we learned that very little is known about this. However, this lack of knowledge regarding sexual incentives applies to those activating approach to the opposite sex exactly as much as it applies to incentives causing approach to the own sex. One of the many adverse consequences of the focus of human sexual behavior research on homosexuality and sexual dysfunctions is that we know very little about fundamental aspects of ordinary sexual behaviors. This lack of knowledge becomes irritatingly evident when we turn to behaviors preceding actual copulation, such as approach behaviors and associated mechanisms. An attentive reader would already have noted that I have remarked, on innumerable previous occasions, that we know far more about copulatory reflexes than about the events leading up to the display of these reflexes. Again, this is equally valid for same-sex as for opposite-sex behaviors.

Another issue, intimately related to the preceding, is the role of learning for a stimulus' sexual incentive properties. In Chapter 7, I showed that all known learning mechanisms may modify sexual behaviors and that classical conditioning may transform any neutral stimulus into a sexual incentive. Although I do regret that we do not have more empirical data with regard to learning with sexual reinforcement, it is possible that we do not need more data than we already have. It is established beyond reasonable doubt that the performance of sexual responses functions as reinforcement in a way similar to food and drink. Likewise, it is established beyond reasonable doubt that food and drink can reinforce all sorts of learning. Our alimentary and drinking habits are acquired through learning and the incentive stimuli directing choice and intake of food and drink are also learned. There is no reason to believe that sex would be any different.

Considering sexual incentives as learned does not mean that we can control that learning any more than we control the learning of alimentary and drinking habits. Many humans have acquired eating habits they consciously would like to change, for example. For most humans, it is usually impossible to explain how these habits were acquired in the first place, although some invent elaborated stories with the purpose of *a posteriori* justification. Once acquired, eating and drinking habits are extremely difficult to modify. In my own case, I would like to reduce my total intake of food, thereby reducing the size of my belly. Despite years of efforts, the result is absent. Everyone has heard many stories about failed dieting efforts and the many failures are not always a consequence of a lack of will. Indeed, if our will were free to decide, all dieting programs and all chemicals purportedly making us thinner would not be necessary. As I have insisted upon many times before in this chapter, being attracted to a particular incentive is an entirely deterministic process over which we have no more control than we have over the size of our suprachiasmatic nucleus or anterior commissure.

A change of food habits may be beneficial both to personal and public health. A change of drinking habits might also be beneficial for personal and public health. The high incidence of cirrhosis in France, for example, could be reduced by convincing the French to drink less wine and more grapefruit juice. On the contrary, a change of the sex of our partner would not have the slightest consequence neither for individual nor for public health. Someone could argue that HIV infection is more common among people having sex with their own sex and therefore the abandonment of that habit would improve public health. This might be true in some cases, but again the improvement would be marginal compared to that potentially caused by improved eating habits. My point here is that there is no particular reason for changing sexual incentives even if that were possible. This statement applies both at the individual and at the societal level. Thus, the question that needs to be asked is not whether it is possible to change sexual incentives from one's own to the opposite sex, but rather whether such a change would be *desirable* or not. Since the answer to this question is negative, the question of whether such a change is possible or not becomes irrelevant.

Although the last few paragraphs are not directly related to the content or empirical bases for the incentive motivational approach to same-sex preferences, they are directly relevant for the issue of willfulness. Considering that the main rationale for much of the research on same-sex preferences was the pursuit of an involuntary cause, it seems important to make clear that willfulness or the possibility of conscious choice are not in any way enhanced by an incentive motivational approach. This point is made with the moralists in mind. To all others, it is sufficient to repeat that the sex of the partner is a strictly individual and completely inconsequential event. The mechanisms deciding the preferred sex may not even be worth searching. Directing our attention to them may seem like a waste of time and effort when we compare their possible impact with the many other problems of some importance that mankind is facing and that require solution. Despite this apparent disdain for research on same-sex preferences, I refuse to end this chapter without a summary of some research supporting the incentive motivational approach.

6 Empirical support for an explanation of same-sex preferences in incentive motivational terms

When discussing human sexual incentives in Chapter 4, I mentioned that sexual incentive stimuli in the form of pornographic movies are efficient for enhancing genital blood flow, the classical indicator of sexual arousal. The genital response is an example of the visceral reactions produced by incentive stimuli through their actions on the central motive state. Provided that the responsiveness of the central motive state is kept constant, it can be assumed that the magnitude of the genital response is determined by the incentive stimulus' efficiency. In all the studies that were mentioned in Chapter 4, the incentive stimuli depicted either individuals of the sex opposite to the experimental subject, or heterosexual activities. Furthermore, these studies included only subjects that declared themselves to be heterosexual. Please excuse me for continuing to use a term that I dismissed myself, but I am using the terminology employed by the authors of the studies I have mentioned and that I will mention. Rather than using incentive stimuli of the sex opposite to the subjects, some scientists have varied the incentives and included depictions of same-sex activities (two individuals of the same sex engaged in copulatory behaviors and of the same sex as the experimental subject (same-same-sex) or of the sex opposite to the experimental subject (opposite same-sex) in addition to heterosexual activities. Furthermore, some studies included not only heterosexuals but also male and female homosexuals, vulgarly called gays and lesbians, as experimental subjects.

I will describe one of the early studies (Wincze and Qualls, 1984) in some detail. The subjects' sexual preferences were determined through an interview and ratings on the famous Kinsey Scale and only individuals whose *current* sexual behavior was exclusively homosexual were included. The italics on '*current*' are not there by accident. It is most important to notice that no assumptions as to stability of same-sex preference are made here. Of importance are only these preferences at the moment of test. The female subjects' genital arousal was assessed by vaginal photoplethysmography and the male subjects' arousal by measurement of penile circumference. All subjects were exposed to a series of film fragments. One film illustrated the coastline of Nova Scotia, which was considered as a neutral stimulus. Another film fragment showed a man and a woman engaged in copulatory activities, a third film fragment depicted two men displaying copulatory behaviors, a fourth film showed two women in the midst of sexual activities and the fifth and last film depicted copulatory activities within a group of men and women. These films were presented in counterbalanced order to the subjects. When the changes in vaginal blood flow in response to the various film fragments were analyzed in the homosexual women, it was found that the film fragment depicting sex within a group of men and women produced the largest response. Then followed the film fragment with sex between two women, the fragment showing sex between a man and a woman and, finally, the film illustrating sex between two men, in decreasing order of magnitude. The difference between the group sex, heterosexual sex and lesbian sex movies was not large. The gay movie had a much inferior capacity to enhance vaginal blood flow. These very interesting data show that the presence of a woman (group sex, heterosexual sex, lesbian sex) is the critical variable determining the vaginal response. The sex of that woman's sexual partner has only a minor influence. Here, the obvious conclusion is that visual stimuli produced by and emitted from the female body are powerful sexual incentives for the lesbians. Visual stimuli produced by and emitted from a man's body are less efficient. With regard to the gay men, the film fragment producing the largest increase in penile circumference was the one depicting sexual activities between two men. That film was followed by the group sex movie, which in turn was followed by the heterosexual movie. The film showing two women engaged in sexual activities produced a penile response that was not much different from baseline. Again, film fragments where a man is present are efficient enhancers of genital responses in gay men, while a film showing only women having sex is inefficient for activating sexual arousal. The conclusion we can draw from the gay men's responses is quite similar to that drawn from the lesbian's responses: visual stimuli produced by and emitted from a man have sexual incentive properties for gay men, while stimuli produced by and emitted from the female body have no or very weak sexual incentive properties.

The kind of results obtained in the study described above has been reported in several other studies (e.g. Freund, 1963; Mavissakalian *et al.*, 1975; Sakheim *et al.*, 1985; Freund *et al.*, 1989; Chivers *et al.*, 2004; Rieger *et al.*, 2005). It seems that stimuli emitted by men are powerful sexual incentives for homosexual men and heterosexual women, but not for heterosexual men or lesbians. On the contrary, stimuli emitted by women are powerful sexual incentives for lesbians and heterosexual men, but not for heterosexual women or homosexual men. Thus, exactly as has been proposed many times, sexual preference is determined by the relative incentive value of the available incentives. Measurements of the sexual arousal response of enhanced genital blood flow give an objective indicator of a stimulus' sexual incentive value, making it possible to base the preceding conclusion on experimental facts rather than on speculations.

The mechanisms determining the incentive value of a specific stimulus remain obscure. One possibility that immediately comes to mind is that the central nervous system is predisposed to respond to certain stimuli with enhanced genital arousal. This kind of expression is quite popular in some areas of psychology. The predisposition may be based on structure or on functional organization. All those suggesting a 'biological' basis for same-sex preferences can argue that the central nervous predispositions differ between same-sex preferring and opposite-sex preferring individuals. This is certainly a possibility that must be considered seriously. In fact, it seemingly explains the causes underlying the results of experiments like those mentioned above. There are, however, many serious problems with the notion of predispositions and one of them is that of stability. If 'predisposition' were understood as the momentaneous central nervous responsivity to certain stimuli, it would pose no problem. On the contrary, if it were understood as a constant response tendency in the central nervous system, it would become highly problematic. Most, if not all, of the biological explanations of same-sex preferences share the belief that such preferences are stable throughout the adult lifespan. Consequently, the predispositions must also be stable. We have already seen abundant evidence to the contrary, making the notion of constant predisposition untenable when faced with experimental and other data. By abandoning the notion of stable predisposition and replacing it with momentaneous predispositions, we can avoid one crucial problem with the notion of predispositions. However, other problems remain.

Although it may be convenient to say that the impact of an incentive is determined by momentaneous predispositions of the central nervous system, the explanatory force of this statement is not impressive. Indeed, the very notion of predisposition offers little in terms of cause–effect relationships. To propose that something was caused by a predisposition is nothing more than pronouncing a string of words without any empirical content until that predisposition has been materialized by a neurobiological fundament. So far, this fundament has not been found, neither for opposite- nor for same-sex preferences. On the other hand, if we propose that the impact of an incentive is determined by prior experience with it, we have provided a piece of useful information. We know that experience affects behavior through several mechanisms of learning and negative alliesthesia. Learning mechanisms, with the possible exception of some aspects of social learning, are well understood. What we need to know in order to explain the value of a sexual incentive is simply the individual's reinforcement history with that incentive and all social learning of relevance to that incentive.

We can advance one step more and maintain that we do not care about the individual's learning history because the result of that history is manifest in the individual's response to the incentive stimulus. The sum of history of reinforcement plus social learning regarding the incentive is the determinant of the present response and if we already have the sum we do not get more illuminated by figuring out its components. This is exactly equivalent to one of the examples of scientific explanation I provided in Chapter 1. I am referring to Einstein's second law of gravity, which can be used to describe perfectly the path of a planet, including predictions of its position in a distant future. However, there is no mention of how the planet got in motion. This is entirely irrelevant, because the result of that motion is what we can observe and what we should be concerned about. Speculations about the origin of motion are futile. Regarding the possibility of prediction of the individual's future response to an incentive, it is evident that this can be done as long as we can predict future reinforcement contingencies and the future social learning related to that incentive.

Even though we can consider that an individual's reinforcement history and social learning with regard to a particular incentive is expressed in the present response to the incentive and that the history therefore is irrelevant, a curious mind may persist in asking what this history may look like. A critical mind may go so far as to ask for some empirical data suggesting or showing that reinforcement history and social learning indeed can determine the sexual incentive value of a stimulus. The answer to the critical mind was partly given in Chapter 7, where I mentioned some studies showing that a neutral stimulus may be transformed into a sexual incentive through classical conditioning. The operation of social learning is less understood, as pointed out in Chapter 7, but the insightful analysis presented by Gagnon and Simon (2002) gives some hints. An answer to the demand made by the curious mind requires some more caution. We have no recorded complete history of sexual reinforcement starting in childhood and ending at some arbitrary point in adult life and we have no complete description of all social learning that may have affected the impact of a particular stimulus. I doubt that such a history or description ever will be possible to obtain.

A state of ignorance is not typical for potential sexual incentives. It is typical for most of the incentives of importance for human behavior. For example, the complex incentive of the performance of a Wagner opera at the Bayreuther Festspiele attracts people from all over the world. Most of those succeeding in getting a ticket have been on the waiting list for many years, while some others have paid an exorbitant price on the black market. This shows that the incentive value of these opera performances is high. Despite its considerable importance for human well-being and the European opera culture, we do not know much about how these performances acquired this high incentive value. Likewise, there are people following a bullfighter all over Spain throughout the long fighting season, at great monetary expense and with a considerable investment of time and effort. We do not know how this bullfighter has acquired his most respectable incentive value. Some pages ago, I mentioned the mysteries surrounding the incentive value of stimuli essential for survival, like food or drink. We do not know how a particular taste acquires its incentive value and, far less, its relative incentive value. How come, for example, that for some people the incentive value of the taste of a Laphroig 20-year old single malt whisky is far superior to that of a 30-year old Tallisker?

Having made clear that our knowledge of how a stimulus acquires sexual incentive properties of a particular intensity is not more limited than our knowledge of most other potential incentives, I will summarize the few facts and observations we have concerning the issue.

An extensive study (van Wyk and Geist, 1984) of the role of early sexual experiences for future sexual preferences was performed on a subset of the interview data obtained by Kinsey and collaborators. A very large number of variables were entered into many regression analyses and their capacity to predict adult homosexuality as evaluated by the Kinsey scale was determined. Among the significant predictors for homosexuality in men we find 'being masturbated by the same sex' and 'observing masturbation' when these men were prepubertal. Other important events occurring prior to puberty and predicting adult homosexuality was sexual interactions with adult men. The number of such experiences, the variety of sexual techniques involved, the attainment of orgasm and the total duration of sexual relations with adult men were all predictors of adult homosexuality. On the contrary, post-pubertal heterosexual activities were all negatively related to adult homosexuality. These extremely interesting data show that the amount of sexual interaction with other young boys or adult men, including the attainment of orgasm, before puberty predicts future homosexuality positively. Likewise, the amount of sexual interaction with women after puberty predicts future homosexuality in a negative way.

The frequent sexual interactions with other boys and men allow stimuli emitted by these boys and men to become conditioned sexual incentives. The only requisite for that is that the sexual activities engaged in were reinforcing and there is no reason to doubt that this was the case. Regarding data obtained with future female homosexuals, it was again found that being masturbated by another girl when prepubertal predicts adult homosexuality. On the contrary, intense prepubertal sexual activity, including oral, anal and vaginal penetration, with boys was negatively related to adult homosexuality. Thus, sexual interaction with boys seems to favor heterosexuality. On the contrary, if the first contact with adult male genitals occurred during a prepubertal sexual encounter, then it was more likely that the girl would become homosexual. Likewise, the larger the difference in age between the girl and the male partner, the more likely it was that the girl would turn into a homosexual. After puberty, the amount of heterosexual activity as well as the absence of orgasm when having such activity were related to future homosexuality.

When interpreting these data in terms of conditioning of sexual incentives, we immediately find out that when potentially reinforcing same-sex events, such as masturbation, were performed by another girl, the likelihood for future homosexuality increased. On the contrary, reinforcing sexual interactions with boys reduced that likelihood. An interesting result is the positive relationship between prepubertal exposure to male genitals, in the course of physical sexual interaction on one hand, the age difference between the girl and the owner of these genitals on the other and future homosexuality. This fact suggests that the male genitals had become a negative incentive when the age difference was large. It is quite likely that physical sexual interaction with a much older man provokes a strong reaction of fear or guilt or some other aversive state in a prepubertal girl. That aversive experience can easily become associated to a salient stimulus in the situation, the man's genitals. They would thereby turn into a negative incentive, activating the response of withdrawal. Considering that the data employed in this study were collected between 1938 and 1963, a time when sexual encounters between a young girl and a much older man must have been regarded as still more outrageous than it is today, the proposal of a fear or guilt reaction does not seem entirely unwarranted. Independently of this, the data reported in the van Wyk and Geist (1984) study are easily interpreted in terms of classical conditioning of initially neutral stimuli to sexual incentives. Interpretations in other terms are quite difficult, if not impossible.

The results reported in the van Wyk and Geist (1984) study were, as I already mentioned, based on regression analyses. A significant regression means in essence that the predictor variable covaries with the predicted variable. Covariation may be a result of accidental events, or it may be an expression of a cause–effect relationship. Whether one or the other of these alternatives applies to a particular regression analysis is impossible to say. Moreover, the data used for the regression analyses come from interviews where the subjects orally reported their past and present experiences. The veracity of these reports is completely unknown. It is not necessary to suppose that false information was obtained only because the interviewees consciously lied. Nevertheless, in addition to the danger of consciously lying subjects, it is well known that memories are a most unreliable source of information. Their correspondence to facts is frequently only remote. The double uncertainty, caused by the origin of the data on one hand and the regression analyses on the other, means that all the arguments exposed in my discussion of the van Wyk and Geist (1984) paper are no more than informed speculations, at best. I must admit that I included a discussion of this paper under substantial doubt. The reason convincing me to mention it was merely that there are so few data available concerning early sexual experience and future sexual preference that the luxury of choice is almost unavailable.

In addition to the van Wyk and Geist (1984) report based on the Kinsey *et al.* data, there is an extensive interview study of about 700 homosexual men and almost 300 homosexual women living in the San Francisco bay area (Bell *et al.*, 1981). The explicit purpose of these interviews was to gather information relevant for the development of sexual preference, from childhood until late adolescence. In addition to interviewing the homosexual men and women, identical interviews were made with more than 300 heterosexual men and with 140 heterosexual women. The inclusion of a heterosexual 'control' group is quite unusual in studies of homosexuality and this fact makes the Bell *et al.* (1981) study stand out. A comparison of interview responses between homosexual and heterosexual individuals gives a lot more valuable information than a simple description of responses within groups of homosexuals, which unfortunately is the commonly used procedure.

Before summarizing the observations of particular relevance to the issues of how neutral stimuli may be transformed into sexual incentives and how the incentive value is determined by experience, a few caveats need to be made. The sexual preference of the interviewees in the Bell *et al.* (1981) study was determined on the basis of a Kinsey scale score obtained in connection with the interview. Those with a score of 2 or more were classified as homosexuals and those with a score of less than 2 were classified as heterosexuals. The classification as homosexual or heterosexual was probably correct at the moment of the interview, but we cannot know if it had been stable over any extended period of time. This is problematic, as it has been in all studies employing the concept of homosexuality. Another factor limiting the value of all information based on interviewees' recollections of events in a distant past is the uncertain relation between what they remember and what actually occurred in this distant past. Most of the subjects in the Bell *et al.* (1981) study were about 35 years old at the moment of the interview. This means that they were asked to remember events that had passed somewhere between 25 and 15 years before. As I pointed out only a few paragraphs ago, it is not impossible that unconscious modifications of memory give rise to recollections of events that are quite different from the original event. It is also possible that these unconscious modifications of memory somehow are influenced by our current sexual preferences. We all have a tendency of trying to make our memories of even the remote past coherent with our present life circumstances. Finally, the elegant statistical methods employed by Bell and collaborators are most adequate for detecting coincidences in a data set, but they say nothing about causal relationships. With these caveats in mind, we can turn to the data.

When looking at the recollections of childhood and adolescent experiences in currently homosexual men, we find a few differences between these and heterosexual men that might support the incentive motivational approach. Before the age of 19, only 26% of the currently homosexual men had either masturbated a woman or been masturbated by a woman, while the corresponding figure for heterosexual men was 58%. Likewise, only 18% of homosexual men had achieved orgasm by rubbing their body against a woman's body. The figure was 54% for heterosexual men. Many heterosexual men, 34%, engaged very often in heavy petting with women, while few, 12%, of the homosexuals did so. While 32% of the homosexual men had performed penile–vaginal intercourse before the age of 19 years, 62% of the heterosexuals had done so. Thirty-seven per cent of heterosexuals had experienced cunnilingus or fellatio against 17% of homosexuals.

When comparing the incidence of homosexual behaviors before 19 years of age between homo- and heterosexual men, it was found that 84% of homosexuals had masturbated or been masturbated by another man, while only 21% of heterosexuals had had this experience. About 70% of homosexual men had fellated or been fellated by another man. Only 15% of heterosexuals had been so. Finally, 39% of homosexuals had obtained orgasm by rubbing their body against another man, but only 5% of heterosexuals had experienced the same. These data clearly show that homosexual men had experienced far less sexual interactions with females than heterosexual men had. An obvious consequence is that the possibility of associating sexual reward with female stimuli was much lower for homosexuals than for heterosexuals. On the contrary, homosexual men had experienced far more experiences with other men than heterosexual men had. Thus, the possibility for homosexuals to associate sexual reward with male stimuli was far above that of the heterosexuals. These data, then, coincide perfectly with the hypothesis outlined some pages ago. Learning theory has established that the number of reinforcements is a critical determinant of the amount of learning, or habit strength. The stronger the association between a rewarding event and a stimulus is, the larger is the conditioned incentive value of that stimulus. Consequently, the larger the number of sexual rewards obtained in the presence of male stimuli, the larger the incentive value these stimuli acquire. The differential reinforcement history of male versus female stimuli can explain how the homosexual became a homosexual and how the heterosexual became a heterosexual. This latter question is often ignored, but it is just as intriguing as the question of how homosexuality arises.

Rather than choosing the parsimonious and elegant explanation in terms of incentive learning proposed above, Bell *et al.* (1981) interpret their data as suggesting that the current male homosexuals as children and adolescents already lacked interest in women. Thus, an incipient homosexuality was already present and caused them to avoid women and approach men. To my humble judgment, this reasoning is to beg the question and enter into a circular argument. The current state, which they wish to explain as a result of childhood and adolescent experiences, suddenly becomes the cause of these childhood and adolescent experiences.

The data obtained in the interviews with homo- and heterosexual women were not identical to those obtained in men. Homosexual women had generally more sexual experiences with other women, but there was not much difference between homo- and heterosexual women with regard to their experiences with men.

However, homosexual women rated the reward value of heterosexual encounters as lower than that of homosexual encounters. The enhanced number of sexual experiences with other women, and their larger reward value, can easily explain how stimuli emitted by women have higher sexual incentive value than stimuli emitted by men. As was the case with male homosexuals, Bell *et al.* (1981) prefer to interpret their data in terms of incipient homosexuality. This appears to be equally absurd for women as it was for men.

The results of the Bell *et al.* (1981) study prompt me to propose a hypothesis accounting for the acquisition of same-sex preferences as well as for opposite-sex preferences in terms of differential numbers of reinforcements. We already know that sexual activity ending in orgasm is reinforcing and rewarding. It is also likely that sexual activities ending short of orgasm are rewarding. Thus, most of the sexual experiences an individual has had can be supposed to have been rewarding and reinforcing. Even though the number of sexual experiences is not necessarily identical to the number of sexual rewards and reinforcements, we can assume that the larger the number of experiences, the larger the number of rewards. It does not seem unreasonable to assume that the number of sexual experiences with the opposite sex is normally distributed in a given population. We can also assume that the number of sexual experiences with the same sex is normally distributed. The mean for the distribution of same-sex experiences is probably much lower than the mean of opposite-sex experiences because social pressure, or social learning, favors the latter. Nevertheless, the individuals on the far right of the normal distribution of same-sex experiences would have a larger number of such experiences than a part of those to the left on the normal distribution of opposite-sex experiences. This group of individuals would then constitute those attributing a larger sexual incentive value to members of the same sex than to members of the opposite sex. The fact that the mean differs in the two normal distributions explains how the incidence of same-sex preference is lower than the incidence of opposite-sex preference.

There are certainly events that may modify the simple scheme outlined in the preceding paragraph. Some sexual activities are more likely than others to lead to orgasm and those leading to orgasm have probably a larger reward value than those not doing so. This means that the number of sexual experiences does not correlate exactly with the number and/or intensity of sexual reward and reinforcement. However, in a large population this should be only of minor importance. Furthermore, some individuals may have a strongly aversive sexual experience, dramatically reducing the incentive value of stimuli present during that experience. Although Bell *et al.* (1981) did not find any evidence for a role of such experiences in the development of homosexuality, they could have some influence in some individuals. Nevertheless, this kind of experience would not be a major factor in a large population.

To end this section, I will present some anecdotal data from a most exotic place, New Guinea, which might seem to challenge the hypothesis presented in the preceding paragraphs. I have not abandoned my skeptical attitude to anecdotes, but

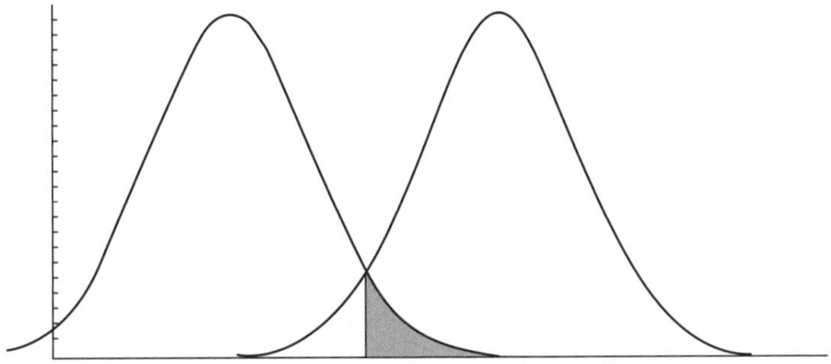

FIGURE 9.1 Distribution of the number of same-sex and opposite-sex reinforcements in an imaginary population of humans. The distribution to the right shows the number of opposite-sex sexual reinforcements. The mean of that distribution is much superior to that of the distribution to the left, showing the number of same-sex sexual reinforcements. This difference is due to social learning, which strongly favors the obtention of opposite-sex sexual reinforcements. Those within the dotted area have received a larger number of same-sex than opposite-sex reinforcements and will therefore display a preference for their own sex. All others will display preference for the opposite sex.

this is the only example I have been able to discover where someone has tried explicitly to analyze the acquisition of sexual preferences in terms of social learning. This is interesting, because one of the basic principles of social learning is that the individual does not need to experience reinforcement in order to learn. However, I will use the following account of sexual behavior in New Guinea for illustrating that an interpretation of the anecdotal evidence in terms of reinforcement is equally possible as an interpretation in terms of social learning. It is important to keep in mind that the following paragraphs are an account of something that *may* have occurred. My summary is entirely based on a paper by Baldwin and Baldwin (1989). Most of the original anthropological accounts were published in a book some years ago (Herdt, 1987).

Sambia is a rather unknown tribe, at least to me, living somewhere on New Guinea. A peculiarity of that tribe is that all boys are isolated from women and housed in a kind of clubhouse from the age of 7–10 until marriage. Before that age, the boys have interacted with their parents and siblings in a way similar to other boys throughout the world. In the male clubhouse, the boys are told that they need to ingest sperm in order to mature and become men capable of impregnating women. The most efficient way to ingest fresh sperm is to fellate other boys or adult men. It appears that not all boys engage in fellatio voluntarily. In that case, they are coerced. The boys have to persist with frequent fellatio for a few years. When puberty arrives, their role changes from fellator to recipients of fellatio. The preceding fellator period does not seem to be associated with any sexual reinforcement since the boys never experience orgasm or even genital stimulation while fellating. On the contrary, being fellated to ejaculation should be associated with reinforcement and reward. In fact, it is said that many boys experience pleasure while being fellated. This period approaches an end in the late teens or early twenties, when the boys get married, always to a very young girl. Sambian marriage is somewhat unusual, though, since the husband and wife do not start living together immediately. In fact, they have to wait until the girl has had her first menstruation. Although not living together, wife and husband may interact in several ways during the waiting period. It seems that one typical interaction consists of the man demanding his wife to fellate him. He can also be fellated by younger boys, whenever he should feel a need for that. At menarche, husband and wife start to live together. From now on, the boy is expected to show only heterosexual behavior with his wife and avoid all sexual activities with other men. This seems to occur, since the incidence of adult homosexuality among the Sambia has been estimated to be similar to the one found in Western societies. This assertion must be most uncertain, though, since the anthropologists cannot have any objective data to substantiate it.

The rather long period the boys spent in the clubhouse and interacted sexually with other boys and adult men instead of with women should have offered many opportunities for conditioning of sexual incentives. When experiencing the reinforcing event of being fellated, for example, the boys should associate stimuli emitted by the other boy or the man with the pleasure of fellatio. Since there were no women available for the boys, no association between sexual activities and stimuli emitted by females could have been established. When adult, then, the conditioning in boyhood should manifest itself as a preference for men. So far as the anthropologists have ascertained, this is not the case. There may be two reasons for this. First, the boys had spent years being the fellator, sometimes with coercion, and always without reinforcement and reward. That should be enough for forming associations between male genitals and an aversive event, the act of fellatio. When assuming the role of the fellated and experiencing reinforcement and reward, the aversive nature of the act should sooner or later be lost and eventually be replaced by an experience of positive affect, making it possible to form associations between that affect and the stimuli provided by the fellating boy. This may well be the case. However, after entering marriage the boy is also fellated by his wife, making it possible for him to form associations between stimuli emitted by a female and sexual reward. The result should be that both male and female stimuli are sexual incentives. Nevertheless, the boy now being man will be heterosexual, with rare exceptions. This can easily be explained by social learning. During early childhood he saw his parents in a heterosexual relationship, and in the clubhouse he was told that he should eat sperm in order to become able to impregnate women. Indeed, Sambian society is based on a rather traditional heterosexual couple. The virtues of this social structure are inculcated into the children during early childhood and during the years in the clubhouse, despite the boys' engagement in frequent homosexual activities. Furthermore, these activities are explicitly understood as a step on the road to adulthood and heterosexual marriage. Another important aspect is that the boys are told many stories about the pleasures of sex with women.

At the same time, they are taught that heterosexual activities are strictly forbidden before and outside of marriage. These two messages do not seem to be much different from what most Western boys will hear. The socially transmitted information regarding the pleasures of sex with women may allow the boys to form mental representations of heterosexual activities and these mental representations may function as incentives. At the same time, the teachings about homosexual activities as something typical of maturing boys but unworthy of adult men may cause a devaluation of the possible incentive value of such activities and consequently of sexual incentive stimuli emitted by men.

The combined result of sexual reinforcement and the socially determined mental representations of homo- and heterosexual activities is apparently enough for assuring that the vast majority of adult Sambia men will be predominantly heterosexual. Despite the social encouragement of homosexual activities during part of childhood and adolescence, it appears that the social construction of sexuality is based on the same principles as that occurring in Western societies, as eloquently analyzed by Gagnon and Simon (2002). All this is based on the supposition that the anthropologists' accounts of the behavior of the Sambians are true. The veracity of that supposition is unknown, particularly since the source of the anthropological information is one single research group.

The studies I have presented above do not include any direct, observational data of relevance to sexual learning in the human. As already suggested, such data will perhaps never be available. The uncertainties with regard to the empirical foundations of the accounts given above imply that we should regard these accounts as *possible* descriptions of how sexual learning may take place. Two kinds of learning, fundamental with regard to the processes by which neutral stimuli acquire sexual incentive properties, were illustrated. The first was classical conditioning. Although this process may be highly important, it is not enough for understanding how external stimuli or mental representations of such stimuli act on the central motive state. The actions of both external and internally generated incentive stimuli are modified by social learning. Some stimuli may get the incentive value, as established by conditioning, enhanced through social learning. Other stimuli may get their incentive value, as established by classical conditioning, reduced because of social learning. Still other stimuli may acquire sexual incentive properties entirely by social learning. Despite these apparent complexities, the mechanisms of classical conditioning and social learning determine the sexual incentive value of a stimulus. Since preference is no more than relative incentive value, these processes also determine sexual preferences in the human.

7 The human is not simply another mammal

In human sexual behavior, conditioning is by itself not enough for explaining how stimuli acquire or lose sexual incentive properties. Social learning must absolutely be added as a fundamental contributor to this process. Independently of the learning mechanisms involved, it is clear that human sexual incentives are learned. There is a general agreement between scientists working in the field of human sexual behavior on this point. In non-human mammals, the existence of unlearned sexual incentives is firmly established. This extremely important difference between humans and other mammals has probably been overlooked by the neuroscientists searching for a biological explanation of homosexuality.

The causes for this costly confusion reside probably in the fact that the vast majority of the scientists looking for brain differences between homosexuals and heterosexuals are unfamiliar with the field of human sexuality and still more so with the field of learning theory. They are normally trained in one of the biomedical or purely biological disciplines and have frequently some basic notions of rat and other non-human sexual behaviors. Since most of the sexual incentive stimuli important for rodents and many other non-human mammals are unconditioned, meaning that they have incentive properties not acquired through learning, it is easy to generalize this fact to the human and imagine that many sexual incentives are also unconditioned in our species. If a stimulus' incentive properties are independent of learning, then these properties must automatically be given to it by the central nervous system. In old-fashioned language, we would say that this kind of stimulus is an inborn incentive. Something in our genes has made the central nervous system identify the stimulus as a sexual incentive, automatically connecting it to the central motive state and there generate appropriate motor output. Within this line of reasoning, it is quite obvious that a preference for one's own sex may be caused by the genes determining the incentive properties of a stimulus. A search for a homosexuality gene becomes completely logical within this context.

The unusual event that a stimulus emitted by an individual of one's own sex functions as a superior sexual incentive can also be explained by some unusual differentiation of the central nervous system. Consequently, a search for structural differences between homosexuals and heterosexuals seems most reasonable. Such differences may be caused by the homosexuality gene or by endocrine accidents during sexual differentiation. Finally, since the inborn incentive properties of a stimulus reasonably can be expected to last for the entire life, the problem of stability disappears.

All this is very nice. It is a pity, though, that human sexual incentives are not inborn. This little detail has escaped the attention of those believing that human sexuality is no different from that of a rat. Since it is very different, much of the logical basis for the search of biological determinants of stimuli functioning as human sexual incentives is lost.

In fact, my proposal is that those searching for a 'biological' basis for same-sex preferences have not fully realized the fundamental differences between human and non-human sexual behaviors. One additional consequence of this, not mentioned before, is the notion that human sexual activities necessarily involve, and usually are limited to, penile–vaginal intercourse as is the case in most non-human mammals. The peculiarity of same-sex behaviors is that they cannot include penile–vaginal intercourse. Therefore, they require an explanation, while penile–vaginal intercourse, as the normative behavior, does not. However, while penile–vaginal sex certainly is the norm in non-human mammals, it is not so in the human. The notion that other sexual motor patterns are unusual, and therefore need to be explained, is also based on a generalization from non-human mammals. As soon as we accept that penile–vaginal intercourse is only one of many motor patterns leading to sexual reward, same-sex preferences and same-sex copulatory relationships become completely natural. In fact, they are in no more need of explanation than opposite-sex preferences are. The rationale for the pursuit of oddities in the central nervous system of those engaging in same-sex behaviors disappears. Their sexual behaviors are determined by exactly the same mechanisms as those preferring the opposite sex. These considerations also make the entire concept of homosexuality unnecessary in scientific thought. I already exposed several other reasons for the lack of usefulness of that concept and the one given here only adds some extra weight to these reasons.

The inquiry into same-sex behaviors made throughout this chapter is completely unrelated to moral or religious conceptions of these behaviors. For men of faith, the theological arguments advanced by Saint Augustine or religious scholars of other confessions have not lost their validity. The moral or religious arguments are based on faith and, as such, they are untouchable by science. This chapter, with its intention of being strictly scientific, has hopefully refrained from making any moral judgments.

8 A note on homosexual identity

The many arguments advanced against the very notion of homosexuality as a scientific concept and the insistence on saying that the difference between same-sex and opposite-sex preferences is trivial, may seem to contradict many claims of some psychologists and many lesbian and gay groups, who talk about a gay or lesbian identity, or about gay and lesbian subcultures, etc. However, I see no contradiction at all. Humans form groups of all kinds and frequently identify with these groups. For example, at the time when the temperance movement was influential in Scandinavia, sociologists talked about a temperance identity. Any human group, particularly groups that are, or perceive themselves as, marginalized for one reason or another has a tendency to facilitate the development of some kind of group identity. There may be several factors favoring the need for the creation of a gay and lesbian identity, such as the status of 'deviant', police persecution, social rejection, and so on. In any case, the subject of identity is of no relevance for the discussion here. It is a socially determined process, without any conceivable importance for the development of sexual preferences. For those interested in the subject there are some papers that may be worth reading and which substantiate the above statements regarding identity as a social construction quite independent from sexual preferences (Johns and Probst, 2004; Kaufman and Johnson, 2004; Sibalis, 2004; Nash, 2006).

Summary

1 The search for an endocrine, genetic or structural basis for same-sex preferences has yielded conflicting results.
2 The category of homosexual is frequently understood as corresponding to a stable characteristic of the individual. There is not much empirical evidence in support of this notion.
3 By explaining homosexuality as caused by innate or developmental biological alterations, the condition is not a consequence of a conscious, free choice. God or Nature rather than the homosexual must be blamed for it.
4 The notions of homosexuality and the euphemism sexual orientation should be replaced with the notion of sexual preference.
5 Preference is determined by the relative incentive value of all available incentive stimuli. The stimulus with the largest incentive value is the preferred stimulus. Volitions do not in the slightest contribute to this process. Furthermore, the relative incentive value of a particular stimulus is determined by the other incentives available at the moment. Preference may, therefore, vary from one situation to another.
6 The incentive value of a stimulus is partly determined by the individual's reinforcement history with regard to that stimulus. Social learning may either enhance or reduce the incentive value of a particular stimulus. Additional experiences, or new social learning experiences, may at any moment change the incentive value of a stimulus. Therefore, preference does not need to be stable over time.
7 Whether the incentive stimulus with the largest incentive value of available incentive stimuli is emitted by a man or a woman is a rather uninteresting question.
8 In non-human mammals, most sexual incentives are unconditioned. They have acquired their incentive properties through the mechanisms of evolution and do not require learning. In the human, sexual incentives are conditioned. Learning is a necessity.

References

Ainslie, G. (2005). Précis of *Breakdown of Will*. *Behav Brain Sci* **28**, 635–673

Allen, L.S. and Gorski, R.A. (1992). Sexual orientation and the size of the anterior commissure in the human brain. *Proc Nat Acad Sci USA* **89**, 7199–7202

Altemeyer, B. (2001). Changes in attitudes toward homosexuals. *J Homosexual* **42**, 63–75

Bailey, J.M. and Pillard, R.C. (1991). A genetic study of male sexual orientation. *Arch Gen Psychiat* **48**, 1089–1096

Bailey, J.M., Pillard, R.C., Neale, M.C. and Agyei, Y. (1993). Heritable factors influence sexual orientation in women. *Arch Gen Psychiat* **50**, 217–223

Baldwin, J.D. and Baldwin, J.I. (1989). The socialization of homosexuality and heterosexuality in a non-Western society. *Arch Sex Behav* **18**, 13–29

Beech, J.R. and Mackintosh, I.C. (2005). Do differences in sex hormones affect handwriting style? Evidence from digit ratio and sex role identity as determinants of the sex of handwriting. *Pers Individ Differ* **39**, 459–468

Bell, A.P., Weinberg, M.S. and Hammersmith, S.K. (1981). *Sexual preference. Its development in men and women*. Indiana University Press: Bloomington

Bullough, V.L. (1979). *Homosexuality: a history*. New American Library: New York

Burg, B.R. (2002). *Gay warriors: a documentary history from the ancient world to the present*. New York University Press: New York

Byne, W. and Parsons, B. (1993). Human sexual orientation: the biologic theories reappraised. *Arch Gen Psychiat* **50**, 228–239

Byne, W., Tobet, S., Mattiace, L.A. *et al*. (2001). The interstitial nuclei of the human anterior hypothalamus: an investigation of variation with sex, sexual orientation, and HIV status. *Horm Behav* **40**, 86–92

Cameron, P. and Cameron, K. (2002). What proportion of heterosexuals is ex-homosexual? *Psychol Rep* **91**, 1087–1097

Chivers, M.L., Rieger, G., Latty, E. and Bailey, J.M. (2004). A sex difference in the specificity of sexual arousal. *Psychol Sci* **15**, 736–744

De Cecco, J.P. and Parker, D.A. (1995). The biology of homosexuality: Sexual orientation or sexual preference? *J Homosexual* **28**, 1–27

de Jong, W. (1980). The stigma of obesity: the consequence of naive assumptions concerning the cause of physical deviance. *J Health Soc Behav* **21**, 75–87

Ernulf, K.E., Innala, S.M. and Whitam, F.L. (1989). Biological explanation, psychological explanation, and tolerance of homosexuals: a cross-national analysis of beliefs and attitudes. *Psychol Rep* **65**, 1003–1010

Freund, K. (1963). A laboratory method for diagnosing predominance of homo- or hetero-erotic interest in the male. *Behav Res Therap* **1**, 85–93

Freund, K., Watson, R. and Rienzo, D. (1989). Heterosexuality, homosexuality, and erotic age preference. *J Sex Res* **26**, 107–117

Gagnon, J.H. and Simon, W. (2002). *Sexual conduct: the social sources of human sexuality*, 2nd edn. AldineTransaction: New Brunswick

Galton, F. (1869). *Hereditary genius: an inquiry into its laws and consequences*. McMillan: London

Hamer, D.H., Hu, S., Magnuson, V.L., Hu, N. and Pattatucci, A.M.L. (1993). A linkage between DNA markers on the X chromosome and male sexual orientation. *Science* **261**, 321–327

Hawkes, G. (2004). *Sex and pleasure in Western culture*. Polity Press: Cambridge

Hegarty, P. (2002). It's not a choice, it's the way we're built: symbolic beliefs about sexual orientation in the US and Britain. *J Community Appl Soc Psychol* **12**, 153–166

Herdt, G.H. (1987). *The Sambia: ritual and gender in New Guinea*. Harcourt Brace: Fort Worth

Humphreys, L. (1970). *Tearoom trade. Impersonal sex in public places*. Aldine Publishing Company: Chicago

Hurtazo, H.A., Paredes, R.G. and Ågmo, A. (2003). Inactivation of the medial preoptic area/anterior hypothalamus (MPOA/AH) reduces sexual incentive motivation in male rats. *Soc Neurosci Abstr* 404.6

Johns, D.J. and Probst, T.M. (2004). Sexual minority identity formation in an adult population. *J Homosexual* **47**, 81–90

Kallman, F. (1952). Comparative twin study of the genetic aspects of male homosexuality. *J Nerv Ment Dis* **115**, 283–298

Kaufman, J.M. and Johnson, C. (2004). Stigmatized individuals and the process of identity. *Sociol Q* **45**, 807–833

Kendler, K.S., Thornton, L.M., Gilman, S.E. and Kessler, R.C. (2000). Sexual orientation in a US national sample of twin and nontwin sibling pairs. *Am J Psychiat* **157**, 1843–1846

Kindon, H.A., Baum, M.J. and Paredes, R.G. (1996). Medial preoptic/anterior hypothalamic lesions induce a female-typical profile of sexual partner preference in male ferrets. *Horm Behav* **30**, 514–527

Kinsey, A.C. (1941). Criteria for a hormonal explanation of the homosexual. *J Clin Endocrinol* **1**, 424–428

Klar, A.J.S. (2004). Excess of counterclockwise scalp hair-whorl rotation in homosexual men. *J Genet* **83**, 251–255

Kraemer, B., Noll, T., Delsignore, A., Milos, G., Schnyder, U. and Hepp, U. (2006). Finger length ratio (2D:4D) and dimensions of sexual orientation. *Neuropsychobiology* **53**, 210–214

Kruijver, F.P.M., de Jonge, F.H., van den Broek, W.T., van der Woude, T., Endert, E. and Swaab, D.F. (1993). Lesions of the suprachiasmatic nucleus do not disturb sexual orientation of the adult male rat. *Brain Res* **624**, 342–346

Lalumière, M.L., Blanchard, R. and Zucker, K.J. (2000). Sexual orientation and handedness in men and women: a meta-analysis. *Psychol Bull* **126**, 575–592

Lancaster, R.N. (2006). Sex, science, and pseudoscience in the public sphere. *Identities-Glob Stud Cult Power* **13**, 101–138

Landén, M. and Innala, S. (2002). The effect of a biological explanation on attitudes towards homosexual persons. A Swedish national sample study. *Nord J Psychiatr* **56**, 181–186

Lasco, M.S., Jordan, T.J., Edgar, M.A., Petito, C.K. and Byne, W. (2002). A lack of dimorphism of sex or sexual orientation in the human anterior commissure. *Brain Res* **936**, 95–98

Lester, D.H., McLaughlin, S., Cohen, R. and Dunn, L. (1977). Sex deviant handwriting, femininity, and homosexuality. *Percept Mot Skills* **45**, 1156

LeVay, S. (1991). A difference in hypothalamic structure between heterosexual and homosexual men. *Science* **253**, 1034–1037

Lillebø, S. (2006) Steng saunaene. Klassekampen (October 5), p. 7. Oslo

Lippa, R.A. (2003a). Are 2D:4D finger-length ratios related to sexual orientation? Yes for men, no for women. *J Personal Soc Psychol* **85**, 179–188

Lippa, R.A. (2003b). Handedness, sexual orientation, and gender-related personality traits in men and women. *Arch Sex Behav* **32**, 103–114

Mavissakalian, M., Blanchard, E.B., Abel, G.C. and Barlow, D.H. (1975). Responses to complex erotic stimuli in homosexual and heterosexual males. *Br J Psychiat* **126**, 252–257

Mbugua, K. (2003). Sexual orientation and brain structures: a critical review of recent research. *Curr Sci* **84**, 173–178

Mustanski, B.S., DuPree, M.G., Nievergelt, C.M., Bocklandt, S., Schork, N.J. and Hamer, D.H. (2005). A genomewide scan of male sexual orientation. *Hum Genet* **116**, 272–278

Nash, C.J. (2006). Toronto's gay village (1969–1982): plotting the politics of gay identity. *Can Geogr-Geogr Can* **50**, 1–16

Paredes, R.G. and Baum, M.J. (1995). Altered sexual partner preference in male ferrets given excitotoxic lesions of the preoptic area anterior hypothalamus. *J Neurosci* **15**, 6630

Paredes, R.G., Tzschentke, T. and Nakach, N. (1998). Lesions of the medial preoptic area/anterior hypothalamus (MPOA/AH) modify partner preference in male rats. *Brain Res* **813**, 1–8

Pillard, R.C. and Weinrich, J.D. (1986). Evidence of familial nature of male homosexuality. *Arch Gen Psychiat* **43**, 808–812

Reeves, R. (1993). Breeding a new revolution. *Atlanta J Atlanta Constitut* (July 16), p. A9

Rice, G., Anderson, C., Risch, N. and Ebers, G. (1999). Male homosexuality: absence of linkage to microsatellite markers at Xq28. *Science* **284**, 665–667

Rieger, G., Chivers, M.L. and Bailey, J.D. (2005). Sexual arousal patterns of bisexual men. *Psychol Sci* **16**, 579–584

Risch, N., Squires-Wheeler, E. and Keats, B.J.B. (1993). Male sexual orientation and genetic evidence. *Science* **262**, 2063–2064

Sakheim, D.K., Barlow, D.H., Beck, J.G. and Abrahamson, D.J. (1985). A comparison of male heterosexual and male homosexual patterns of sexual arousal. *J Sex Res* **21**, 183–198

Sibalis, M. (2004). Urban space and homosexuality: the example of the Marais, Paris' 'gay ghetto'. *Urban Stud* **41**, 1739–1758

Skinner, B.F. (1971). *Beyond freedom and dignity*. Alfred A. Knopf: New York

Spencer, C. (1995). *Homosexuality. A history*. Fourth Estate: London

Swaab, D.F. and Hofman, M.A. (1990). An enlarged suprachiasmatic nucleus in homosexual men. *Brain Res* **537**, 141–148

van Wyk, P.H. and Geist, C.S. (1984). Psychosocial development of heterosexual, bisexual and homosexual behavior. *Arch Sex Behav* **13**, 505–544

von Krafft-Ebing, R. (1886). *Psychopatia sexualis: eine klinisch-forensische Studie*. Enke: Stuttgart

Whitley, B.E. (1990). The relationship of heterosexuals' attributions for the cause of homosexuality to attitudes toward lesbians and gay men. *Pers Soc Psychol Bull* **16**, 369–377

Wilcox, S.A. (2003). Cultural context and the conventions of science journalism: drama and contradiction in media coverage of biological ideas about sexuality. *Crit Stud Media Comm* **20**, 225–247

Wincze, J.P. and Qualls, C.B. (1984). A comparison of structural patterns of sexual arousal in male and female homosexuals. *Arch Sex Behav* **13**, 361–370

Xiao, K., Kondo, Y. and Sakuma, Y. (2004). Sex specific effects of gonadal steroids on conspecific odor preference in the rat. *Horm Behav* **46**, 356–361

CHAPTER 10

Some comments on the concepts of 'normal' or 'functional' sexual behavior

1 Introduction

In the preceding chapters of this book, we have discussed several basic aspects of sexual behaviors in mammals. From this chapter and until the end of this book, we will focus our attention on something frequently called sexual dysfunctions and/or abnormal sexual behaviors. To begin with, we need to make clear of what the difference between normal, abnormal, functional and dysfunctional consists.

Many experts on human sexual behavior do not use the expressions 'functional' or 'dysfunctional' very often. They prefer the terms 'normal' and 'abnormal'. Some might consider 'normal' and 'functional' as synonyms. Others would argue that they are fundamentally different, because the word 'functional' refers to something having a desired or appropriate effect, while 'normal' is thought to refer to behaviors coinciding with a norm of some kind. On the contrary, 'dysfunctional' refers to something *not* having the desired or appropriate effect and 'abnormal' refers to behaviors that do *not* fall within some arbitrary norm. There is some vagueness in the preceding statement, but we do not need to care about it for the moment. It should be succinct enough as a basis for the discussion of these terms. The next few pages will try to give us some notion about the meanings attributed to the concept of normal. Then we will turn to the concept of functional.

2 The notion of normal versus abnormal

The noun 'normal' is often understood as meaning 'habitual', 'common', 'regular' or 'typical'. Although rarely defined in the clinical literature, the habitual meaning of 'normal' seems to be 'close to average'. Anything far from the average is correspondingly called 'deviant'. One of those believing that others believed in this conception of normality was Kinsey. He was convinced that when people learned how common homosexuality or extramarital sex was, they should immediately stop considering these behaviors as deviant and instead accept them as normal. In retrospect, it may perhaps be ascertained that he was right, although the instantaneous change in attitude he expected in reality required several decades to appear.

The idea that everything close to average is normal and everything far from it is deviant is deeply rooted in the human mind, including that of most clinicians and many scientists. In the context of basic bodily functions, like blood pressure or blood concentrations of glucose or cholesterol, large deviations from the average can legitimately be considered abnormal and they can justify treatments with the purpose to reduce the deviation and bring values back to the average. Clinical experience shows that values far from average sooner or later will lead to disease or enhance the likelihood of disease. Without any doubt, in the case of essential physiological functions, treatments reducing deviations from average are often beneficial to the individual's health. Within the medical sciences, then, the notion of normality as everything not too far from average seems to be justified.

The basis for the respectable medical use of the concept of normality is the fact that most, if not all, physiological functions, such as heart rate, blood pressure, urine production and so on, are similar in all human beings. There may be variations associated with age or sex and there may be some variations associated with ethnic origin but, basically, all physiological parameters vary in a quite narrow range. Consequently, they are generally similar in all individuals of the same age and sex. Any large deviation from average in any physiological parameter can reasonably be called abnormal, while all individuals with these parameters close to average are called normal. As used in medicine, the concept of normality poses no serious problem. This assertion does not exclude intense discussions among clinicians with regard to the magnitude of deviation from average required for employing the label abnormal or pathological, and consequently deemed as indicative of a need for corrective treatment.

When we turn to sexual behaviors, the notion of normality as everything close to average becomes problematic. We can start with a simple illustration of this. A large Swedish survey has determined the average frequency of penile–vaginal intercourse to be 7.3 per month for people between 26 and 30 years old (Lewin *et al.*, 1998). Unfortunately, no data on variation around the average, like the standard deviation or range, are reported. This omission reduces the usefulness of the reported data. By the way, if a student paper contained the same omission, the poor student would be punished with a very low grade. Anyway, we imagine someone having 30 penile–vaginal intercourses per month, one per day including weekends. This is rather far from the average. In fact, a person with a blood glucose concentration four times above average would be in a critical condition. On the contrary, the person having intercourse once a day would probably be in splendid health, although he is far from the average. Likewise, a person with no sexual activity would suffer no health problem as pointed out in Chapter 1. Thus, large variations in the frequency of penile–vaginal intercourse are innocuous to an individual's health. Considering this, it is not evident that *any* deviation from average frequency of penile–vaginal intercourse can be called abnormal. Deviations to lower frequencies would rather be regarded as virtuous by some. At the same time, an extremely high frequency of intercourse might interfere with other essential activities and be a cause of personal problems for the individual. However, the same would be true for any activity practiced with an exceedingly high intensity. Psychologists might wish to characterize the practice of an activity with an intensity such that it interferes with other essential activities as a compulsion or as sex addiction. In both these cases, there is an underlying disorder expressing itself in intense sexual activity. The latter is a symptom of a potential disorder, not the disorder itself. It is doubtful whether variations in the frequency of penile–vaginal intercourse ever should be called abnormal. We will discuss this issue at length in subsequent chapters.

The idea that no frequency of expression of a sexual behavior by itself is abnormal can be applied to cunnilingus, fellatio, anal sex, and so on. Consequently, variations in the frequency of execution of sexual behaviors cannot by themselves be regarded as abnormal, even if they should be far from the average frequency. Exceptional frequencies may, however, be symptoms of other psychiatric disorders.

The notion of normality as something close to average in the statistical sense poses problems in addition to those mentioned in the preceding paragraph. One concerns the origin of the average, in other words the population for which it is estimated. I will illustrate this with an uninteresting example. On the Irish coast, there is a community where sexual behaviors are most peculiar. Anthropologists have described that the only sexual activity known is penile–vaginal intercourse with the woman supine. Obviously, sexual intercourse is unthinkable outside of marriage. During intercourse, the partners are fully dressed, but the clothes are discretely arranged so that the genitals can come into contact. The notion of female orgasm is unknown to this human group and the frequency of copulation is estimated to be very low (Messenger, 1971). On the other side of the earth, in the middle of the Pacific, there is a group of islands known as the Cook Islands. One of them is Mangaia. Again, anthropologists have described some of the sexual habits of the Mangaians. From the end of puberty until marriage, boys and girls are encouraged to engage in as much and as varied sexual activity as possible. One purpose with this is to permit them to acquire the experience needed for maximizing sexual enjoyment during the years of marriage expected to follow adolescence. Both boys and girls are reported to experience several orgasms per night, perhaps with different partners (Marshall, 1971). The average number of orgasms per unit time is quite high. The point of this example is that being normal on the Irish coast is something very different from being normal on Mangaia.

The fact that the concept of normality is unstable in space is perhaps not so disturbing. Who cares about normality on Mangaia, anyway? However, even within a limited region, for example France, sexual behaviors may vary quite dramatically between one geographical area and another. The sexual behaviors displayed by the young intellectuals in Paris' Saint Germain des Prés are probably rather different from those displayed by the peasants of the Massif Central. Likewise, the average frequency of fellatio on New York's Upper East Side may be very different from that found in Western Harlem. The first case would illustrate differences between geographical areas within a country, while the second case illustrates that substantial differences may be found between one neighborhood and another within the same borough. In addition to this, it is not unusual that several subcultures coexist in the same geographical area. This is perhaps not very common in the countryside of la France profonde or on the plains of the Midwest, but in all large cities it is the rule rather than the exception. The average frequency of any behavior, like the number of visits to a gay bar per week, or number of sexual partners per hour, may show major variations between subcultures at a given geographical location.

All notions of normality as everything close to average suffer from the problem illustrated in the preceding two paragraphs: the average of a specific behavior may vary between one place and another and between one group and another at the same place. In order to determine if the behavior of someone is normal, we have first to determine from where this someone comes and to which subculture this someone belongs. This makes the notion of normality unstable, or relativistic, as some like to say. What is normal for one individual is abnormal for another. Moreover, the average frequency of a particular behavior may vary over time. Some experts suggest, for example, that the frequency of cunnilingus has increased over the last few years because of the social learning offered by means of pornographic movies. The current average frequency would have been abnormal only a few years ago. Thus, the concept of normality is not only unstable in space and between cultures, but also in time. A concept with such a variable meaning has no place in science and probably not in clinical practice either.

The concept of normality does not always refer to something close to average in the statistical sense of the word. A particular behavior pattern, for example, cannot be 'average'. It is impossible to maintain that penile–vaginal intercourse has a value of 2, while anal intercourse has a value of 3, and fellatio a value of 1. Behavior patterns cannot be assigned meaningful quantities, making it impossible to calculate an 'average behavior'. However, among the many possible sexual behavior patterns we could determine the frequency of performance of each of them and then propose that the one with the highest frequency is the average. Statistically, we would talk of mode rather than average. Behavior patterns that are performed with a frequency not far from the mode could be considered as normal while those performed with a frequency far below the mode could be regarded as abnormal. It was probably this kind of 'normality' that Kinsey had in mind. He also had a generous appreciation of how large the deviation from the average frequency should be before a behavior could be called abnormal.

The notions of normality and abnormality presented in the preceding paragraph are essentially meaningless. The frequency of some behaviors is extremely low, yet they are usually not considered abnormal. When Edmund Hillary climbed Mount Everest, he executed a most exceptional behavior. However, few people would consider his behavior as abnormal. The large deviance from common behavior patterns did not impede the queen to knight him for his exploits. Among sexual behaviors, some are probably quite unusual, albeit far from the level of exceptionality of Sir Edmund's behavior. For example, amid the many sexual activities grouped together under the name of BDSM (bondage and discipline, domination and submission, sadism and masochism) are a couple that seem to be quite infrequent, at least as far as I can judge. One of those is often called erotic spanking, meaning that one of the partners spanks the other for mutual or own pleasure, or for the pleasure of the spankee. Erotic spanking may be performed early in sexual interactions, or it may be used to enhance the quality of orgasm. The specific motor patterns involved are variable, as always in the human. One of the more demanding motor patterns consists of having the spankee slung over the spanker's shoulder. This artistic arrangement allows the spanker to paddle the spankee with his or her free hand. In my innocence, I imagine that this sexual behavior pattern is extremely infrequent, at least among those equipped with modest gymnastic capacities.

The fact that the behavior pattern is infrequent would, according to some, mean that it is abnormal. However, erotic spanking in the position described above is probably far more frequent than Sir Edmund Hillary's ascent on the Everest, so infrequency is not a convincing reason for considering the behavior abnormal. Then it could be added that spanking is not only infrequent, but it may also cause injury. The risk for injury associated with infrequency could then be sufficient reason for calling it abnormal. This argument is not particularly strong, since the probability of injury and even death during the Everest climb certainly was far above that associated with spanking. Independently of how we exhaust ourselves in a search for arguments for deeming erotic spanking as abnormal while maintaining that climbing the Everest is normal, we will not succeed.

As we have seen during several paragraphs by now, the notion of normality is void of empirical content. It is rather an arbitrary value statement, based on our own convictions of what is *acceptable* or *unacceptable* behavior. When we judge something as abnormal, what we actually do is consider it as unacceptable to our personal value system, or to our prejudice. Exactly as I did some paragraphs ago, I maintain that the concept of normality is unsuitable for scientific use.

A most interesting solution to the relativistic nature of the concept of normality was suggested many years ago (Shoben, 1964). Rather than considering anything close to average as normal, any behavior that reflects the unique attributes of the human animal should be considered normal. Shoben sided with the philosopher Cassirer (Cassirer, 1944) and considered the capacity of symbolization as the foremost of the uniquely human attributes. That capacity allows us to employ propositional language, which can be used as a vehicle for dealing with things not physically present. Such 'things' are recollection of events from the past and projections of our experience into events that have not yet occurred as well as events that have not and will not occur. The latter kind of events is frequently called imagination or fantasy. Throughout this book I have included all these uses of propositional language in the notion of 'mental representation'. Exactly as Cassirer did and exactly as I have already mentioned many times, I consider mental representations of past, future and non-existent events as something exclusive to the human. The second uniquely human attribute mentioned by Shoben is dependency on others. We need someone to feed us when infants, we need someone to teach us the appropriate use of symbols and all intricacies of human language and we need to be introduced into the mysteries of human social interactions with the assistance of others. I do not know if the dependency on others is uniquely human or not, but I do not doubt that a strong argument could be built for maintaining that it is. I will uncritically accept it as such for the purpose of the present discussion.

The two unique characteristics of the human can, according to Shoben (1964) be used as criteria for normality. As long as a human has the capacity to use symbols, meaning language, in an appropriate way, and as long as a human accepts the dependency on others, then the human should be considered as normal. Language allows us to predict the consequences of our acts in terms of rewarding or aversive consequences for ourselves and for others that may be affected by our actions. The dependency on others should make us act dependably in relation to those around us and to make us acknowledge our need for them. This approach to normality is void of references to average and consequently also from the relativism associated with that. It is universally applicable. Personally, I find the idea of trying to identify the uniquely human and then employ that as criteria for being a normal human as most attractive. Despite the attractiveness of the approach, however, there are certainly some problems with its application.

In the case of sexuality, Shoben's notion of normality fits nicely. Any sexual behavior that can be predicted to lead to reward for all partners and which is devoid of aversive consequences must be regarded as normal. Some sexual behaviors, particularly those included in the vulgar concept of kinky sex, may have some apparently aversive consequences. If the participants in these activities correctly predict the aversive consequence and find it outweighed by the rewarding consequence, then we would have to accept these activities as normal. Examples could be sadism or bondage. If the participants in such activities find that predicted reward is larger than predicted pain or other discomfort, then sadism or bondage must be accepted as normal. Likewise, a woman suffering from dyspareunia may predict that the pain suffered from penile–vaginal intercourse is less important than the predicted reward for the partner. If this woman engages in penile–vaginal intercourse, we would have to accept that her behavior is normal. These simple illustrations of the application of Shoben's criteria to sexual behaviors suggest, at least to me, that the concept may be very useful.

As a prelude to the section on 'functional' and 'dysfunctional', it is convenient to see how we can relate normality in Shoben's sense to notions of functionality. Many of the popular sexual dysfunctions, like impotence or female sexual arousal disorder, would not be considered abnormalities according to the definition provided by Shoben. Rather than judging this as a major problem, I consider it as insignificant. There is, *a priori*, no reason to consider sexual dysfunction as more abnormal than sexual function, or vice versa. As mentioned, none of the sexual dysfunctions constitutes a threat to health and they are all part of the human condition. This suggests that we need to keep the concept of 'normal' separate from the concept of 'functional'. That is exactly what I think we should. The words 'normal' and 'abnormal' are always loaded with many connotations. To most people, there is something disturbing with the word 'abnormal' while the word 'normal' is reassuring. 'Abnormal' is frequently associated with words like aberrant, irregular, contrary to some rule or principle, and similar. It contains some implicit moral judgment. This is not the case with 'dysfunction'. It is a far more neutral word, which might inspire compassion rather than opprobrium. Talking about abnormal sexual behavior has nothing to do with talking about dysfunctional sexual behavior.

The arguments exposed above may seem to contradict my earlier statement that it is entirely legitimate to consider, for example, a blood glucose concentration far from average as abnormal. There is, in fact, no contradiction. An abnormal blood glucose concentration may cause death and is therefore a threat to human life, not a normal part of it. All illnesses potentially causing death or serious injury to the organism may never be considered as normal, not even if the illness is unique to the human.

3 Sexual function and dysfunction

The expression 'functional' can be given an understandable sense without too much effort. However, first we must eliminate the biological sense of 'functional' from our discussion. To a biologist, a functional sexual behavior means that it leads to fertilization. This notion is completely unacceptable when analyzing sexuality from a behavioral point of view, as noted already in Chapter 1. An example of the uselessness of the biological sense can be taken from the preceding chapter: homosexuality is dysfunctional to a biologist since it cannot be associated with fertilization. So is cunnilingus, fellatio, penile–vaginal intercourse between individuals using contraceptives, and so forth. In fact, the biological sense of 'functional' makes most human sexual behaviors dysfunctional.

After having established the inutility of the biological sense of the concepts of function and dysfunction, we need to find a convincing replacement. I suggest that we tentatively define 'functional sexual behaviors' as *any behavior pattern involving genital stimulation and leading to sexual reward for the individual executing the behavior or for the partner*. This definition is almost identical to the one I proposed for sexual behavior in the introduction to Chapter 1. It is not complete, though. Rape, copulatory behavior with an infant and extreme sadism would all be included in this definition and most sensate people would not consider such behaviors as functional. In the sense of the definition they are functional for the rapist, the pedophile and the sadist, because they probably obtain sexual reward from these activities. It appears that we need to add something to the definition of functional and, after adding that something, it would be like this: *Any* socially accepted *behavior pattern* in response to socially accepted incentives *involving genital stimulation and leading to sexual reward for the individual executing the behavior or for the partner*. We see immediately that the addition of 'socially accepted' excludes the rapist, the pedophile, the sadist and all others showing socially repulsive sexual behaviors. The only problem with this definition is that 'socially accepted' is an arbitrary concept. A few years ago, homosexuality or fellatio would have been dysfunctional according to this definition. Furthermore, the acceptability of specific sexual behaviors and sexual incentives may vary from one subculture to another or from a geographical location to another within the same society. In addition, there may be large variations between societies at a given time and between different time points within a given society. An ideal definition should have universal validity and temporal stability. Considering that, the definition proposed above has a serious deficiency. By the way, we recognize these arguments from our discussion of normality. Nevertheless, I have tried to figure out a scientifically sound definition not including arbitrary elements but, despite having spent many sleepless nights thinking about it, I have not succeeded. Perhaps we have no choice but to accept that the meaning of 'functional sexual behavior' is unclear.

Dysfunction can be regarded as an opposite to or an absence of function. The easiest way to define dysfunctional sexual behaviors would then be to maintain that *'all sexual behaviors that are not functional are dysfunctional'*. A disadvantage with this kind of extensional definition is that it tells us nothing about sexual dysfunction. It tells us what it is not but not what it is. An intentional definition could be *'the inability to obtain sexual reward through genital stimulation'*. That definition would include many of the conditions classified as sexual dysfunctions in the diagnostic manuals. Others, like impotence, might be difficult to include under this definition. It is not impossible to obtain sexual reward without erection, but it is impossible to obtain it through penile–vaginal or penile–anal intercourse. Now, since these are only two of several motor patterns leading to sexual reward through genital stimulation, it is not obvious that the absence of these motor patterns should be considered a sexual dysfunction. The afflicted individual would be able to obtain sexual reward by other means.

In addition to impotence, there is another condition, also considered a sexual dysfunction according to recent editions of the diagnostic manuals, named hypoactive sexual desire disorder. That disorder is not characterized by an inability to obtain sexual reward through genital stimulation. It consists of a lack of interest in obtaining such rewards. A lack of interest for engaging in behaviors leading to an unnecessary reward is usually not considered dysfunctional. In fact, to attribute the notion of dysfunctionality to a lack of interest in the execution of copulatory behaviors is quite recent. Independently of this, the absence of sexual behaviors cannot be considered a dysfunction of these behaviors. They are simply not expressed and that should be the end of it. However, this does not seem to be correct. In order to apply the diagnostic label of 'hypoactive sexual desire disorder' to an individual, that individual must consider that the lack of sexual interest is a problem and the clinician must agree. If an individual considers a lack of desire as a problem, we can propose that this individual in fact desires to have desire. In other words, the individual regrets that sexual incentives do not activate the desire he/she desires to have. If we accept this reasoning, then we also must accept that hypoactive sexual desire disorder indeed is a sexual dysfunction. These examples show that the definition provided above is unsatisfactory. A better alternative is perhaps *'the inability to respond to sexual incentives with the responses of approach and genital arousal'*. This definition includes the conditions of impotence, female arousal disorder and hypoactive sexual desire disorder.

Some of the problems with the initial definitions were solved by the modified version presented at the end of the preceding paragraph. Despite the improvement, disorders like dyspareunia or orgasmic disorder would be excluded. Furthermore, in many cases we are expected not to respond to sexual incentives with approach and arousal. On a nudist beach, it would be considered as most inappropriate to approach and get sexually aroused by other nudists, although it is well known that a nude body of the preferred sex is a powerful incentive for approach and arousal in many other situations. There are many other occasions where sexual approach to other individuals with sexual incentive properties would be dysfunctional. We would have to add something to the preceding definition in order to further improve it. For example we could state: *'Sexual dysfunction is the inability to respond to sexual incentives with the responses of approach and genital arousal and to obtain sexual reward through genital stimulation when so desired'*. This modified definition would immediately allow for the inclusion of orgasmic disorder since the obtention of sexual reward was added. However, dyspareunia rests problematic. Sexual reward can probably not be achieved by penile–vaginal intercourse if penetration is painful, so if sexual behaviors were limited to that, there would be no reward and the condition would be covered by the definition. However, sexual reward can easily be obtained by other means, such as cunnilingus. Thus, dyspareunia would continue to be excluded from the dysfunctions. There are more problems with the new definition. *'When so desired'* had to be added in order to exclude all situations where the responses of approach and arousal would be inappropriate, as in the case of the nudist beach. The underlying assumption is that no decent human desires to be sexually aroused and initiate sexual approach on occasions where social norms prohibit this. However, as we have learned from Saint Augustine, sexuality is not controlled by the will. Thus, people may involuntarily desire to approach and get aroused by someone on the nudist beach, but at the same time social norms may inhibit the full expression of these responses. This example makes it clear that 'when so desired' refers to an unobservable, internal state. The only way to know if someone desires to get sexually aroused or not is to request that someone introspect and tell us the result of the introspection. This seems to be a scientifically unacceptable way to proceed, although frequently used in clinical practice. Maybe we have to accept it with the argument that we are now analyzing clinical entities.

Even if we accept the inclusion of the non-scientific expression 'when desired', the paraphilias raise a serious problem. The pedophile or fetishist desire to obtain sexual reward is not, then, included under the definition. We would have to add 'socially acceptable' in order to include them. Then it would look like this: *'Sexual dysfunction is the inability to respond to socially acceptable sexual incentives with the responses of approach and genital arousal and to obtain sexual reward through genital stimulation when so desired'*. With this addition, we can also include most of the paraphilias. Exhibitionism, however, rests tricky. An exhibitionist obtains some form of sexual reward by observing other people's reaction to the exposition of his/her genitals. Some exhibitionists masturbate when exposing themselves, thus satisfying the criterion of genital stimulation. Others do not and some do not even show signs of sexual arousal. The latter cannot be included within the definition as stated above. One solution to this inconvenience is to maintain that exhibitionism, at least those forms where genital stimulation and sexual arousal are absent, is not a sexual dysfunction. I will tentatively suggest that this is the most reasonable solution. Our definition has already become quite cumbersome and it would not be helpful to add the qualifications needed to be able to include exhibitionism. By the way, I will present more arguments for not considering exhibitionism a sexual dysfunction at all in a couple of pages.

The final definition contains one arbitrary element, '*socially acceptable*' and one subjective element, '*when so desired*'. There is no easy way to eliminate these elements without also eliminating some well established, contemporary sexual dysfunctions. I also think that the concept of sexual dysfunction inevitably must contain arbitrary elements. Some years ago, masturbation was regarded as a very dangerous sexual perversion (the label dysfunction had not become popular at that time) and so was homosexuality. Indulgence in the former led to mental retardation, general weakness and eventually to a premature death, while the latter was the cause of long post-mortem visits to hell and, if it occurred within the British Empire, to immediate hanging. Likewise, the notion of a female sexual hypoactive desire disorder would have been ludicrous only a few decades ago and still is in many circles. Within a few years some new, presently unheard of, sexual dysfunctions will appear and some of the now popular will have disappeared in the oblivion of history. It is probable that the arbitrariness inherent in the definition I proposed above, and still propose, is a significant advantage rather than a deficiency. A fundamentally arbitrary concept should not be made look like an absolute truth by providing an impeccable definition.

Having said all that, it can probably be concluded that the terms 'functional' and 'dysfunctional' are no better than the terms 'normal' and 'abnormal'. None of these expressions can be given a meaningful, universally acceptable empirical content. In connection with some future language reform they could just as well be thrown into the garbage can.

4 A short note on non-human mammals

The only normality criterion that may be applied to non-human mammals is probably that of close to average. This criterion seems to be used rather frequently in the scientific community. It is not unusual to read about rats displaying a 'subnormal level of sexual behavior' or 'having abnormally short ejaculation latencies'.

These statements mean that the rats displayed fewer mounts, intromissions and ejaculations than the average of the control rats, and that the average of the ejaculation latency was much below that of the control group, respectively. This is perhaps a legitimate use of the concept of 'normal', although it certainly would be preferable simply to describe the rats' behavior in comparison to other rats without employing any reference to normality or abnormality.

It is difficult to have any idea of what may constitute functional sexual behavior in non-human mammals. The discarded biological definition could come in handy here, but there is no reason to believe that a rat or a rhesus monkey cares more about fertility than a human does. However, it is almost certain that all non-human mammalian sexual behavior leads to sexual reward. Considering that any behavior pattern probably is socially acceptable among non-human mammals, we could argue that their sexual behavior, whenever displayed, is functional.

It is evident that my final definition of dysfunctional human sexual behavior cannot be applied to non-human mammals for several reasons in addition to the one given in the preceding paragraph. Among those is the fact that we cannot easily know if an animal would like to have desire while being unable to respond to sexual incentives. It is easy to determine whether an animal performs expected responses to an incentive or not, but there is no way to know whether an absence of a response is against the animal's desire or not. Furthermore, the expression '*socially acceptable incentive*' has a most uncertain empirical content with regard to non-human mammals. I have no idea of the sexual incentives which should be considered as socially acceptable for a marmoset or a rat. Thus, the notions of socially acceptable behavior patterns and socially acceptable incentives, which were found to be indispensable in the definition of human sexual dysfunction, are senseless when considering non-human sexual behaviors.

Since the definition of dysfunctional found acceptable for human use is inapplicable to non-human mammals, we need another under the condition that something possible to regard as dysfunctional sexual behavior exists in these mammals. Indeed it does. I mentioned somewhere that a small proportion of rats, and certainly also of most other mammals, will not show any sexual behavior in the usual laboratory conditions and some are also unable to sire offspring even when given prolonged opportunity to do so. They seem to suffer from a condition similar to hypoactive sexual desire disorder in the way that sexual incentives are unable to activate sexual motivation. So far as we know, there is no evident disturbance in these animals' endocrine system and no observable peculiarities in their central nervous system. In that way, they are similar to humans suffering from hypoactive sexual desire disorder. It is not impossible that we should consider the sexually inactive, unmanipulated mammals as dysfunctional. Since the definition we found acceptable for use in humans cannot be applied to other mammals, we need an alternative. A suitable one could be the unacceptable proposal I made first in our discussion of human dysfunctional sexual behavior, '*the inability to respond to sexual incentives and to obtain sexual reward through genital stimulation*'. That definition obviously applies to sexually inactive rats and other mammals. There are perhaps non-human sexual dysfunctions in addition to the inability to respond to sexual incentives. If there are, I must confess that I know nothing about them. However, I assume that they are covered by the definition proposed here.

The most sagacious conclusion from these simple considerations is probably that it is easier to define dysfunctional than functional sexual behaviors in non-human mammals. The notion of normality can only be used in the statistical sense, close to average.

References

Cassirer, E. (1944). *An essay on man: an introduction to the philosophy of human culture*. Yale University Press: New Haven

Lewin, B., Fugl-Meyer, K., Helmius, G., Lalos, A. and Månsson, S.A. (1998). *Sex i Sverige. Om sexuallivet i Sverige 1996*. Folkhälsoinstitutet: Stockholm

Marshall, D.S. (1971). Sexual behavior on Mangaia. In *Human sexual behavior. Variations in the ethnographic spectrum* (D.S. Marshall and R.C. Suggs, eds), pp. 103–162. Basic Books: New York

Messenger, J.C. (1971). Sex and repression in an Irish folk community. In *Human sexual behavior. Variations in the ethnographic spectrum* (D.S. Marshall and R.C. Suggs, eds), pp. 3–37. Basic Books: New York

Shoben, E.J. Jr (1964). Toward a concept of the normal personality. In *The study of abnormal behavior* (M. Zax and G. Stricker, eds), pp. 12–21. Macmillan: New York

CHAPTER

11

Hypoactive sexual desire disorder

1 Introduction

We have, throughout this book, constantly reminded the reader that, in traditional Western society, sexual activity was legitimate only when procreation was the aim. This opinion is, as we know, still held in many groups. We also know that Saint Augustine favored the idea that once a suitable number of children had been begotten, complete abstinence from sexual activity was the ideal. An absence of sexual desire would certainly be helpful in attaining that ideal. Likewise, in many groups in which sexual activity was not allowed, low or no sexual desire would be most useful. Among these groups we find unmarried men and women, clergymen of the Holy Roman Catholic and Apostolic Church, members of religious orders, and many others. The search for low sexual desire took sometimes quite drastic expressions, as when Origen castrated himself with the aid of a sharp stone, or when the Valesii not only castrated themselves but also their visitors, as I mentioned in Chapter 5. Over centuries, it was probable that the presence of sexual desire was of far more concern than its absence.

Within certain groups, the presence of sexual desire is still problematic and many creative ways have been invented for reducing the likelihood that sexual desire manifests itself in sexual activity. Other collectives are less obsessed with the social control of sexual behaviors and acknowledge that these behaviors belong to the individual domain. Moreover, the influence of the Augustinian principles has become reduced and perhaps even disappeared in many contemporary societies. Although still officially imposed by the Holy Roman Catholic and Apostolic Church and many protestant groups, members of the Catholic Church or of the protestant groups do not seem to adhere to these rules. In Europe, they are respected only by members of some marginal fundamentalist organizations. In parts of North America, religious fundamentalism appears to be far more extensive than in Europe, but it is certainly only a matter of time before these groups collapse. Nevertheless, once liberated from Saint Augustine's restrictive ideals, absence of sexual desire may be transformed from virtue to problem.

In 1980, the 3rd edition of the *Diagnostic and Statistical Manual of Mental Disorders* (American Psychiatric Association, 1980) included a category called 'inhibited sexual desire'. In a later edition of the manual, the name was changed to 'hypoactive sexual desire disorder' (American Psychiatric Association, 1987). In the *International Statistical Classification of Diseases and Related Health Problems*, 1989 revision (ICD-10) (World Health Organization, 1992), the name used is 'lack or loss of sexual desire'. In the preceding edition, the ICD-9 from 1977, there was no such category. Thus, low sexual desire turned into an internationally accepted problem in the period between 1977 and 1989.

2 Diagnostic criteria

The 4th edition of the *Diagnostic and Statistical Manual of Mental Disorders* (DSM-IV; American Psychiatric Association, 1995) describes human sexual interaction in the same way as van de Velde (1965) did 70 years ago (I remind the reader that van de Velde's book was originally published in 1926), although the terminology has been modified. Instead of talking about sexual interaction as consisting of a series of phases as van de Velde did, the more fancy denomination 'sexual response cycle' is used. Moreover, rather than calling the first phase of sexual interaction prelude, it is now called the desire phase. It is defined in this way: 'this phase consists of fantasies about sexual activity and the desire to have sexual activity' (American Psychiatric Association, 1995, p. 505).

In order to receive the diagnosis 'hypoactive sexual desire disorder' (code F.520 both in the ICD-10 and DSM-IV), a person must satisfy three criteria:

1 Persistently or recurrently deficient (or absent) sexual fantasies and desire for sexual activity. The judgment of deficiency or absence is made by the clinician, taking into account factors that affect sexual functioning, such as age and the context of the person's life.
2 The disturbance causes marked distress or interpersonal difficulty.
3 The sexual dysfunction is not better accounted for by another Axis 1 disorder (except another sexual dysfunction) and is not due exclusively to the direct physiological effects of a substance (e.g. a drug of abuse, a medication) or a general medical condition. (Reprinted with permission from the *Diagnostic and Statistical Manual of Mental Disorders*. Copyright 2000. American Psychiatric Association).

The disorder may be life-long, with the start at the onset of sexual functioning, according to DSM-IV. I assume that 'onset of sexual functioning' is a euphemism for puberty, but I am not completely sure. Considering that it is an American text, it may also mean onset of marriage. There is also an 'acquired type' of the disorder, meaning that the afflicted individual had satisfactory sexual desire during some period of life. Furthermore, hypoactive sexual desire may be limited to particular situations, to particular kinds of stimulation or to particular partners. If so, it is then considered to be of the 'situational type'. If the lack of desire is omnipresent, then the disorder is of the generalized type. Finally, a good diagnosis should include some mention of possible cause and here the DSM-IV generously offers two alternatives, psychological factors and combined factors. Whenever a general medical condition and drug use or abuse can be excluded, the cause is considered to be psychological. When the clinician suspects that a general medical condition or drug use or abuse *contribute* to the disorder, it is due to combined factors. However, if a general medical condition or drug use or abuse is sufficient to account for the lack of sexual desire, then the diagnosis 'sexual dysfunction due to a general medical condition' or 'substance induced sexual dysfunction' should be used.

There are several important things to learn from this. First, the absence of interest in sexual activities does not by itself constitute a problem. The absence must cause marked distress or interpersonal difficulties, or perhaps both at the same time. Adding the criterion of distress or interpersonal difficulties excludes, for example, Saint Augustine from the diagnosis. He would have been enchanted rather than distressed by a complete absence of sexual desire. Another important thing is that low sexual desire associated with somatic or mental disease pertaining to axis 1 should be excluded from the label of hypoactive sexual desire disorder. Axis 1 includes major psychotic states like schizophrenia, depression and bipolar disorder as well as anxiety disorders and attention deficit hyperactivity disorder (ADHD). These exclusion criteria are sometimes forgotten, as we will see later. In any case, Origen could not have been diagnosed with hypoactive sexual desire disorder for two reasons. Absence of sexual desire did presumably not cause him distress or interpersonal difficulties and the cause was a general medical condition, albeit self-inflicted. Iatrogenic hypoactive sexual desire, caused by opiate use or medication with SSRIs, is also excluded.

The subcategory situational hypoactive sexual desire is also most interesting. A man lacking sexual desire in hotel rooms, but experiencing strong desire at home, should be diagnosed as suffering from this desire disorder, provided that the absence of desire in hotel rooms causes the man distress or interpersonal difficulties. It is quite unlikely that the situational type of the disorder is due to a general medical condition, so that exclusion criterion can be excluded. The man could obviously take drugs only when staying at hotels, so this possibility should be checked by the clinician before pronouncing a final diagnosis. Still more interesting is the situational type where certain 'types of stimulation' do not activate sexual desire. It would seem that these 'types of stimulation' must be expected to produce sexual desire in most people, but that is not made explicit. My interpretation of this rather cryptic statement is that certain stimuli having sexual incentive properties for most people are inefficient in the individual suffering from situational hypoactive sexual desire disorder. The most interesting situational type is the one where a particular individual fails to induce sexual desire. Again, I assume that this individual either did induce sexual desire at some moment or that it induces sexual desire in most other individuals. Alternatively, the failure of the individual to induce sexual desire causes personal distress or interpersonal difficulties independently of whether this individual earlier produced sexual desire or is attractive to most other people. The essential point here is that a person suffering from situational hypoactive sexual desire disorder may experience desire in response to some individuals but not in response to others and that this latter causes distress or interpersonal difficulties.

We will return to several of the points made above when we arrive at the section on etiology. For the moment, our attention should be focused elsewhere.

3 Epidemiology

Several studies have revealed an amazingly high prevalence of hypoactive sexual desire disorder. An American nationwide survey of 1622 women between 18 and 59 years of age found prevalence to be around 30%, independent of age. The corresponding value for a sample of 1249 men was 15%, again independent of age (Laumann *et al.*, 1994, 1999). In the Valencia-Castellon area of Spain, 37% of the women asked and 25% of the men asked complained of low sexual desire (Arnal *et al.*, 1995). There was no age effect in men, but in women there was a continuous increase between 14 and 40 years of age. Thereafter, prevalence remained stable. In Denmark, a representative sample revealed that 11% of Danish women and 3% of Danish men consider themselves to have low sexual desire (Ventegodt, 1998), independently of age. In Sweden, 34% of women between 18 and 74 years old reported a sexual desire sufficiently low to constitute a problem. In men, the figure was 16% (Fugl-Meyer and Sjögren-Fugl-Meyer, 1999). The Swedish study belongs to the rather small group of surveys where it was explicitly determined whether the respondents considered their level of sexual desire as problematic. A positive answer was interpreted to mean that it caused distress or interpersonal difficulties. Thereby, the Swedish study could determine not only the presence of low sexual desire, but also do so in relation to the diagnostic criteria employed in DSM-IV and ICD-10. Another interesting feature of the Swedish study was that the prevalence of low sexual desire increased with increasing age in both men and women.

A reasonably recent, large multinational study determined the prevalence of low sexual desire in men and women between 40 and 80 years of age in 29 countries. The results of that study are very similar to those reported above, with a prevalence of 30–40% among women and 10–20% among men. The exception was southeast Asia, including Indonesia, Malaysia, Philippines, Singapore and Thailand. There, the prevalence among women was as high as 43.3%, and that of men was 28%. Neither the authors of the paper nor myself have any explanation to offer for this unusually sad state of affairs. With regard to the influence of age, it was found that it had none (Laumann *et al.*, 2005).

A few general conclusions can be drawn from the data presented in the preceding paragraphs. First, the incidence of low sexual desire in women is at least double of that in men. This is a very consistent finding that has been replicated in all epidemiological studies. I know of no contradictory data with regard to this point. Second, the vast majority of studies fails to detect any effect of age. There are some rare exceptions to this, but those exceptions did not prevent an informed reviewer from concluding that the prevalence of sexual dysfunctions remains constant with age in women (Hayes and Dennerstein, 2005). I cannot but agree with this conclusion.

The epidemiological data show that low sexual desire is the most common of all sexual dysfunctions in women. It is frequently followed by orgasmic disorder and sexual arousal disorder (deficient vaginal lubrication). In men, erectile dysfunction is less common than desire disorders when young, while the reverse is true in old men. Premature ejaculation is the most common dysfunction in young men and almost certainly followed by low sexual desire. The prevalence of premature ejaculation declines with increasing age and, at the time impotence is common, premature ejaculation is rare in those still potent. Nevertheless, while low sexual desire appears to be the major sexual dysfunction in women, it does not appear to be so in men of any age. These epidemiological data justify the fact that female hypoactive sexual desire disorder has received far more attention than the corresponding state in men.

The fact that self-reported low sexual desire is more frequent in women than in men attracts my attention. There is no evident reason for women to suffer from that problem more frequently than men. However, the prevalence of depression or of anxiety disorders is also at least double in women than in men, without any evident reason. Furthermore, it is known that morbidity in general is higher in women than in men (Verbrugge, 1985; Bogner and Gallo, 2004), and there is no generally accepted explanation for that. Different incidence of disease in men and women is, then, quite common and not at all special for disorders of sexual desire. There may be one hitherto unknown factor that can explain all these sex differences, making it futile to try to invent some credible explanation for the particular case of sexual desire. At the same time, each of the many differences may have a unique cause. Since there is no way to know, I will allow myself to launch some speculations regarding a possible explanation for the high prevalence of low sexual desire in women compared to men.

In Western society, conventional wisdom holds that men should be more sexually active than women. In fact, the notion that women engage in sexual activity only for the pleasure of their male companion disappeared quite recently. A lot of questionnaire data support the popular wisdom of lower sexual motivation in women than in men (Baumeister *et al.*, 2001). These data are easily explained by factors unrelated to differences in sexual motivation. In fact, there is no convincing reason, besides social stereotypes, to believe in any sex difference in motivation between men and women. However, the belief is strongly rooted and certainly influences much of our thinking. One result of that belief is that it is socially acceptable for men to express their sexual urges, while women are supposed to show more constraint in such expressions. A woman having no interest in sex should, in several ways, coincide with the ideal of a woman. The lack of desire would, thus, be completely unproblematic. If the woman has a male partner, it is not certain that the partner would find the lack of sexual desire in his companion as unproblematic as she does. Since men are allowed to express their sexual impulse, it is not impossible that the partner makes explicit his inconformity with the situation. It then turns into a problem and the woman's condition satisfies the criterion of 'personal distress or interpersonal difficulty'. Perhaps it is not even necessary for the male partner to express anything at all for making the woman perceive an interpersonal difficulty. If she is a strong believer in the myth of male hypersexuality she may, by her own initiative, imagine that the situation is unpleasant for the partner, thereby making an interpersonal problem of it.

A man with a lack of sexual desire may react in a different way. Again, according to social stereotype he knows that women have little interest in sex and he might convince himself that his absent desire is actually something pleasant for the partner. It will not be perceived as a cause for interpersonal difficulties. Moreover, the female partner is not expected to show her eventual sexual urges that she might have and she probably does not show them. Accordingly, the man may be totally unaware of the distress he is causing his partner and he sees no problem. This reasoning may appear extremely simplistic, perhaps even vulgar, to a cultivated reader. Tragically enough, social stereotypes are more influential than many cultivated women and men would like to admit. Admittedly, there are no data directly supporting my argument but, at the same time, there are no data contradicting it. Since I cannot find any other reasonable explanation for the sex difference in the prevalence of hypoactive sexual desire disorder, I have no choice but to consider mine the best available.

There are also people without a stable partner suffering from low or absent sexual desire. In their case, the explanation proposed above is useless. No epidemiological study has directly addressed the prevalence of low desire in people without a stable partner, making it impossible to know if there is also a sex difference in this group.

4 Etiology

4.1 General

There are many speculations and few data with regard to the causes of hypoactive sexual desire disorder. The name originally given to it, inhibited sexual desire disorder, suggested that the underlying sexual motivation was intact, but some processes blocked its expression. Such a distinction between a basic, unaffected sexual motivation system and the blockade of its expression has important consequences for treatment. If we believe in this hypothesis, a successful treatment should remove the inhibitions imposed on this intact motivation rather than trying to enhance it. On the contrary, if we suspect that the cause of the lack of desire resides in absent motivation, then we would need to find a way of enhancing that motivation rather than removing inhibitions, particularly since there are no inhibitions to remove, according to the suspicion determining our choice of treatment. In view of the therapeutic implications of the conflicting hypotheses, it becomes quite important to determine which of them is correct.

A substantial amount of clinical data supports the hypothesis of active inhibitory processes as causes of hypoactive sexual desire disorder. The most eloquent exponent of this hypothesis is undoubtedly Helen Kaplan. In several books, she has summarized an impressive clinical experience (e.g. Kaplan, 1979, 1995). She suggests that intrapsychic sexual conflicts and neurotic interactions with the partner are the deeper causes of disorders of sexual desire. A common feature of her cases is that the partner suffering from the disorder perceives the other partner as ugly, rejecting, hostile, brute and ignorant, just to take a few adjectives from the many case histories reported by Kaplan. To the clinician, however, the partners rarely gave the same impression. Most of them are described as handsome, attractive, considerate, soft-spoken and similar. In an incentive motivational language, it seems that the partners' perceived incentive value was not particularly high from the viewpoint of the other partner. Furthermore, few of the clinical cases described suffered from life-long hypoactive sexual desire disorder. On the contrary, at some earlier point most of them had maintained a satisfactory sexual relationship with their current partner. This implies that something had happened in the course of the relationship that had transformed the partner from a positive incentive to a neutral or even a negative incentive.

Factors unrelated to sexuality may perhaps reduce the partner's sexual incentive value. There is little research on the issue, but it has been suggested that something called poor marital adjustment could contribute to the development of hypoactive sexual desire. The data on this issue are mixed, though. Some studies report marital adjustment in couples where one member complains of low desire to be at the same level as in couples without such complaints (e.g. Schiaivi *et al.*, 1992), while others have found poor marital adjustment in such couples (e.g. Trudel *et al.*, 1993). A recent study evaluated the relationship between severity of hypoactive sexual desire disorder assessed by the Sexual Interest and Desire Inventory-Female (SIDI-F; Sills *et al.*, 2005) and marital adjustment assessed by the Marital Adjustment Scale (Locke and Wallace, 1959). No significant relationship was found (Clayton *et al.*, 2006). These data are of considerable relevance, since both scales have been shown to be reliable and probably also valid. It would appear, then, that alterations in general marital adjustment cannot be considered a universal cause of low sexual desire. These negative data do not exclude the possibility that poor adjustment is a contributing factor in some couples. However, even if poor marital adjustment were associated with hypoactive sexual desire, it is not entirely clear which is the cause of which. An unwanted absence of sexual activity, or imposed celibacy, may have adverse effects on marital adjustment. Rather than being the cause of low desire, poor adjustment may be a consequence of it. The absent or weak relationship between general marital adjustment/satisfaction and hypoactive sexual desire disorder makes it unlikely that extrasexual elements contribute to the reduction of a partner's sexual incentive value to any sizeable extent. Furthermore, since these potential extrasexual elements are entirely unknown, I will simply ignore them and focus the attention on the quality of the sexual relationship.

4.2 An incentive motivation/learning account of the origin of hypoactive sexual desire disorder

In the next few paragraphs, I will discuss the development of acquired hypoactive sexual desire disorder in incentive motivational and learning terms. With the purpose of keeping the arguments as comprehensible as possible, I will consider that the sex of the partner suffering from hypoactive sexual desire disorder is a woman. With slight modifications, the arguments can equally well be applied to men. I will also assume that the woman lives in a stable relationship. This assumption is made because most of the clinical descriptions available refer to people living in such relationships.

At the beginning of the relationship, the partner was a sexual incentive. Approach behaviors were activated and copulatory activities ensued. As we know, copulatory activities usually, but not always, end in orgasm. We also know that the frequency of orgasm is an important determinant both of sexual activity and satisfaction (Gebhard, 1966; Bentler and Peeler, 1979; Lief, 1980). It also determines the intensity of motivation (Arafat and Cotten, 1974). These old data have been confirmed in a quite elegant report showing that women rate the copulatory behavior pattern most frequently associated with the obtention of orgasm, cunnilingus, as the most satisfying. At the opposite end of satisfaction they rate anal intercourse and fellatio. These behaviors rarely lead to orgasm in women (Hurlbert *et al.*, 1993a). Knowing that orgasm is the most reinforcing and rewarding of all sexual events, we maintain that the preceding data establish, beyond doubt, that the frequency of orgasm is an important determinant of the partner's sexual incentive value. This conclusion should not come as a surprise by now.

Having established that the degree of satisfaction with sexual activities as well as the partner's incentive value are determined by the likelihood of obtaining orgasm, I now proceed to suggest that the degree of dissatisfaction with sexual activities is directly proportional to the likelihood of *not* obtaining orgasm. The rationale is straightforward. When attracted by the sexual incentive and after completed approach, copulatory activities are frequently initiated. If not made long before, at the beginning of copulatory interaction the partners can make the reasonable prediction that orgasm and the associated intense positive affect sooner or later will follow. They produce a mental representation of that positive affect and a strong expectancy of its obtention is created. If it does not arrive, the expectancy of reward turns out to be false. The absence of an expected reward has been shown to induce a state of negative affect, an aversive state, both in humans and other mammals (see Amsel, 1962). An association will be formed between the most salient stimulus at the moment, the partner, and the aversive state induced by the frustrated expectancy. This association will reduce the positive incentive value of the partner. If this frustration of expectancy should be repeated many times, then the partner's positive incentive value may disappear and he will become a neutral stimulus. It is also possible that the association between the partner and aversive events continues beyond the point of neutrality, turning the partner into a negative incentive activating the response of withdrawal.

There are many examples in the clinical literature of women suffering from hypoactive sexual desire disorder who display active withdrawal responses from stimuli that would be expected to function as sexual incentives. Even the mentioning of anything sexual may 'bring tears to the eyes and outbursts of rage', as so eloquently described by Helen Kaplan in one of her case histories (Kaplan, 1979, p. 136).

In addition to the frustrating, hence aversive, event of lack of expected orgasm, other aspects of the sexual interaction may reduce the partner's incentive value. A man insisting on anal intercourse as the sole copulatory pattern may not only deprive the woman of the experience of orgasm, but he may also cause her physical pain when penetrating the rectum. It is also possible that the woman finds anal intercourse repugnant for other reasons, for example religious conviction or an obsession with hygiene. Many of these considerations may also apply to fellatio and other forms of copulatory motor patterns. Even the traditional and unobjectionable penile–vaginal intercourse may never end in orgasm if the partner constantly supplies too little sensory stimulation for one reason or another and it may become an aversive event if the partner's behavior is sufficiently inconvenient. There are undoubtedly many ways in which a partner can lose the sexual incentive properties and either become a neutral stimulus or a negative incentive.

There is no direct empirical support for the hypothesis concerning the origin of low or absent sexual desire described above. Some indirect support is obtained from a study where it was found that couples where one member suffered from hypoactive sexual desire disorder reported a lower intensity of pleasure derived from sexual activities than that reported by couples without such a diagnosis. Moreover, the sexual behavior repertoire was more limited in couples with low sexual desire than in other couples (Trudel *et al.*, 1995). It could be argued that these differences were not a cause of low sexual desire but the consequences of it. However, it appears that the reduced experience of sexual reward in couples with hypoactive sexual desire disorder precedes the onset of the disorder (Trudel *et al.*, 1997), suggesting that it indeed may be a cause and not a consequence. There are no other data that can be considered as relevant for the explanation of low sexual desire in incentive motivational terms. The reason for this is presumably that few, if any, clinicians are used to thinking of sexual behaviors in incentive motivational terms. Helen Kaplan, for example, is of psychoanalytical inclination with a very weak interest in learning processes and motivation theory. Most other clinicians share this lack of interest in learning and motivation theory and prefer to pursue their research in far more sophisticated directions. Due to this unfortunate state of affairs, the reasoning presented above is nothing more than speculation.

Despite the absence of supportive empirical data, the incentive motivational approach offers some important advantages in relation to other explanations of hypoactive sexual desire disorder. One is that it is part of a theoretical framework making it possible to organize our thoughts and potential data in a systematic way. I have already referred to the importance of this on many occasions, and I will not insist any further. Another is that the futile discussion concerning possible inhibitory mechanisms impeding a basic motivation from being expressed versus intrinsic deficiencies in this basic motivation can be reformulated in testable terms. By accepting that sexual behaviors, including motivation or desire, are determined either by external stimuli or mental representations of such stimuli *and* the central motive state, we can know where to look for explanations for the disorder. The central motive state requires the exposure to appropriate amounts of gonadal hormones, as we learned in Chapter 5, and some brain structures, as we learned in Chapter 6. Both these requirements can be empirically evaluated. Furthermore, the central motive state corresponds probably to the mysterious concept of basic or intrinsic motivation employed in the discussions of hypoactive sexual desire disorders.

The incentive motivational explanation of hypoactive sexual desire disorder may account for some clinical facts that are most difficult to explain in other terms. One particularly evident case is the variant of situational hypoactive sexual desire disorder where the patient lacks sexual interest in one person, usually the partner, while being attracted to and performing sexual responses with other persons. The husband who is not attracted to his wife, and even actively avoids sexual approaches from her, while going to a brothel three times per week for having and wanting to have sex with prostitutes is an excellent example. The wife that has no interest whatsoever in having sex with her husband, but happily copulates with an occasional lover in a shabby hotel room is another.

Besides the lack of desire for a specific partner, there are other variants of situational hypoactive sexual desire disorder that are equally easy to account for in the context of the incentive motivation model, but which are representing difficulties for other models. One such variant is lack of desire at a particular place, like the conjugal bedroom. A person might be completely uninterested in having sex with her partner in their own bedroom, but that person may experience intense desire to have sex with that same partner in the guest bedroom at her grandfather's farm in Machtolsheim outside Tübingen. The incentive motivational model would state that some stimulus or set of stimuli in the home bedroom had acquired negative incentive properties by being part of the context in which aversive experiences had occurred. Since this stimulus/set of stimuli is absent from the bedroom at the farm, no negative incentive activates the response of withdrawal thereby allowing the sexual incentive properties of the partner to manifest themselves. An explanation of hypoactive sexual desire as a result of neurotic reactions to sexuality would have problems explaining this kind of situational low desire. Likewise, an assumption of lacking intrinsic motivation would also fail.

4.3 Some rat data supporting the incentive motivation/learning model and a note on the relationship between desire and motivation

The observant reader may have noticed that I make mixed use of the terms desire and motivation in this chapter. This is not a mistake, but a deliberate strategy having the purpose of creating confusion. After suffering from that confusion for several pages, I calculate that the reader will be more receptive to my clarification of the issue. In the clinical literature, the term motivation is rarely used in the sense it has in experimental psychology. In experimental psychology, the term desire is never used. Yet the meaning of 'motivation', in experimental psychology, is very similar or identical to the meaning 'desire' has in the clinical literature. I do not know why clinicians have chosen to employ the term desire rather than motivation. It would be unfair to suggest that clinicians are ignorant of psychological terminology, so that explanation must be rejected. In any case, I have convincingly argued for the equivalence of the terms 'motivation' and 'desire' elsewhere (Ågmo *et al.*, 2004). Briefly, according to the Merriam-Webster online dictionary (http://www.m-w.com/cgi-bin/dictionary) the word desire has several meanings: *1. Conscious impulse toward something that promises enjoyment or satisfaction in its attainment; 2. Sexual urge or appetite; 3. A usually formal request or petition for some action.* The first two meanings may be of particular relevance here. 'An impulse towards something that promises enjoyment or satisfaction' can easily be translated to terms used in scientific psychology. A 'something' with these properties is exactly what is called a positive incentive in motivation theory (Bindra, 1974, 1976, 1978). Moreover, the 'conscious impulse toward' and 'urge and appetite' are normally signs of motivation, again as the term is used in psychology. Combining positive incentive and motivation we find that desire is equivalent to incentive motivation. If we also take the word 'sexual' from the second meaning in Merriam-Webster, we end up with the term 'sexual incentive motivation' as equivalent to the term 'sexual desire'.

In contrast to the proposal that sexual desire and sexual motivation are equivalent concepts, it has been argued that they need to be separated (Hurlbert *et al.*, 2000). Sexual desire was defined as 'a cognitive process to either approach or avoid sexual activity', while sexual motivation is 'a behavioral process to either approach or avoid sexual activity' (Hurlbert *et al.*, 2000, p. 326). Most motivation theorists would disagree with this distinction, though, since motivation is not *behavior* but a *hypothetical factor causing behavior* (Bindra, 1974). The same factor may also determine cognitive activities. For this reason, the suggested distinction between motivation and desire seems artificial and unfounded in psychological theory.

As a courtesy to clinicians, I try to use the term 'desire' when talking about the human and the term 'motivation' when talking about other mammals. There is one exception, though. When I treat points of theoretical consequence, I try to use the term 'incentive motivation'. 'Desire' would be too vulgar in the context of psychological theory.

After this excursion into the fascinating world of conceptual confusion, we can now approach some rat data relevant for the topic of learned modifications of sexual incentive motivation. It is known that the association of an aversive event with the execution of copulatory behavior eventually will lead to the disappearance of the behavior. In a series of ingenious studies, male rats were injected with LiCl shortly after ejaculation. This injection causes the stomach to be upset for a short while and is believed to constitute an aversive experience. After several pairings of copulation until ejaculation with the LiCl injection, copulatory behavior was extinguished (Peters, 1983; Koch and Peters, 1987; Peters *et al.*, 1989). Disappearance of copulatory behavior was not due to reduced blood androgen concentration (Koch

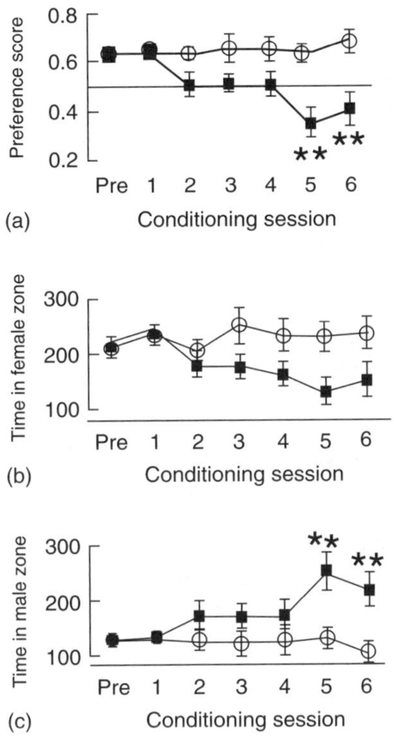

FIGURE 11.1 Preference score (a), time (in seconds) spent in the female incentive zone (b) and time spent in the male incentive zone (c) in male rats tested before conditioning (Pre) and after 6 conditioning sessions where LiCl was administered either contingent (conditioned group, □) or non-contingent (control, ○) on copulation. Data are mean ± sem. **, different from control, $P < 0.01$. There were 12 animals per group. (Reprinted from Ågmo, 2002. Copyright 2002, with permission from Elsevier.)

and Peters, 1992). Based upon these observations, I did a funny little experiment some years ago (Ågmo, 2002). Male rats were allowed to copulate until ejaculation with females scented with capelin oil. This is an oil from a North Atlantic fish slightly aversive to the human nose but perfectly neutral to the rat. One minute after ejaculation, some males received an injection of LiCl while the controls received saline. After a few pairings the LiCl treated rats ceased to copulate. In a test for sexual incentive motivation, these males showed a complete loss of interest in the scented female. In fact, rather than approaching her, they avoided her. The scented female had been transformed into a negative incentive, activating withdrawal behaviors. This is exactly what we see in humans suffering from hypoactive sexual desire disorder. They frequently withdraw from sexual incentives.

Some data from this experiment are shown in Figure 11.1. The experimental setup employed for quantifying the intensity of sexual incentive motivation is shown in Figure 11.2.

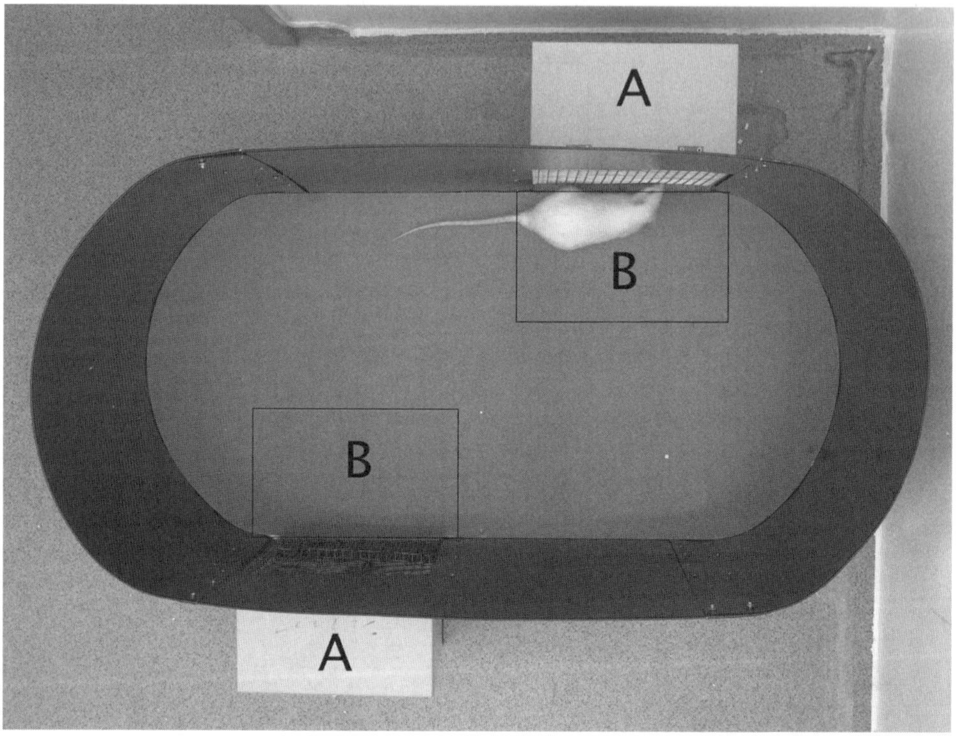

FIGURE 11.2 Design of the apparatus used for evaluating sexual incentive motivation. The incentive animal cages (marked A) are removable and can easily be exchanged within an experimental session. Outside these cages, a virtual zone is defined (marked B). The time the experimental subject spent in the zone adjacent to a sexually receptive female compared to that spent in the zone adjacent to another male during a 10 min test is the measure of sexual incentive motivation. Many data show that approach to a male represents social motivation while approach to a sexually receptive female represents social + sexual motivation. The difference between the incentives is, then, dependent on the intensity of sexual motivation. (Reprinted from Spiteri and Ågmo (2006). Copyright 2006, with permission from Elsevier.)

When the scented female was replaced by another female, the LiCl treated males approached that female as much as control animals did. The aversion was specific to the stimuli associated with the aversive event. In the human, we would call this situational hypoactive sexual desire disorder. When given the possibility to copulate with the unscented female, most of the LiCl treated males did so. However, their behavior was less intense than that of the control males. This shows that the inhibition of copulatory behavior also was situational to a large degree. In any case, the results of this little experiment show that aversive experiences associated with sexual interactions can transform a sexual incentive into a negative incentive, causing avoidance behavior.

4.4 Potential endocrine causes of hypoactive sexual desire disorder

It is difficult to imagine that deficiency in some gonadal hormone could account for the situational variants of hypoactive sexual desire disorder. It may also be difficult to explain the acquired variant in terms of endocrine alterations, unless we assume that the onset of these endocrine alterations coincide with the onset of hypoactive sexual desire disorder. On the contrary, an endocrine explanation may be very suitable for explaining the life-long variant. However, it is most important not to forget that hypoactive sexual desire caused by a general medical condition is excluded from the diagnosis of hypoactive sexual desire disorder. Whenever reduced sexual desire can be attributed to diseases in the endocrine glands, castration or ovariectomy or any other medical condition, another diagnosis must be given. On the contrary, the combined type accepts organismic conditions as *contributing* factors, acting in conjunction with unknown psychic entities.

In Chapter 5, I made clear that adequate blood concentrations of androgens are necessary for maintaining a functional central motive state in both men and women. Therefore, it is not startling that several studies have determined blood androgen concentrations in women diagnosed with hypoactive sexual desire disorder. Early studies failed to detect any difference between women reporting an adequate level of desire and those diagnosed with hypoactive desire disorder (Stuart *et al.*, 1987; Schreiner-Engel *et al.*, 1989). In contrast, a group of women with life-long hypoactive sexual desire disorder were found to have a lower free testosterone blood concentration than controls (Riley and Riley, 2000). This is a very interesting observation, suggesting that the life-long variant of desire disorder can be attributed to reduced reactivity of the central motive state because of insufficient androgen stimulation. Later studies have extended these findings. Premenopausal, regularly cycling women with self-reported low libido (a common clinical expression referring to low sexual motivation) were found to have lower blood testosterone concentrations than a control group judged to have normal (probably meaning average) libido (Guay, 2001). An additional study from the same group replicated these findings and expanded them to postmenopausal women. Unfortunately, no control group was used. Instead, reduced androgen concentration was defined as any concentration of free testosterone below 2 pg/ml in women under 50 years of age, and as below 1.2 pg/ml for women over 50 years. These values are purportedly far below the means generally reported in the literature. Results showed that 70% of the women selected because of decreased sexual desire had free testosterone concentrations below the criterion. The proportion of women with below-criterion free testosterone concentration did not differ between premenopause and menopause. It may also be noted that total testosterone and dehydroepiandrosterone sulfate were reduced in women with low desire (Guay and Jacobson, 2002). Similar data were reported in another study of women complaining of low sexual desire. In that latter study, a control group of healthy women of about the same age as the patients was included. Half of the women in each group was premenopausal and the other half was in menopause. Total testosterone, free testosterone and dehydroepiandrosterone sulfate were reduced in women with low desire compared to controls (Turna *et al.*, 2005).

The last few studies mentioned above coincided in reporting reduced androgen concentrations in women with low desire. They also coincided in sloppy descriptions of the diagnostic criteria used for considering a woman as suffering from low sexual desire. All we are told is that their chief complaint was decreased sexual desire. In the Turna *et al.* (2005) study, we are also told that the women had presented symptoms for at least 6 months. Moreover, all subjects participating in that study answered the Female Sexual Function Index. Results revealed a significant difference between patients and controls, suggesting that there indeed was a difference between patients and controls with regard to sexual desire. Nevertheless, in none of the latter studies mentioned here was the diagnosis based on the criteria given in the DSM-IV or ICD-10. Among other things, we do not know if the patients suffered from acquired or life-long hypoactive sexual desire. This could be of some importance, strangely enough. Likewise, we are not told whether the disorder is generalized or situational. It appears that the clinical researchers do not realize the crucial role of a complete and correct diagnosis before taking blood samples and determining hormone concentrations.

We will now turn to some observations contradicting the notion of reduced androgen concentration in women with low sexual desire. In a study employing women with loss of libido assessed by a psychosexual counselor and a group of healthy controls of about the same age, no difference between women with low desire and the controls was found with regard to blood concentrations of total testosterone, free testosterone, sex-hormone binding globulin, estradiol, dihydrotestosterone and dehydroepiandrosterone sulfate (Nyunt *et al.*, 2005). The criteria for being considered a patient were less clear in this study than in any of those mentioned above, strongly reducing the value of its results. Nevertheless, they coincide with results reported by van Anders *et al.* (2005). These researchers measured testosterone in saliva rather than in blood, but that has become a popular procedure and it may perhaps be acceptable. The patients were assessed through a careful clinical interview. No difference in saliva testosterone concentration between patients and controls was found.

The conclusion that may be drawn from the studies having assessed alterations in androgen concentrations in women suffering from low desire in comparison to women not suffering from it is uncertain. About half of the studies report no difference while the other half does. That is exactly what we should expect if results were random. The most logical conclusion here would be that there is no endocrine difference between healthy women and women with low desire. This may be true, but due to the sloppy diagnostic criteria employed in almost all studies, such a conclusion is probably unjustified with regard to hypoactive sexual desire disorder as defined by the diagnostic manuals. Perhaps the only reasonable conclusion is that self-reported low sexual desire is not necessarily associated with reduced blood androgen concentrations. Furthermore, it is impossible to determine whether the central motive state shows a reduced sensitivity to sexual incentives because of deficient stimulation by androgens or not.

The preceding discussion was limited to studies in women. Male hypoactive sexual desire disorder has attracted far less attention. So little, in fact, that there are no recent data concerning blood androgen concentrations in men complaining of low sexual desire. An old review suggested that men suffering from low desire had blood androgen concentrations within the normal range (LoPiccolo and Stock, 1986). This is not unlikely, since there are data showing that symptoms of androgen deficiency, including reduced sexual interest, appear at blood concentrations far below the normal range or at the bottom of that range. There are large, stable interindividual variations in androgen sensitivity in men. This means that some men report symptoms of deficiency only when blood androgen concentration are very low while others report such symptoms already when concentration approaches the lower normal range (Kelleher *et al.*, 2004). A most modest relationship between scores on some obscure questionnaire evaluating sexual desire (among other things) and blood total testosterone concentration was reported in a German study (Beutel *et al.*, 2005) of adult and old men visiting andrology clinics for evaluation of 'aging-male' symptoms, whatever that may mean. There was no correlation between free testosterone and desire. The significant correlation between total testosterone and desire was 0.188, meaning that 1.39% of the interindividual variation in desire can be attributed to variations in testosterone concentration. This figure does not seem to be particularly impressive and may justify the suggestion that low sexual desire is not a consequence of low blood androgen concentrations made by LoPiccolo and Stock (1986) some 20 years ago.

The fact that the normal variations in blood testosterone concentrations are not associated with sexual desire does in no way exclude that men suffering from hypoactive sexual desire disorder may have lower blood androgen concentrations than men not suffering from it. The lack of empirical studies in men with an appropriate diagnosis obliges me to refrain from any more tangible conclusion.

4.5 Other potential causes

The presence of the premenstrual syndrome, oral contraceptives, hysterectomy, ovariectomy, perimenopausal or menopausal states has been invoked as possible causes for low sexual desire. There is no evidence that any of them is a significant etiological factor (Warnock, 2002). They may be important in some exceptional cases, though. Generalized anxiety has been proposed as a factor contributing to low sexual desire in men and women, and comorbidity such as depression and diabetes have also been suggested to be a factor (Bozman and Beck, 1991; Trudel, 1991; Spector *et al.*, 1993; van Minnen and Kampman, 2000). As we know, all these states can be included in the combined variant of hypoactive sexual desire disorder. There is no doubt that all these factors, and probably many others, can play a role in the manifestation of low desire. It is far more doubtful whether they can be the original cause.

Treatment with the SSRI antidepressants is frequently associated with sexual side effects, including reduced desire (e.g. Aldrich *et al.*, 1996; Gregorian *et al.*, 2002). We know that iatrogenic low sexual desire is excluded from the diagnosis of hypoactive sexual desire disorder, but that knowledge does not impede a quite intense interest for the search of a treatment for SSRI-induced low sexual desire. That does not concern us now, but the fact that drugs enhancing serotoninergic neurotransmission can produce a state similar to hypoactive sexual desire disorder could easily lead to the hypothesis that alterations in serotoninergic systems are at the origin of the disorder. Abundant rat data show that enhanced serotoninergic neurotransmission reliably inhibits all aspects of sexual behavior (see e.g. Ahlenius *et al.*, 1980; Vega-Matuszczyk *et al.*, 1998). Strangely enough, this rather evident hypothesis has not even been proposed. Consequently, it is not easy to emit any kind of judgment as to its veracity. I hope, though, that it will be proposed and subjected to test in a not too distant future.

4.6 A curious observation of some consequence

Two most remarkable studies were published quite recently. They were remarkable first because they included groups of women diagnosed with hypoactive sexual desire disorder or female sexual arousal disorder *according to the criteria established in DSM-IV*. This is certainly admirable, but the diagnosis was not complete. There is no indication whatsoever neither with regard to life-long or acquired disorder, nor to generalized or situational disorder. These omissions are most unfortunate. The second reason making these studies remarkable is that they included an objective measure of sexual arousal in response to a sexual incentive. This is exceptional indeed. Changes in vaginal blood flow, determined by photoplethysmography, were the dependent variable. A pornographic movie segment depicting precopulatory, heterosexual behaviors followed by penile–vaginal intercourse was the sexual incentive stimulus. The final remarkable aspect of these studies is the result. It turned out that women diagnosed with hypoactive sexual desire disorder or sexual arousal disorder enhanced their vaginal blood flow in response to the pornographic movie as much as the healthy controls did. The women were also asked to rate their arousal by responding to a questionnaire. Again, there was no difference between women with low desire and healthy controls (Meston and McCall, 2005). In the second study, some women with orgasmic disorder, dyspareunia or vaginismus were included in addition to women diagnosed with hypoactive sexual desire disorder or sexual arousal disorder. Again, the dysfunctional females showed an enhancement of vaginal blood flow during exposure to the pornographic movie similar to that found in healthy controls. Also, the women in this study were asked to fill out a questionnaire evaluating sexual arousal. Now it turned out that the dysfunctional women had a lower score than the functional women (Meston, 2006). Since the effects of the pornographic movie on questionnaire responses is somewhat variable we may just as well ignore them.

If the observations regarding vaginal blood flow responses to pornographic movie fragments were confirmed, we would be able to draw some amazing conclusions. For example, the impact of the sexual incentive employed on the central motive state must be identical in women with and without sexual desire disorder. This can be deduced from the similar intensity of the visceral response displayed in both groups. Furthermore, the incentive value of the movie fragments must be similar in both groups. Otherwise there would have been some difference in intensity of response. Thus, there can be no deficiency in 'intrinsic' motivation between these women, and there can be no difference in response to a new sexual incentive. Knowing that all women were in a stable heterosexual relationship, we may perhaps conclude that they must have suffered from a hypoactive sexual desire disorder of the situational type, in which the partner had lost his sexual incentive properties or in which the home environment somehow functioned as a negative incentive. It is utterly regrettable that no complete diagnosis was offered in the reports. Otherwise we could have known whether this conclusion is correct or not.

An ancient study of sexual arousal in men suffering from low desire, impotence, or premature or retarded ejaculation yielded results somewhat different from those mentioned in the preceding paragraph. The penile response to a pornographic audiotape was smaller in the patient group than in healthy controls of about the same age (Heiman and Rowland, 1983). The interest of this observation in the context of disorders of sexual desire is limited, since more than half of the participating men complained of impotence. The responses of the two subjects with low desire probably disappeared among those of the nine impotent men. I mention this study only because it is all we have with regard to men. By doing so I expect to avoid all possible accusations for promoting sexual inequality.

4.7 Conclusion

To date, the etiology of hypoactive sexual desire disorder is badly understood. In fact, there is no generally accepted hypothesis as to its origin. A factor contributing to our ignorance is the sloppy diagnostic procedures employed in most of the clinical research. This seems to be a long tradition and there is no hope that it will change in the near future. Despite the paucity of empirical data, I would not be surprised if it some day was confirmed that the life-long variant of the disorder is related to a deficient androgen production. With regard to the acquired variant, I can find no better explanation than the incentive motivation/learning proposal made in the beginning of this section. We will soon see that it receives ample support from the therapeutic procedures most successful for the treatment of hypoactive sexual desire disorder.

5 Treatment

5.1 Psychotherapeutic procedures

As so many times before, I need to point out that there is an impressive discrepancy between the efforts invested in finding a suitable treatment for hypoactive sexual desire disorder in women and that invested in searching for treatments of the disorder in men. I do not know if the real reason for this is known, but a feminist would certainly sustain that male malevolence is at the root. Since most researchers are men, and since most sufferers of the disorder are women, it is quite natural that the male scientists want to find a cure for an annoying state in their wives. Be it how it may with this, we have to admit that there is no known psychotherapeutic treatment program with controlled outcome data for low sexual desire in men. Despite my very best intentions I am, therefore, obliged to limit the following account to treatments for female hypoactive desire disorder.

There is no psychotherapeutic treatment for female hypoactive sexual desire disorder among the therapies with proven efficacy. However, a therapeutic procedure named Orgasm Consistency Training is included on the American Psychological Association's list of probable efficacious treatments (Chambless *et al.*, 1998). I will begin with a rather thorough examination of this therapy.

The basis of the treatment program is a four-step procedure developed, practiced and described by LoPiccolo and Friedman (1988). As almost every treatment program aimed at sexual dysfunctions, the LoPiccolo and Friedman program is heavily influenced by the work of Masters and Johnson (1970). It is also, in one way or another, influenced by a series of speculations as to the cause of low sexual desire. The number of potential causes listed by LoPiccolo and Friedman (1988) is impressive indeed, and goes far beyond my previous discussion of etiology. The potential causes are religious orthodoxy, anhedonic or obsessive-compulsive personality, problems with gender identity or object choice, sexual phobias and aversions, fear of loss of control over sexual urges, masked sexual deviation, fear of pregnancy, depression, side effects of drugs, concerns about aging, lack of attraction to the partner, poor sexual skills in the partner, marital conflict, fear of closeness, couple differences regarding point of optimal closeness, passive-aggressive solution to power imbalance, and inability to fuse feelings of love and sexual desire. This long list of potential causes stems from clinical experience rather than from some kind of controlled studies. It probably reflects the results of the therapists' intellectual efforts to understand their patients. Such an understanding can be useful for directing the therapeutic process. For example, if the therapist considers religious orthodoxy to be the main cause of the sexual desire problems, then the most rational approach to therapy would be to attack the orthodoxy and nothing else. Similarly, if the therapist assumes that the partner's unsatisfactory skills are responsible for the patient's low desire, then the most efficient treatment would be to have these skills improved. Indeed, it is a general rule in psychotherapeutic and medical practice that the most efficient treatment is the elimination of the cause of the disturbance.

Since the list of potential causes is very long and varied, and since many patients unquestionably present multiple potential causes, an efficient treatment program has to address this multiplicity in an efficient way. This is one of the important aspects of the LoPiccolo and Friedman (1988) program. However, the Orgasm Consistency Training includes additional elements, as we will see.

The first phase of the original LoPiccolo and Friedman program consists in enhancing sensory awareness. The aim is to make the patients recognize when they are experiencing feelings like anxiety, pleasure, anger and disgust by attending to bodily cues. This is accomplished by asking questions such as: 'How do you know you are feeling happy? Describe the sensations in your body at the time of the feeling.' The therapist may also provoke mental representations of intense sexual activities while observing the patients' emotional and physical reactions. The patients can then be asked to examine the experienced emotions. After some practice of sensory awareness, the therapist explains his/her hypothesis with regard to the cause or causes of the patient's low desire. Whether the hypothesis is true or not is probably of minor importance, so long as the patient can be made to believe that it is true. By being given an explanation for the problem, worries and anxiety about it normally become much reduced. As soon as the patient has been convinced of a cause, an efficient treatment starts to appear possible, and the simple knowledge of the fact that a solution is possible and perhaps even likely is a powerful tranquilizer. In the third phase, irrational thoughts about sexuality are altered or removed. The purpose of this is to make the patient able to approach sexually charged situations or even sexual activity without experiencing emotions of fear or repulsion. An aid to that is training the patient to cope with sexual fears, fears of losing control, feelings of guilt, and so on. For example, the patients may be asked to write down 15–20 individually adapted coping statements. They are then instructed to read and think about these statements every day for a while.

The fourth phase is behavioral training. This phase consists of assertion training, communication training, stress reduction interventions and drive induction, among other things. Of particular interest is the drive induction. Patients are encouraged to go to movies with sexual content, read books with sexual content and to watch pornographic videos. They are also assigned tasks including expressions of physical affection, like hugging, kissing and patting. This kind of behavior is usually infrequent among patients suffering from low sexual desire, because they fear that the partner might interpret such behaviors as invitations to sexual activity, which they want to avoid.

The program is completed in 15–20 weekly sessions. A follow-up study established that several of the positive effects of the program lasted for at least one year (Schover and LoPiccolo, 1982). Since none of the usual control groups were included in the study of treatment effect and duration of effect, there is no way of knowing if the treatment in itself was effective, or if any treatment would have produced the same beneficial effect, or if no treatment would have been equally good.

The Orgasm Consistency Training program is an add-on to the program described above. The rationale for adding it stems from another hypothesis as to the cause of hypoactive sexual desire, namely that 'most cases of female sexual dysfunction are situational or transient conditions determined by sexual technique and skill deficits rather than the result of enduring the pejorative individual problems' (Hurlbert, 1993, p. 44). It may be noted that this view on low sexual desire coincides very nicely with the incentive motivation/learning hypothesis of its origin presented in the preceding section. Stating that deficits in sexual technique or skills are crucial factors is exactly the same as stating that lack of sexual reinforcement and reward is the crucial factor. Sexual techniques and sexual skills can be regarded as deficient only if they fail to provide sexual reward or if they provide aversive experiences instead of reward, or conceivably if they provide a combination of both.

The first part of the Orgasm Consistency Training program consists of exercises of directed masturbation. This is a technique frequently employed in the treatment of anorgasmia (Kohlenberg, 1974), but it has also been successfully used in a variety of other conditions, including low sexual desire. At the beginning of training, the patient and the partner are told that masturbation is very common, socially acceptable, etc. The purpose of this is to make them assume a relaxed and positive attitude towards masturbation. Then the patient is instructed to observe her own genitals in a mirror for a minimum of 5 minutes per day in order to get familiarized with the anatomy and appearance. The partner is instructed to do the same, allowing the couple to talk about their exploratory activities and share their discoveries. After one week of daily mirror exercise, the patient and the partner are asked not only to look at their genitals in the mirror but also to touch them with the goal of finding out which areas offer the most pleasant sensation. This should go on for at least 5 minutes every day. The third week, demands were increased from simply touching the genitals to the application of more intense stimulation. Now a maximum amount of pleasure should be obtained by exploring different methods of stroking and applying pressure. Again, this should last for at least 5 minutes every day. The fourth week, the duration of the most pleasurable genital stimulation should be no less than 20 minutes. For the comfort of the patients and their partners, they were encouraged to use jelly to avoid tissue irritation. During the fifth and last week, the patient and the partner should stimulate themselves together. The man should show the woman (only heterosexual couples have been included in this program) how he obtains maximal pleasure and the woman should do likewise. Furthermore, it was suggested that the couple should engage in mutual genital stimulation.

Throughout the 5 weeks, the couple was instructed to have at least three sexual encounters per week. These encounters should focus on obtaining sensate pleasure from touching, also the genitals if the couple felt for that. Vaginal penetration was also allowed if the couple desired it. In that case, rear entry or the lateral position was recommended since these positions are supposed to permit continuous caressing of the genitals while copulating.

In The Orgasm Consistency Training, the emphasis is gradually shifted from masturbation to explicit sensate focus exercises and techniques to improve voluntary male self-control. Sensate focus is the classical Masters and Johnson procedure (1970), adopted by almost every sexologist for almost every purpose. It consists of making the patient concentrate attention on sensual pleasure obtained from sensory stimulation and on the kind of stimulation that causes maximal pleasure. In the original version, the patient and the partner were instructed to remove their clothing, once they were in privacy, and one of them should then assume the role of touch provider and the other that of touched object. These roles were changed whenever judged convenient. Any part of the body, except the genital area, could be touched. In later stages, touching the genitals was also permitted. The use of creamy lotions, with and without fragrance, was encouraged, both in order to avoid tissue irritation and in order to stimulate olfactory systems in addition to the tactile system. Besides making the patient and the partner discover how to provide pleasurable stimulation to the other, the procedure is thought to facilitate communication within the couple. In the Orgasm Consistency Training program, sensate focus exercises are implicitly introduced quite early in the treatment, just as Masters and Johnson (1970) did.

Another feature of the later phases of the program is the improvement of male self-control. This means that the male partner should learn to delay his ejaculation until the woman has experienced at least one orgasm. In fact, as part of the treatment, the couple signs a contract where one of the clauses establishes that the woman is entitled to orgasm in every copulatory encounter, and that her orgasm should precede that of the man. This, in fact, is nothing more than an application of the old gentlemanly principle *the ladies come first*. It is permissible for the woman to obtain her orgasm in any way she likes, including but not limited to self-masturbation, masturbation by the partner, and cunnilingus. Finally, they are taught how to enhance the likelihood of obtaining orgasm through penile–vaginal intercourse. Employment of the coital alignment technique is suggested (Eichel *et al.*, 1988). In this technique, the woman lies on her back with the man positioned between her legs. In contrast to what usually occurs in this the most classical of all copulatory motor patterns, the man should locate himself as far forward as possible. The couple should assure intense clitoral stimulation by moving both pelvises together, in short, rocking strokes. The man initiates downward strokes, while the woman is supposed to initiate upward strokes. An important part of this is that the frequency of movement should be kept constant throughout the intercourse, meaning that any increase as orgasm approaches should be avoided.

After completion of the Orgasm Consistency Training program, including its three parts–cognitive behavioral modifications, directed masturbation training, and the training in how to assure orgasm at most if not every sexual encounter – the low sexual desire should have been replaced by a level of desire judged adequate by the patient. This seems to be the case. Treatment efficacy studies (Hurlbert, 1993) have shown that the Orgasm Consistency Training program as described above is superior to the standard program (which here means a somewhat extended LoPiccolo and Friedman program). Not only did the positive treatment effect last for at least 6 months, but the former patients also continued to improve between the end of treatment and the last follow up. These observations were confirmed in another follow-up study. Here, the control group consisted of patients on the waiting list. It is somewhat unclear whether these patients remained on the waiting list at follow up, but I assume they did not receive any treatment. The patients were divided in two groups, one consisting of couples only and one where there were only heterosexual women. Interestingly, the couples only group was superior to the waiting list group at the end of treatment. However, the women only group did not show any consistent improvement. At the 6 months follow-up, the couples only group was superior to both the women only and the waiting list control group (Hurlbert *et al.*, 1993b).

The results of the Hurlbert *et al.* (1993b) study are quite interesting, because they suggest that psychotherapeutic treatment of the patient without the concomitant treatment of the partner is of little use. This is not particularly surprising, since one of the aims of the Orgasm Consistency Training program is to assure that the woman complaining of low desire experiences orgasm at each sexual encounter. The obtention of orgasm is strongly dependent on the partner's behavior. Indeed, it could be maintained that a deficient sexual behavior on the part of the male partner is the main cause for the woman's low desire. However, Hurlbert does not want to put the blame on any specific partner. He prefers talking of deficient sexual skills and techniques in general. However, the fact that improvement is marginal when the patient alone participates in the therapeutic program suggests that the male partner carries a large part of the responsibility for the hypoactive desire experienced by his female partner. Furthermore, the positive results obtained in the Orgasm Consistency Training program directly support the incentive motivation/learning hypothesis. The characteristic feature of this program is that it tries to assure that sexual activities always are associated with reward. Moreover, an implicit supposition is that the disorder is caused by absence of sexual reward, exactly as suggested by the incentive motivation/learning hypothesis. This assumption is actually made explicit in the Hurlbert (1993) paper. An overview of studies evaluating the coital alignment technique went still further when suggesting that the 'hidden epidemic' of hypoactive sexual desire disorder, as well as other sexual disorders, may be a consequence of inadequate copulatory motor patterns leading to premature ejaculation and female coital anorgasmia. It was even suggested that '. . .the physical dynamics of the sex act may be more the etiology of common sex problems than factors of mental pathology rooted in childhood trauma' (Pierce, 2000, p. 266).

For some unknown reasons, the orgasm consistency training described in the preceding paragraphs has not attracted much attention from clinical researchers. No extensive evaluation of its efficacy in addition to the rather small studies mentioned above has been performed, making it difficult to determine its practical utility as a routine treatment. Despite this, it is the only psychotherapeutic treatment for hypoactive sexual desire disorder for which we have any data satisfying minimal methodological criteria (O'Donahue *et al.*, 1997).

There is no psychotherapeutic program for the treatment of male hypoactive sexual desire disorder with established efficacy (Baucom *et al.*, 1998; O'Donohue *et al.*, 1999).

5.2 Pharmacological treatments

Hormones

Men Men complaining of low sexual desire because of hypogonadism have been successfully treated with androgens (Bancroft, 1984; Burris *et al.*, 1992). The effectiveness of this treatment is not surprising. We learned already in Chapter 5 that adequate functioning of the central motive state requires stimulation of androgen receptors in the brain and that this stimulation is insufficient in men with very low blood androgen concentrations, as is the case in hypogonadal men. However, since it is quite likely that most men suffering from hypoactive sexual desire disorder have blood testosterone concentrations within the normal range, it is not evident that they would benefit from treatment with testosterone. Nevertheless, it would be interesting to have empirical data with regard to the effects of supplementary testosterone in such men. It seems that most clinicians regard the possibility of such an effect remote, at least too remote for trying treatments with androgen. There is, to my knowledge, no study of the effects of testosterone treatment in men diagnosed with hypoactive sexual desire disorder. In eugonadal men suffering from impotence, though, testosterone treatment has been tried. This is of no relevance here, but the men's sexual desire was also evaluated in some of these studies. The results of one such study are summarized in a very amusing way: there were marked, but statistically non-significant increases in median frequency of reported sexual desire (Schiavi *et al.*, 1997). An unusually creative way of saying that there was no effect. Another study of grown-up eugonadal men (mean age about 60) compared the blood testosterone concentrations between one group reporting high levels of sexual desire, one reporting moderate levels, and a third reporting low desire. There was no difference, neither in total nor in free testosterone concentration (Ansong and Punwaney, 1999). Similar data were obtained in a later study (Köhn *et al.*, 2004). Although none of the subjects participating in these studies had been diagnosed as suffering from hypoactive sexual desire disorder according to the diagnostic criteria of the DSM-IV or ICD-10, the results speak against any role for testosterone in the control of sexual desire in eugonadal men. Thus, the clinicians' reluctance to test testosterone as a treatment for low sexual desire in men is well founded.

Women In women, the situation is slightly different. First, the data with regard to reduced testosterone concentrations in subjects suffering from low desire are inconclusive. Second, there are some suggestions that enhanced blood concentration of androgens may be beneficial for sexual desire, even in women having the androgen concentration within the normal range. This leaves the possibility for improvement with supplementary androgens open. Consequently, several clinical trials have been performed. In one, women with low sexual desire were injected with 100 mg of testosterone cyprionate, a long-acting testosterone ester, once a month. Both premenopausal and menopausal women were included. How they had been diagnosed is slightly unclear and whether they fulfilled the habitual criteria for hypoactive sexual desire disorder is impossible to determine. Controls were women who came in for routine annual pelvic examinations. They did not receive any injection. This was, then, an open study. Treatment lasted for 3 months. Evaluations of sexual desire were performed before treatment and at the end of the 3-month treatment period. There was no change in desire in the control group.

However, the testosterone-treated group showed a marked and significant improvement. At the pretreatment evaluation, the sexual desire score in the women with low desire was far below that of controls, while there was no difference between the groups at the end of treatment evaluation. As could be expected, blood testosterone concentrations had more than doubled between the pre- and end of treatment blood samples. While both groups had similar blood testosterone concentrations before treatment, the concentration in the treated group was much higher than in the control group at the end of treatment (van Anders *et al.*, 2005). These data are suggestive, but the group sizes were small, there was no untreated low sexual desire control group and the study was open. I would be very cautious before concluding anything at all from this study.

Another small clinical trial, this time including a placebo control and double-blind design, showed effects of testosterone administration similar to those mentioned above (Goldstat *et al.*, 2003). Only premenopausal women were included in this study and testosterone was administered in the form of a cream. Hypoactive sexual desire disorder was diagnosed with the aid of a questionnaire, which is not exactly the most reliable way. Nevertheless, the beneficial effects of testosterone were substantial, suggesting that testosterone indeed is helpful in premenopausal women with baseline androgen blood concentrations within the normal range.

There are several clinical trials of androgen therapy in menopausal women. A carefully conducted, double-blind, placebo controlled study revealed that the addition of androgens to estrogen replacement therapy in menopausal women complaining of low sexual desire had a beneficial effect (Lobo *et al.*, 2003). This observation has been confirmed in another study, this one including several hundred spontaneously menopausal women suffering from hypoactive sexual desire disorder. They were all on estrogen replacement therapy. In addition, some were given a testosterone patch twice weekly for 24 weeks, while others received a placebo patch for an identical period of time. The improvement from baseline was significantly larger in women given testosterone than in those given placebo (Shifren *et al.*, 2006). Unfortunately, this extensive study suffers from major problems. One is the unusually sloppy diagnostic procedure. A home-made questionnaire consisting of five questions was the basis for classifying a woman as suffering from low sexual desire. The relationship between the label the authors allow themselves to employ, hypoactive sexual desire disorder, and their diagnostic criterion is far from clear. According to my own judgment, appropriate diagnostic criteria *must* be applied as soon as the official clinical category name is used. This is not the only problem, though. The procedure for evaluating sexual functioning, including desire, leaves a lot to be desired. It appears to consist of some kind of diary where the subject is requested to record sexual activities during a 7-day period. Some obscure questionnaires were used in addition. These shortcomings reduce the informative value of this study quite a lot. Nevertheless, since it coincides with the results from other clinical trials of testosterone administration, we may accept its conclusion under some reserve. The conclusion is: 'testosterone patch treatment increased the frequency of satisfying sexual activity and sexual desire' (Shifren *et al.*, 2006, p. 770). Similar studies in surgically menopausal women report essentially the same results (Braunstein *et al.*, 2005; Buster *et al.*, 2005). Since these studies share all the weaknesses of the one described above, their informative value is not overwhelming.

We can easily see that menopausal women with reduced sexual desire have been more attractive than fertile women for studies of the effects of androgens. Nevertheless, the van Anders *et al.* (2005) study included both premenopausal and menopausal women and no difference in effects of testosterone between these women was reported. Furthermore, the data from the Goldstat *et al.* (2003) trial with transdermal testosterone clearly suggest a positive effect in premenopausal women. In fact, all existing data point to a beneficial effect of testosterone treatment in premenopausal as well as in menopausal women.

Hormones other than androgens have not been proposed to be useful in the treatment of female hypoactive sexual desire disorder. Estrogen replacement therapy is regularly employed in menopausal women, but no beneficial effects of estrogens on sexual desire have been reported. This coincides with the large amount of data showing that androgens rather than estrogens are important for sexual behaviors in women. We can now conclude this subsection by stating that androgen therapy probably is useful in the treatment of female hypoactive sexual desire disorder while it does not seem to have any beneficial effect in men.

Psychopharmacological treatments

The encouraging results reported in the clinical trials with androgens contrast sharply to the lack of progress with regard to psychopharmacological approaches. There is currently no established drug treatment for hypoactive sexual desire disorder, neither in men nor in women. Recent reviews of the treatment strategies for this disorder have focused on hormone treatments (Köhn, 2006; Nijland *et al.*, 2006; Segraves and Woodard, 2006). No new data on the effects of drugs acting in the central nervous system seem to have appeared during the last few years, although the pharmaceutical industry is investing a considerable effort in search for effective drugs. A few data do suggest that some psychopharmacological agents may be effective, but these data do not seem to appear convincing enough for launching more extensive clinical evaluations. Nevertheless, I will mention some of them.

Men A group of men suffering from impotence was treated with the α_2-adrenergic antagonist yohimbine for 8 weeks. The drug was given orally in a dose of 30 mg per day. Before and during the treatment period, the men recorded several aspects of sexuality in a diary and at the end of treatment they participated in an interview. Although the main purpose of this double-blind drug trial was to evaluate the effects of yohimbine on erection, data with regard to sexual desire were also obtained. Yohimbine had no effect on desire (Vogt *et al.*, 1997). This study is not particularly relevant to the issue of drug treatment of masculine hypoactive sexual desire disorder, since none of the participants was diagnosed with it. Furthermore, the procedure for determining level of sexual desire was quite primitive. The two reasons for mentioning it were, on one hand to illustrate the dearth of data and, on the other hand, to bring our attention to the carelessness of some review writers. The Vogt *et al.* (1997) data were recently mentioned as evidence for a positive effect of yohimbine in men with low sexual desire (Köhn, 2006).

Clinicians have occasionally suggested that dopamine agonists could be an effective treatment option. The rather extensive experience obtained with apomorphine employed as a centrally acting stimulant of erection has clearly demonstrated that the compound does not enhance desire (reviewed in Giuliano and Allard (2002)). Again, these observations are not directly relevant to the problem of hypoactive sexual desire disorder, since the impotent patients using apomorphine do not necessarily suffer from this disorder. It could, obviously, be reasoned that if the drug indeed were an efficient enhancer of desire, it should also be so in impotent men. Although no firm conclusion can be drawn, it would appear that apomorphine is not particularly efficient for stimulating sexual desire. In sum, the very limited information available on psychopharmacological treatments of male desire disorders suggest that neither of the potential candidates, yohimbine and apomorphine, is promising. A major breakthrough in our understanding of sexual pharmacology seems to be required before any alternative can appear.

For the pleasure of those fond of botany, I want to mention a small trial with a Peruvian plant, *Lepidium meyenii*, called Maca by the locals. The root has a reputation of being aphrodisiac and healthy in many ways and it is available worldwide in the form of encapsulated powder. A group of 30 men without any sexual problem ate the plant for 12 weeks and men in another group received placebo. Measures of sexual desire were based on self-report. More than 40% of the men eating Maca reported increased desire after 8 and 12 weeks, while none of the placebo-eaters did. In fact, ratings of sexual desire were significantly increased from 4 weeks of treatment onwards. Blood concentration of testosterone was unaffected (Gonzales *et al.*, 2002). These data obtained in healthy men cannot immediately be extended to men suffering from hypoactive sexual desire disorder, but it would seem that the plant should be the subject of a clinical trial.

The scant interest in finding drugs enhancing sexual desire in men probably has practical explanations. Prevalence of hypoactive sexual desire disorder is lower in men than in women, making a treatment for the former less commercially attractive than a treatment for the latter. The difference in the number of potential clients is considerable. In view of this, it is quite natural to give priority to a search for treatments for women with low sexual desire. Other explanations for the bias against men are probably of social origin. Men now and then commit the crime of rape, while women rarely do it. Men are sporadically accused of exhibitionism, pedophilia or other unsuitable sexual activities. There are many reasons to believe that at least some of these behaviors are related to high levels of sexual desire. In the case of rape, this is most uncertain or outright false, but many uninformed people hold the belief that rape is even a crime of passion. Imagine that a man ingests a drug known to have a stimulating effect on sexual desire and then is arrested for pedophilia while under the influence of that drug. Or still worse, a rapist could be caught after that he had taken the drug and coincidentally raped someone. Both these criminals would probable allege that the drug rather than themselves was responsible. Because of considerations of this kind, be they true or false, there is not much enthusiasm for supporting research with the purpose of finding a drug treatment for male hypoactive sexual desire disorder.

Women Women rarely perpetrate sexual crimes. This fact removes one of the concerns associated with the search for drugs enhancing sexual desire. I already mentioned the epidemiological data favoring women. It is no surprise, then, that some research has been done with the explicit aim of finding a drug treatment for female hypoactive sexual desire disorder. Most of this research is of recent date. In a review published in 1996, Waldinger (1996) concluded that there did not exist any controlled clinical study of drug therapy for hypoactive sexual desire disorder. Since then, the situation has changed. The first methodologically sound evaluation of a drug treatment was made in a nicely performed study (Piletz *et al.*, 1998). Nine premenopausal women, diagnosed as suffering from hypoactive sexual desire disorder, were treated either with yohimbine (16.2 mg daily) or placebo in randomized order. After one baseline menstrual cycle, treatment with yohimbine or placebo started on the first day of menses of the following cycle. At the end of this cycle, treatments were switched. The diagnosis of hypoactive sexual desire disorder was made according to the criteria established in the DSM-III-R (American Psychiatric Association, 1987). An experienced clinical psychologist was in charge of diagnosis. This is a most commendable aspect of this study. We have seen so many examples of slapdash diagnostic procedures that it is a real pleasure to find a study where it has been done correctly. Besides the women diagnosed with low desire, a group of healthy women were used as control. The effectiveness of the yohimbine treatment with regard to stimulating noradrenergic transmission (like other adrenergic α_2 antagonists, yohimbine enhances noradrenaline release) was determined by measuring concentration of 3-methoxy-4-hydroxyphenylglycol in blood. A clear effect was obtained. On the contrary, the drug failed to modify sexual desire. Even though the sample size was small, this carefully performed clinical evaluation of yohimbine inspires some confidence. It is interesting to note that the negative results of this trial coincide with the results of the yohimbine trial in men mentioned earlier. Perhaps yohimbine should be put on the list of drugs with proven inefficacy.

Apomorphine has been subjected to a clinical trial in a peculiar group of premenopausal women. All 62 participants had a recurring deficiency of sexual fantasies and/or thoughts and did not experience any clitoral or vaginal sensation and no lubrication in response to sufficient sexual stimulation. It is evident that these women suffered from sexual arousal disorder in addition to a possible low desire. During the first phase of this study, the women were instructed to take a pill about 30 to 45 minutes before having sex. The pill should be kept under the tongue rather than swallowed, in order to reduce the emetic properties of apomorphine. In the second phase, the women were assigned to two different doses of apomorphine or to placebo. Now the pill should be taken once every afternoon regardless of whether they had planned to have sex or not. The results from the first phase showed that apomorphine was ineffective with regard to desire and arousal, while having some positive effect on the incidence of orgasm at a dose of 3 mg but not at a dose of 2 mg. During phase 2, apomorphine improved all aspects of sexual functioning at both doses. Moreover, the larger dose was superior to the smaller (Caruso *et al.*, 2004). These results are certainly encouraging, but they suffer from a rather serious problem, and that is the double diagnosis of low desire and sexual arousal disorder in all the women participating in the trial. Knowing that apomorphine can promote erection in men and other animals, it is not far fetched to assume that it also can stimulate genital blood flow in women. The primary target of apomorphine may well be the lacking arousal. Improved lubrication and enhanced clitoral erection may make sexual activities more rewarding, and increased reward will, as we know, enhance future desire. I will not take the merit away from these data, but a clinical trial in women suffering exclusively from hypoactive sexual desire disorder is badly wanted.

Bupropion is a mixed dopamine-noradrenaline reuptake inhibitor. It is an efficient antidepressant and rumors claim that it is sporadically used as an anti-smoking pill. As an antidepressant, bupropion has a reputation of having very small or no side effects on sexual function. This contrasts to the common occurrence of sexual side effects of many other antidepressants, notably the SSRIs. Not only is bupropion almost devoid of sexual side effects, but it has also been successfully used for ameliorating the inhibitory effects of SSRIs in depressed women (Ashton and Rosen, 1998; Gitlin *et al.*, 2002; Vanderkooy *et al.*, 2002). As many as 20% of women (and men) taking the SSRI antidepressants manifest sexual malfunction, including reduced sexual desire. The diagnosis of hypoactive sexual desire disorder is excluded according to the criteria established in the DSM-IV or the ICD-10, but anyway it is interesting that bupropion reduces iatrogenic low sexual desire. Sufficiently interesting, in fact, for testing the drug in women not on SSRIs. Another observation drawing attention to bupropion was that some women treated with the drug for depression reported enhanced sexual desire (Modell *et al.*, 1997).

A group of premenopausal women were diagnosed with hypoactive sexual desire disorder using criteria similar to those of the DSM-IV. Following a baseline period during which placebo was given, the women received bupropion for 8 weeks (Segraves *et al.*, 2001). At several time points during this period, they were evaluated with 'investigator-rated scales of sexual desire and sexual functioning'. Potential readers were not judged worthy of knowing which these scales were. There was no change during the placebo period. When the results of the treatment period are presented, we are immediately faced with a terrible confusion. First, it turns out that 29 of the 51 patients who entered the study responded while the rest did not. Of the 29 patients who completed the study, 38% responded. With 'respond' is meant that the subjects had improved much or very much according to the Clinical Global Impressions Scale for Improvement of Illness. I have no idea of what that scale scales, but we can leave that. Second, an attrition rate of about

50% is quite astounding, but what is more astounding is that throughout the rest of the results, there is no way of knowing whether the data stem from the responders or from all the participants. The next unanswered question is which responders and which participants. Those entering, those completing, or perhaps a salad of both? Despite the dramatic deficiencies in the report, this miserable study might suggest that bupropion has some beneficial effect.

A later, more extensive study (Segraves *et al.*, 2004), from the same group confirms their habit of writing confusing reports. Premenopausal women, diagnosed with hypoactive sexual desire disorder, were assigned either to placebo or bupropion. Their sexual function was, as always, rated by the investigator, but they were also asked to fill out the Changes in Sexual Functioning Questionnaire (Clayton *et al.*, 1997). This time we are not told the proportion of responders and nonresponders, but we are told that sexual desire was unaffected by the drug. However, orgasm completion scores were improved, whatever that might mean. Arousal was also enhanced. Nevertheless, it would seem that the effects of bupropion are quite irregular and quite modest.

Drug treatment for female hypoactive sexual desire disorder is undeniably in its infancy. The candidates evaluated so far do not seem to be particularly promising. Exactly as was the case with pharmacological treatments for men, we are in need of some breakthrough. It is uncertain if we can expect that to occur in the near future. Our knowledge of the pharmacology of sexual behavior is extremely limited and the only reliable data we have, from humans and other mammals, applies to inhibitory drug effects.

6 General conclusion concerning hypoactive sexual desire disorder

The most common of all sexual dysfunctions is one of the least known. Taking into account that the disorder was created only some 20 years ago, this state of profound ignorance is perhaps what we should expect. Moreover, it is not a life-threatening disease and funding research regarding its causes and cures is not a top priority for any funding agency. However, speculations concerning the causes of low sexual desire and its treatment have existed long before the diagnostic category was invented. None of these speculations seems to have been very successful, since they are all forgotten or hidden in the mind of some nostalgic clinician. My explanation for this sad state of affairs is that sexual desire has not been analyzed in an appropriate theoretical context. It has certainly been the subject of much disordered thought, but the futility of the brain-storming approach should have become widely apparent.

The incentive motivation/learning approach outlined in this section can provide a framework sufficiently strong for organizing thought and for generating testable hypotheses. It is the only explanation that can account for the many variants of the disorder, including the life-long variant. We have simply to assume that the afflicted individual for some reason or another never got to experience sexual reward and that social learning did not compensate for this. Perhaps social learning was actually the cause for not acquiring experience with sexual incentives and activities. Religious orthodoxy is one of the many unusual factors now and then suggested to be causally related to low or absent desire. If social learning has taught us that sex is ugly and sinful, and that everything making us think of it should be avoided, then the step to low sexual desire cannot be too long. Furthermore, within most orthodox groups, low sexual desire is not a problem but the highest of virtues.

Psychotherapeutic treatment seems to be the best available option at present. The Orgasm Consistency Training program is the only treatment that has reported some lasting success, but more controlled outcome studies are required before it can enter into the category of established, efficacious treatments. Its focus on assuring the regular obtention of sexual reward coincides, accidentally without doubt, with the incentive motivation/learning model proposed. Androgens may be effective in women, but their usefulness in men is questionable. There is no currently available psychopharmacological agent with reliable effect.

The activation of sexual desire is a result of an interaction between incentive stimuli and the central motive state. A highly sensitive central motive state can be expected to react even to weak sexual incentives. Accordingly, any agent that can enhance the sensitivity of the central motive state can also enhance the impact of incentives. A drug having the capacity to increase the sensitivity of the central motive state responding to sexual incentives without affecting any other central motive state could perhaps represent an efficient treatment. Such a drug would make it possible to treat the patient in a very effortless way. An alternative with higher likelihood of success in female hypoactive sexual desire disorder may be androgen therapy. Supplementary androgens would simply enhance the action of endogenous androgens on the central motive state. In that way, their effect would be physiological rather than pharmacological. However, androgens have many peripheral effects that are not desirable in women. Perhaps a specific *a*ndrogen *r*eceptor *m*odulator (SARM) with actions only in the central nervous system would solve this problem. There is no such compound at the moment, making this hypothetical treatment dependent on important advances in androgen receptor pharmacology. These are likely to come soon, though, since the SARMs have attracted much attention in recent years. An excellent review of this exciting field of research is available (Cadilla and Turnbull, 2006).

Even if the ideal psychopharmacon or SARM were to be discovered some day, a serious dilemma remains. A chemically induced, enhanced reactivity of the central motive state may indeed make the patient respond to sexual incentives with desire. This desire may also be the cause of approach behaviors and sexual interaction. Thereby the problem of low desire might seem to be solved. However, the solution will not be lasting if the sexual interaction is not followed by reward. The incentive motivation/learning hypothesis would predict that if the reinitiated sexual interaction were followed by an aversive experience rather than reward, then the chemically induced enhanced reactivity to sexual incentives would eventually be neutralized by continued conditioning. On the other hand, if the low sexual desire was independent of learning but caused by an unusually unresponsive central motive state, then the chemical treatment would constitute a final solution. Here we have a nice example illustrating that the etiology we believe in determines which kind of treatment we predict to be the most efficient.

The psychotherapeutic approach is far more costly in time and probably also in money than drug or hormone treatments. Moreover, it requires the participation of the partner. Although the partner may be the cause of the problem, as pointed out before, it is not certain that she/he wants to be the cure. This poses a problem, since we know that the absence of a cooperative partner makes psychotherapeutic treatment less efficient. In these cases, treatment with chemicals may be the best and perhaps the only solution.

Summary

1 Hypoactive sexual desire disorder is a rather recent addition to the list of diagnostic categories included in manuals like DSM-IV and ICD-10. This reflects social changes that have transformed low sexual desire from a virtue to a problem.
2 The etiology is unknown. Endocrine changes do not seem to be involved and no other organismic cause has been found.
3 An incentive motivation/learning model is proposed. Lack of sexual reward following sexual interaction or aversive experiences during that interaction may, through classical conditioning, transform sexual incentives (stimuli emitted by the partner) into neutral or even aversive stimuli. These stimuli can no longer activate sexual desire.
4 When exposed to sexual incentive stimuli in the form of a pornographic movie, women with hypoactive sexual desire disorder showed enhancement of vaginal blood flow of the same magnitude as healthy women. If true, these data demonstrate that the central motive state is exactly as responsive to sexual incentives in women with hypoactive sexual desire disorder as in healthy women. The women with low desire employed in these studies must belong to the category situational hypoactive sexual desire disorder.
5 The Orgasm Consistency Training psychotherapeutic program has reported substantial success and patients continued to improve at a 6 months follow-up. More controlled clinical studies are required before this program can be considered as truly efficacious. Treatment of both members in the couple appears to be of great importance.
6 In men, treatment with androgens does not seem to have any consistent effect on low sexual desire. On the contrary, there is a substantial amount of data suggesting that androgen therapy is efficient in both premenopausal and menopausal women diagnosed with hypoactive sexual desire disorder.
7 Clinical trials with the adrenergic α_2-antagonist yohimbine in both men and women have failed to find a positive effect. The non-selective dopamine agonist

apomorphine seems to be ineffective in men, but some positive data have been obtained in women suffering from hypoactive sexual desire disorder combined with sexual arousal disorder. The mixed noradrenaline/serotonin reuptake inhibitor bupropion has been reported to be efficient in iatrogenic hypoactive sexual desire disorder, but its effect in other cases is unclear.

8 Specific androgen receptor modulators may become the chemical treatment of choice, if and when compounds with appropriate actions on the androgen receptor are made available.

References

Ågmo, A. (2002). Copulation-contingent aversive conditioning and sexual incentive motivation in male rats: evidence for a two-stage process of sexual behavior. *Physiol Behav* **77**, 425–435

Ågmo, A., Turi, A.L., Ellingsen, E. and Kaspersen, H. (2004). Preclinical models of sexual desire: Conceptual and behavioral analyses. *Pharmacol Biochem Behav* **78**, 379–404

Ahlenius, S., Larsson, K. and Svensson, L. (1980). Further evidence for an inhibitory role of central 5-HT in male rat sexual behavior. *Psychopharmacology* **68**, 217–220

Aldrich, A.P., Cook, M.D. and Pedersen, L.R. (1996). Retrospective review of selective serotonin reuptake inhibitor-induced libido disturbance in women. *Clin Drug Invest* **11**, 353–359

American Psychiatric Association (1980). *Diagnostic and statistical manual of mental disorders*, 3rd edn. American Psychiatric Association: Washington, DC

American Psychiatric Association (1987). *Diagnostic and statistical manual of mental disorders* (DSM-III-R), 3rd edn, revised. American Psychiatric Association: Washington, DC

American Psychiatric Association (1995). *Diagnostic and statistical manual of mental disorders*, 4th edn. American Psychiatric Association: Washington, DC

Amsel, A. (1962). Frustrative nonreward in partial reinforcement and discrimination learning: Some recent history and a theoretical extension. *Psychol Rev* **69**, 306–328

Ansong, K.S. and Punwaney, R.B. (1999). An assessment of the clinical relevance of serum testosterone level determination in the evaluation of men with low sexual drive. *J Urol* **162**, 719–721

Arafat, I. and Cotten, W. (1974). Masturbation practices of males and females. *J Sex Res* **10**, 293–307

Arnal, A., Llario, R. and Gil, M.D. (1995). Estudio epidemiológico sobre la prevalencia de disfunciones sexuales en la comunidad valenciana. *Psicothema* **7**, 95—104

Ashton, A.K. and Rosen, R.C. (1998). Bupropion as an antidote for serotonin reuptake inhibitor-induced sexual dysfunction. *J Clin Psychiat* **59**, 112–115

Bancroft, J. (1984). Testosterone therapy for low sexual interest and erectile dysfunctions in men: a controlled study. *Br J Psychiat* **14**, 146–151

Baucom, D.H., Shoham, V., Mueser, K.T., Daiuto, A.D. and Stickle, T.R. (1998). Empirically supported couple and family interventions for marital distress and adult mental health problems. *J Consult Clin Psychol* **66**, 53–88

Baumeister, R.F., Catanese, K.R. and Vohs, K.D. (2001). Is there a gender difference in strength of sexual drive? Theoretical views, conceptual distinctions, and a review of relevant evidence. *Pers Soc Psychol* **5**, 242–273

Bentler, P.M. and Peeler, W.H. (1979). Models of female orgasm. *Arch Sex Behav* **8**, 405–423

Beutel, M.E., Wiltink, J., Hauck, E.W. *et al.* (2005). Correlations between hormones, physical, and affective parameters in aging urologic outpatients. *Eur Urol* **47**, 749–755

Bindra, D. (1974). A motivational view of learning, performance, and behavior modification. *Psychol Rev* **81**, 199–213

Bindra, D. (1976). *A theory of intelligent behavior*. Wiley: New York

Bindra, D. (1978). How adaptive behaviour is produced: a perceptual-motivational alternative to response reinforcement. *Behav Brain Sci* **1**, 41–52

Bogner, H.R. and Gallo, J.J. (2004). Are higher rates of depression in women accounted for by differential symptom reporting? *Soc Psychiatr Psychiatr Epidemiol* **39**, 126–132

Bozman, A.W. and Beck, J.G. (1991). Covariation of sexual desire and sexual arousal: the effects of anger and anxiety. *Arch Sex Behav* **20**, 47–60

Braunstein, G.D., Sundwall, D.A., Katz, M. *et al.* (2005). Safety and efficacy of a testosterone patch for the treatment of hypoactive sexual desire disorder in surgically menopausal women – A randomized, placebo-controlled trial. *Arch Intern Med* **165**, 1582–1589

Burris, A.S., Banks, S.M., Carter, C.S., Davidson, J.M. and Sherins, R.J. (1992). A long-term, prospective study of the physiologic and behavioral effects of hormone replacement in untreated hypogonadal men. *J Androl* **13**, 297–304

Buster, J.E., Kingsberg, S.A., Aguirre, O. *et al.* (2005). Testosterone patch for low sexual desire in surgically menopausal women: A randomized trial. *Obstet Gynecol* **105**, 944–952

Cadilla, R. and Turnbull, P. (2006). Selective androgen receptor modulators in drug discovery: Medicinal chemistry and therapeutic potential. *Curr Top Med Chem* **6**, 245–270

Caruso, S., Agnello, C., Intelisano, G., Farina, M., Di Mari, L. and Cianci, A. (2004). Placebo-controlled study on the efficacy and safety of daily apomorphine SL intake in premenopausal women affected by hypoactive sexual desire disorder and sexual arousal disorder. *Urology* **63**, 955–959

Chambless, D.L., Baker, M.J., Baucom, D.H. *et al.* (1998). Update on empirically validated therapies, II. *Clin Psychol* **51**, 3–16

Clayton, A.H., McGarvey, E.L., Clavet, G.J. and Piazza, L. (1997). Comparison of sexual functioning in clinical and nonclinical populations using the changes in sexual functioning questionnaire (CSFQ). *Psychopharmacol Bull* **33**, 747–753

Clayton, A.H., Segraves, R.T., Leiblum, S. *et al.* (2006). Reliability and validity of the sexual interest and desire inventory-female (SIDI-F), a scale designed to measure severity of female hypoactive sexual desire disorder. *J Sex Marital Ther* **32**, 115–135

Eichel, E.W., Eichel, J.D. and Kule, S. (1988). The technique of coital alignment and its relation to female orgasmic response and simultaneous orgasm. *J Sex Marital Ther* **14**, 129–141

Fugl-Meyer, A.R. and Sjögren-Fugl-Meyer, K. (1999). Sexual disabilities, problems and satisfaction in 18–74 year old Swedes. *Scand J Sexol* **2**, 79–105

Gebhard, P.H. (1966). Factors in marital orgasm. *J Soc Issues* **22**, 88–95

Gitlin, M.J., Suri, R., Altshuler, L., Zuckerbrow-Miller, J. and Fairbanks, L. (2002). Bupropion sustained release as a treatment for SSRI-induced sexual side effects. *J Sex Marital Ther* **28**, 131–138

Giuliano, F. and Allard, J. (2002). Apomorphine SL (Uprima®): preclinical and clinical experiences learned from the first central nervous system-acting ED drug. *Int J Impot Res* **14**, S53–S56

Goldstat, R., Briganti, E., Tran, J., Wolfe, R. and Davis, S.R. (2003). Transdermal testosterone therapy improves well-being, mood, and sexual function in premenopausal women. *Menopause-J N Am Menopause Soc* **10**, 390–398

Gonzales, G.F., Cordova, A., Vega, K. *et al.* (2002). Effect of Lepidium meyenii (MACA) on sexual desire and its absent relationship with serum testosterone levels in adult healthy men. *Andrologia* **34**, 367–372

Gregorian, R.S., Golden, K.A., Bahce, A., Goodman, C., Kwong, W.J. and Khan, Z.M. (2002). Antidepressant-induced sexual dysfunction. *Ann Pharmacother* **36**, 1577–1589

Guay, A.T. (2001). Decreased testosterone in regularly menstruating women with decreased libido: a clinical observation. *J Sex Marital Ther* **27**, 513–519

Guay, A.T. and Jacobson, J. (2002). Decreased free testosterone and dehydroepiandrosterone-sulfate (DHEA-S) levels in women with decreased libido. *J Sex Marital Ther* **28**, Suppl. 1, 129–142

Hayes, R. and Dennerstein, L. (2005). The impact of aging on sexual function and sexual dysfunction in women: a review of population-based studies. *J Sex Med* **2**, 317–330

Heiman, J.R. and Rowland, D.L. (1983). Affective and physiological sexual-response patterns: the effects of instructions on sexually functional and dysfunctional men. *J Psychosom Res* **27**, 105–116

Hurlbert, D.F. (1993). A comparative study using orgasm consistency training in the treatment of women reporting hypoactive sexual desire. *J Sex Marital Ther* **19**, 41–55

Hurlbert, D.F., Apt, C. and Rabehl, S.M. (1993a). Key variables to understanding female sexual satisfaction–an examination of women in nondistressed marriages. *J Sex Marital Ther* **19**, 154–165

Hurlbert, D.F., White, L.C., Powell, R.D. and Apt, C. (1993b). Orgasm consistency training in the treatment of women reporting hypoactive sexual desire: an outcome comparison of women-only groups and couples-only groups. *J Behav Ther Exp Psychiat* **24**, 3–13

Hurlbert, D.F., Apt, C., Hurlbert, M.K. and Pierce, A.P. (2000). Sexual compatibility and the sexual desire-motivation relation in females with hypoactive sexual desire disorder. *Behav Modificat* **24**, 325–347

Kaplan, H.S. (1979). *Disorders of sexual desire and other new concepts and techniques in sex therapy*. Simon and Schuster: New York

Kaplan, H.S. (1995). *The sexual desire disorders. Dysfunctional regulation of sexual motivation*. Brunner/Mazel: New York

Kelleher, S., Conway, A.J. and Handelsman, D.J. (2004). Blood testosterone threshold for androgen deficiency symptoms. *J Clin Endocr Metab* **89**, 3813–3817

Koch, P.C. and Peters, R.H. (1987). Suppression of adult copulatory behaviors following LiCl-induced aversive contingencies in juvenile male rats. *Dev Psychobiol* **20**, 603–611

Koch, P.C. and Peters, R.H. (1992). Plasma testosterone concentration in adult male rats following acquisition of copulation-illness associations. *Physiol Behav* **51**, 873–876

Kohlenberg, R.J. (1974). Directed masturbation and treatment of primary orgasmic dysfunction. *Arch Sex Behav* **3**, 349–356

Köhn, F.M. (2006). Gibt es eine Pharmakotherapie bei Libidostörungen des Mannes? *Urologe* **45**, 999–1001

Köhn, F.M., Schuppe, H.C., Schill, W.B. and Elbers, J. (2004). Treatment of late onset hypogonadism with a new formulation of testosterone undecanoate – baseline data of a multicenter study. *Andrologia* **36**, 174

Laumann, E.O., Gagnon, J.H., Michael, R.T. and Michaels, S. (1994). *The social organization of sexuality. Sexual practices in the United States*. University of Chicago Press: Chicago

Laumann, E.O., Nicolosi, A., Glasser, D.B. *et al.* (2005). Sexual problems among women and men aged 40–80 y: prevalence and correlates identified in the Global Study of Sexual Attitudes and Behaviors. *Int J Impot Res* **17**, 39–57

Laumann, E.O., Paik, A. and Rosen, R.C. (1999). Sexual dysfunction in the United States: prevalence and predictors. *J Am Med Assoc* **281**, 537–544

Lief, H.I. (1980). Current thinking of the orgasm experience. *Med Aspects Hum Sexual* **14**, 55–62

Lobo, R.A., Rosen, R.C., Yang, H.M., Block, B. and van der Hoop, R.G. (2003). Comparative effects of oral esterified estrogens with and without methyltestosterone on endocrine profiles and dimensions of sexual function in postmenopausal women with hypoactive sexual desire. *Fertil Steril* **79**, 1341–1352

Locke, H.J. and Wallace, K.M. (1959). Short marital adjustment scale and prediction test: their reliability and validity. *Marriage Family Living* **21**, 251–255

LoPiccolo, J. and Friedman, J.M. (1988). Broad-spectrum treatment of low sexual desire: integration of cognitive, behavioral, and systemic therapy. In *Sexual desire disorders* (S.R. Leiblum and R.C. Rosen, eds), pp. 107–144. Guilford Press: New York

LoPiccolo, J. and Stock, W.E. (1986). Treatment of sexual dysfunctions. *J Consult Clin Psychol* **54**, 158–167

Masters, W.H. and Johnson, V.E. (1970). *Human sexual inadequacy*. Little, Brown & Co.: Boston

Meston, C.M. (2006). The effects of state and trait self-focused attention on sexual arousal in sexually functional and dysfunctional women. *Behav Res Ther* **44**, 515–532

Meston, C.M. and McCall, K.M. (2005). Dopamine and norepinephrine responses to film-induced sexual arousal in sexually functional and sexually dysfunctional women. *J Sex Marital Ther* **31**, 303–317

Modell, J.G., Katholi, C.R., Modell, J.D. and DePalma, L. (1997). Comparative sexual side effects of SSRIs and bupropion. *Clin Pharmacol Ther* **61**, 476–487

Nijland, E., Davis, S., Laan, E. and Schultz, W.W. (2006). Female sexual satisfaction and pharmaceutical intervention: A critical review of the drug intervention studies in female sexual dysfunction. *J Sex Med* **3**, 763–777

Nyunt, A., Stephen, G., Gibbin, J. *et al.* (2005). Androgen status in healthy premenopausal women with loss of libido. *J Sex Marital Ther* **31**, 73–80

O'Donahue, W., Dopke, C.A. and Swingen, D.N. (1997). Psychotherapy for female sexual dysfunction: a review. *Clin Psychol Rev* **17**, 537–566

O'Donohue, W.T., Swingen, D.N., Dopke, C.A. and Regev, L.G. (1999). Psychotherapy for male sexual dysfunction: a review. *Clin Psychol Rev* **19**, 591–630

Peters, R.H. (1983). Learned aversions to copulatory behaviors in male rats. *Behav Neurosci* **97**, 140–145

Peters, R.H., Blythe, B.L., Koch, P.C. and Kueker, C.A. (1989). Copulation-illness association in male rats: Lithium chloride dose and delay manipulations. *Behav Neurosci* **103**, 117–123

Pierce, A.P. (2000). The coital alignment technique (CAT): an overview of studies. *J Sex Marital Ther* **26**, 257–268

Piletz, J.E., Segraves, K.B., Feng, Y.Z., Maguire, E., Dunger, B. and Halaris, A. (1998). Plasma MHPG response to yohimbine treatment in women with hypoactive sexual desire. *J Sex Marital Ther* **24**, 43–54

Riley, A. and Riley, E. (2000). Controlled studies on women presenting sexual drive disorder: I. Endocrine status. *J Sex Marital Ther* **26**, 269–283

Schiaivi, R.C., Karstaedt, A., Schreiner-Engel, P. and Mandeli, J. (1992). Psychometric characteristics of individuals with sexual dysfunctions and their partners. *J Sex Marital Ther* **18**, 219–230

Schiavi, R.C., White, D., Mandeli, J. and Levine, A.C. (1997). Effect of testosterone administration on sexual behavior and mood in men with erectile dysfunction. *Arch Sex Behav* **26**, 231–241

Schover, L.R. and LoPiccolo, J. (1982). Treatment effectiveness for dysfunctions of sexual desire. *J Sex Marital Ther* **8**, 179–197

Schreiner-Engel, P., Schiari, P.C., White, D. and Ghizzani, A. (1989). Low sexual desire in women: the role of reproductive hormones. *Horm Behav* **23**, 211–234

Segraves, R. and Woodard, T. (2006). Female hypoactive sexual desire disorder: History and current status. *J Sex. Med* **3**, 408–418

Segraves, R.T., Clayton, A., Croft, H., Wolf, A. and Warnock, J. (2004). Bupropion sustained release for the treatment of hypoactive sexual desire disorder in premenopausal women. *J Clin Psychopharmacol* **24**, 339–342

Segraves, R.T., Croft, H., Kavoussi, R.J. *et al.* (2001). Bupropion sustained release (SR) for the treatment of hypoactive sexual desire disorder (HSDD) in nondepressed women. *J Sex Marital Ther* **27**, 303–316

Shifren, J.L., Davis, S.R., Moreau, M. *et al.* (2006). Testosterone patch for the treatment of hypoactive sexual desire disorder in naturally menopausal women: results from the INTIMATE NM1 Study. *Menopause-J N Am Menopause Soc* **13**, 770–779

Sills, T., Wunderlich, G., Pyke, R. *et al.* (2005). The Sexual Interest and Desire Inventory-Female (SIDI-F): Item response analyses of data from women diagnosed with hypoactive sexual desire disorder. *J Sex. Med* **2**, 801–818

Spector, I.P., Leiblum, S.R., Carey, M.P. and Rosen, R.C. (1993). Diabetes and female sexual function: a critical review. *Ann Behav Med* **15**, 257–264

Stuart, F.M., Hammond, D.C. and Pett, M.A. (1987). Inhibited sexual desire in women. *Arch Sex Behav* **16**, 91–106

Trudel, G. (1991). Review of psychological factors in low sexual desire. *Sex Marital Ther* **6**, 261–272

Trudel, G., Aubin, S. and Matte, B. (1995). Sexual behaviors and pleasure in couples with hypoactive sexual desire. *J Sex Educ Ther* **21**, 210-216

Trudel, G., Boulos, L. and Matte, B. (1993). Dyadic adjustment in couples with hypoactive sexual desire. *J Sex Educ Ther* **19**, 31–36

Trudel, G., Fortin, C. and Matte, B. (1997). Sexual interaction and communication in couples with hypoactive sexual desire. *Scand J Behav Ther* **26**, 49–53

Turna, B., Apaydin, E., Semerci, B., Altay, B., Cikili, N. and Nazli, O. (2005). Women with low libido: correlation of decreased androgen levels with female sexual function index. *Int J Impot Res* **17**, 148–153

van Anders, S.M., Chernick, A.B., Chernick, B.A., Hampson, E. and Fisher, W.A. (2005). Preliminary clinical experience with androgen administration for pre- and post-menopausal women with hypoactive sexual desire. *J Sex Marital Ther* **31**, 173–185

van de Velde, T.H. (1965, originally published in 1926). *Ideal marriage. Its physiology and technique.* William Heinemann Medical Books: Westport

van Minnen, A. and Kampman, M. (2000). The interaction between anxiety and sexual functioning: A controlled study of sexual functioning in women with anxiety disorders. *Sex Relationship Ther* **15**, 47–57

Vanderkooy, J.D., Kennedy, S.H. and Bagby, R.M. (2002). Antidepressant side effects in depression patients treated in a naturalistic setting: a study of bupropion, moclobemide, paroxetine, sertraline, and venlafaxine. *Can J Psychiat* **47**, 174–180

Vega-Matuszczyk, J., Larsson, K. and Erikson, E. (1998). Subchronic administration of fluoxetine impairs estrous behavior in intact female rats. *Neuropsychopharmacology* **19**, 492–498

Ventegodt, S. (1998). Sex and quality of life in Denmark. *Arch Sex Behav* **27**, 295–307

Verbrugge, L.M. (1985). Gender and health: An update on hypotheses and evidence. *J Health Soc Behav* **26**, 156–182

Vogt, H.J., Brandl, P., Kockott, G. *et al.* (1997). Double-blind, placebo-controlled safety and efficacy trial with yohimbine hydrochloride in the treatment of nonorganic erectile dysfunction. *Int J Impot Res* **9**, 155-161

Waldinger, M.D. (1996). Use of psychoactive agents in the treatment of sexual dysfunction. *CNS Drugs* **6**, 204–216

Warnock, J.K. (2002). Female hypoactive sexual desire disorder: Epidemiology, diagnosis and treatment. *CNS Drugs* **16**, 745–753

World Health Organization (1992). *International Statistical Classification of Diseases and Related Health Problems, 1989 Revision*. World Health Organization: Geneva

CHAPTER

12

Hyperactive sexual desire and the paraphilias

1 Introduction

While low sexual desire disorders have attracted a lot of attention from the scientific and clinical communities for some time, problems with high desire have been and are still far less popular. However, high sexual desire in women and men has been a theme in literature from antiquity onwards. Female hypersexuality is usually given the poetic name of nymphomania, or *furor uterinus*, madness of the womb. The noun nymphomania is constructed from the middle French word *nymphe* and the latinized version of the greek word μανια, mania. In Greek mythology, a nymph is any member of a low class of female divinities. They are actually spirits, living in the sea, in rivers, mountains, trees and so on. Some were loyal attendants on a particular god. For some reason, they must have got the reputation of being most sexually active. In contemporary English, the word nymph is often thought to mean a prostitute or any other woman regarded as a means of sexual gratification. The second part of the word nymphomania means, as any cultivated person knows, grossly excessive desire. I gather that nymphomania then must mean 'excessive desire for a nymph'. In men, hypersexuality is called satyriasis. Satyrs were, in ancient Greece, small human-like creatures with exaggerated appetites who accompanied Dionysus. They were later transformed into a mixture of human and goat. Satyrs are thought to be equipped with strong sexual urges and antique painters had the habit of portraying them with an erect penis. –asis is an ancient Greek and Latin suffix added to nouns or verbs to make them refer to a disease. Satyriasis is, then, a disease in which men reproduce the intense sexual urges of the satyrs, normally without becoming half goat.

Historians consider both nymphomania and satyriasis as social constructions, a kind of reflection of typical Western preconceptions of sexuality (Berrios and Kennedy, 2002; Berrios, 2006). I do agree that the categories themselves are no more meaningful than the categories of homosexual or heterosexual, other social constructions. In fact, what we consider as an unusually high level of sexual activity is arbitrary. It would also be completely capricious to consider an arbitrarily determined level of sexual activity a disorder or a disease, in the same way as it was capricious to consider homosexuality a disease. As I discussed in Chapter 10, the very notions of normal and functional and, consequently, also of abnormal, are intrinsically arbitrary. Even though it might be most convenient to limit the use of the notions of nymphomania and satyriasis as well as that of hypersexuality to the realms of the fine arts, society or the medical profession may decide that they should be considered as disorders. In fact, respectable scientists have eloquently argued for the need to include hyperactive sexual desire disorder as a diagnostic category in the future DSM-V and ICD-11 (Vroege *et al.*, 1998). This suggestion seems to have partially prospered even before it was made, because the latest versions of the ICD-10 do include such a category.

The paraphilias, with such conditions as pedophilia, exhibitionism and fetishism among others, have been the objects of much attention in most Western societies for many years. Old ladies or young schoolgirls frightened by a completely inoffensive exhibitionist make headlines in local newspapers and appear at prime time on local TV shows with their terrified eyes staring at us. Revelations of pedophilia among the clergy will be broadcast not only on local TV stations, but also on all national chains and at all times. Expressions of paraphilia are widely regarded as a threat to social order and personal security. Then again, the interest in a scientific understanding of these unusual behaviors is far less intense than the outcry for longer prison terms for those displaying them.

In this chapter, I will review the question of high sexual desire, including the proposed diagnostic criteria, epidemiology and treatment. Then I will turn to the paraphilias and provide a similar review. As always, I will profit from these evaluations of current knowledge to propose that the incentive motivation/learning model may provide a helpful theoretical framework for the analysis of etiological factors as well as treatment alternatives.

2 Hyperactive sexual desire

2.1 The search for diagnostic criteria

In the DSM-IV, including the 2000 version called DSM-IV-TR (American Psychiatric Association, 2000), there is no diagnostic category at all for persons expressing high levels of sexual activity. On the contrary, there is a rather extensive discussion of several of the paraphilias. In the ICD-10, the 2006 online edition (http://www.who.int/classifications/apps/icd/icd10online/) includes a category labeled excessive sexual drive. There are two subheadings, nymphomania and satyriasis. Like the DSM-IV, the ICD-10 supplies a long list of paraphilias.

The ICD-10, in the 2006 version mentioned above, is not very explicit with regard to the diagnostic criteria corresponding to the label 'excessive sexual desire'. Actually it says nothing. Its companion 362 pages publication entitled *The ICD-10 Classification of Mental and Behavioural Disorders: Clinical descriptions and diagnostic guidelines* does not provide much more information (World Health Organization, 1992). These guidelines are quite eloquent in relation to many other sexual disorders, but when arriving at the diagnostic category F52.7, excessive sexual drive, it simply states:

> Both men and women may occasionally complain of excessive sexual drive as a problem in its own right, usually during late teenage or early adulthood. When the excessive sexual drive is secondary to an affective disorder (F30–F39) or when it occurs during the early stages of dementia (F00–F03), the underlying disorder should be coded.
>
> Includes: nymphomania, satyriasis
>
> (World Health Organization, 1992, p. 194)

Another publication accompanying the ICD-10, the 248 pages *The ICD-10 Classification of Mental and Behavioural Disorders: diagnostic criteria for research* is not more talkative. It simply states: 'No research criteria are attempted for this category. Researchers studying this category are recommended to design their own criteria' (World Health Organization, 1993, p. 120). Rumors have it that the experts could not come to any consensus on the criteria for excessive sexual drive and it was therefore decided to leave the label and subheadings without any criteria at all. Another solution to the lack of consensus would have been to simply omit the diagnostic category, but for some unknown reason this solution must have been rejected.

Judging from the short statement concerning excessive sexual drive reproduced above, any individual articulating complaints about such a state could be classified under category 52.7. If the complaining individual were a woman, the diagnostic category would be nymphomania, if it were a man, then he would be suffering from satyriasis. A minor problem here is that the notion of 'excessive' is extremely unclear. It is, in fact, close to meaningless. First, we would have to know who determines whether the sexual drive is excessive or not. Is it the complaining individual, the clinician, a questionnaire? Second, we would have to know in relation to what the sexual drive is excessive. According to Saint Augustine, almost all expressions of sexual drive are excessive. According to some porn film directors, excessive sexual drive must be quite extreme. All this is unclear.

The lack of clarity concerning the empirical content of the term 'excessive' is not the only problem with the diagnostic category we are discussing here. The notion of 'drive' is also ambiguous. In old behaviorist literature, drive was the term used for motivation. It always referred to a bodily necessity generated by deprivation. The existence of conditioned drives was suggested, but never proved for anything except fear. As a scientific term, 'drive' was abandoned many years ago. Nevertheless, in popular writings and in the writings of those unfamiliar with contemporary behavior theory it may still appear. Although obsolete, expressions like 'hunger drive' may have some meaning, since the mental state called hunger somehow is related to an organismic necessity. 'Sex drive', on the contrary, is an unacceptable expression since we now know that sex is not at all linked to any bodily need. In any case, the term 'drive', as employed in the ICD-10, is probably equivalent to what we now would call motivation. We already know that motivation goes under the label of desire in most of the clinical literature. Using a contemporary language, then, it would be appropriate to call the diagnostic category 52.7 'excessive sexual desire'.

The preceding two paragraphs should have exposed enough compelling arguments for regarding the notion of excessive sexual drive as a notion void of empirical content. To me, this diagnostic category appears just as ridiculous as the subcategories nymphomania and satyriasis. In sexual matters, the otherwise respectable experts of the World Health Organization appear to have been left behind long ago, still remaining in the late baroque when histories of nymphs and satyrs were popular.

In contrast to experts engaged in huge bureaucracies, scientists have made reasonable suggestions as to how criteria for hyperactive sexual desire disorder could be formulated. These scientists have abstained from proposing mythological subcategories taken from ancient Greece, perhaps because of a weaker interest in Hellenistic history than in sound science. We will now look at a few of the efforts made with the aim of finding an acceptable rationale for a diagnostic category of hyperactive sexual desire and to determine clinically useful criteria for diagnosis.

Quite clear criteria for considering a person to suffer from hyperactive sexual desire were proposed some years ago (Kafka, 1997). The person should have a minimum total sexual outlets of seven per week and that frequency of sexual outlet should have been sustained for at least 6 months during some period in life after the age of 15 years. The notion of total sexual outlet has already been explained, but I remind the forgetful reader that all sexual activities leading to orgasm are included in it. The rationale for establishing the criteria was clear-cut. The minimum duration of 6 months was chosen since the DSM from its third edition onwards requires that duration for the diagnosis of paraphilia. In addition, a frequency measure with a rather long period, as the one used here, would be most variable if a short time span were used as a basis for estimating it. The magical number seven was chosen because the median total sexual outlet is around two per week in men between adolescence and age 85 as determined in a large interview study and only 7.6% of young Americans (from adolescence to 30 years) have a total sexual outlet of more than seven per week (Kinsey *et al.*, 1948). A cutoff criterion of seven ensures that those included in the category have a total sexual outlet far above the median and that they belong to a small, high-end segment of the population.

The proportion of young men having a total sexual outlet above seven reported in the Kinsey *et al.* (1948) study deviates somewhat from that found in a more recent survey. As many as 15% of American men between 18 and 50 had daily sexual activity leading to orgasm, meaning a total sexual outlet above seven (Janus and Janus, 1993). However, this survey has been criticized on several grounds and there is no particular reason to give it more weight than the old Kinsey data. Later studies of the intensity of sexual behaviors in the American population (e.g. Laumann *et al.*, 1994) preferred to estimate sexual activity as the frequency of penile–vaginal intercourse, of masturbation or of any other particular sexual behavior rather than the total sexual outlet, making it difficult to determine the amount of sexual behaviors actually displayed. However, as far as can be determined, the data reported in these later studies do not deviate much from the Kinsey *et al.* (1948) data. The criterion used by Kafka (1997) seems to derive from a rather solid statistical base concerning average intensity of sexual behavior.

Here we have a splendid example of the employment of deviation from average as a sign of abnormality, or at least something requiring a label leading thoughts to a disorder. We can agree or disagree with that kind of reasoning. Personally I strongly disagree. Despite the disagreement, it must be recognized that the criteria suggested by Kafka (1997) are clear and, as such, respectable. It may be worthwhile mentioning that Kafka's aim was to investigate whether there was any relationship between high sexual desire and paraphilia. When asking a group of men seeking treatment for paraphilia or paraphilia-related disorders to estimate their total sexual outlet, he found that 72% of those men had an outlet above seven and they had had it for more than 10 years. This coincidence is not any evidence in favor of an association between paraphilia and hypersexuality, although Kafka (1997) makes that suggestion.

The usefulness of a category of hyperactive sexual desire disorder has been questioned in an amusing Swedish study (Långström and Hanson, 2006). After criticizing earlier efforts made in order to justify such a category as well as the criteria proposed for inclusion in it, the researchers suggest that large amounts of impersonal sex might be indicative of something worth calling hypersexuality. Items representing impersonal sex are masturbation, pornography use, the number of sexual partners, sexual activities outside but simultaneous with the current relationship, more than one simultaneous partner, preference for casual sex over a stable partner and the participation in group sex. Based on questionnaire answers from a nationwide sample, the respondents were classified as hypersexual if they belonged to the top 5–10% with regard to frequency of the behaviors on the impersonality list just mentioned. For men, a frequency of masturbation above 15 per month, use of pornography more than 31 times last year and more than three partners last year were all indices of hypersexuality. Likewise, ever having had sex with another person while married/cohabitating, having more than one current stable partner, a preference for casual sex and ever having participated in group sex were also indicators of hypersexuality. For women, the requirements were less stringent. The frequency of masturbation needed to be five or more per month, pornography use must have occurred more than four times during the last year and the number of partners must have been two or more during last year. Among men, 12.1% were characterized as hypersexual by applying these criteria, while 7% of the women satisfied them. When comparing these women and men with those reporting low levels of sexual activity, it turned out that high sexual activity was associated with a higher tendency to expose genitals to a stranger, to spy on others' sexual activities, to smoke and drink more alcohol, and to engage more frequently in illegal drug use, just to mention some examples.

The requirements for being considered hypersexual in Sweden do not seem to be remarkably demanding. For a man, reading a pornographic magazine once a week and masturbating every other day, plus having shown one or more infidelities in a lifetime appears enough. For a woman, masturbating once a week and watching a pornographic movie once every three months, plus having been unfaithful to the partner or ever having engaged in group sex is sufficient. In a statistical sense, all these behaviors are far from average, and therefore acceptable criteria of abnormality for those defining 'normal' as being close to average. The fact that there is some relationship between high impersonal sexual activity and behaviors judged inconvenient by contemporary society, like smoking and drinking alcohol, justifies, according to Långström and Hanson (2006), the use of the label excessive when referring to these activities. They show some modesty, though, when remarking that 'the current study did not identify the causes of hypersexuality or whether it is a distinct disorder' (p. 50).

Interestingly enough, Långström and Hanson (2006) also report data showing that personal sex, meaning penile–vaginal intercourse with a stable partner, is associated with sexual satisfaction and satisfaction with life in general. The higher the frequency of intercourse, the larger the satisfaction. On the contrary, a high frequency of masturbation (impersonal sex) is associated with low sexual satisfaction and low satisfaction with life in general.

The Långström and Hanson (2006) report is an excellent example of how scientists' prejudice determines the questions asked and the answers provided. Their list of 'impersonal' sexual activities coincides beautifully with dominant Swedish stereotypes regarding unacceptable sexual behavior patterns. All those on the list are, in fact, repulsive to the social democratic ideal of decency, which has replaced protestant fundamentalism as the social norm. For example, they include masturbation under the label of impersonal sex. This seems extremely strange to me. It is difficult to imagine any activity more personal than self-masturbation. However, masturbation is socially regarded as inferior to penile–vaginal sex. In the same way, watching a pornographic movie together with your partner may be a quite personal experience, but since pornography has been made socially suspect among the respectable, educated Swedish middleclass through the efforts of an unholy alliance between the religious right and feminists, any use of it is badly viewed (Östergren, 2006). Furthermore, the fact that socially unacceptable sexual activities are related to other unacceptable behaviors, according to the social democratic ideal, like smoking and drinking alcohol comes as no surprise. Those not sharing the dominant ideal of decency can be expected to show a host of behaviors disliked by the guardians of morality.

In sum, even those fond of the idea that deviations from average represents abnormality must accept that the criteria chosen by Långström and Hanson (2006) are particularly unfortunate. The basic notion of 'impersonal sex' refers to most sexual activity other than penile–vaginal intercourse within a stable, devoted couple. I really do not know why, but Saint Augustine comes to mind. What Långström and Hanson (2006) have done, actually, is to characterize people not respecting dominant social norms, be it with regard to sexual activities or with regard to the relationship to smoking and alcohol consumption, as hypersexual. This does not seem to be a promising approach.

The two examples provided so far do not give any convincing reason for the usefulness of a diagnostic category of hyperactive sexual desire. The diagnostic criteria proposed seem quite arbitrary. They are not only based on the notion of normality as average, but they are also contaminated by a substantial amount of prejudice. Scientifically, they have no value and in clinical practice the situation is the same.

It is undeniably so that some individuals show a rather intense preoccupation with sexuality. They may spend hours every day masturbating while looking in pornographic magazines, viewing on-line pornography or participating in chat rooms with sexual content. They may also spend hours in public places such as bars, bookstores or bathhouses trying to pick up occasional sexual partners. These behaviors may be performed at great expense, causing financial problems. Since the time investment is heavy, it may be impossible to keep a job, which exacerbates the financial problems. Some of those able to hold a job may spend most of the working day downloading pornographic pictures on their employer's hard drive. They may also fill their own or their employer's computer with child pornography or other illegal content. Typical for some of these individuals is that the risks associated with their sexual behavior are of no or little concern. These risks range from financial ruin, unemployment, contagion with sexually transmitted diseases, penal pursuit and possibly a prison term and, eventually, a complete loss of social esteem. Conditions of this kind have been called sexual addiction or compulsive sexual behavior (Coleman, 1990, 1991; Goodman, 1992).

There is a substantial disagreement with regard to the utility of these labels. This disagreement has lasted for at least 20 years (Coleman, 1986), without any solution appearing. Some suggest that compulsive sexual behavior is a manifestation of disorders in the impulsive/compulsive spectrum (Raymond *et al.*, 2003) and others maintain that addiction to sex is not essentially different from other kinds of addiction (Goodman, 2001). A thoughtful review of the concept of sexual addiction concluded by stating that the empirical usefulness of the concept remains to be established. The relevant literature consists of conjecture and theoretical considerations, largely based on clinical observation rather than on systematic research (Gold and Heffner, 1998). More recently, another review considered that neither addiction nor compulsion were adequate labels for cases of intense sexual preoccupations and behaviors. Instead, terms like 'out of control sexual behavior' or 'unregulated sexual behavior' are preferable until some more solid knowledge has been accumulated (Bancroft and Vukadinovic, 2004).

Hyperactive sexual desire, sexual addiction or compulsive sexual behavior, all refer to some state of heightened sexual activity. As we have seen, all these notions are problematic in several ways. The few studies that tried to establish a set of usable criteria for distinguishing hyperactive sexual desire from euactive (euactive is a neologism I just created to avoid the unsuitable word 'normal') sexual desire are based on the notion of normal as close to average. This might be legitimate if the deviation from average has some clinical significance. In the case of the criteria proposed by Kafka (1997), it is extremely difficult to see this significance. He suggests that hyperactive sexual desire may be related to the presence of paraphilia. Even if this were to turn out to be correct, then so what? There are already several diagnostic categories for paraphilia. The addition of a new diagnostic category of hyperactive sexual desire would not be of any help for those suffering from that disorder. On the contrary, if it were established that high sexual desire *causes* paraphilia, then a treatment of desire could be of clinical utility. However, hypotheses concerning the etiology of paraphilia do not, in the slightest, justify the creation of a new diagnostic category, unless that new category is to replace the existing diagnostic category of paraphilia. This proposal seems entirely unwarranted, since available data seem to suggest that the vast majority of those expressing high sexual desire are not paraphiliacs.

Sexual addiction and compulsive sexual behavior are, so far as I and many others can see, manifestation of other underlying behavioral disorders. These disorders happen to manifest themselves in unusually intense sexual preoccupations, but they could in all probability have taken other forms of expression. Treatment aimed at removing the underlying disorder would be the most efficient solution in these cases.

2.2 A proposal for diagnostic criteria

It is not self-evident that it is worth the trouble to establish diagnostic criteria for a state that might be completely unproblematic for the individual. This was the implicit basis for my extensive criticism of the category of 'excessive sexual drive' and its subcategories. In contrast, if the state causes personal distress or interpersonal difficulties, then it might be worthy of consideration. In the case of hypoactive sexual desire disorder, the diagnostic criteria require both absence of desire for sexual activities *and* the experience of personal distress or interpersonal difficulties caused by this absence. The absence of sexual desire is by itself no problem. We even know that it may represent the highest of virtues. It becomes a problem only when the afflicted individual perceives it as such. With regard to the opposite condition, hyperactive sexual desire, it appears that the preoccupation has been concentrated on finding an objective limit above which the intensity of sexual behavior must be located in order to be considered as problematic. This was the case in the Kafka (1997) study and in the Långström and Hanson (2006) report. In none of these studies, though, were the subjects asked if their level of sexual activity caused them problems. In the Kafka (1997) study, the subjects had sought attention for paraphilia or paraphilia-like problems, but we do not know if they ever associated that with high levels of sexual activity. In the Långström and Hanson (2006) report, there is no indication whatsoever that those included in the sample ever considered their level of sexual activity as problematic.

We saw that the rationale for and the results of the effort to establish a quantitative criterion for high sexual activity or hypersexuality were questionable on many grounds. Perhaps the insistence on a quantitative cut-off point is completely off the point, so to say. We also learned that the notions of sexual compulsion or addiction can best be forgotten and lumped together with obsessive/compulsive or addictive disorders. In consequence, we do not have to care any further about them.

Forgetting about an objective limit of total sexual outlet or frequency of solitary masturbation or whatever, we can propose three simple diagnostic criteria for hyperactive sexual desire disorder:

1 A persistent or recurrent high level of sexual activity or persistent or recurrent feelings of a need for a high level of sexual activity.
2 The disturbance causes marked personal distress or interpersonal difficulty.
3 According to the clinician's judgment, the high level of sexual activity is causally related to the distress or interpersonal difficulties.

An exclusion criterion must be added. If that the patient satisfies the diagnostic criteria for any of the paraphilias, that diagnosis should be given.

This proposal is probably lacking in many ways. For example, the notion of a high level of sexual activity is very vague. I assume that the efforts aimed at establishing a cut-off point had a goal to avoid this vagueness. However, the alternative is, as we have seen, to propose a minimum level of sexual activity based on deviations from average. In Chapter 10, I mentioned the many deficiencies inherent in this practice. These deficiencies carry a cost that is higher than that of a certain imprecision, I would believe. Another crucial point in my reasoning is that the clinician should use her/his judgment to determine whether the high level of sexual desire is at the root of the patient's distress or not. This introduces an intuitive element, which some might consider inconvenient. At the same time, there is an intuitive element in most diagnoses of psychic problems and I cannot see why it should be more devastating here than elsewhere. Finally, it could be maintained that we do not need the diagnostic category at all more than we need the absurd labels of nymphomania or satyriasis that I criticized not long ago. However, if a diagnostic category can help some humans solve a behavior problem reducing their quality of life, then the label might be justified. Application of the third criterion should ensure that the source of the patient's problem indeed is the hyperactive sexual desire. The correct identification of the origin of the problem is a requisite for successful treatment. Whether a treatment is available or not is another question, but that problem can potentially be solved. However, if we fail to identify the origin, then treatment is frequently destined to failure.

2.3 Epidemiology and etiology

It is impossible to make any kind of informed speculation as to the prevalence of hyperactive sexual desire since there has been no agreement around a definition of it. We can repeat that Kinsey *et al.* (1948) found that 7.6% adult of men had seven or more sexual outlets per week, and we could try to find similar data for women. However, that would be of little use, since the amount of sexual activity *per se* is not the crucial element. Rather than presenting irrelevant data, I prefer to claim that we first need an acceptable and accepted set of diagnostic criteria for hyperactive sexual desire disorder, under the condition that the clinical community finds that of interest, obviously. Then we need decent population based studies. Until such studies have been performed, there is nothing more to say than that we do not know.

The etiology must also remain unknown. Most scientists would consider it absurd to try to describe the etiology of a disorder before the disorder has been discovered. However, just as speculations as to the causes of hypoactive sexual desire preceded the creation of the diagnostic category, we might speculate about causes for an unusually high sexual activity. What we cannot speculate about at present is how a high sexual activity turns into a problem.

A classical hypothesis suggested that the intensity of sexual activity is correlated with blood concentrations of gonadal hormones. This is partly correct. In Chapter 5, we saw some examples illustrating that the absence of gonadal hormones, more specifically testosterone, strongly reduces sexual activity in women and men. We also saw, in Chapter 11, that androgen treatments to women suffering from hypoactive sexual desire had beneficial effects. Thus, there is no doubt that low blood concentration of testosterone is associated with low sexual activity. By extension, it could be assumed that high blood testosterone concentrations should be associated with high sexual activity. In many popular writings we are told that this is the case. On the contrary, in the case of the human male, the scientific literature abounds with evidence showing that once above a minimum concentration, further increases in testosterone have no effect. Treatment of eugonadal men with large testosterone doses have generally failed to affect sexual behaviors (Anderson *et al.*, 1992; Bagatell *et al.*, 1994a; Yates *et al.*, 1999) and variations within the physiological range are not associated with variations in the intensity of sexual activities (Buena *et al.*, 1993). Indeed, there are data showing that the physiological blood concentration of testosterone is above that needed for maintaining sexual behaviors at an adequate level (Bagatell *et al.*, 1994b). There is not much reason to believe, then, that enhanced blood concentrations of androgens are related to unusually high sexual activity in men.

Data from non-human animals support this notion. In a classic study, the intensity of copulatory behavior in a group of male guinea-pigs was determined. The animals were then castrated and all were treated with a constant dose of testosterone. The males having displayed a high level of copulatory behavior before castration continued to do so when castrated and treated with testosterone while animals displaying a low level of sexual activity before castration also did so when given the constant testosterone dose (Grunt and Young, 1952). In rats, the individual differences in the intensity of copulatory behavior between males were not altered when these males were treated with very large testosterone doses (Larsson, 1966). Likewise, sexually inactive or sluggish male rats do not have lower blood testosterone concentrations than active rats and they do not enhance their performance in response to additional testosterone (Damassa *et al.*, 1977). Data similar to those reported for guinea-pigs and rats have also been obtained in rabbits (Ågmo, 1976). The most likely interpretation of these results is that relevant receptors are saturated already at physiological hormone concentrations, making it impossible for a further increase in hormone availability to have additional effects. If the individual differences are dependent on gonadal hormones at all, these differences must have been established during the organizing action of hormones around birth. Alternatively, the differences are due to interindividual variations in efficiency of the transcription activating capacity of the hormone-receptor complex, or of post-transcriptional events. Another possibility is that unknown, hormone-independent central nervous mechanisms determine the reactivity to sexual incentives. My bet would be on the last alternative. In any case, the non-human data give some experimental support to the notion that the intensity of sexual behavior in males is not determined by the availability of testosterone as long as that availability is above a certain limit, probably located close to the lower end of the physiological range.

In women, the situation is unclear. Since testosterone has peripheral, masculinizing effects, which most women do not like, large testosterone doses are not normally given to women. Nevertheless, those with hypoactive sexual desire disorder who responded favorably to testosterone therapy had their basal blood testosterone concentrations within the physiological range, suggesting that an increase within or slightly above that range indeed may enhance sexual activity in women. There are also some data suggesting a relationship between the concentration of circulating androgens and the intensity of sexual desire in women (e.g. Davis, 2000; Davis and Tran, 2001) even when androgen concentrations are within the physiological range. This stands in sharp contrast to data obtained in men, where such a relationship is not found. One explanation is that the much lower concentration of androgens found in women compared to men is insufficient for saturating androgen receptors. Thus, an increase in testosterone availability would enhance the effects of androgen receptor stimulation.

Whether testosterone concentrations above the normal range are causally related to an unusually high intensity of sexual activity in women is not completely known. However, some women suffer from supraphysiological blood androgen concentrations because of Cushing's syndrome, other endocrine diseases or ovarian tumors. One of the most evident manifestations of high testosterone concentrations is hirsutism, growth of bodily hair in a man-like fashion. An interesting but preliminary study was performed in women diagnosed with polycystic ovary syndrome or idiopathic hirsutism undergoing treatment with the androgen receptor antagonist cyproterone acetate for alleviation of symptoms. An assessment of these women's sexual desire before the onset of treatment revealed that it was somewhat lower than that of women in a healthy control group (Conaglen and Conaglen, 2003). This observation speaks against an association between high testosterone concentrations and high sexual desire. Interestingly, the patient group manifested reductions in sexual desire as a consequence of the androgen antagonist treatment, confirming that androgen receptor stimulation indeed is important for female sexual desire. Studies of sexual desire in women suffering from Cushing's syndrome are rare and, because of the alterations of other adrenal steroids, such studies would not be conclusive. Although limited, the available information suggests that supraphysiological testosterone concentrations are not causally related to high levels of sexual desire.

It appears that endocrine factors cannot account for unusually high levels of sexual desire. This does not exclude a heightened reactivity of the central motive state as a cause. Little is known about the central nervous mechanisms determining that reactivity and speculations on the matter seem unwarranted. Instead of searching for explanations in terms of an endogenous factor enhancing reactivity of the central motive state, explanations can, as always, be sought in the individual's own experience. For example, we know that the frequency with which a person engages in sexual activities is related to the satisfaction obtained from them (Giambra and Martin, 1977; Wilson and Lang, 1981; Långström and Hanson, 2006). The reward value of sexual activity is, then, a determinant of its frequency of occurrence. The incentive motivation/learning hypothesis poses that the more frequently sexual reward is obtained, the more likely is the formation of association between that reward and stimuli present in the context. These stimuli will acquire sexual incentive properties and, in the future, they will activate sexual incentive motivation and sexual activity. The stronger the association between the sexual reward and a stimulus is, the larger is its capacity to induce sexual motivation. The larger the number of stimuli that have been associated with sexual reward, the larger the probability of being exposed to one of these stimuli. Taking into account these facts, it is no wonder that frequent sexual reward enhances the likelihood of sexual behaviors. This description of the relationship between sexual reward and sexual activity may look like a positive feedback system where more sex will be a cause of still more sex. However, there are several constraints that will control the system, even if in principle it is actually founded on positive feedback. Negative alliesthesia may reduce the impact of sexual incentives and situational circumstances may limit the exposure to such stimuli or prevent their effects. I already gave an example in which nude female breasts on a beach in Normandy are not working as a sexual incentive while they may be most efficient in the bedroom.

Learning as a factor in intense sexual activities has not been much explored, but there is some indirect evidence in support of the notion. Some people spend a lot of time interacting with computers and some specialize in sites providing pornographic materials ranging from nude pictures through depictions of intercourse to bestiality or sadistic violence. Fully interactive services are also available, offering mutual masturbation accompanied by image, sound and text among other things (Cooper *et al.*, 1999). Amid enthusiasts, it is not unusual to spend 15–20 hours per week connected to electronic pornographic materials. A quite interesting hypothesis concerning the causes of this behavior has been advanced (Putnam, 2000). Through social learning and mental representations, an individual connects to a pornographic site and exposes her/himself to pornographic materials on the screen. As we know from Chapter 4, this kind of material produces arousal and it is not unlikely that this arousal is associated with some reward. It is also reinforcing, assuring that the behavior pattern leading to it will be repeated. Because of operant learning, the frequency of visits to pornographic sites will be increased. A salient stimulus in this situation is obviously the screen itself, the keyboard and other paraphernalia around the screen. All these stimuli will become associated with the sexual incentives exposed on the screen and pouring out from the loudspeakers and will eventually acquire the function of positive incentives, actually sexual incentives. After some amount of conditioning, the sight of the computer screen and the keyboard will induce sexual arousal and activate the operant responses needed for connection to the pornographic site. In terms of the incentive motivation/learning model, the screen and the keyboard, in their quality as conditioned sexual incentives, activate arbitrary approach responses bringing the individual in contact with the sexual stimulus.

This extremely interesting hypothesis would be still more interesting if supported by empirical data. An experiment teaching psychology students, the favorite research subjects, to connect to pornographic Internet sites while penile tumescence and vaginal blood flow were monitored could easily be arranged. The students could be asked to connect for an hour per day for a couple of weeks. After that, changes in penile tumescence and vaginal blood flow in response to exposure to the screen and keyboard could be determined. Possible spontaneous increases in connection time could also be evaluated and a lot of additional data illustrating progress of learning could be collected. I wonder why an experiment so simple and so illuminating has not been performed already.

An elegant study evaluating the change in Internet usage among 'compulsive Internet users', that is people staying connected for more than 16 hours a week, during a period of 1 year found that the best predictor for increase in use was the time spent connected to pornographic sites (Meerkerk *et al.*, 2006). This is exactly what would have been predicted from the hypothesis outlined above. Just as with other sexual activities, circumstantial constraints impose an upper limit on electronic pornography consumption.

Although the arguments exposed in the preceding two paragraphs do not constitute any proof that the incentive motivation/learning model is the best explanation for unusually high levels of sexual activity, they are at least suggestive. I would also like to emphasize that sexual activity on the Internet may be illustrative of the working of sexual motivation and of a subset of sexual behaviors relieved from many habitual social constraints. It is most often performed in privacy (although some people may visit pornographic Internet sites from computers in public places, and even masturbate while doing so), there is no external control of the activities watched (unless the user is under police surveillance), it is cheap (compared to the cost of hiring a prostitute, it is very cheap) and it is always available. Under these circumstances, it is most likely that many external constraints are relaxed and the individual can allow her/his sexual incentive motivation to manifest itself in an almost natural way. Internalized moral restraints will still be at work, but since most people are hypocrites and only respect their presumed morals when observed by others, their influence will probably be quite limited.

To summarize this section on etiology, it can be stated that endocrine alterations are unlikely sources of high sexual activity. Learning may have some role.

2.4 Treatment

Since hyperactive sexual desire does not exist as a diagnostic category, I can present no trial where patients suffering from it have been included. The excessive sexual drive notion of the ICD-10 has not shown itself to be more successful in generating clinical research than in its conceptualization. Not a single study has included patients from that category. We are, then, in a situation where relevant empirical data are completely absent.

It would not be particularly useful to analyze all the means by which sexual activity can be reduced. Already, in Chapter 5, we learned that reducing androgen concentrations in blood or blocking androgen receptors were efficient for reducing sexual behaviors, including desire, in men and women. There is no reason at all why this should not also apply to those having an unusually high level of desire. Thus, treatment for the potential disorder of hyperactive sexual desire should be easily accomplished by any means reducing the cellular actions of androgens. If this treatment was not attractive for some, there is a reliable alternative immediately available. A huge amount of data from non-human mammals, as well as an abundant clinical literature, shows that enhanced serotoninergic neurotransmission leads to inhibition of sexual behaviors and sexual motivation (e.g. Ahlenius *et al.*, 1980; Matuszczyk *et al.*, 1998; Vega-Matuszczyk *et al.*, 1998; Montejo *et al.*, 2001; Clayton *et al.*, 2002; Bonierbale *et al.*, 2003).

Reduced androgen availability, as well as SSRIs, acts on the central motive state, causing it to be less reactive to sexual incentives. Another treatment option would be to reduce the incentive value of the incentive stimuli. How this could be accomplished is not entirely evident. Blocking the sexual reward induced by orgasm might be possible through the administration of pharmacological agents. If this were done systematically, the stimuli associated with the situation would lose their incentive value by extinction of learning. It is also possible that the pharmacologically caused absence of expected orgasm would create frustration and an aversive state, which would facilitate extinction and eventually transform the positive incentive into a negative one. However, multiple drug treatments would be required. In fact, this procedure seems unnecessarily convoluted compared to direct inhibition of the central motive state. Furthermore, in contrast to the solidly established inhibitory effects of the SSRIs or reduction of androgen availability, there is no clinical experience with an extinction procedure like the one outlined here. If the high sexual desire disorder manifested itself in exaggerated use of Internet pornographic sites, a different kind of extinction procedure could be envisaged. Nevertheless, it would probably require far more time and effort and represent a much more risky approach than treatments directed towards the central motive state.

The treatment of hyperactive sexual desire disorder does not seem to pose any serious problem at all. Efficient pharmacological methods for reducing the reactivity of the central motive state exist already and they could be applied to this disorder whenever required.

3 The paraphilias

3.1 General

Paraphilia is, as I already mentioned or at least insinuated, the display of sexual arousal and/or behaviors in response to stimuli not judged appropriate as sexual incentives or not judged appropriate in the specific situation. An example of the former would be fetishism, a response of sexual arousal to an object considered inappropriate, like a black leather boot. An example of the latter could be exhibitionism, where a man exposes his genitals in response to a woman, usually considered an acceptable sexual incentive, in an unsuitable situation, like in a public place or in other contexts where social norms consider exposure of genitals as inappropriate. In addition, some behavior patterns executed in response to appropriate sexual incentives are included among the paraphilias. Sadistic behaviors performed as part of copulatory interaction between a man and a woman, perfectly appropriate sexual incentives for each other, is an example of this kind of paraphilia. In this chapter, I will give short descriptions of the main clinical features of exhibitionism, fetishism and pedophilia. I have chosen these paraphilias since I find them the most illustrative of the difficulties in understanding the etiology and treatment of paraphilia. Furthermore, pedophilia is a subject of much social concern, while exhibitionism and fetishism have an aura of the exotic.

3.2 Exhibitionism

Diagnostic criteria and general description

The diagnosis of exhibitionism, according to the DSM-IV, can be given if the following criteria are satisfied:

A. Over a period of at least 6 months, recurrent, intense sexually arousing fantasies, sexual urges, or behaviors involving the exposure of one's genitals to an unsuspecting stranger
B. The fantasies, sexual urges, or behaviors cause clinically significant distress or impairment in social, occupational, or other important areas of functioning (Reprinted with permission from the *Diagnostic and Statistical Manual of Mental Disorders*. Copyright 2000. American Psychiatric Association)

These criteria are quite clear. It must immediately be observed that no overt behavior is required according to these criteria. Mental representations of exposing genitals to strangers are sufficient as long as they produce sexual arousal. The concept of 'sexual urge' is quite mysterious. I assume it refers to the conscious experience of sexual incentive motivation, like the term limerence I discussed extensively in Chapter 4 does. It adds nothing of value to the criteria, but it does not cause major disturbance either. As always, the notions of distress and impairment are crucial. If the patient does not experience distress or impairment because of the wish to expose genitals, then no diagnosis should be given.

The typical exhibitionist is a man. Exhibitionism among women is very unusual and there are only a couple of cases reported (e.g. Hollender *et al.*, 1977; Grob, 1985). The onset of the urge to expose the genitals to unsuspecting women occurs usually between 20 and 30 years of age (Mohr *et al.*, 1964). However, exhibitionism among adolescent boys is not unknown and there are also cases where the first episode occurs after the age of 60. Somewhat more recent data have confirmed the age of onset as located between ages 20 and 30. In a group of men diagnosed as exhibitionists according to the DSM-IV, the mean age at which the wish to expose themselves appeared was 23.4 years (Grant, 2005). Of these men, only 40% were married. This contrasts to older data reporting that 74% of exhibitionists were married (Mohr *et al.*, 1964). In any case, it seems that there is no relationship between availability of sexual partners and episodes of exhibition. There are even individuals who expose their genitals in a public place shortly after having had sexual intercourse.

The exact stimulus activating exposure is variable. It is almost always a stranger, but the age may vary. The majority of exhibition episodes seem to occur in response to adult women, but a substantial proportion of episodes involve children and adolescents. There are exhibitionists exposing only in response to adult women and those only to children or adolescents. A minority does not make age distinctions. A majority of episodes of exposure occurs to one woman, but some occur to two or several women. It is not known whether exhibitionists systematically expose themselves to a single woman or to two or more, or if circumstances determine the number of potential observers.

The places chosen for exposure are quite varied. Many exhibitionists like to use their car, others expose themselves on the street or in public parks. Public buildings are sometimes chosen. Occasionally, exhibitionists expose themselves through a window, either from his home or from a store. The place and mode of exposure seems to be determined by circumstantial or cultural factors. In North America, the car is popular, while in England public parks are the favorite place.

Most exhibitionists get an erection and some masturbate when exposing themselves. They are said to achieve an intense experience of sexual satisfaction. Others do not show any genital response and have no conscious experience of satisfaction. A particular individual may show the first kind of behavior on some occasions and the second kind on others. A common feature of most, if not all, exhibitionists is that the reaction of the woman or women to whom they expose is crucial. An old study (Ellis, 1933) describes three principal reactions:

1 The woman or women get frightened and run away.
2 The woman or women get indignant and abuse the exhibitionist.
3 The woman or women gets pleased or amused, and laugh or smile.

The word woman should be replaced by the word girl whenever adequate in the preceding sentences. It has been suggested that the last reaction provides the most intense satisfaction to the exhibitionist, but this suggestion is not based on empirical data.

Exhibitionists are considered to be entirely inoffensive. They do not seek any direct sexual contact and do not represent any danger to the woman or women whose presence activates the exposure response. Nevertheless, some women get frightened. It has been suggested that the reaction to an exhibitionist is dependent on the woman's own psychological health and attitudes towards sexual matters (Mohr *et al.*, 1964).

The preceding account of exhibitionism is heavily influenced by the points of view of Mohr *et al.* (1964). Although now more than 40 years old, these points of view are still held by the majority of those working in the field of exhibitionism. Recent research has focused more on the potential causes and treatments of exhibitionism than on characterizing the exhibition episodes. For example, the external stimulus eliciting the behavior, as well as the exact circumstances under which the exhibition occurs, have been ignored. This means that we know almost nothing about the stimulus control of the genital exhibition. Likewise, the precarious knowledge we have concerning the potential rewarding and reinforcing consequences of genital exposure is based on anecdotal evidence. It appears that research on exhibitionism has been erratic and not guided by any theoretical framework within which to understand sexual behaviors.

Etiology

Alterations in blood androgen concentration seem an unlikely cause for the behavior of exposing genitals to unknown women without having any further sexual interaction in mind. In fact, exhibitionists are not different from non-exhibitionists with regard to testosterone concentration (Langevin *et al.*, 1979b).

Among the many suggestions as to causes of exhibitionism and other paraphilias is hypersexuality, as mentioned in a preceding section of this chapter (Kafka, 2003). The idea of exhibitionists' hypersexuality was actually launched quite early (East, 1924), but experimental support has been weak. A simple way to test the hypothesis of hypersexuality would be to expose exhibitionists to pornographic movies and compare the intensity of sexual arousal with that of a group of controls. If exhibitionists indeed are hypersexual, their response to a sexual incentive should be larger than that found in 'normal' controls. This was not the case, though. The enhancement of penile tumescence in response to pornographic movie fragments depicting women did not differ between groups. Like controls, exhibitionists did not respond to a male stimulus and not to pictures of children. It has also been suggested that exhibitionists get aroused by looking at their own genitals, while non-exhibitionists do not. One group of four exhibitionists was exposed to slides of male nudes, female nudes and themselves clothed or nude. While responding with penile tumescence to female nudes, they did not respond to pictures of themselves. Some exhibitionists were also exposed to audiotapes describing genital exposure to women responding either with disgust and embarrassment, with fear or making no response at all. A control group was also included. There was no difference in penile response to the different situations between exhibitionists and controls (Langevin *et al.*, 1979b). This very interesting series of experiments succeeded in destroying many of the hypotheses proposed to account for exhibitionism.

Another experiment exposed exhibitionists, other sex offenders (in the scientific literature as well as in the courts, exhibitionists are always offenders) and 'normal controls' to slides depicting landscapes, fully dressed women in public areas, young boys in erotic or explicit situations, adult men in similar situations, young girls and adult women in erotic or explicit situations. The penile response to these slides was recorded. All groups responded with erection to pictures of adult females, somewhat less to young girls and very little to other stimuli. The only difference between groups was found in the response to a fully dressed female. Exhibitionists responded more than the other two groups (Fedora *et al.*, 1986). The authors give that detail a lot of attention, but they forget to mention that the exhibitionists' response to the dressed woman was no larger than their response to a landscape. Looking at the data, it turns out that both the 'normal control' and 'other offender' groups showed an unusually low degree of tumescence when looking at the dressed woman. Both groups showed, probably by simple coincidence, penile circumference slightly below baseline while the exhibitionists had a circumference slightly above baseline. None of the deviations from baseline appears to be significant. The fact that the exhibitionists differed from the other groups is completely uninteresting, since none of the groups, including the exhibitionists, showed any penile response to the dressed woman. Thus, the authors interpreted their data like the devil interprets the bible. What they have shown is something contrary to what they say they have shown, namely that exhibitionists and other men respond equally to sexual incentives.

Among the many other likely and unlikely explanations put forth for exhibitionism is the proposal that exhibitionists actively avoid sexual intercourse, because they confer some aversive quality to it. As a substitute, they take pleasure in exposing their genitals. This notion is based on the unproved supposition that exhibitionists indeed avoid intercourse. A very nice experimental test of this hypothesis consisted of exposing a group of exhibitionists and a control group to audiotapes describing the several aspects of human sexual interaction, starting with the search for a partner, approach behaviors, precopulatory behaviors, and intercourse. The penile response to these descriptions was registered. The largest response was obtained, not surprisingly, when listening to the description of intercourse. Then followed the precopulatory phase, the approach phase and the search phase, in decreasing order. A neutral auditory message had no effect. There was no difference in response between exhibitionists and controls in any of these phases of sexual interaction (Freund *et al.*, 1984). Thus, there is no reason whatsoever to imagine that exhibitionists find sexual intercourse aversive and therefore are forced to show their genitals to unknown women instead. The fact that this hypothesis was found to be false when subjected to empirical test does not necessarily mean that it was bad. Many logically well-founded hypotheses suffer the same fate when confronted with reality. However, the particular hypothesis shown to be false in the Freund *et al.* (1984) study was already absurd at its conception. It was based on the Freudian and ethological notion that motivation obeys the first law of thermodynamics: it cannot be destroyed. Thus, if not given outlet, motivation will accumulate, just as vapor accumulates in a pressure-cooker. The pressure of the motivation will eventually be so strong that rather unlikely behaviors are activated, just as the safety valve will let some pressure escape from the pressure-cooker. As we know, motivation is not equal to physical energy. The idea that an abstract concept like motivation is subjected to the same laws as physical energy is by itself extraordinary. Anyway, sexual motivation requires the presence of an incentive in order to be activated.

There is no solid evidence showing that exhibitionists respond to sexual incentives in a different way from non-exhibitionists. In addition to the studies evaluating sexual arousal in response to different incentives described above, there are a couple of others I have not mentioned. The results of these studies suggest that the majority of exhibitionists respond to sexual incentives in a similar way to that of non-exhibitionists. It must also be mentioned that all studies of penile responses have been criticized on the grounds that it is possible to fake these responses (Marshall and Fernandez, 2003). However, it is not easy to find any reason for faking in the studies mentioned above and in those not mentioned. We know that similar studies in men and women preferring to have sex with their own sex give reliable results in the sense that genital response to the own sex is larger than that to the opposite sex. We do not know if the subjects in these studies consciously tried to fake their responses, but since data were as expected this possibility was not even considered. One thing is that people partially *can* control their genital response when asked to do so. A completely different thing is if they spontaneously do so in studies where they are not asked to exercise control over their responses. In the absence of evidence to the contrary, there is not much reason to suspect that exhibitionists fake their responses more than other experimental subjects do. To the best of my judgment, the data reported in the studies of exhibitionists' genital responses are just as reliable indicators of sexual arousal as data from other studies.

Exhibitionists do not seem to differ reliably from non-exhibitionists in any way except in the habit of exposing genitals, and perhaps also in the habit of looking at other people having sex (voyeurism), or getting sexual pleasure from rubbing their body against that of an unknown, fully dressed female (frotteurism), or from making obscene phone-calls to unsuspecting women (Freund *et al.*, 1983). These other behaviors seem to have a higher frequency among exhibitionists than among the general population. Because of that, Freund and colleagues liked to call them 'courtship disorders', since they all somehow concern the initial phases of sexual interaction.

A substantial comorbidity with axis 1 disorders, as well as frequent personality disorders, have been reported in exhibitionists (e.g. Grant, 2005) and some childhood experiences may contribute (Lee *et al.*, 2002). These observations, and those mentioned in the preceding paragraph, suggest that exhibition of genitals in an unusual context rarely is the only extravagant behavior shown by exhibitionists. Such a complex behavioral syndrome cannot easily be explained by the incentive motivation/learning hypothesis. It seems though, that the mental representation of the act itself must be the crucial incentive. Otherwise the exhibitionists should have shown some peculiar reaction to the incentives employed in the studies mentioned earlier. In fact, the only incentive not tested was mental representation of exhibitionist acts. We know from Chapter 1 that mental representations can be powerful sexual incentives, so there is no reason why they could not be so in exhibitionists. Explanations in terms of external incentives and classical conditioning or operant learning do not seem feasible here.

Other paraphilias, like voyeurism and frotteurism, can easily be explained in simple incentive motivational terms. Watching other people having sex is arousing as we have seen many examples of, and it is probably also rewarding. If not, it would be difficult to explain the success of the pornography industry. Rubbing the body against that of a woman provides tactile stimulation and we know that certain kinds of tactile stimulation are sexually arousing, and we may assume that it is also rewarding. Those engaging in voyeurism or frotteurism do so because they obtain a sexual reward. The sexual reward in exhibitionism is more difficult to describe. Some exhibitionists masturbate while exposing their genitals and that will obviously produce sexual reward. The problem is why they choose such an impractical context for masturbating. Standing in a corner in a public building ready to lower the pants when a woman passes by in order to obtain sexual reward by masturbation seems like an extreme waste of time and effort when the same sexual reward could be obtained by masturbation while comfortably seated on a couch at home. The fact that the exhibitionist sometimes chooses the former alternative must mean that it offers a superior reward, particularly since the public masturbation carries substantial costs in the form of social opprobrium and possible penal action. Perhaps a peculiar constellation of mental representations enhances the reward value of the act of masturbation in the presence of unknown women to such a degree that the costs are outweighed. The pleasure of breaking social conventions, of attracting attention by unknown women, the suspense surrounding the act, and many other elements may join the sensual pleasure of masturbation into a kind of supreme reward. The mental representation of that supreme reward would be the irresistible incentive activating the behavior. This, obviously, is pure speculation. In addition to being superior to solitary masturbation, exhibition must afford a reward larger than intercourse. Many of the exhibitionists are married and display regular sexual activity with their wife, yet they occasionally add on an episode of exhibition.

The fact that some exhibitions fail, either because of unexpected discovery and arrest, or because of lack of erection and/or of opportunity of masturbation assures that reward and reinforcement is intermittent. All those familiar with learning theory know that intermittent reinforcement enhances the resistance to extinction.

The etiology of exhibitionism is essentially or perhaps even totally unknown. All efforts that have been made have met with limited success. Nevertheless, it is probably the result of a combination of psychic factors, some of which may be considered as disorders, vastly variable between individual exhibitionists. The consequence of this combination is that the individual creates mental representations of the exhibitionist act. These mental representations function as a powerful incentive impelling the individual to express the behavior. All evidence suggests that the occurrence of episodes of exhibition is outside the control of the will, in the same way as all sexual acts are according to Saint Augustine.

Treatment

Psychotherapy There is no established psychotherapeutic treatment for exhibitionists. Several approaches have been tried, but none has been the object of controlled studies of efficacy. The best documented psychotherapeutic treatment is probably that described by Maletzky and McGovern (1991). It consists of a mixture of aversive conditioning, usually with an olfactory aversive stimulus, aversive behavioral rehearsal, plethysmographic biofeedback, masturbatory reconditioning, cognitive therapy and empathy training. The amount of each ingredient is adjusted for each patient according to the therapist's judgment. Most of the therapeutic procedures are applied in a one-to-one context, but some group sessions are also included. Over a 25-year period, 1604 exhibitionists were treated and, of those, 957 participated in a follow-up evaluation 5 years after end of treatment. Treatment failure was defined as any self-report of genital exhibition or commitment of a sexual crime. Using these criteria, the failure rate was 12.8% at the follow up. Some patients were followed over a longer time, a few for as long as 25 years. Failure rate increased only marginally from the 5-year follow-up, suggesting that the treatment had provided a definitive cure of the problem. For patients who terminated the therapeutic program prematurely, relapse was substantially higher (Maletzky and Steinhauser, 2002). These promising data would have been still more promising if a control group receiving no treatment had been used. Nevertheless, the results of the Maletzky program suggest that psychotherapeutic treatment consisting of a combination of learning procedures (classical conditioning and social learning, biofeedback) and cognitive training is efficacious.

Aversive classical conditioning procedures have been used for treating undesirable sexual behaviors for many years. Habitually, it consists of presenting the inappropriate stimulus followed by an aversive event like electric shock, an unpleasant odor or a drug producing nausea (Maletzky, 1980; Laws and Osborn, 1983; Rice *et al.*, 1991; Abel and Osborn, 1996). For a time, it was employed for the treatment of the disease of homosexuality. A picture of a nude male was presented to a male homosexual and an electric shock or a putrid odor appeared shortly after. To a pedophile, a picture of a young boy was shown, followed by the aversive stimulus. Through classical conditioning, the inappropriate stimulus should become associated to the aversive event, leading to a reduction of its positive incentive properties. After an adequate number of pairings, the incentive would have been transformed into a neutral stimulus. The logic behind this procedure is perfectly coherent and the expected association and subsequent loss of incentive value of the stimulus did occur as expected (Quinsey and Marshall, 1983). However, controlled studies evaluating its efficacy are lacking (Quinsey *et al.*, 1993) and the procedure has been the victim of some objections. It is now rarely used.

Rather than employing techniques of classical conditioning, social learning may be used for making exhibition less attractive. This is the procedure currently applied in the Maletzky program. It was described many years ago (Wickramasekera, 1976) and consists of either making the patient expose himself to people who know him or watching a videotape of someone else in the same situation. The people watching the exhibitionist are instructed to display a set of adverse reactions, turning the episode into a strongly aversive event. Watching the videotape or being part of the situation is supposed to teach the exhibitionist that the act of exhibition is followed by aversive consequences through social learning. In this case, it is not the stimulus provoking the act that is treated, it is the act itself. This seems most reasonable since the stimulus control of exhibitionism is unknown. I even suggested that there is no stimulus control. Instead, the behavior is activated by mental representations of the act and the pleasure caused by it. Reducing the incentive value of the act, by associating it with aversive rather than pleasurable consequences, seems far more promising than manipulating the value of non-existent external incentives. In fact, a small sample treated with this procedure without any other therapeutic interventions showed no relapse in a follow up 7 years after the end of treatment. As always, no control group was included. Nevertheless, the efficacy of this procedure alone may account for all the successes reported by the Maletzky program.

Masturbatory reconditioning is another interesting learning procedure employed in the Maletzky program. There are four variants of the procedure, but the reasoning behind them is quite similar, as we will see. The basic principle applied is classical conditioning and the basic suppositions are that most masturbatory behaviors are associated with sexual fantasies, that is mental representations of sexual acts, and that these mental representations influence the expression of sexual behaviors (Marquis, 1970). One of the variants of masturbatory reconditioning is called thematic shift. The patient is instructed to masturbate in the company of his habitual mental representations. In the case of an exhibitionist, these are thought to include elements of genital exposure. At the point of orgasm, the patient should shift his mental representation from the socially unacceptable ones to more appropriate mental representations. The idea is that these last representations should become associated with the reward state induced by orgasm. The strengthening of these associations should be assured by the well-established reinforcing properties of orgasm. Another variant consists of making the patient alternate between acceptable and unacceptable mental representations between masturbatory events. On one day, unacceptable mental representations should accompany masturbation to orgasm, while on another day acceptable mental representations should accompany masturbation. The reasoning behind this variant is slightly obscure in terms of learning. However, the idea was to ensure that the patient had the correct mental representations at the correct moment. In the thematic shift variant, the patient himself must shift fantasy when approaching orgasm and that might be difficult to accomplish. By instructing him (exhibitionists are always men) to keep to one single kind of mental representation during the masturbatory episode, this problem was avoided. The reasons for including masturbation accompanied by unacceptable mental representations at all are difficult to understand, but therapists are not always familiar with fundamental principles of learning theory. However, Maletzky (1985) suppressed the sessions with unacceptable mental representations and made his patients masturbate with appropriate mental representations only. This variant is called directed masturbation. The fourth variant, satiation, is rather interesting from a theoretical point of view. Here, the patient is instructed to continue to masturbate for a prolonged period beyond ejaculation. During the post-orgasmic period, he is instructed to activate his favorite mental representations of the unacceptable act (Marshall and Lippens, 1977). As we know, prolonged exposure to a reward leads to negative alliesthesia. Thus, the mental representations occurring during post-ejaculatory masturbation should become associated to the negative alliesthesia by simple contiguity.

In an attempt to evaluate the success of the masturbatory reconditioning procedures, it was found that the thematic shift and alternation variants were of doubtful usefulness, while the directed masturbation and satiation variants might be effective (Laws and Marshall, 1991). As always, the final conclusion was that more controlled studies are needed before their efficaciousness could be considered as firmly ascertained.

In addition to procedures including elements of learning, there are several case studies of more exotic therapeutic approaches (e.g. Epstein and Deyouts, 1983; Paul *et al.*, 1999; Chopin-Marcé, 2001). Obviously, there are no convincing data with regard to duration of successful treatment results. None of these alternative procedures offers any hope of constituting realistic treatment options.

A feature common to all studies of psychotherapeutic treatments is that their possible side effects are completely ignored. We know that the majority of exhibitionists display socially acceptable sexual behaviors in addition to the episodes of exhibition. Whether the treatment of exhibitionism affects these other sexual behaviors or not does not seem to be of much concern for those interested in eliminating the specific undesirable behavior. If the elimination of episodes of exhibitionism causes a general deterioration of sexual functioning, it is most unlikely that the quality of life of the exhibitionist has been improved. If this is so, the utility of the treatment must be seriously questioned. The absence of data concerning possible side effects of the treatments makes it impossible to express any opinion with regard to their clinical utility.

Pharmacotherapy While some of the psychotherapeutic approaches described above *may* have specific effects on exhibitionistic behaviors, the pharmacological agents sometimes employed in the treatment of exhibitionism have in common that they reduce all aspects of sexual functioning. Basically, two techniques have been used. One consists of reducing the activity of androgen receptors, either by compounds lowering blood concentrations of androgens or by androgen receptor antagonists. The other technique is to administer psychopharmacological agents known to have adverse effects on sexual functioning.

Among the drugs employed for lowering blood androgen concentrations is a progestin, medroxyprogesterone acetate. An early study showed that the drug had some positive effect (Langevin *et al.*, 1979a). However, dropout rate was high, suggesting that the patients disliked the general reduction of sexual behaviors. A later study of paraphiliacs with varied diagnoses confirmed that acceptable sexual behaviors were reduced as much as the unacceptable ones, with the possible difference that unacceptable behaviors disappeared somewhat faster (Kravitz *et al.*, 1995). Blood testosterone concentration was also much reduced, suggesting that reduced androgen receptor stimulation indeed is the basis for the effects of medroxyprogesterone.

Androgen antagonists (e.g. Bradford and Pawlaka, 1993; Czerny *et al.*, 2002) or GnRH agonists (e.g. Briken *et al.*, 2000, 2001, 2003) have been successfully used, but the sexual side effects are similar to those reported for medroxyprogesterone. All sexual behaviors are affected in addition to the unacceptable behavior.

SSRIs have also been used. Their efficaciousness seems to be beyond doubt, but so are their side effects (Greenberg *et al.*, 1996; Greenberg and Bradford, 1997). I have mentioned many times that the SSRIs have solidly recognized inhibitory effects on all aspects of sexual behavior.

In an attempt to solve the problem of generalized inhibition of sexuality, a kind of treatment scheme with progressively increasing intensity of treatment has been proposed. The mildest treatment, psychotherapy, should be tried first. If not working, then an SSRI should be tried instead. If the unacceptable behavior still persisted, then an androgen antagonist or medroxyprogesterone should be tried. If that failed, then GnRH agonists should be employed. If even that is not enough, the GnRH agonist should be combined with the administration of an androgen receptor antagonist (Hill *et al.*, 2003). Perhaps this is a good suggestion. The idea that a treatment should not be more violent than required for reducing or eliminating the undesired behavior is basically sound, but it is not self-evident that the cost associated with the high-intensity treatments outweighs the benefit obtained from them.

In the absence of a specific pharmacological treatment for exhibitionism, and for all paraphilias for that matter, the choice of the pharmacological approach does not seem to be recommendable.

3.3 Fetishism

Diagnostic criteria, epidemiology and general description

Fetishism is another popular paraphilia. The diagnostic criteria, again according to the DSM-IV (American Psychiatric Association, 1995, p. 540) are straightforward:

A. Over a period of at least 6 months, recurrent, intense sexually arousing fantasies, sexual urges, or behaviors involving the use of nonliving objects (e.g. female undergarments).

B. The fantasies, sexual urges, or behaviors cause clinically significant distress or impairment in social, occupational or other important areas of functioning.
C. The fetish object is not limited to articles of female clothing used in cross-dressing (as in Transvestic Fetishism) or devices designed for the purpose of tactile genital stimulation (e.g. a vibrator). (Reprinted with permission from the *Diagnostic and Statistical Manual of Mental Disorders*, Copyright 2000. American Psychiatric Association).

From the diagnostic criteria of exhibitionism, we recognize the notions of urge, minimum duration, fantasies in addition to behavior, distress and impairment. The only new thing here is that we are provided with some extremely trivial examples of sexual incentive stimuli that can be regarded as fetishes. Point C seems just as unnecessary as the mention of urge.

It appears that any object can acquire sexual incentive properties and activate some sexual behavior pattern, usually approach to the object and solitary masturbation in its presence. However, in reality it appears that most fetishes are associated with the human body. Garments, often used to cover it, and boots or shoes are typical fetishes. Although excluded from the DSM-IV diagnostic criteria, parts of the human body may acquire all the properties of a fetish. Some examples of specific fetishisms are given in Table 12.1. Some of them fall outside the diagnostic criteria, but in the clinical literature they all appear under the label of fetish.

The prevalence of fetishism is unknown. About 18% of a sample of healthy individuals (healthy means here that they had not sought treatment for sexual disorders) reported having fetishistic fantasies (Gosselin and Wilson, 1980). This is certainly not the same as being a fetishist, according to the DSM-IV criteria. Others have

TABLE 12.1 A list of a small selection of paraphilias, subcategory fetishism

1	Abasiophilia: sexual attraction to people who use leg braces or other orthopaedic appliances
2	Acousticophilia: sexual arousal from certain sounds
3	Acrotomophilia: sexual attraction to amputees
4	Agalmatophilia: sexual attraction to statues or mannequins or immobility
5	Algolagnia: sexual pleasure from pain
6	Amaurophilia: sexual arousal by a partner whom one is unable to see due to artificial means, such as being blindfolded or having sex in total darkness.
7	Andromimetophilia: love of women dressed as men
8	Aquaphilia: arousal from water and/or in watery environments, including bathtubs or swimming pools
9	Aretifism: sexual attraction to people who are without footwear
10	Trichophilia: sexual arousal from hair
11	Urolagnia: sexual attraction to urine
12	Vorarephilia: sexual attraction to being eaten by, and/or eating, another person or creature
13	Xenophilia: sexual attraction to foreigners (in science fiction, can also mean sexual attraction to aliens)
14	Zoophilia: emotional or sexual attraction to animals
15	Zoosadism: the sexual enjoyment of causing pain and suffering to animals

determined the prevalence of fetishism among those seeking treatment for psychiatric disorders. It turned out to be 0.8% of patients treated during a 20-year period (Chalkley and Powell, 1983). These figures are not based on representative samples, but it seems reasonable to propose that the proportion of individuals finding fetishism a problem and seeking treatment for it is low. It must also be mentioned that fetishism often appears together with other paraphilias. The most frequently associated with fetishism are pedophilia and transvestism (Abel and Osborn, 1992).

Some have argued that fetishism should not be considered a disorder at all (Reiersol and Skeid, 2006). The fact of becoming sexually aroused in response to unusual stimuli is a private matter, particularly since fetishism, as defined in the DSM-IV, does not involve any human except the fetishist himself (like exhibitionism, fetishism is most unusual in women). A reason for continuing to consider it a disorder could be that fetishism may cause personal distress and/or occupational and other problems, as mentioned in the second criterion. However, there are innumerable private events causing distress and occupational problems lasting more than 6 months for which there is no diagnostic category. My prediction is that fetishism has a rather short life-expectancy as a diagnostic category. It will probably soon follow the category of homosexuality into the history of arbitrary diagnoses.

Etiology

Psychoanalysts have offered splendid accounts of the origin of fetishism. None of these accounts is particularly helpful. For example, Freud himself wrote, in a 1927 essay entitled 'Fetischismus' (translated and reprinted in Freud, 1961), that the origin of fetishism was the traumatic discovery sooner or later made by most boys that the mother has no penis. This discovery causes castration anxiety. In order to reduce anxiety, the boy substitutes his mother's missing genital with any object he finds convenient for that purpose. This object acquires sexual significance and, through the combined work of the defense mechanisms and the drives of the *id*, the fetishistic behavior will manifest itself in one form or another. Although intellectually attractive, the Freudian account is of little help in habitual clinical practice.

The most popular beliefs concerning the etiology of fetishism hold that it is a result of classical conditioning. Somehow, sexual reward came to be associated with a salient stimulus that accidentally was present at the moment the sexually rewarding and reinforcing event occurred. This is a perfectly reasonable account. The only problem with it is that we do not have any direct empirical evidence for it. Nevertheless, there is probably no other way to explain how a substantial diversity of inanimate objects has acquired sexual incentive properties. Furthermore, we learned in Chapter 7 that any object *can* acquire sexual incentive properties through classical conditioning. There is a funny experiment in the quail that illustrates this very nicely. Just before a female became available for copulatory interaction, an artificial object was presented. That object was the conditioned stimulus, while the female quail was the unconditioned stimulus. The female rapidly activated the unconditioned responses of approach and copulation. After several pairings, the artificial object also activated the responses of approach and, in some of the male quails, also copulatory behavior leading to ejaculation. The anatomy of the quail is such that no vagina is required for the execution of copulatory behavior. It so happens that the male has no penis. During copulation, he grabs the back of the female's head or neck feathers, then he mounts and arches back so that his cloaca comes in contact with the female's cloaca. Cloacal contact is typically associated with ejaculation and the end of copulatory interaction. Provided that the male has the opportunity to grab an object at the top and sustain himself on the object, something similar to cloacal contact is possible. Therefore, some quails succeeded in obtaining ejaculation with the artificial object. After conditioning, the males were never again exposed to the female quail but to the artificial object. Those males that regularly achieved ejaculation with the object did not show any signs of extinction. Those who did not ejaculate lost their interest in the artificial object after a couple of exposures. The fact that the males who copulated with the object obtained sexual reward on each exposure to it prevented extinction. In reality, there were no extinction sessions since reward always was obtained. The authors concluded that their quail data constituted a bird equivalent to human fetishism (Köksal *et al.*, 2004). Some of the results from the experiment are illustrated in Figure 12.1.

The very elegant experiment described in the preceding paragraph suggests that conditioning can account for continued sexual responses to an inanimate object. Also the human fetishist frequently obtains reward when exposing himself

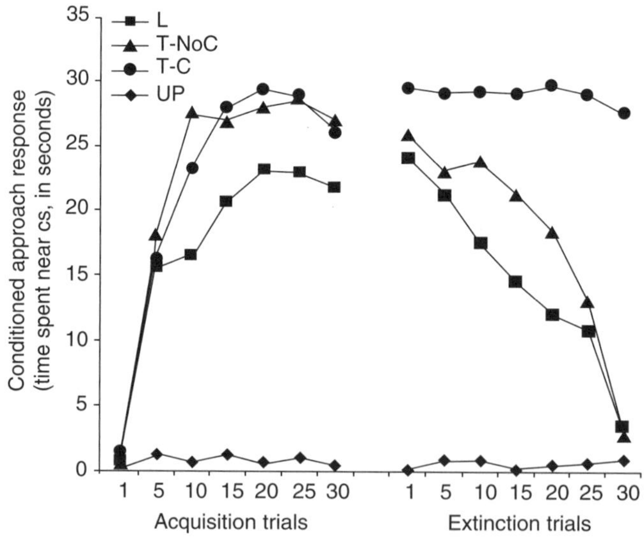

FIGURE 12.1 Mean seconds spent in the CS zone during the 30 second CS periods for groups L, T-C, T-NoC and UP during acquisition and extinction in Experiment 1 (sexual conditioning). The CS was a light for group L and a terrycloth objects for groups T-C, T-NoC and UP. Subjects in group T-C copulated with the terrycloth CS, whereas those in groups L and T-NoC did not show conditioned copulatory behavior. (From Köksal *et al.*, 2004. Copyright 2004. Reproduced with permission from Elsevier.)

to the fetish, provided that he masturbates to ejaculation or engages in some other rewarding and reinforcing activity in its presence. Thus, there will be no reduction of the fetish's incentive properties.

A role for social learning in the development of fetishism cannot be excluded. By direct observation or through verbal descriptions of other individuals getting sexually aroused by arbitrary objects, we may produce mental representations of those objects as sexually arousing. Perhaps mental representations of an object as sexually arousing are not sufficient for turning it into a fetish, but they may direct our attention to similar objects in situations where sexual reward is obtained, for example through solitary masturbation. Social learning could explain the cultural differences with regard to the most popular fetishes.

The learning hypothesis outlined above fits very nicely with the incentive motivation/learning model of sexual behaviors.

Treatment

Psychotherapy Controlled clinical studies of psychotherapeutic treatments of fetishism have not been performed. In fact, this disorder seems to have attracted little attention among research oriented psychotherapists. Although psychoanalysts have offered entertaining accounts of possible etiology, they have been far less active in treatment, at least with regard to publishing reports concerning the results of psychoanalytic management of fetishism. The majority of the few reports available on the subject are focused on learning-based therapies (Wise, 1985). As we know, the main hypothesis concerning etiology poses that neutral stimuli have become conditioned sexual incentives through classical conditioning. An efficient therapeutic procedure must, then, have as aim the extinction of that conditioning. In most learning situations, the habitual way to extinguish a conditioned response is repeatedly to present the conditioned stimulus without any ensuing reward. The efficiency of this approach was nicely demonstrated in the quail study mentioned above. The quails not obtaining reward when exposed to the previously conditioned stimulus rapidly extinguished their approach behaviors. On the contrary, those that continued to obtain reward did not show signs of extinction.

Applying simple learning principles, the most efficient treatment for fetishism would be repeated exposure to the fetish in the absence of sexual reward. A disadvantage with this simple procedure is that extinction would require a large number of exposures, if it will take place at all. The strength of the association between the fetish and sexual reward is certainly considerable, making a large number of nonrewarded exposures necessary. However, since the conditioned arousal induced by the fetish may have some rewarding properties in itself, extinction may never occur. Thus, the safest way to reduce the fetish's incentive value would be to transform it into a conditioned stimulus predicting an aversive event. This is exactly what has been done in a couple of case studies. Presentations of the fetish have been followed by electric shock, a procedure euphemistically called 'faradic aversion'. Two fetishists were treated with this procedure, and their penile tumescence in response to the fetish was monitored. The results were promising and the patients reported improvement at follow up (Marks and Gelder, 1967). There are many other case studies that employed aversive conditioning of different kinds. An excellent review of these studies and a detailed description of one case are available in a book chapter (Junginger, 1997). Since the total number of treated patients is very low, and because no controls were included, it is impossible to determine whether any of the conditioning procedures are efficacious or not. Thus, there is presently no recognized psychotherapeutic treatment of fetishism. Perhaps the demand for such treatments is so limited that clinical research efforts are not justified.

Pharmacotherapy Very little is known about pharmacological treatments of fetishism. Studies of agents blocking androgen receptors or reducing blood androgen concentrations are lacking. In fact, I have been unable to find any such study explicitly evaluating these kinds of drugs in patients diagnosed as fetishists. They usually lump all paraphilias together, and the main concern is always those paraphilias regarded as societal problems, essentially pedophilia. Since fetishism is not usually a criminal offense, individuals suffering from it are rarely if ever included in clinical trials. Nevertheless, as was the case with exhibitionism, the cost in terms of reduced general sexual functioning of treatment reducing androgen receptor stimulation does probably not justify its use for alleviating a minor complaint like fetishism. Unless the patient unequivocally demands such a treatment for one reason or another, obviously.

Psychopharmacological treatments have been employed now and then. There are a few case studies reporting favorable effects with the SSRIs (e.g. Lorefice, 1991; Clayton, 1993; Rubenstein and Engel, 1996), which is not surprising. Again, the cost is that also unproblematic sexual behaviors are reduced. More recently, an anticonvulsant, topiramate, modulating voltage-dependent sodium and calcium ion channels, has been employed in a case of fetishism. The drug potentiates GABA neurotransmission and blocks kainate/AMPA glutamate receptors. Fetishistic behavior disappeared after 4 weeks of treatment with topiramate (Shiah *et al.*, 2006). The frequency of masturbation in absence of the fetish was also reduced, suggesting that sexual behavior in general and not only fetishism was affected. This coincides with another case study of a man displaying symptoms of hyperactive sexual desire. All sexual behaviors disappeared after 6 weeks of treatment (Fong *et al.*, 2005). This does not come as a surprise, since rat data show that compounds facilitating GABAergic neurotransmission are inhibitory to sexual behavior (Ågmo and Paredes, 1985; Paredes *et al.*, 1997). It appears that no selective drug exists for the treatment of fetishism. Available pharmacological data suggest that a compound either is ineffective or blocks all aspects of sexual behavior in addition to fetishism. Such a violent treatment does not seem to be justified except in extreme cases.

3.4 Pedophilia

Of all the paraphilias, pedophilia is certainly the one causing most concern to society. This means that research on this disorder is much more abundant than on exhibitionism or fetishism. However, there are serious limitations with regard to the research effort. The most serious one is that research has been concentrated on pedophiles who have been the object of criminal prosecution.

Diagnostic criteria, epidemiology and general description

According to the DSM-IV, the criteria that need to be satisfied before pronouncing the diagnosis of pedophilia are:

A. Over a period of at least 6 months, recurrent, intense sexually arousing fantasies, sexual urges, or behaviors involving sexual activity with a prepubescent child or children (generally age 13 or younger).
B. The fantasies, sexual urges, or behaviors cause clinically significant distress or impairment in social, occupational or other important areas of functioning.
C. The person is at least 16 years of age and at least 5 years older than the child or children in Criterion A. (Reprinted with permission from the *Diagnostic and Statistical Manual of Mental Disorders*. Copyright 2000. American Psychiatric Association.)

Many elements in these criteria are identical to those found in the diagnostic criteria for exhibitionism and fetishism. It is only the object of sexual attentions that has changed. The criteria also contain some specifications that are important. One concerns the age of the incentive, which must be younger than 13, another concerns the age of the patient, that must be over 16, and the third concerns the difference between incentive and patient, that must be a minimum of 5 years. All these age and time limits are arbitrary, essentially reflecting dominant social conventions. Nevertheless, they are unambiguous and as such useful. In addition to the three criteria mentioned above, pedophiles are also categorized according to the sex of the partner. There are pedophiles attracted exclusively to boys, exclusively to girls and those that are nonselective. More elegant terms have been used, for example heterosexual pedophilia, homosexual pedophilia and undifferentiated pedophilia. Those belonging to the last group are believed to be a minority of the pedophiles. Somewhere between 60 and 70% of all cases of pedophilia are supposed to be heterosexual (Mohr *et al.*, 1964).

Penile–vaginal intercourse is rarely the sexual behavior of choice for a heterosexual pedophile. In fact, it appears to occur essentially with partners around or above puberty. Most occasions of pedophile penile–vaginal intercourse have been reported to be consensual (Vetter, 1972). The most common behaviors displayed by a pedophile is a kind of childish behavior, with looking at the partner's genitals, or making the partner look at their own genitals, fondling or being fondled by the partner being the most common behaviors. Masturbation during this sex play is not uncommon. The homosexual pedophile likes to masturbate his partner and sometimes be masturbated by him. Fellatio is also quite common, while anal intercourse is rare.

Pedophiles may be of all ages, from adolescence to senescence. The mean age was estimated to be 37 years in one study (Fisher, 1969). About half of them were married. Other data suggest a trimodal age distribution of pedophiles, with the largest peak in early adolescence, followed by a slightly smaller peak in the mid-30s, and a still smaller peak in the mid-50s. Of the pedophiles over 20 years, 75% were married (Mohr *et al.*, 1964). Other characteristics like intelligence, personality traits, occupation, education and similar do not appear to distinguish pedophiles from the general population. Although there have been numerous suggestions concerning all kinds of abnormalities among pedophiles, such as deviant responses on the Minnesota Multiphasic Personality Inventory, deficiencies in empathy, signs of psychopathy, or addiction to sex, just to mention a few of the many disturbances attributed to them, none of these suggestions resists careful scrutiny (Okami and Goldberg, 1992; Marshall *et al.*, 1995; Marshall and Hall, 1995). It seems that a typical pedophile is not substantially different from others except for the tendency to respond to socially unacceptable sexual incentives.

The most frequent partners for pedophilic activities are persons known to the pedophile, such as own children, friends' children or a neighbor's children. This is particularly the case for heterosexual pedophiles. Among the homosexual pedophiles, it is not unusual to procure the services of a child prostitute. This means that the proportion of partners unfamiliar to the pedophile is somewhat higher than is the case for heterosexual pedophiles. The age of the partner is usually between 6 and 12. A majority of the partners consent to sexual interaction with the pedophile, and some are even the instigator of such interaction (Virkkunen, 1975).

The role of the pedophile's partner is not of primary concern in this chapter. However, there are many myths and misconceptions regarding the 'victim'. I have systematically avoided the term victim and replaced it with the neutral term partner. This was not done by accident. There are many data showing that the kind of sexual activities pedophiles normally engage in are quite childish, similar to the sex play occurring between children. It is not certain that the participating young child interprets these activities as sexual at all (Gagnon and Simon, 2002) and it is still more unlikely that she/he spontaneously assumes the role of victim. Nevertheless, adults have a tendency to react strongly to discoveries or accounts of such activities and they immediately consider the child as the victim. Moreover, it is not uncommon for adults to imagine the most extravagant consequences of interaction with a pedophile. In view of the apprehension associated with this, I feel obliged to make some short mention of the partner, if not for anything else just to be irritating.

The real consequences of having participated in pedophilic activities are difficult to evaluate. It has been proposed that at least children who willingly become partners are in a problematic situation before the event, making it still more difficult to determine possible adverse consequences. However, a couple of recent, extensive studies have shown that the sequels of childhood sexual experiences with pedophiles are slight (Bauserman and Rind, 1997; Rind *et al.*, 1998). In fact, they may even be positive (Rind, 2003; Sbraga and O'Donohue, 2003). Proposing that the partner in pedophilic activities does not suffer grave psychic consequences is contrary to dominant social prejudice. As a reward for daring to present scientific data offending that prejudice, the congress in a country proclaiming itself as a leader of the free world issued an official condemnation (Rind *et al.*, 2000) of the Rind *et al.* (1998) article. This nicely illustrates that science is never independent of or above the society in which it is exercised. However, societies frequently change with time and the reactions to the pedophile and his partner may also change. To end this short discussion of the partner with a proposal for such a change, I want to mention a provocative point of view on pedophilia that was expressed in an intellectually most stimulating paper a few years ago (Mirkin, 1999). It was argued that the current strong social disapproval of pedophilia is similar to the disapproval of homosexuality expressed a few decades ago and that the fears and myths once associated with homosexuality now have been transferred to pedophilia. It was predicted that the disorder of pedophilia will share the destiny of the disorder of homosexuality and disappear within a not too distant future.

Etiology and epidemiology

Many factors have been proposed to account for pedophilia. There are explanations in psychoanalytic terms, in terms of disordered personality, of emotional immaturity, of hypersexuality and many others. Although some of these explanations may have considerable literary value, they do not seem to correspond to any tangible reality. This minor problem should not be an obstacle for reading some of these accounts. Personally, I enjoy the psychoanalytic line of thought, brilliantly illustrated in a classic paper (Cassity, 1927). For a more recent variant of psychoanalytic thought, an erudite article of Charles Socarides was published in 1959 (Socarides, 1959). It is still excellent reading. For those preferring an approach based on a salad of psychological notions like emotional congruence, disinhibition, and blockage, there is an exuberant model available (Finkelhor, 1984). These kinds of explanations will not be considered here. None of them has a sound empirical base and any suggestion that they have scientific value would be nothing more than a dramatic manifestation of a lack of modesty.

I already mentioned that the pedophiles' personality and intellectual abilities are found to be no different from non-pedophiles when appropriately controlled studies are performed. Some maintain that their childhood experiences may be different from those of non-pedophiles (e.g. Marshall *et al.*, 1993; Lee *et al.*, 2002), particularly with regard to having been victim of sexual abuse (e.g. Cohen *et al.*, 2002), but this is difficult to judge. I have mentioned the uncertainties inherent in recollections of childhood events before. These uncertainties become accentuated in the situation of a pedophile. Almost all pedophiles included in published studies have been convicted for a sexual offense, and many have been given harsh sentences. In this situation, unconscious transformations of childhood memories are very likely to occur. The more tragic the offender makes his childhood, the easier it is to inspire compassion and pass guilt on to unhappy events, like an unloving father, a violent mother, an alcoholic sister, or a perverted landlord. Those claiming that pedophiles have suffered all this and much more have a substantial burden of evidence before they can convince us that the pedophiles' recollections coincide with actual childhood events. Also parents may unconsciously transform their memories when asked to recollect the childhood of their son, who happened to be condemned to a long prison term for pedophilia some time ago. Finally, some experts claim that the probability of arrest and subsequent sentence for sexual offenses is dependent on the individual's social class. Family structure and child-rearing procedures in low income families may be quite different from the middle class ideals to which they are compared, and eventually described as deviant by the scientists. All these reasons make descriptions of differences in childhood experiences between pedophiles and non-pedophiles particularly risky and even if we assume a modest influence of the factors mentioned above, we will find that pedophiles are no different from others.

It seems most unlikely that the causes for the attraction to children or adolescents are to be found in factors unrelated to sexuality. Instead, I suggest that children, for some reason or another, have acquired sexual incentive properties for those expressing pedophilic behavior while they do not have such properties for others. Adult individuals, of the same or of the opposite sex, are also sexual incentives for the pedophile exactly as they are for almost everyone else. This simple suggestion can easily be subjected to experimental test. We could, for example, expose pedophiles to pornographic movies depicting sexual interaction between adults. To these movies, the pedophiles should show exactly the same response as non-pedophiles, that is an erection. Both groups could then be exposed to a pornographic movie depicting sexual interactions between a child and an adult. In this case, we would predict that the pedophile would respond with erection whereas the non-pedophile would not. In fact, according to social expectations, the non-pedophile should respond with disgust to the latter movie. The predictions concerning differential erection response in pedophiles and non-pedophiles have been tested in a number of studies. However, none of them has employed the most appropriate stimulus material. We know from Chapters 4 and 9 that the most efficient sexual incentives for enhancing genital blood flow are pornographic movies depicting sexual interaction between members of the preferred sex. Moving pictures produce a larger and more reliable response than slides, and slides are more efficient than audiotapes describing sexual activities. Despite this solid knowledge, studies of sexual arousal in pedophilia employ usually slides or audiotapes. I do not know why this is the case, but it seems that clinical researchers are slow to follow developments in basic research. Notwithstanding the regrettable choice of stimulus material, data are consistent. Pedophiles and non-pedophiles show a similar response to slides depicting adult incentives or audiotapes describing sexual interaction between adults. On the contrary, slides depicting young incentives or audiotapes describing sexual interactions with children caused a larger response in pedophiles than in controls (e.g. Avery-Clark and Laws, 1984; Marshall *et al.*, 1986; Wormith, 1986; Grossman *et al.*, 1992). This appears to be a quite consistent finding, which has been confirmed in a large number of studies. Contradictory observations are rare, and can easily be explained by deficiencies in methods. The fact that a child of the preferred sex causes a larger response in pedophiles than in control subjects unequivocally shows that it is an efficient sexual incentive for pedophiles, and less so for others.

Some non-pedophiles show responses of similar magnitude to adult and child incentives (Hall *et al.*, 1995), suggesting that the age of the incentive is not a critical determinant of its incentive value in all individuals. Regrettably, the exact age of the girls depicted on the stimulus slides was not given. All we are told is that they were prepubertal. I interpret this as meaning that these girls lacked pubic hair and developed breasts as well as the body shape of an adolescent. Nevertheless, they may have displayed some stimuli carrying the sexual incentive properties of an adult woman, such as hairstyle, way of looking, and similar. Since we do not know the features of an adult woman's body, shown on a slide, that carry her sexual incentive property, it is entirely possible that some men recognize the crucial feature also on a picture of a prepubescent girl. An elegant study employing slides depicting women of different ages suggests that this explanation is not impossible. Non-pedophilic men show some increase in their penile response to slides of young girls when age increases from 5 to 13. The increase is much larger when the age of the incentive jumps from 13 to 20. There is no change in response magnitude when the age of the incentive increases from 20 to 24 (Marshall *et al.*, 1988). Nevertheless, these data indicate that young girls are not devoid of sexual incentive properties for non-pedophiles. However, their mean magnitude of response is well below that of the pedophiles.

The issue of an erectile response to stimuli depicting or describing children in non-pedophiles is not of much scientific concern. Individual variation is legion, and some overlap between groups is not problematic as long as the means are reliably different. A non-pedophilic man showing a larger response to a picture of a child than to that of an adult woman does not invalidate the observation that an overwhelming majority of pedophiles does show a larger response to the child than most non-pedophiles. In contrast, when the erectile response to a picture of a child is employed as a clinical tool for distinguishing pedophiles from others, then there is a problem. In the clinic, the procedures for determining penile tumescence carry the graceful name of phallometry. Because some non-pedophilic men indeed respond to the child, there will be some false positives. Being classified as a pedophile while not being one may have devastating consequences if the classification is used in court. It might actually influence the verdict. It may also be used in prisons, for determining release on probation, and by child care workers when deciding custody. All these uses of penile responses to sexual incentives are completely grotesque. Adequate use of a research instrument for scientific studies is most desirable, but taking it out from the laboratory and making it instrumental in decisions of vast importance for the fate of a human being appears to be extremely doubtful. A well formulated critique against phallometric assessment has been published elsewhere (Marshall and Fernandez, 2000, 2003).

After this digression we shall now return to children as sexual incentives for pedophiles. Two questions must be answered. The first is how children acquired the property of sexual incentives. It is obvious that it must be through learning and that is the opinion among most of those working with pedophilia. It is also obvious that the learning must be in the forms of classical conditioning and operant learning. The child will become a sexual incentive because of classical conditioning, associating sexual reward with stimuli emitted by a child. The performance of sexual acts must be controlled by operant learning. Somehow, the pedophile must have learned that certain motor patterns executed in the presence of the child or involving the child lead to sexual reward. Exactly how this learning occurs is a matter of speculation. The factual, albeit not theoretical, impossibility of determining the precise circumstances of learning was discussed in Chapter 9 and there is no reason to do it again.

The second question we have to answer with regard to children as sexual incentives concerns the capacity of these incentives to trigger a response. We already know that a sexual incentive does not necessarily activate approach behaviors and ensuing sexual responses. I have mentioned examples illustrating that the incentive properties of a stimulus are dependent on the context in which the stimulus appears. One of these examples was the sight of a half naked woman on a beach in Normandy. Although most young men would show a sexual arousal response in the form of erection to the picture of a half naked woman presented in the laboratory, and perhaps also in response to the woman on the beach, very few of them would approach her. Because of social learning these young men know that a nude female torso on the beach is not an acceptable sexual incentive, and that knowledge almost always impedes the display of approach behaviors. This is also the case with children as sexual incentives. Even though pictures of children may have sexual incentive properties for some men when presented in the laboratory, and perhaps also in the world outside the laboratory, these incentive properties should not activate any approach behaviors since social learning has taught us that such behaviors are inappropriate. In the case of the pedophile we have, consequently, to explain the way in which social learning has failed. At this moment, we do not have enough data even for attempting to propose a hypothesis. There is one observation that might provide some clue, though. In a study of sexual delinquents, among those a group of pedophiles, impulsivity was slightly related to sexual delinquency (Blocher *et al.*, 2001). This suggests that social learning may have less capacity to control behavior in pedophiles, and not that social learning by itself is deficient. More studies aimed at exploring these important issues are urgently needed.

We are now ready for a conclusion concerning the etiology of pedophilia. Unusual psychic events during childhood or during adult life cannot explain the interest in engaging in sexual activities with children. I have not mentioned factors like gonadal hormones or brain development for the simple reason that there is no evidence whatsoever for any anomaly. After excluding endocrine or intrapsychic events, we are left with an explanation in terms of the incentive motivation/learning model. That explanation is, probably, the only one that can accommodate the variations in onset of pedophilic behavior. We know that there are three main age groups showing pedophilic behavior, adolescents, those around 35, and those around 55. A developmental, organic or psychic, explanation cannot account for this distribution among ages. Adult psychopathology could, but no known pathology has an age distribution that coincides with pedophilia, so this explanation is most unlikely.

Learning, however, can occur at any moment. A likely moment is shortly after puberty, when the central motive state sensitive to sexual incentives is activated by gonadal hormones. At this moment, learning with sexual reward should be likely to occur. Interestingly, the adolescence period coincides with the main peak of pedophilic behavior. An additional contributor to the adolescence peak of pedophilia may be the availability of partners. Many adolescents have siblings in the preferred age range, and friends with siblings in that range. In the mid-30s, many people have children between 6 and 12, and their presence offer possibilities for the formation of associations. It may also be so that children come to mind in mental representations of sexual activities. Furthermore, couples with children frequently have friends with children of about the same age. We already know that frequent partners for pedophiles are their own or their friends' children. In the mid-50s, grandchildren of an appropriate age are not unusual. Again, the same kind of associations and mental representations possibly occurring with the own children now appear with the grandchildren. This is a very simple and most likely explanation for the age distribution of pedophilic behavior.

The prevalence of pedophilia is unknown. Data on the prevalence of abused children have become abundant, but such data do not necessarily coincide with the prevalence of pedophilia. One pedophile may abuse many children, and many cases of pedophilic sexual activities are not perceived as abuse by the partner, and probably not reported as such on the typical questionnaires employed for estimating prevalence. Thus, there is a very confused relationship between the number of reported cases of abuse and the number of individuals satisfying the criteria for the diagnosis of pedophilia. Some like to look at conviction rates for sexual offenses, but again the number of convictions may be very different from the number of pedophiles according to established diagnostic procedures. In view of this uncertainty, any estimate of the prevalence of pedophilia is prone to be false. Prudence obliges me to abstain from giving any figure at all.

Treatment

Psychotherapy Many therapeutic approaches have been used in the treatment of pedophilia. They include psychoanalytically oriented therapies (e.g. Juda, 1986), group therapies of all kinds (e.g. Cabeen and Coleman, 1961), and behavioral therapies. The latter have generally focused on the sexual incentive properties of inappropriate stimuli and have combined the exposure to such stimuli with aversive events (e.g. Maletzky, 1980; Laws and Osborn,1983; Quinsey and Marshall, 1983; Rice *et al.*, 1991). It was expected that the inappropriate stimulus would lose its sexual incentive properties because of classical conditioning. The rationale for this kind of therapy was outlined some pages ago and there is no need to repeat it here. The success of the behavioral therapies was frequently determined from the penile response to inappropriate and appropriate stimuli. It was not unusual to find that the magnitude of response to the inappropriate stimuli became reduced as a result of treatment and that was reason enough for considering it efficacious.

In the case of pedophilia, treatment success can be measured in a way rather different from those used for evaluating success in the treatment of most other disorders. Since pedophilia is a criminal behavior, those engaging in it, and who are discovered, are brought before justice. The vast majority of the patients treated for pedophilia has, in fact, been convicted before treatment. An easy way to evaluate the success of a treatment is to determine the incidence of recidivism among the patients. However, there is a widespread opinion holding that sexual offenses are under-reported, making recidivism rates doubtful estimates of success. Cognizant of these problems, a review of treatment success determined that sex offenders having been subjected to psychotherapeutic treatments did not show lower rates of recidivism than offenders whose only treatment had been the legal sanctions imposed on them (Furby *et al.*, 1989). In this review, no distinction was made between different kinds of sexual offenses and between different kinds of treatments. However, this was simply because no pattern could be detected among all offenses and among all treatments. The fact that recidivism rate was not visibly affected by any treatment does not exclude that they have had some beneficial effects on the quality of life in some individuals.

A comment must be made with regard to the apparent weak success of aversive conditioning procedures. If children are sexual incentives for pedophiles, then a reduction of their incentive value should inevitably reduce the incidence of pedophilic behavior. Pictures of children or audiotapes describing sexual interaction with children had actually suffered a reduction of incentive value because of aversive conditioning, since many of the studies revealed a reduced sexual arousal to these stimuli following the conditioning procedure. How come, then, that recidivism was not clearly reduced in these pedophiles? There are at least two likely explanations. One is that the conditioning did not generalize from audiotapes or pictures to real children. A behaving child is much more than what can ever be captured on a slide. The child emits many stimuli in addition to body shape. Stimuli like specific kinds of movements, the auditory quality of voice, the content of verbal messages, the glare, and many more. If any of these stimuli contribute to a child's sexual incentive properties, aversive conditioning to a picture may not generalize at all to a living child. The same argument can obviously be used with regard to generalizations from an audiotape to a living child. A second explanation can be that the aversive conditioning is much less resistant to extinction than the conditioning of a child's sexual incentive properties is. Considering these explanations, it is not at all surprising that aversive conditioning as performed is quite inefficient. If for a moment we compare these conditioning procedures with those implicitly forming the basis for Hurlbert's successful Orgasmic Consistency Training for the treatment of hypoactive sexual desire disorder, we find immediately a crucial difference. In the Hurlbert procedure, a living partner constitutes the stimulus to which sexual reward is conditioned, and not a picture of the partner.

After this rather negative account of psychotherapeutic treatments for pedophilia, I must immediately mention that things have changed after the 1989 review mentioned in the preceding paragraph. One treatment program is now on the list of probably efficacious treatments for sexual offenders (Chambless *et al.*, 1998). I will dedicate a couple of lines to a description of it. A much more elegant account has been published elsewhere (Marshall, 1999).

Treatment is offered to individuals convicted of a sexual offense. The program has three levels, two of which are intramural. One is for those considered to have the largest needs (based on reoffense risk and security status among other elements), another is for those with moderate needs, and the third level is common to all. The highest level treatment lasts for 6 months, with daily 3-hour sessions. The other in-prison treatment lasts for 4 months, and consists of three sessions of 3 hours each per week. A shorter alternative is offered to those considered as low-need, low-risk offenders. This alternative lasts 3 months and there are only two 3-hour sessions per week. The third treatment level, localized to a community-based clinic, starts after release from prison. This program includes all kinds of sexual offenders, like rapists, exhibitionists, and pedophiles. The latter kind of offenders are frequently assigned to the short alternative of level 2.

After a careful assessment, the patients are assigned to the treatment level considered adequate. The content of the treatment program is varied. It includes denial/minimization correction, to make the patient assume responsibility for his acts. It also includes corrections of distorted perceptions. The patient frequently perceives the partner rather than himself as at fault, and this misconception is modified. He is also trained in feeling empathy for the partner. Of more interest is the effort to eliminate deviant mental representations. Patients are taught to resist them, for example by applying covert sensitization. The virtues of masturbatory reconditioning are also explained. This procedure was described exhaustively in a preceding section. Both the directed masturbation and the satiety variants are recommended.

The program includes some other elements, which are not essential for the present discussion. It is clear that the treatment program tries to modify several aspects that might lie behind pedophilic behavior. One of these aspects is the incentive properties of children and another is the capacity of social learning to control behavior. Stimulation of a sense of empathy with the partner and the correction of misperceptions of fault will serve both these ends. Empathy is somehow based on the thought that the child is the victim and, once being convinced that this is the case, the pedophile must realize that he himself is the victimizer, something unacceptable according to social rules. The inculcation of this probably strengthens the capacity of socially learned norms to control the incentive stimuli's impact on the central motive state or the likelihood that an activated central motive state expresses itself in motor patterns leading to approach. The directed masturbation may help to reduce the incentive value of children. The combination of these two effects of the treatment program, enhanced control over the central motive state and reduced incentive value of children, may be enough for eliminating pedophilic behavior. Socially acceptable sexual behaviors should, at least in principle, not be affected. It must be noted that pharmacological treatment may be included in this program if there are no signs of success with the behavioral/cognitive procedures.

Pharmacotherapy As was the case with pharmacological treatments of the other paraphilias, treatments for pedophilia are not specific. They all reduce the reactivity of the central motive state to all sexual incentives, appropriate as well as inappropriate. In fact, the treatments employed for pedophiles are exactly the same as those used for the treatment of exhibitionism or fetishism. The only notable difference is that society considers pedophilia as far more offensive than exhibitionism or fetishism, making it easier to justify treatments abolishing all sexual behaviors. The SSRIs, fluoxetine, sertraline and fluvoxamine, have been used and were reported to be efficient (Greenberg *et al.*, 1996; Greenberg and Bradford, 1997). Androgen receptor antagonists (Bradford and Pawlak, 1993) and drugs reducing blood testosterone concentration like medroxyprogesterone (Meyer *et al.*, 1992; Gottesman and Schubert, 1993) and leuprolide (Briken *et al.*, 2000, 2001, 2003; Czerny *et al.*, 2002; Schober *et al.*, 2005) have likewise reported to be efficacious. Reviews of the effects of pharmacological treatments conclude unanimously that the efficacy of these treatments is established beyond reasonable doubt (Rosler and Witztum, 2000; Hill *et al.*, 2003).

The disadvantages of pharmacological treatments of pedophilia are similar to those discussed above with regard to exhibitionism and fetishism. I will not repeat the arguments here, neither will I mention the need for a treatment that specifically reduces pedophilic behavior while leaving acceptable sexual behaviors intact.

4 General conclusion

Hyperactive sexual desire and the paraphilias represent aspects of human sexual behavior that are morally questionable in contemporary Western society. We saw very clear examples of that when we reviewed the efforts that have been made with the aim of establishing criteria for hypersexuality. A frequency of sex far above average was, by itself, enough for talking about hypersexuality. Despite the absence of a rational criterion for distinguishing this condition from something I called eusexuality, I suggested that hyperactive sexual desire could be given a diagnostic category based on the same principle as other categories of sexual disorders, namely that the condition is experienced as a problem of sufficient magnitude for causing marked distress. The rationality of this principle is open to discussion. There are many situations in life, lasting for more than 6 months, that cause marked distress or interpersonal difficulty without benefiting from a diagnostic category in the DSM-IV or ICD-10. I assume that social prejudice and professional concerns about potential markets determine, to a substantial degree, which problems are to be selected as requiring a diagnostic label and corresponding category. Science has only a marginal influence in that process.

The arbitrariness of the existence of diagnostic categories becomes still more evident when we turn to the paraphilias. As was mentioned, some social groups suggest that some of the paraphilias on the official diagnostic lists should be removed, for example fetishism, sadism and masochism. This will certainly occur sooner or later, in the same way as homosexuality was removed from the list when society was ripe for that move. New diagnoses will be added, just as hyperactive sexual desire disorder was added a couple of years ago. All diagnostic categories of sexual dysfunctions, perhaps except the arousal disorders, anorgasmia and premature ejaculation, are social constructions and not entities of nature.

Summary

1 Hyperactive sexual desire is an ambiguous concept. Its meaning is most variable and always arbitrarily determined. Several definitions have been tried, but none is convincing. Dominant social prejudice appears to be the main determinant.
2 The causes for unusually intense sexual activity are essentially unknown. If treatment is required, there are plenty of efficient pharmacological alternatives available.
2 Exhibitionism is difficult to explain and to treat without affecting other sexual behaviors. An incentive motivation/learning model might be helpful, but its application requires some suppositions.
3 Fetishism is normally explained in terms of learning and the incentive motivation/learning model can easily account for it. Some psychotherapeutic approaches may selectively reduce fetishistic behaviors. Pharmacological treatments affect all sexual behaviors in the same way.
4 Pedophilia is of much concern to contemporary society, despite the fact that most pedophiles engage in quite childish behavior patterns. The fate of the partner, frequently called victim, is only marginally affected by pedophilic experiences.
5 The causes of pedophilia are unknown, but most scientists believe that conditioning in one form or another is involved. The incentive motivation/learning model can account for the development of pedophilia and can explain its distribution over the life span.
6 There is a psychotherapeutic procedure potentially useful for treating pedophilic behaviors without affecting other sexual behaviors. It is based on masturbatory reconditioning combined with cognitive group therapy. The same pharmacological treatments that can be employed for treating hyperactive sexual desire, exhibitionism or fetishism can be and are used in pedophilia. They have, as always, the costly side effect of reducing or eliminating all sexual behaviors.
7 Most sexual disorders are social constructions without any correspondence to entities in nature.

References

Abel, G.G. and Osborn, C. (1992). The paraphilias: the extent and nature of sexually deviant and criminal behavior. *Psychiatr Clin North Amer* **15**, 675–687

Abel, G.G. and Osborn, C.A. (1996). Behavioural therapy treatment for sex-offenders. In *Sexual deviation*, 3rd edn (I. Rosen, ed.), pp. 382–398. Oxford University Press: Oxford

Ågmo, A. (1976). Individual differences in the response to androgen in male rabbits. *Physiol Behav* **17**, 587–589

Ågmo, A. and Paredes, R. (1985). GABAergic drugs and sexual behavior in the male rat. *Eur J Pharmacol* **112**, 371–378

Ahlenius, S., Larsson, K. and Svensson, L. (1980). Further evidence for an inhibitory role of central 5-HT in male rat sexual behavior. *Psychopharmacology* **68**, 217–220

American Psychiatric Association (2000). *Diagnostic and statistical manual of mental disorders*, 4th edn, Text revision. American Psychiatric Association: Washington, DC

Anderson, R.A., Bancroft, J. and Wu, F.C.W. (1992). The effects of exogenous testosterone on sexuality and mood of normal men. *J Clin Endocr Metab* **75**, 1503–1507

Avery-Clark, C.A. and Laws, D.R. (1984). Differential erection response patterns of sexual child-abusers to stimuli describing activities with children. *Behav Ther* **15**, 71–83

Bagatell, C.J., Heiman, J.R., Matsumoto, A.M., Rivier, J.E. and Bremner, W.J. (1994a). Metabolic and behavioral effects of high-dose, exogenous testosterone in healthy-men. *J Clin Endocr Metab* **79**, 561–567

Bagatell, C.J., Heiman, J.R., Rivier, J.E. and Bremner, W.J. (1994b). Effects of endogenous testosterone and estradiol on sexual behavior in normal young men. *J Clin Endocr Metab* **78**, 711–716

Bancroft, J. and Vukadinovic, Z. (2004). Sexual addiction, sexual compulsivity, sexual impulsivity, or what? Toward a theoretical model. *J Sex Res* **41**, 225–234

Bauserman, R. and Rind, B. (1997). Psychological correlates of male child and adolescent sexual experience with adults: a review of the non-clinical literature. *Arch Sex Behav* **26**, 105–141

Berrios, G.E. (2006). Madness from the Womb. *Hist Psychiatr* **17**, 223–235

Berrios, G.E. and Kennedy, N. (2002). Erotomania: a conceptual history. *Hist Psychiatr* **13**, 381–400

Blocher, D., Henkel, K., Retz, W., Retz-Junginger, P., Thorne, J. and Rösler, M. (2001). Symptome aus dem Spectrum des hyperkinetischen Syndroms bei Sexualdelinquenten. *Fortschr Neurol Psychiatr* **69**, 453–459

Bonierbale, M., Lançon, C. and Tignol, J. (2003). The ELIXIR study: evaluation of sexual dysfunction in 4557 depressed patients in France. *Curr Med Res Opin* **19**, 114–124

Bradford, J.M. and Pawlak, M.A. (1993). Effects of cyproterone acetate on sexual arousal patterns of pedophiles. *Arch Sex Behav* **22**, 629–641

Bradford, J.M.W. and Pawlaka, A. (1993). Double-blind placebo crossover study of cyproterone-acetate in the treatment of the paraphilias. *Arch Sex Behav* **22**, 383–402

Briken, P., Berner, W., Noldus, J., Nika, E. and Michl, U. (2000). Therapie mit dem LHRH-agonisten Leuprolinacetat bei Paraphilien und sexuell aggressiven Impulshandlungen. *Nervenarzt* **71**, 380–385

Briken, P., Hill, A. and Berner, W. (2003). Pharmacotherapy of paraphilias with long-acting agonists of luteinizing hormone-releasing hormone: A systematic review. *J Clin Psychiat* **64**, 890–897

Briken, P., Nika, E. and Berner, W. (2001). Treatment of paraphilia with luteinizing hormone-releasing hormone agonists. *J Sex Marital Ther* **27**, 45–55

Buena, F., Swerdloff, R.S., Steiner, B.S. *et al.* (1993). Sexual function does not change when serum testosterone levels are pharmacologically varied within the normal male range. *Fertil Steril* **59**, 1118–1123

Cabeen, C.W. and Coleman, J.C. (1961). Group therapy with sex offenders: description and evaluation of a group therapy program in an institutional setting. *J Clin Psychol* **17**, 122–129

Cassity, J.H. (1927). Psychological considerations of pedophilia. *Psychoanal Rev* **14**, 189–199

Chalkley, A.J. and Powell, G.E. (1983). The clinical description of 48 cases of sexual fetishism. *Br J Psychiat* **142**, 292–295

Chambless, D.L., Baker, M.J., Baucom, D.H. *et al.* (1998). Update on empirically validated therapies, II. *Clin Psycholog* **51**, 3–16

Chopin-Marcé, M.J. (2001). Exhibitionism and psychotherapy: a case study. *Int J Offender Ther Comp Criminol* **45**, 626–633

Clayton, A.H. (1993). Fetishism and clomipramine. *Am J Psychiat* **150**, 673–674

Clayton, A.H., Pradko, J.F., Croft, H.A. *et al.* (2002). Prevalence of sexual dysfunction among newer antidepressants. *J Clin Psychiat* **63**, 357–366

Cohen, L.J., McGeoch, P.G., Gans, S.W., Nikiforov, K., Cullen, K. and Galynker, I.I. (2002). Childhood sexual history of 20 male pedophiles vs. 24 male healthy control subjects. *J Nerv Ment Dis* **190**, 757–766

Coleman, E. (1986). Sexual compulsion vs. sexual addiction: the debate continues. *SIECUS Rep* **14**, 7–11

Coleman, E. (1990). The obsessive-compulsive model for describing compulsive sexual behavior. *Am J Preventive Psychiatr Neurol* **2**, 9–14

Coleman, E. (1991). Compulsive sexual behavior: new concepts and new treatments. *J Psychol Human Sexual* **4**, 37–52

Conaglen, H.A. and Conaglen, J.V. (2003). Sexual desire in women presenting for antiandrogen therapy. *J Sex Marital Ther* **29**, 255–267

Cooper, A., Scherer, C.R., Boies, S.C. and Gordon, B.L. (1999). Sexuality on the Internet: From sexual exploration to pathological expression. *Professional Psychol-Res Practice* **30**, 154–164

Czerny, J.P., Briken, P. and Berner, W. (2002). Antihormonal treatment of paraphilic patients in German forensic psychiatric clinics. *Eur Psychiatr* **17**, 104–106

Damassa, D.A., Smith, E.R., Tennent, B. and Davidson, J.M. (1977). The relationship between circulating testosterone levels and male sexual behavior in rats. *Horm Behav* **8**, 275–286

Davis, S. (2000). Testosterone and sexual desire in women. *J Sex Educ Ther* **25**, 25–32

Davis, S.R. and Tran, J. (2001). Testosterone influences libido and well being in women. *TIPS* **12**, 33–37

East, W.N. (1924). Observations on exhibitionism. *Lancet* **2**, 370–375

Ellis, H. (1933). *The psychology of sex.* William Heinemann: London

Epstein, S.J. and Deyouts, P.L. (1983). Hypnotherapeutic control of exhibitionism: a brief communication. *Int J Clin Exp Hypnosis* **31**, 63–66

Fedora, O., Reddon, J.R. and Yeudall, L.T. (1986). Stimuli eliciting sexual arousal in genital exhibitionists – a possible clinical application. *Arch Sex Behav* **15**, 417–427

Finkelhor, D. (1984). *Child sexual abuse: new theory and research.* Free Press: New York

Fisher, G. (1969). Psychological needs of heterosexual pedophiliacs. *Dis Nerv Syst* **30**, 419–421

Fong, T.W., De La Garza, R. and Newton, T.F. (2005). A case report of topiramate in the treatment of nonparaphilic sexual addiction. *J Clin Psychopharmacol* **25**, 512–514

Freud, S. (1961). Fetishism. In *The standard edition of the complete psychological works of Sigmund Freud, vol. 21, The future of an illusion, civilization and its discontents and other works.* (J. Strachey, ed.). Hogarth Press: London

Freund, K., Scher, H. and Hucker, S. (1983). The courtship disorders. *Arch Sex Behav* **12**, 369–379

Freund, K., Scher, H. and Hucker, S. (1984). The courtship disorders – a further investigation. *Arch Sex Behav* **13**, 133–139

Furby, L., Weinrott, M.R. and Blackshaw, L. (1989). Sex offender recidivism: a review. *Psychol Bull* **105**, 3–30

Gagnon, J.H. and Simon, W. (2002). *Sexual conduct: the social sources of human sexuality*, 2nd edn. AldineTransaction: New Brunswick

Giambra, L.M. and Martin, C.E. (1977). Sexual daydreams and quantitative aspects of sexual activity: some relations for males across adulthood. *Arch Sex Behav* **6**, 497–505

Gold, S.N. and Heffner, C.L. (1998). Sexual addiction: Many conceptions, minimal data. *Clin Psychol Rev* **18**, 367–381

Goodman, A. (1992). Sexual addiction: designation and treatment. *J Sex Marital Ther* **18**, 303–314

Goodman, A. (2001). What's in a name? Terminology for designating a syndrome of driven sexual behavior. *Sex Add Compulsiv* **8**, 191–213

Gosselin, C. and Wilson, G. (1980). *Sexual variations*. Faber and Faber: London

Gottesman, H.G. and Schubert, D.S.P. (1993). Low-dose oral medroxyprogesterone acetate in the management of the paraphilias. *J Clin Psychiat* **54**, 182–188

Grant, J.E. (2005). Clinical characteristics and psychiatric comorbidity in males with exhibitionism. *J Clin Psychiat* **66**, 1367–1371

Greenberg, D.M. and Bradford, J.M.W. (1997). Treatment of the paraphilic disorders: A review of the role of the selective serotonin reuptake inhibitors. *Sex Abuse: J Res Treat* **9**, 349–360

Greenberg, D.M., Bradford, J.M.W., Curry, S. and O'Rourke, A. (1996). A comparison of treatment of paraphilias with three serotonin reuptake inhibitors: a retrospective study. *Bull Am Acad Psychiat Law* **24**, 525–532

Grob, C.S. (1985). Female exhibitionism. *J Nerv Ment Dis* **173**, 253–256

Grossman, L.S., Cavanaugh, J.L. and Haywood, T.W. (1992). Deviant sexual responsiveness on penile plethysmography using visual stimuli: alleged child molesters vs normal control subjects. *J Nerv Ment Dis* **180**, 207–208

Grunt, J.A. and Young, W.C. (1952). Differential reactivity of individuals and the response of male guinea-pigs to testosterone propionate. *Endocrinology* **51**, 237–248

Hall, G.C.N., Hirschman, R. and Oliver, L.L. (1995). Sexual arousal and arousability to pedophilic stimuli in a community sample of normal men. *Behav Ther* **26**, 681–694

Hill, A., Briken, P., Kraus, C., Strohm, K. and Berner, W. (2003). Differential pharmacological treatment of paraphilias and sex offenders. *Int J Offender Ther Comp Criminol* **47**, 407–421

Hollender, M.H., Brown, C.W. and Roback, H.B. (1977). Genital exhibitionism in women. *Am J Psychiat* **134**, 436–438

Janus, S.S. and Janus, C.L. (1993). *The Janus report on sexual behavior*. Wiley: New York

Juda, D.P. (1986). The usefulness of self psychology in understanding and treating a case of homosexual pedophilia. *Dynamic Psychother* **4**, 99–123

Junginger, J. (1997). Fetishism. Assessment and treatment. In *Sexual deviance. Theory, assessment, and treatment* (D.R. Laws and W. O'Donahue, eds), pp. 92–110. The Guilford Press: New York

Kafka, M.P. (1997). Hypersexual desire in males: An operational definition and clinical implications for males with paraphilias and paraphilia-related disorders. *Arch Sex Behav* **26**, 505–526

Kafka, M.P. (2003). Sex offending and sexual appetite: the clinical and theoretical relevance of hypersexual desire. *Int J Offender Ther Comp Criminol* **47**, 439–451

Kinsey, A.C., Pomeroy, W.B. and Martin, C.E. (1948). *Sexual behavior in the human male*. Saunders: Philadelphia

Köksal, F., Domjan, M., Kurt, A. *et al*. (2004). An animal model of fetishism. *Behav Res Ther* **42**, 1421–1434

Kravitz, H.M., Haywood, T.W., Kelly, J., Wahlstrom, C., Liles, S. and Cavanaugh, J.L. (1995). Medroxyprogesterone treatment for paraphiliacs. *Bull Am Acad Psychiatr Law* **23**, 19–33

Langevin, R., Paitich, D., Hucker, S. *et al.* (1979a). Effect of assertiveness training, Provera and sex of therapist in the treatment of genital exhibitionism. *J Behav Ther Exp Psychiatr* **10**, 275–282

Langevin, R., Paitich, D., Ramsay, G. *et al.* (1979b). Experimental studies of the etiology of genital exhibitionism. *Arch Sex Behav* **8**, 307–331

Larsson, K. (1966). Individual differences in reactivity to androgen in male rats. *Physiol Behav* **1**, 255–258

Laumann, E.O., Gagnon, J.H., Michael, R.T. and Michaels, S. (1994). *The social organization of sexuality. Sexual practices in the United States*. University of Chicago Press: Chicago

Laws, D.R. and Marshall, W.L. (1991). Masturbatory reconditioning with sexual deviates: an evaluative review. *Adv Behav Res Ther* **13**, 13–25

Laws, D.R. and Osborn, C.A. (1983). How to build and operate a behavioral laboratory to evaluate and treat sexual deviance. In *The sexual aggressor: Current perspectives on treatment* (J.G. Greer and I.R. Stuart, eds), pp. 293–335. Van Nostrand Reinhold: New York

Lee, J.K.P., Jackson, H.J., Pattison, P. and Ward, T. (2002). Developmental risk factors for sexual offending. *Child Abuse Neglect* **26**, 73–92

Lorefice, L.S. (1991). Fluoxetine treatment of a fetish. *J Clin Psychiat* **52**, 41

Långström, N. and Hanson, R.K. (2006). High rates of sexual behavior in the general population: Correlates and predictors. *Arch Sex Behav* **35**, 37–52

Maletzky, B. and McGovern, K.B. (1991). *Treating the sexual offender*. Sage: Newbury Park

Maletzky, B.M. (1980). Self-referred versus court-referred sexually deviant patients: success with assisted covert sensitization. *Behav Ther* **11**, 306–314

Maletzky, B.M. (1985). Orgasmic reconditioning. In *Dictionary of behaviour therapy techniques* (A.S. Bellack and M. Hersen, eds), pp. 157–158. Pergamon Press: New York

Maletzky, B.M. and Steinhauser, C. (2002). A 25-year follow-up of cognitive/behavioral therapy with 7,275 sexual offenders. *Behav Modificat* **26**, 123–147

Marks, I.M. and Gelder, M.G. (1967). Transvestism and fetishism: clinical and psychological changes during faradic aversion. *Br J Psychiat* **113**, 711–729

Marquis, J. (1970). Orgasmic reconditioning: changing sexual object choice through controlling masturbation fantasies. *J Behav Ther Exp Psychiatr* **1**, 263–271

Marshall, D.S. and Lippens, K. (1977). The clinical value of boredom: a procedure for reducing inappropriate sexual interests. *J Nerv Ment Dis* **165**, 283–287

Marshall, W.L. (1999). Current status of North American assessment and treatment programs for sexual offenders. *J Interpers Viol* **14**, 221–239

Marshall, W.L., Barbaree, H.E. and Butt, J. (1988). Sexual offenders against male-children – Sexual preferences. *Behav Res Ther* **26**, 383–391

Marshall, W.L., Barbaree, H.E. and Christophe, D. (1986). Sexual offenders against female-children: sexual preferences for age of victims and type of behavior. *Can J Behav Sci-Rev Canad Sci Comport* **18**, 424–439

Marshall, W.L. and Fernandez, Y.M. (2000). Phallometric testing with sexual offenders: Limits to its value. *Clin Psychol Rev* **20**, 807–822

Marshall, W.L. and Fernandez, Y.M. (2003). Sexual preferences – Are they useful in the assessment and treatment of sexual offenders? *Aggr Violent Behav* **8**, 131–143

Marshall, W.L. and Hall, C.G.N. (1995). The value of the MMPI in deciding forensic issues in accused sexual offenders. *Sexual Abuse: J Res Treat* **7**, 205–219

Marshall, W.L., Hudson, S.M. and Hodkinson, S. (1993). The importance of attachment bonds in the development of juvenile sex offending. In *The juvenile sex offender* (H.E. Barbaree, W.L. Marshall and S.M. Hudson, eds), pp. 164–181. Guilford Press: New York

Marshall, W.L., Hudson, S.M., Jones, R. and Fernandez, Y.M. (1995). Empathy in sex offenders. *Clin Psychol Rev* **15**, 99–113

Matuszczyk, J., Larsson, K. and Erikson, E. (1998). The selective serotonin reuptake inhibitor fluoxetine reduces sexual motivation in male rats. *Pharmacol Biochem Behav* **60**, 527–532

Meerkerk, G.J., Van den Eijnden, R.J.J.M. and Garretsen, H.F.L. (2006). Predicting compulsive Internet use: It's all about sex! *Cyberpsychol Behav* **9**, 95–103

Meyer, W.J., Collier, C. and Emory, E. (1992). Depo-Provera treatment for sex offending behavior: an evaluation of outcome. *Bull Am Acad Psychiatr Law* **20**, 249–259

Mirkin, H. (1999). The pattern of sexual politics: Feminism, homosexuality and pedophilia. *J Homosexual* **37**, 1–24

Mohr, J.W., Turner, R.E. and Jerry, M.B. (1964). *Pedophilia and exhibitionism. A handbook.* University of Toronto Press: Toronto

Montejo, A.L., Llorca, G., Izquierdo, J.A. and Rico-Villademoros, F. (2001). Incidence of sexual dysfunction associated with antidepressant agents: a prospective multicenter study of 1022 outpatients. *J Clin Psychiat* **62** Suppl. 3, 10–21

Okami, P. and Goldberg, A. (1992). Personality-correlates of pedophilia – are they reliable indicators. *J Sex Res* **29**, 297–328

Östergren, P. (2006). *Porr, horor och feminister*. Natur och Kultur: Stockholm

Paredes, R.G., Karam, P., Highland, L. and Ågmo, A. (1997). GABAergic drugs and sociosexual behavior. *Pharmacol Biochem Behav* **58**, 291–298

Paul, R.H., Marx, B.P. and Orsillo, S.M. (1999). Acceptance-based psychotherapy in the treatment of an adjudicated exhibitionist: A case example. *Behav Ther* **30**, 149–162

Putnam, D.E. (2000). Initiation and maintenance of online sexual compulsivity: implications for assessment and treatment. *Cyberpsychol Behav* **3**, 553–563

Quinsey, V.L., Harris, G.T., Rice, M.E. and Lalumiere, M.L. (1993). Assessing treatment efficacy in outcome studies of sex offenders. *J Interpers Viol* **8**, 512–523

Quinsey, V.L. and Marshall, W.L. (1983). Procedures for reducing inappropriate sexual arousal: an evaluation review. In *The sexual aggressor. Current perspectives on treatment* (J.G. Greer and I.R. Stuart, eds), pp. 267–289. Van Nostrand Reinhold: New York

Raymond, N.C., Coleman, E. and Miner, M.H. (2003). Psychiatric comorbidity and compulsive/impulsive traits in compulsive sexual behavior. *Comprehens Psychiat* **44**, 370–380

Reiersol, O. and Skeid, S. (2006). The ICD diagnoses of fetishism and sadomasochism. *J Homosexual* **50**, 243–262

Rice, M.E., Harris, G.T. and Quinsey, V.L. (1991). Evaluation of an institution-based treatment program for child molesters. *Can J Prog Eval* **6**, 111–129

Rind, B. (2003). An elaboration on causation and positive cases in child sexual abuse. *Clin Psychol Sci Practice* **10**, 352–357

Rind, B., Bauserman, R. and Tromovich, P. (2000). Science versus orthodoxy: anatomy of the congressional condemnation of a scientific article and reflections on remedies for future ideological attacks. *Appl Prevent Psychol* **9**, 211–225

Rind, B., Tromovitch, P. and Bauserman, R. (1998). A meta-analytic examination of assumed properties of child sexual abuse using college samples. *Psychol Bull* **124**, 22–53

Rosler, A. and Witztum, E. (2000). Pharmacotherapy of paraphilias in the next millennium. *Behav Sci Law* **18**, 43–56

Rubenstein, E.B. and Engel, N.L. (1996). Successful treatment of transvestic fetishism with sertraline and lithium. *J Clin Psychiat* **57**, 92

Sbraga, T.P. and O'Donohue, W. (2003). Post hoc reasoning in possible cases of child sexual abuse: Symptoms of inconclusive origins. *Clin Psychol Sci Practice* **10**, 320–334

Schober, J.M., Kuhn, P.J., Kovacs, P.G., Earle, J.H., Byrne, P.M. and Fries, R.A. (2005). Leuprolide acetate suppresses pedophilic urges and arousability. *Arch Sex Behav* **34**, 691–705

Shiah, I.S., Chao, C.Y., Mao, W.C. and Chuang, Y.J. (2006). Treatment of paraphilic sexual disorder: the use of topiramate in fetishism. *Int Clin Psychopharmacol* **21**, 241–243

Socarides, C.W. (1959). Meaning and content of pedophiliac perversion. *J Am Psychoanal Assoc* **7**, 84–94

van Anders, S.M., Chernick, A.B., Chernick, B.A., Hampson, E. and Fisher, W.A. (2005). Preliminary clinical experience with androgen administration for pre- and postmenopausal women with hypoactive sexual desire. *J Sex Marital Ther* **31**, 173–185

Vega-Matuszczyk, J., Larsson, K. and Erikson, E. (1998). Subchronic administration of fluoxetine impairs estrous behavior in intact female rats. *Neuropsychopharmacology* **19**, 492–498

Vetter, H.J. (1972). *Psychology of abnormal behavior*. Ronald Press: New York

Virkkunen, M. (1975). Victim-precipitated pedophilia offences. *Br J Criminol* **15**, 175–180

Vroege, J.A., Gijs, L. and Hengeveld, M.W. (1998). Classification of sexual dysfunctions: Towards DSM-V and ICD-11. *Comprehens Psychiat* **39**, 333–337

Wickramasekera, I. (1976). Aversive behavior rehearsal for sexual exhibitionism. *Behav Ther* **7**, 167–176

Wilson, G.D. and Lang, R.J. (1981). Sex differences in sexual fantasy patterns. *Pers Individ Differ* **2**, 343–346

Wise, T.N. (1985). Fetishism – etiology and treatment: a review from multiple perspectives. *Comprehens Psychiat* **26**, 249–257

World Health Organization (1992). *The ICD-10 classification of mental and behavioural disorders: clinical descriptions and diagnostic guidelines*. World Health Organization: Geneva

World Health Organization (1993). *The ICD-10 classification of mental and behavioural disorders: diagnostic criteria for research*. World Health Organization: Geneva

Wormith, J.S. (1986). Assessing deviant sexual arousal: physiological and cognitive aspects. *Adv Behav Res Ther* **8**, 101–137

Yates, W.R., Perry, P.J., MacIndoe, J., Holman, T. and Ellingrod, V. (1999). Psychosexual effects of three doses of testosterone cycling in normal men. *Biol Psychiatr* **45**, 254–260

CHAPTER

13

On the omissions

Throughout this book, many arguments that could have been made in support for a proposition were not made. Sometimes arguments that could have been made against a proposition were not made. A good example of the former may be the proposal made in Chapter 1 that natural selection has not exercised a strong pressure for making reproductive behaviors as efficient as possible. I never mentioned the most convincing argument of all, namely that selective pressure has been so low that the vast majority of vertebrate species reproduce sexually with one-sexed individuals. This reproductive strategy reduces the reproductive potential of a species to half compared to what could have been obtained by making all individuals two-sexed. This solution would have preserved all the advantages of sexual reproduction without the enormous costs it carries. Biologists have desperately tried to find an explanation for this strange fact by suggesting, in a most creative way, all kinds of disadvantages associated with hermaphroditism. However, none of these disadvantages has been shown to be a necessary consequence of hermaphroditism, and none of them appears to outweigh the enormous costs.

In the same Chapter 1, I could have made a much more profound analysis of the influences of Saint Augustine and other religious scholars on contemporary views on sexuality and I could have traced the roots of much of their reasoning further back in history. In Chapter 2, my presentation of incentive motivation theory can be regarded as extremely superficial and my description of copulatory behaviors can be regarded as lacking. Literally hundreds of papers have not been cited and innumerable intricacies of rat sex were never mentioned. Still worse, information regarding other mammalian species was scant and superficial. In Chapter 3, I only covered a fraction of the experimental studies concerning the role of the vomeronasal organ and I almost completely forgot to make any mention of the neurophysiology of the main olfactory system. I could continue like this for hours, but the few illustrations offered hitherto have made it clear that not a single subject treated in this book has been treated exhaustively enough.

Only a few of the sexual dysfunctions or disorders have been covered. Notable omissions are the sexual arousal disorders, so named in women, but called erectile disorder in men. Both conditions are characterized by a lack of enhanced genital blood flow in response to sexual incentives. I decided to omit these disorders because we have talked so much about sexual arousal in other chapters and the mechanisms controlling it have been analyzed in some detail. Again, the physiology of genital circulation was not adequately covered and the anatomical features of the genital vascular systems were not mentioned at all. This, in fact, would have become indispensable for a reasonable thorough discussion of the arousal disorders. Without these aspects, I could only have concluded that the impact of sexual incentives on the central motive state and the ensuing activation of visceral responses were deficient. Everyone could have figured out that by her/himself, making it quite unnecessary to mention it. The application of the incentive motivational/learning model to disorders was illustrated with several examples and I could feel no need for presenting more. The essential reasoning was already made clear and extensively illustrated in these examples. All the many alternative approaches to deficient sexual arousal have been presented in a huge number of popular and scientific books and research papers and there is no need to repeat these once more.

I have also omitted vaginismus, dyspareunia, premature ejaculation and anorgasmia. In all these cases, I had reasons slightly different from those mentioned in the preceding paragraph. Although all these disorders in principle can be treated with behavioral techniques, conditioning to be exact, we do not have many good data with regard to this. In this situation, there were three options available. One was to present an entirely speculative account, the second was to limit coverage to all that which is told in many other books already, and the third was simply to omit these subjects from discussion. The first option seemed intellectually unsatisfactory and the second seemed quite boring. I chose the third. As any choice, it can be criticized, but the same would have occurred if I had chosen option 1, or option 2 for that matter.

Most of the omissions are, as the ones I mentioned above, a consequence of choice. I alone am responsible for these choices. Some omissions are a consequence of ignorance. I alone am responsible for that. We may also turn the cake around and say that many inclusions are unfortunate. I alone am responsible for this and I disagree. All material found in this book is included, not with the purpose of making the book long, but with the purpose of making arguments clear and comprehensible and in order to illustrate the empirical or logical support required for turning the arguments into something approaching fact rather than speculation.

I have conscientiously tried to avoid any mention of purpose when discussing non-human behavior, while at the same time making frequent reference to it when mentioning human behavior. As was mentioned in Chapter 1, the human capacity of mental representation of future events makes the notion of purpose legitimate. However, it was not until Chapter 10 that I introduced Cassirer's brilliant analysis of the peculiarities of the human, particularly our extraordinary capacity for the use of symbols. This capacity takes its foremost expression in human language. With the aid of language, we can create all kinds of marvelous mental representations that would be completely impossible in the absence of language. This uniquely human capacity forces us to look at human behavior as somewhat different from that of all other animals. I have mentioned the extraordinary power of mental representations as incentives hundreds of times throughout the book, and that is not an accident. I am completely convinced that most of the differences between human and non-human behavior can be explained by our possession of language. When we come to sexual behaviors, language has combined with an unusual flexibility of motor systems, turning human sexual behavior into a most variable activity.

This fact, of outstanding importance, should perhaps have been stressed still more than it has. My impression is that most of my colleagues have not fully understood the deep chasm separating non-human and human sexual behaviors. Not at all with regard to the central nervous or endocrine mechanisms involved, which certainly are quite similar, neither with regard to the basic behavioral mechanisms, which certainly also are quite similar, but with regard to the motor patterns employed for the obtention of sexual reward and the possibility of mental representations of future sexual reward through completely novel motor patterns. One of the many consequences is, as repeated many times throughout this book, that penile–vaginal intercourse is only one of the large number of potential motor patterns.

Another fact frequently forgotten is the role of learning as a determinant of sexual incentives and sexual motor patterns. Although an entire chapter was dedicated to that issue, I feel that still more insistence on learning would have been useful. Particularly the role of social learning should have been emphasized more. We will never be able to understand the essentials of human sexual behaviors if we do not realize the fundamental role of social learning. At the same time, operant learning can profoundly affect the specific motor patterns we employ and this is another fact that could have been treated in more detail and illustrated with more experimental data. In sum, the roles of human language and of learning have not been omitted from this book, which should be evident even for the most distracted reader, but I fear that they have not been sufficiently included. Now it is too late to do anything about that.

Among the basic notions behind all reasoning in this book is to regard human and non-human sexual behaviors as an activity aimed at the obtention of an ephemeral positive affect. I have not added anything more to it. I do not think it is desirable to add anything more to it. With regard to non-human animals, many might agree. However, when turning to the human, the romantics or the moralists would in all likelihood add that sex is much more than that. It may be an expression of the esoteric concept of love, or of the emotional state of affection, or a means for acquiring control over someone, or for advancing the career, or for obtaining a revenue. Most of these ideas are nothing more than irrelevant social constructions, or prejudice, and some are observations on the context where sexual activities are executed. None of these observations has any bearing on our aim, understanding sexual behavior. It is evident that a human can use mental representations of a future positive affect obtained through the payment for execution of copulatory motor patterns, in the same way as a carpenter expects payment and ensuing positive affect as reward for the execution of different motor patterns. Sex as a source of revenue or as a means for making a career are expressions of human civilization and quite uninteresting for our knowledge of the mechanisms determining sexual behaviors.

Considering sexual behaviors as a means for obtaining positive affect, and nothing more than that, is a kind of revolution. It challenges many of the established myths and most of the old moral principles. Saint Augustine, however, would probably have reacted mildly. He was convinced that the search for any pleasure was a distraction from the ways of God. This conviction made him eat no more than absolutely necessary for survival and to drink no more than necessary for avoiding death from dehydration. These teachings have been forgotten in societies where obesity is a main problem of public health. A fat man does not provoke moral outcries because of his overindulgence in the pleasures of the table. He is not even expected to associate his eating with feelings of love or affection towards the cow or the lamb he is ingesting. Perhaps the man may explain his behavior with reference to the fact that eating is an activity necessary for survival, even if you should happen to weigh 164 kg. I am not sure that Saint Augustine would have believed him.

The irrational fears of the consequences of considering sexual activities as an entirely innocent way of enhancing the quality of human life will sooner or later disappear, in the same way as Saint Augustine's asceticism and the Valesii's habit of autocastration disappeared long ago. The modifications in social learning that will be associated with that disappearance will be profound and most interesting. They are already well under way in several European countries and within a couple of centuries they will probably move westwards.

Index